Advanced Engineering Calculus

A Collection of Problems with Solutions

Advanced Engineering Calculus

A Collection of Problems with Solutions

Veselin Jungić

Simon Fraser University, Canada

World Scientific

NEW JERSEY · LONDON · SINGAPORE · BEIJING · SHANGHAI · HONG KONG · TAIPEI · CHENNAI · TOKYO

Published by

World Scientific Publishing Co. Pte. Ltd.

5 Toh Tuck Link, Singapore 596224

USA office: 27 Warren Street, Suite 401-402, Hackensack, NJ 07601

UK office: 57 Shelton Street, Covent Garden, London WC2H 9HE

Library of Congress Control Number: 2025001959

British Library Cataloguing-in-Publication Data
A catalogue record for this book is available from the British Library.

Front cover art by Listiarini Listiarini and Veselin Jungic ©

ADVANCED ENGINEERING CALCULUS
A Collection of Problems with Solutions

ISBN 978-981-98-0908-0 (hardcover)
ISBN 978-981-98-1158-8 (paperback)
ISBN 978-981-98-0909-7 (ebook for institutions)
ISBN 978-981-98-0910-3 (ebook for individuals)

For any available supplementary material, please visit
https://www.worldscientific.com/worldscibooks/10.1142/14208#t=suppl

Desk Editors: Soundararajan Raghuraman/Cian Sacker Ooi

Typeset by Stallion Press
Email: enquiries@stallionpress.com

To my sons, my best teachers.

Preface

The purpose of this collection of problems is to serve as a supplementary learning resource for students who are studying what is broadly referred to as advanced engineering calculus. This book covers the following areas: Fourier series and transforms, ordinary differential equations, difference equations, Laplace transforms, and vector analysis. Further, with this collection, I seek to provide another teaching resource for colleagues who teach those areas of advanced calculus.

The book can also be used as a refresher, or as a quick reference, by all of those students, instructors, and engineers who have taken advanced calculus in the past but may have forgotten parts of the subject or need to further their understanding in certain topics of the above-listed areas.

The collection contains 440 problems, many of them with multiple parts. The problems are sorted by topic, and almost all of them are accompanied by detailed solutions. The majority of the problems were used as exam questions in courses that I taught, going back to the 1980s [15]. My selection of problems was guided by classical texts [3, 13, 16, 25, 26, 30], standard textbooks [8, 20, 33], and modern resources [9, 11, 12, 18, 27, 28, 34, 37]. In particular, I was strongly influenced by Professor Dobrilo Tošić's work [35].

Each chapter starts with a summary of relevant mathematical notions and their properties. My expectation is that the reader is familiar with the topics covered in a standard North American sequence of calculus courses: differential, integral, and multivariable calculus. These topics include the mathematical notions of a function, limits, derivatives (including partial derivatives), indefinite, definite, improper, and multiple integrals, and power series. Occasionally,

I ask the reader to recall facts and techniques from linear algebra and complex analysis. It goes without saying that the reader is expected to be proficient with various algebraic techniques that are part of the secondary school mathematics curriculum.

This is a reader-friendly book, with a range of problems, from routine to not-so-routine. Since the purpose of this text is to serve as a supplementary learning and teaching resource, the reader will occasionally need the help of a *more knowledgeable other*: a peer, another book or article, or modern technology.

Still, this is a mathematical text. Precise and consistent mathematical notation and vocabulary are used throughout the book. Checking given properties before applying certain techniques, as a crucial step in mathematical thinking, is one of the main general messages of this book. The book clearly communicates some of the rather advanced calculus ideas and their applications, such as representing a function in a *convenient* and *meaningful* way, being a differential or a difference equation, the Fourier series, the Fourier or the Laplace or the z-transform, depending on the user's need.

The collection includes exercises that establish many general properties, facts, or techniques related to the topics of interest, including the properties of the Laplace transformation, the facts about the form of the Fourier series for certain classes of functions, and techniques for solving various types of differential and difference equations.

This book is written at the time of *two-eyed mathematical seeing*, a term that refers to learning to see from one eye with the strengths of mathematical rigour and from the other eye with the strengths of modern technology, and to using both these eyes together for the benefit of all [7]. Since, in my experience, a modern engineering student sees mathematics with their *technology eye* wide open, the intention of this collection is to remind the reader that they need to keep their *rigour* eye open too. Consequently, the reader is invited to try to solve problems *by hand* and experience the happiness, pride, and joy of *doing* mathematics manually, step-by-step.

No project such as this can be free from errors and incompleteness. I would be grateful to everyone who points out any typos or errors, or sends any other suggestions on how to improve this manuscript.

Veselin Jungić
Department of Mathematics, Simon Fraser University

Contents

Preliminaries

0.1 Symbols

$\emptyset$	empty set
$\{a, b, c\}$	set containing a, b, and c
$\{x : F(x)\}$	set of all x such that $F(x)$
$x \in A$	x belongs to A
$x \notin A$	x does not belong to A
$\mathbb{N}, \mathbb{N}_0, \mathbb{Z}, \mathbb{Q}, \mathbb{R}, \mathbb{C}$	sets of natural numbers, whole numbers, integers, rationals, reals, and complex numbers, respectively
$A \cup B$ and $A \cap B$	union and intersection of sets A and B, respectively
$A \backslash B$	$\{x : x \in A \text{ and } x \notin B\}$
$A \subset B$	A is a subset of B
$[a, b]$	$\{x \in \mathbb{R} : a \leq x \leq b\}$
$[a, b)$	$\{x \in \mathbb{R} : a \leq x < b\}$
(a, b)	$\{x \in \mathbb{R} : a < x < b\}$
$f : A \to B$	function from A to B (often, we will only write *a function f*)
$x \mapsto f(x)$	function that maps x to $f(x)$
$y = f(x)$	function that maps x to y
y' or $f'(x)$ or $\dfrac{dy}{dx}$	first derivative of the function $y = f(x)$

$\dfrac{\partial f}{\partial x_i}$	first partial derivative of the function $\omega = f(x_1, x_2, \ldots, x_n)$ with respect to the variable x_i, $1 \le i \le n$
$\int f(x)\,dx$	antiderivative of the function $y = f(x)$
$a \approx b$	a and b are close to each other by some given criterion
π	number pi, the ratio of a circle's circumference to its diameter
e	number e, also known as Euler's number or Napier's constant
$\sum_{i=1}^{n} a_i$	sum $a_1 + a_2 + \cdots + a_n$
$\{a_n\}_{n \in \mathbb{N}_0}$	sequence a_0, a_1, a_2, $\ldots$
$\mathbb{E}^2$ and $\mathbb{E}^3$	two-dimensional and three-dimensional Euclidean spaces, respectively
i	imaginary unit

0.2 Identities

If not stated otherwise, x and y are any real numbers.

$$x^k - y^k = (x - y) \sum_{i=0}^{k-1} x^i y^{k-1-i}, \ k \in \mathbb{N}$$

$$\sqrt{x^2} = |x| \qquad\qquad a^x a^y = a^{x+y}, \ a > 0$$

$$\frac{a^x}{a^y} = a^{x-y}, \ a > 0 \qquad\qquad (a^x)^y = a^{xy}, \ a > 0$$

$$\log_a xy = \log_a x + \log_a y, \qquad \arcsin(\sin x) = x, \ x \in \left[-\frac{\pi}{2}, \frac{\pi}{2}\right]$$

$$0 < a \ne 1, x, y > 0 \qquad \sin(\arcsin x) = x, \ x \in [-1, 1]$$

$$\log_a \frac{x}{y} = \log_a x - \log_a y, \qquad a^{\log_a x} = x, 0 < a \ne 1, x > 0$$

$$0 < a \ne 1, x, y > 0 \qquad \sin^2 x + \cos^2 x = 1$$

$$\log_a (a^x) = x, 0 < a \ne 1 \qquad \cos 2x = \cos^2 x - \sin^2 x$$

$$\log_a (x^y) = y \log_a x, \qquad \sin^2 x = \frac{1 - \cos 2x}{2}$$

$$0 < a \ne 1, x > 0$$

$$\sin 2x = 2 \sin x \cos x \qquad\qquad \cosh^2 x - \sinh^2 x = 1$$

$$\cos^2 x = \frac{1 + \cos 2x}{2} \qquad\qquad e^{ix} = \cos x + i \sin x$$

0.3 n-Dimensional Real Space

For $n \in \mathbb{N}$, the set $\mathbb{R}^n = \underbrace{\mathbb{R} \times \cdots \times \mathbb{R}}_{n} = \{(a_1, \ldots, a_n) : a_i \in \mathbb{R}$ for $1 \leq i \leq n\}$, together with the inner operation "+", defined by $(a_1, \ldots, a_n) + (b_1, \ldots, b_n) = (a_1 + b_1, \ldots, a_n + b_n)$, and the scalar multiplication "·", defined by $c \cdot (a_1, \ldots, a_n) = (ca_1, \ldots, ca_n)$, is called the *n-dimensional real space*. In this text, we will mostly consider functions defined on a subset of $\mathbb{R}^n$, for $n \in \{1, 2, 3\}$.

Number Line. Let ℓ be an Euclidean line, and let O and A be two points on ℓ.[1] This splits the line ℓ into two rays with the initial point O: one ray, call it r^+, that contains the point A and another ray, call it r^-, that does not contain the point A. We take the line segment $\overline{OA}$ as the unit of measurement of the distance between two points on the line ℓ. This allows us to establish a one-to-one correspondence between the elements of the set of real numbers and the points on the line:

- The point O, called the *origin*, corresponds to the number 0, and the point A corresponds to the number 1.
- The point X that belongs to the ray r^+ corresponds to the number $x > 0$ if and only if the distance between the point X and the point O is x units.[2]
- The point X that belongs to the ray r^- corresponds to the number $x < 0$ if and only if the distance between the point X and the point O is $-x$ units.

The line ℓ together with the above-described correspondence is called a *number line*.

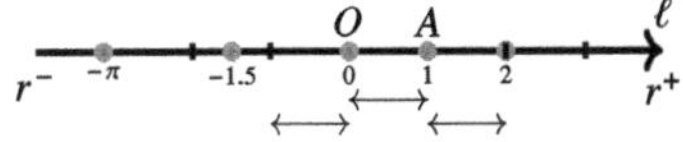

---

[1] Euclid, c. 300 BC, ancient Greek mathematician.

[2] The existence of the point X that corresponds to a rational number $\frac{p}{q}$, $p, q \in \mathbb{N}$, is intuitively clear: divide the line segment $\overline{OA}$ into q mutually congruent subsegments, and starting at O, apply one of those subsegments p times to reach the point X. If x is an irrational number, the existence of the corresponding point X is guaranteed by Dedekind's axiom. Richard Dedekind, 1831–1916, was a German mathematician.

Coordinate System in $\mathbb{E}^2$. In the Euclidean plane $\mathbb{E}^2$, we fix a point O and two mutually perpendicular number lines, both having O as the origin. This establishes, via signed distances between a point and the chosen lines, a one-to-one correspondence between $\mathbb{E}^2$ and $\mathbb{R}^2$. The point O, called the *origin*, and the number lines, called the *coordinate axes*, together with the said correspondence, determine a Cartesian coordinate system in $\mathbb{E}^2$. Commonly, we visualise this coordinate system by a drawing in which, from the viewer's point of view, one coordinate axis is horizontal with its positive ray to the right from the origin, and the other axis is vertical with its positive ray above the origin.

It is common to refer to the horizontal coordinate axis as the *x-axis* and the vertical coordinate axis as the *y-axis*. In this case, the coordinate system determined by the origin O, the x-axis, and the y-axis will be called the *Oxy coordinate system*.

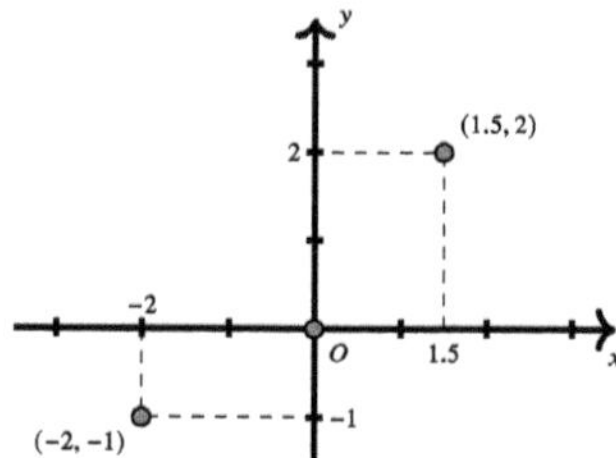

Coordinate System in $\mathbb{E}^3$. In the Euclidean space $\mathbb{E}^3$, we fix a point O, called the *origin*, and three mutually perpendicular number lines, called *coordinate axes*, each having O as the origin. Each pair of coordinate axes determines a *coordinate plane*. This establishes, via signed distances between a point and the coordinate planes, a one-to-one correspondence between $\mathbb{E}^3$ and $\mathbb{R}^3$. The origin and the coordinate axes, together with the said correspondence, determine a Cartesian coordinate system in $\mathbb{E}^3$.

It is common to refer to the coordinate axes as the x-axis, the y-axis, and the z-axis and to orient them in accordance with the *right-hand rule*: if one holds the index finger of their right hand pointing in the positive direction of the x-axis and the middle finger pointing in the positive direction of the y-axis, then the thumb is pointing in the positive direction of the z-axis. In this case, the coordinate system determined by the origin O, the x-axis, the y-axis, and the z-axis will be called the *Oxyz coordinate system*.

This one-to-one correspondence between $\mathbb{R}^n$ and $\mathbb{E}^n$, $n \in \{2,3\}$, is one of the reasons why it is common to refer to elements of $\mathbb{R}^n$ as *points*. We will make an explicit distinction between $\mathbb{R}^n$ and $\mathbb{E}^n$ only when we wish to underline that a particular notion in $\mathbb{E}^n$ does not depend on our choice of the coordinate system.[3]

The *distance* between two points in $\mathbb{R}^n$, $P = (a_1, \ldots, a_n)$ and $Q = (b_1, \ldots, b_n)$, is defined as $d(P,Q) = \sqrt{\sum_{i=1}^{n}(a_i - b_i)^2}$. In $\mathbb{R}^n$, $n \in \{1,2,3\}$, this distance is the length, in the appropriate unit of measurement, of the line segment with the end points P and Q in the underlying Euclidean space.

A set $D \subset \mathbb{R}^n$ is *bounded* if there is $M > 0$ such that, for any $P, Q \in D$, $d(P,Q) < M$.

Interior and Boundary Points and Open and Closed Sets. Let $n \in \mathbb{N}$ and let D be a subset of $\mathbb{R}^n$. We say that $P \in D$ is an *interior point of D* if there is $r > 0$ such that $S(P,r) = \{Q \in \mathbb{R}^n : d(P,Q) < r\} \subset D$. In $\mathbb{R}$, $S(P,r)$ is an open interval centred at P; in $\mathbb{R}^2$, it is the interior of the disc with centre at P and radius r; and in $\mathbb{R}^3$, $S(P,r)$ is the interior of the sphere with centre at P and radius r.

A point $Q \in \mathbb{R}^n$ is a *boundary point of D* if, for any $r > 0$, $S(Q,r) \cap D \neq \emptyset$ and $S(Q,r) \cap (\mathbb{R}^n \backslash D) \neq \emptyset$. A boundary point may or may not belong to the set. For example, the point $Q = (1,0)$ is a boundary point of the set $D = \{(x,y) \in \mathbb{R}^2 : x^2 + y^2 < 1\}$. Observe that any point in D is an interior point of D.

A set $D \subseteq \mathbb{R}^n$ is *open* if each of its points is an interior point of D. The empty set is an open set by definition. A set is closed if its complement is open, or, alternatively, a set is closed if it contains all of its boundary points.[4] If $D \subseteq \mathbb{R}^n$, then the set $\overline{D} \subseteq \mathbb{R}^n$ that consists only of the interior and boundary points of D is the smallest – in the sense of set inclusion – closed set that contains D as a subset. We say that $\overline{D}$ is the *closure of D*.

---

[3] One such notion is a *curve* as a set of points in $\mathbb{E}^3$ with certain properties. Representing the given curve in a particular coordinate system becomes a question of convenience.

[4] The notion of open and closed sets and the reason why the empty set and the set $\mathbb{R}^n$ are considered both open and closed is commonly studied in an Introduction to Topology course. The topology in $\mathbb{R}^n$, and, correspondingly, in $\mathbb{E}^n$, determined by the family of open sets defined as above is called the *Euclidean topology*. See, for example, [2].

For example, the set $D = \{(x, y) \in \mathbb{R}^2 : x^2 + y^2 < 1\}$ is open, the set $E = \{(x, y) \in \mathbb{R}^2 : x^2 + y^2 \leq 1\}$ is closed, and the set $F = \{(x, y) \in \mathbb{R}^2 : 0 < x^2 + y^2 \leq 1\}$ is neither closed nor open in $\mathbb{R}^2$. Observe that $\overline{D} = \overline{F} = \overline{E} = E$.

We say that a set $A \subseteq \mathbb{R}^n$ is a *neighbourhood* of a point $a \in \mathbb{R}^n$ if there is an open set U such that $a \in U \subseteq A$.

We say that a set $A \subseteq \mathbb{R}^n$ is *connected* if there are no two disjoint open sets U and V such that $A = (A \cap U) \cup (A \cap V)$.

A *region* is a non-empty connected open set in $\mathbb{R}^n$.

0.4   Functions

Function. A *function* $f : A \to B$ is a rule that assigns to each element in a set A exactly one element in a set B. The set A is called the *domain* of the function f.[5] We write $A = \mathrm{Dom}(f)$. The set B is called the *codomain* of the function f.

When defining and manipulating functions with non-empty domains, it is common to use the so-called *functional notation*, $y = f(x)$, which is read "y equals f of x," and means that the output y in the codomain correspondents to the input x in the domain by the rule f. In this setting, x denotes the *independent variable*, a variable that represents the elements of the domain, and y denotes the *dependent variable*, a variable that represents the elements of the codomain of the function.

If $f : A \to B$, then the *range* of the function f is the set of all elements y in B so that there is an element x in the domain A such that $y = f(x)$. The range is a subset of the codomain, $\mathrm{Range}(f) = \{y \in B : y = f(x) \text{ for some } x \in A\}$.

For example, consider sets $A = \{-2, -1, 0, 1, 2\}$ and $B = \{0, 1, 2, 3, 4\}$ and the rule f that assigns to each element x in the set A the element in the set B that is equal to x^2. Hence, $f(-2) = f(2) = 4$, $f(-1) = f(1) = 1$, and $f(0) = 0$. The set A is the domain of the function f, and the set B is its codomain. The range of the function f is the set $\{0, 1, 4\} \subset B$.

---

[5] If $A = \emptyset$, f is commonly called the *empty function*.

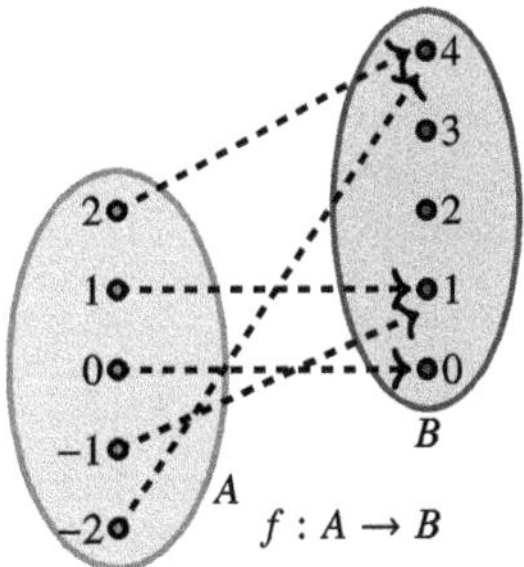

Occasionally, we will use the notation $x \mapsto f(x)$, which is read "x maps to $f(x)$" and means that the output $f(x)$ in the range correspondents to the input x in the domain by the rule f.

Piecewise Defined Function. If $f : A \to B$ and $g : A' \to B$, where $A \cap A' = \emptyset$ are two functions, then the function $F : A \cup A' \to B$, defined by

$$F(x) = \begin{cases} f(x) & \text{if } x \in A, \\ g(x) & \text{if } x \in A', \end{cases}$$

is said to be a *piecewise defined function*.

For example, the *absolute value function* $y = |x| = \begin{cases} -x & \text{if } x \in (-\infty, 0), \\ x & \text{if } x \in [0, \infty), \end{cases}$ is a piecewise defined function.

Extension and Restriction of a Function. If A and B are sets such that $A \subset B$ and if functions $f : A \to C$ and $g : B \to C$ are such that, for all $x \in A$, $f(x) = g(x)$, then we say that g is an *extension of f in the set B*. Equivalently, the function f is a *restriction of g in the set A*.

For example, if $f(x) = 1$, $x \in (0, \pi]$, then $g(x) = \begin{cases} -1 & \text{if } x \in [-\pi, 0], \\ 1 & \text{if } x \in (0, \pi], \end{cases}$ and $h(x) = \begin{cases} -1 & \text{if } x \in [-\pi, 0), \\ 1 & \text{if } x \in [0, \pi], \end{cases}$ are two extensions of f in $[-\pi, \pi]$.

One-to-One Function. We say that a function f is *one-to-one* if different inputs have different outputs. In other words, f is one-to-one if, for any x_1, x_2 in the domain of f, $x_1 \neq x_2$ implies $f(x_1) \neq f(x_2)$ or, equivalently, if $f(x_1) = f(x_2)$ implies $x_1 = x_2$.

For example, the function $f : [0, \infty) \to [0, \infty)$ defined by $f(x) = x^2$ is one-to-one and the function $g : \mathbb{R} \to [0, \infty)$ defined by $g(x) = x^2$ is not.

Inverse Function. Let $f : A \to B$ be a one-to-one function, and let the set B be the range of the function f. The *inverse function* of f, denoted f^{-1}, is the function defined in the domain B such that, for any $x \in A$ and $y \in B$, $f^{-1}(y) = x$ if and only if $f(x) = y$.

For example, the *square root function* $x \mapsto \sqrt{x}$ is the inverse function of the function $f : [0, \infty) \to [0, \infty)$ defined by $f(x) = x^2$.

Composition of Functions. Let $f : A \to B$ and $g : B \to C$ be two functions. The function $F : A \to C$ defined by $F(x) = g(f(x))$, $x \in A$, is called the *composition of f and g* and is denoted by $F = g \circ f$.

For example, if $x \mapsto f(x) = \sqrt{x}$ and $x \mapsto g(x) = x^2$, then $x \mapsto F(x) = (g \circ f)(x) = g(f(x)) = g(\sqrt{x}) = (\sqrt{x})^2 = x$ for $x \in [0, \infty)$. On the other hand, $x \mapsto G(x) = (f \circ g)(x) = f(g(x)) = \sqrt{x^2} = |x|$ for $x \in \mathbb{R}$.

Equation: Let A, B, and C be non-empty sets, and let $f : A \to C$ and $g : B \to C$ be functions. Let $A' = \{x \in A : f(x) \in \text{Range}\ (g)\}$ and $B' = \{x \in B : g(x) \in \text{Range}\ (f)\}$. To solve the equation $f(x) = g(x)$ means to determine the set $S = \{a \in A' \cap B' : f(a) = g(a)\}$. Any element of the set S is called a *solution* to the equation $f(x) = g(x)$. The set $A' \cap B'$ is called a *permissible set* of the equation $f(x) = g(x)$.

Properties of Functions. Let $S \subseteq \mathbb{R}$ be such that $x \in S$ if and only if $-x \in S$. Let $f : S \to \mathbb{R}$. We say that f is an *odd function* if $f(-x) = -f(x)$ for all $x \in S$. We say that f is an *even function* if $f(-x) = f(x)$ for all $x \in S$.

For example, the *sign function*

$$\text{sign}\ (x) = \begin{cases} -1 & \text{if } x < 0, \\ \ \ 0 & \text{if } x = 0, \\ \ \ 1 & \text{if } x > 0 \end{cases}$$

is an odd function.

Any *constant function* $x \mapsto f(x) = c$, $c \in \mathbb{R}$ is an even function.

The *Heaviside step function*

$$H(x) = \begin{cases} 0 & \text{if } x < 0, \\ 1 & \text{if } x \geq 0 \end{cases}$$

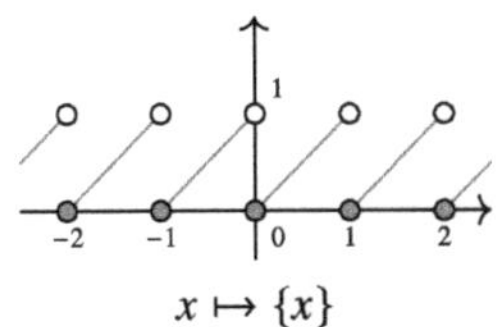

$$x \mapsto H(x)$$

is neither odd nor even.[6]

Let $S \subseteq \mathbb{R}$. A function $f : S \to \mathbb{R}$ is said to be *periodic* if there is a real number $T \neq 0$ such that, for any $x \in S$, $x + T \in S$ and $f(x + T) = f(x)$. The number T is called a *period* of the function f. The least positive period, if it exists, is called the *fundamental period* of the function f.

For example, the *fractional part function* $x \mapsto \{x\}$, a function that is defined as the difference between a real number and its greatest integer value, is periodic with the fundamental period $T = 1$.

$$x \mapsto \{x\}$$

Homogeneous Functions. Let $n \in \mathbb{N}$ and let $D \subseteq \mathbb{R}^n$. A function $f : D \to \mathbb{R}$ is called a homogeneous function with the degree of homogeneity $k \in \mathbb{R}$ if, for any $t \in \mathbb{R}\backslash\{1\}$ such that $tD = \{(tx_1, \ldots, tx_n) : (x_1, \ldots, x_n) \in D\} \subset D$, $f(tx_1, \ldots, tx_n) = t^k f(x_1, \ldots, x_n)$.

For example, the function $f(x, y) = y\sqrt{x} + x\sqrt{x}$, $(x, y) \in \mathbb{R}^+ \times \mathbb{R}$, is homogeneous with the degree of homogeneity $k = \frac{3}{2}$. The function $g(x, y, z) = \frac{xy + xz + yz}{x^2 + y^2 + z^2}$, $(x, y, z) \in \mathbb{R}^3\backslash\{(0, 0, 0)\}$, is homogeneous with the degree of homogeneity $k = 0$.

---

[6] Oliver Heaviside, 1850–1925, English mathematician and physicist.

Continuous and Piecewise Continuous Functions. Let $m, n \in$ $\mathbb{N}$, let $D \subseteq \mathbb{R}^n$, and let $f : D \to \mathbb{R}^m$. We say that the function f is continuous at a point $P \in D$ if for any open set $V \subseteq \mathbb{R}^m$ that contains $f(P)$, there is an open set $U \subseteq \mathbb{R}^n$ such that $P \in U$ and that, for any $Q \in U \cap D$, $f(Q) \in V$. If $m = 1$, $n \in \{1, 2, 3\}$, and P is an interior point in D, this is a restatement of the common $\varepsilon - \delta$, i.e., the limit definition of a function continuous at a point in its domain. By definition, a function may be continuous at a boundary point of its domain that belongs to the domain. A function is continuous in its domain if it is continuous at any point in the domain.

If $D \subseteq \mathbb{R}$ is an open, possibly unbounded interval, the function $f : D \to \mathbb{R}$ is *piecewise continuous* in D if for any bounded set $D' \subseteq D$ the function f is continuous in D' or there is a non-empty finite set $A \subset \overline{D'}$ such that: (a) f is continuous in the set $D' \backslash A$, and (b) for each $a \in A$ one sided limits $\lim\limits_{x \to a^+} f(x)$ and[7] $\lim\limits_{x \to a^-} f(x)$ exist.

For example, the function $f : \mathbb{R} \to \mathbb{R}$ defined by $f(x) = \lceil x \rceil$ is piecewise continuous.

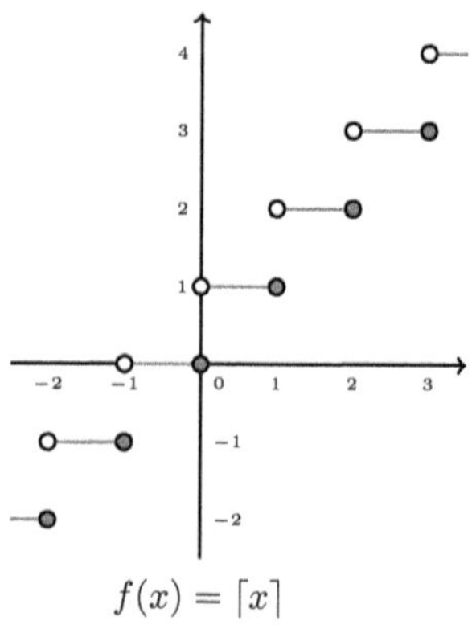

$$f(x) = \lceil x \rceil$$

Intuitively, the notion of a piecewise continuous function captures the idea of a function whose domain can be partitioned locally into finitely many pieces in each of which the function is continuous or it has a continuous extension.

Differentiable Functions. Let $D \subseteq \mathbb{R}$ and let $f : D \to \mathbb{R}$ be a function. We say that f is *differentiable at* $a \in D$ if the limit

---

[7] *or, in the case that $a = \min \overline{D'}$ or $a = \max \overline{D'}$.*

$f'(a) = \lim\limits_{h \to 0} \dfrac{f(a+h) - f(a)}{h}$ exists. The number $f'(a)$ is called the *derivative of f at a.* If D is an open set and if f is differentiable at any point in D, we say that *f is differentiable in the set D.*

Let $n \in \mathbb{N}\backslash\{1\}$, let $D \subseteq \mathbb{R}^n$, and let $f : D \to \mathbb{R}$ be a function. For each $i \in \{1, 2, \ldots, n\}$, the *partial derivative of f with respect to the variable x_i at* $(a_1, \ldots, a_i, \ldots, a_n) \in D$ is the number $\frac{\partial f}{\partial x_i}(a_1, \ldots, a_i, \ldots, a_n)$ defined as the value of

$$\lim_{h \to 0} \frac{1}{h}\left(f(a_1, \ldots, a_i + h, \ldots, a_n) - f(a_1, \ldots, a_i, \ldots, a_n)\right),$$

if this limit exists. The domain of the function $(x_1 \ldots, x_n) \mapsto \frac{\partial f}{\partial x_i}(x_1, \ldots, x_n)$ is the set of all $(x_1, \ldots, x_n) \in D$ for which the limit exists.

The *linearisation* of the function $(x, y) \mapsto f(x, y)$ at $(a, b) \in \mathrm{Dom}(f)$, for which $\frac{\partial f}{\partial x}(a, b)$ and $\frac{\partial f}{\partial y}(a, b)$ exist, is the linear function $L(x, y) = f(a, b) + (x - a)\frac{\partial f}{\partial x}(a, b) + (y - b)\frac{\partial f}{\partial y}(a, b)$. Since for $(\alpha, \beta) \in \mathrm{Dom}(f)$ that is *close* to (a, b), $f(\alpha, \beta) \approx L(\alpha, \beta)$, the number $L(\alpha, \beta)$ is called the *linear approximation* of $f(\alpha, \beta)$, the value of f at (α, β).

Similarly, the linearisation of a function $(x, y, z) \mapsto f(x, y, z)$ at $(a, b, c) \in \mathrm{Dom}(f)$ is $L(x, y, z) = f(a, b, c) + (x - a)\frac{\partial f}{\partial x}(a, b, c) + (y - b)\frac{\partial f}{\partial y}(a, b, c) + (z - c)\frac{\partial f}{\partial z}(a, b, c)$. The number $L(\alpha, \beta, \gamma)$ is called the *linear approximation* of $f(\alpha, \beta, \gamma)$.

The function $(x, y) \mapsto f(x, y)$ is *differentiable* at $(a, b) \in \mathrm{Dom}(f)$ if

$$\lim_{(x,y) \to (a,b)} \frac{f(x, y) - L(x, y)}{\sqrt{(x - a)^2 + (y - b)^2}} = 0,$$

where $L(x, y)$ is the linearisation of the function f at (a, b).

Similarly, the function $(x, y, z) \mapsto f(x, y, z)$ is differentiable at $(a, b, c) \in \mathrm{Dom}(f)$ if

$$\lim_{(x,y,z) \to (a,b,c)} \frac{f(x, y, z) - L(x, y, z)}{\sqrt{(x - a)^2 + (y - b)^2 + (z - c)^2}} = 0.$$

Connected and Simply Connected Sets. A set $D \subseteq \mathbb{R}^n$ is connected if and only if it is *path-connected*, i.e., if for any two points $A, B \in D$, there is a continuous function $f : [0,1] \to D$ such that $f(0) = A$ and $f(1) = B$.[8] The function f is called a *path* from A to B [22].

A connected set $D \subseteq \mathbb{R}^n$ is *simply connected* if for any $A, B \in D$ and any two paths f and g from A to B, the path f can be continuously deformed into the path g while keeping the endpoints A and B fixed, i.e., if there is a continuous function $F : [0,1] \times [0,1] \to D$ such that, for each $x \in [0,1]$, $F(x,0) = f(x)$ and $F(x,1) = g(x)$. Intuitively, a simply connected set is a connected set with no holes.

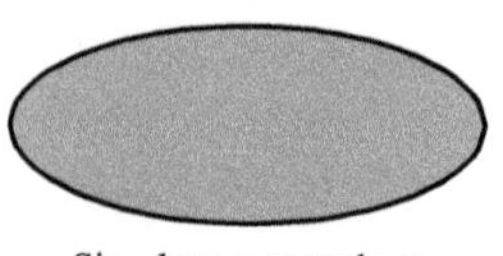

Simply connected set

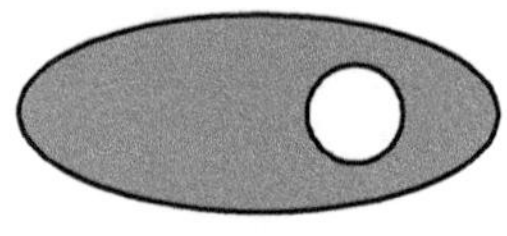

Connected, but not
simply connected set

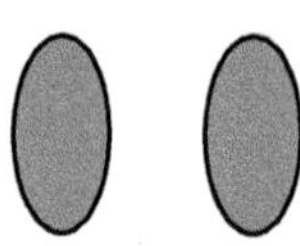

Not connected set

Smooth and Piecewise Smooth Functions. Let $m, n \in \mathbb{N}$. A real-valued function defined in an open set $D \subseteq \mathbb{R}^m$ is n-smooth, or a function of class C^n, if in D it has all partial derivatives up to order n and the derivatives of order n are continuous. For a function of class C^1, we say that it is a *smooth function* [21].

If $D \subseteq \mathbb{R}$ is an open, possibly unbounded, interval, a piecewise continuous function $f : D \to \mathbb{R}$ is *piecewise smooth* if for any bounded interval $D' \subseteq D$, the function f' is continuous in D' or there is a non-empty finite set $A \subset \overline{D'}$ such that: (a) f' is continuous in the set $D'\backslash A$, and (b) for each $a \in A$ one sided limits $\lim\limits_{x \to a^+} f'(x)$ and[9] $\lim\limits_{x \to a^-} f'(x)$ exist.[10]

---

[8] This characterisation corresponds to our intuition of *connectivity* on a line, plane, or three-dimensional space: a set is connected if it is possible to go from one point in the set to any other point in a continuous manner while staying in the set all the time.

[9] *or*, in the case that $a = \min \overline{D'}$ or $a = \max \overline{D'}$.

[10] For the purpose of this text, we have adjusted Rockafellar's definition [29] of a piecewise smooth function $f : D \to \mathbb{R}$, where $D \subseteq \mathbb{R}^n$, $n \geq 1$, by allowing f to be piecewise continuous. Rockafellar's definition requires f to be continuous.

Absolutely and Square Integrable Functions: Let $D \subseteq \mathbb{R}$. A function $f : D \to \mathbb{R}$ is called an *absolutely integrable* function if $\int_D |f(x)| \, dx < \infty$. A function $f : D \to \mathbb{R}$ is called a *square-integrable* function, a *quadratically integrable* function, or an L^2 function if $\int_D (f(x))^2 \, dx < \infty$.

0.5   Vectors

0.5.1 $\vec{\mathbb{E}}^2$ *and* $\vec{\mathbb{E}}^3$

For $A, B \in \mathbb{E}^2$, $A \neq B$, we represent the ordered pair $(A, B) \in \mathbb{E}^2 \times \mathbb{E}^2$ by a *directed line segment* in $\mathbb{E}^2$, i.e., by an arrow with the initial point at A and the tip point at B. We denote this directed line segment by $\overrightarrow{AB}$. We say that two directed line segments $\overrightarrow{AB}$ and $\overrightarrow{CD}$ are having the *same direction* and write $\overrightarrow{AB} \uparrow\uparrow \overrightarrow{CD}$ if one of the following occurs:

- if points A, B, C, and D are collinear, then one of the following three cases is required: (a) if $A = C$, then B and D are on the same side of $A = C$; (b) if $A \neq C$ and if B and C are on the same side of A, then C is in between A and D; or (c) if $A \neq C$ and if A is between B and C, then A and D are on the same side of C.
- if the lines AB and CD are parallel to each other and the points B and D are on the same side of the line AC.

We define the relation "$\sim$" in the set $\mathbb{E}^2 \times \mathbb{E}^2$ by, for any $A, B, C, D \in \mathbb{E}^2$, $(A, B) \sim (C, D)$ if and only if one of the following occurs: (a) $A = B$ and $C = D$; or (b) the directed line segments $\overrightarrow{AB}$ and $\overrightarrow{CD}$ have the same direction and the same length, i.e., $\overrightarrow{AB} \uparrow\uparrow \overrightarrow{CD}$ and $|\overline{AB}| = |\overline{CD}|$.

The relation $\sim$ is an equivalence relation, and it partitions the set $\mathbb{E}^2 \times \mathbb{E}^2$ into equivalence classes. This means that the relation $\sim$ determines a family, commonly denoted by $\mathbb{E}^2 \times \mathbb{E}^2/\sim$, of mutually disjoint sets whose union is $\mathbb{E}^2 \times \mathbb{E}^2$. The elements of $\mathbb{E}^2 \times \mathbb{E}^2/\sim$ are called *vectors in* $\mathbb{E}^2$ and denoted by a symbol with an over arrow, $\vec{v}$ for example. The set $\{(A, A) : A \in \mathbb{E}^2\} \in \mathbb{E}^2 \times \mathbb{E}^2/\sim$ is called the *zero vector*, and it is denoted by $\vec{0}$. If $(A, B) \in \vec{v} \neq \vec{0}$, then

$\vec{v} = \{(C, D) : |\overline{CD}| = |\overline{AB}| \text{ and } \overrightarrow{CD} \uparrow\uparrow \overrightarrow{AB}\}$. The set $\vec{v}$ is completely determined by any of its elements. We denote the set of all vectors in $\mathbb{E}^2$ by $\vec{\mathbb{E}}^2$.[11]

If $(A, B) \in \vec{v}$, then the *magnitude of the vector $\vec{v}$* is defined as $|\vec{v}| = |\overrightarrow{AB}|$. In particular, $|\vec{0}| = 0$.

If $\vec{v}$ and $\vec{w}$ are nonzero vectors such that $(A, B) \in \vec{v}$ and $(C, D) \in \vec{w}$, then $\vec{v}$ and $\vec{w}$ *have the same direction* if and only if $\overrightarrow{AB} \uparrow\uparrow \overrightarrow{CD}$. The zero vector has no direction.

For nonzero vectors $\vec{u}$ and $\vec{v}$, the *measure of the positively oriented angle between $\vec{u}$ and $\vec{v}$*, $\angle(\vec{u}, \vec{v})$, is defined in the following way. Let $X, Y, Z \in \mathbb{E}^2$ be such that $(X, Y) \in \vec{u}$ and $(X, Z) \in \vec{v}$. Then, $\angle(\vec{u}, \vec{v})$ is the measure of the positively oriented angle YXZ, i.e., the angle with the vertex at X, the initial ray XY, and the terminal ray XZ, rotated counterclockwise. Since any translation preserves angles, this definition does not depend on the choice of the point X.

The above definitions are among the reasons why it is common to abridge the notation and simply write $\overrightarrow{AB}$ and say "the vector A, B," meaning the vector $\vec{v}$, the equivalence class that contains the pair (A, B), rather than a single directed line segment.

The set $\vec{\mathbb{E}}^3$, the set of *vectors in $\mathbb{E}^3$*, is defined in a similar way.

0.5.2   *Scalar components of a vector*

In the Cartesian plane, i.e., in the Euclidean plane $\mathbb{E}^2$ equipped with a Cartesian coordinate system Oxy, we represent each vector $\vec{v} \in \vec{\mathbb{E}}^2$ by the directed line segment $\overrightarrow{OA}$, where $(O, A) \in \vec{v}$. If $A = (a, b)$, we write $\vec{v} = \langle a, b \rangle$ and call a and b the *scalar components* of the vector $\vec{v}$. The *length* (i.e., the *magnitude*) of the vector $\vec{v} = \langle a, b \rangle$ is the number $|\vec{v}| = |\langle a, b \rangle| = \sqrt{a^2 + b^2}$.

In the same manner, a vector $\vec{v}$ in the Cartesian three-dimensional space is denoted as an ordered triple, $\vec{v} = \langle a, b, c \rangle$, meaning that for $A = (a, b, c)$, $(O, A) \in \vec{v}$. In particular, $\vec{0} = \langle 0, 0, 0 \rangle$. The *length* of the vector $\vec{v} = \langle a, b, c \rangle$ is the number $|\vec{v}| = \sqrt{a^2 + b^2 + c^2}$.

---

[11]Hence, $\vec{\mathbb{E}}^2 = \mathbb{E}^2 \times \mathbb{E}^2 / \sim$.

Algebraic Operations with Vectors. Let $\vec{u} = \langle u_1, u_2, u_3 \rangle$, $\vec{v} = \langle v_1, v_2, v_3 \rangle$, and let $c \in \mathbb{R}$ be a scalar. Then:

(a) $\vec{u} = \vec{v}$ if and only if $u_1 = v_1$, $u_2 = v_2$, and $u_3 = v_3$.
(b) $\vec{u} + \vec{v} = \langle u_1 + v_1, u_2 + v_2, u_3 + v_3 \rangle$.
(c) $c \cdot \vec{u} = c\vec{u} = \langle cu_1, cu_2, cu_3 \rangle$.

These definitions of *addition* "+" of two vectors in $\mathbb{E}^3$ and *multiplication* "·" of a vector in $\mathbb{E}^3$ by a real number, in view of their geometrical interpretation in terms of directed line segments, do not depend on our choice of the coordinate system.

Properties of Algebraic Operations with Vectors. Let $\vec{u}$, $\vec{v}$, $\vec{w}$, be vectors and a, b be scalars. Then:

(a) $\vec{u} + \vec{v} = \vec{v} + \vec{u}$.
(b) $\vec{u} + \vec{0} = \vec{u}$.
(c) $0 \cdot \vec{u} = \vec{0}$.
(d) $(ab)\vec{u} = a(b\vec{u}) = b(a\vec{u})$.
(e) $(a + b)\vec{u} = a\vec{u} + b\vec{u}$.
(f) $(\vec{u} + \vec{v}) + \vec{w} = \vec{u} + (\vec{v} + \vec{w})$.
(g) $\vec{u} + (-1)\vec{u} = \vec{u} + (-\vec{u}) = \vec{0}$.
(h) $1 \cdot \vec{u} = \vec{u}$.
(i) $a(\vec{u} + \vec{v}) = a\vec{u} + a\vec{v}$.

In other words, $(\vec{\mathbb{E}}^3, +, \cdot)$ is a vector space.

Unit Vectors. A *unit vector* is a vector of length 1. If $\vec{v} \neq \vec{0}$, then $\vec{u} = \frac{1}{|\vec{v}|}\vec{v} = \frac{\vec{v}}{|\vec{v}|}$ is the unique unit vector with the same direction as $\vec{v}$.

Standard Basis Vectors. In the given Cartesian coordinate system in $\mathbb{E}^2$, it is common to take the unit vectors $\vec{\imath} = \langle 1, 0 \rangle$ and $\vec{\jmath} = \langle 0, 1 \rangle$ as the basis vectors of the vector space $(\mathbb{E}^2, +, \cdot)$. Then, for any $a_1, a_2 \in \mathbb{R}$, $\vec{a} = \langle a_1, a_2 \rangle = a_1\vec{\imath} + a_2\vec{\jmath}$. In the given Cartesian coordinate system in $\mathbb{E}^3$, it is common to take the unit vectors $\vec{\imath} = \langle 1, 0, 0 \rangle$, $\vec{\jmath} = \langle 0, 1, 0 \rangle$, and $\vec{k} = \langle 0, 0, 1 \rangle$ as the basis vectors of $(\vec{\mathbb{E}}^3, +, \cdot)$. Then, for any $a_1, a_2, a_3 \in \mathbb{R}$, $\vec{a} = \langle a_1, a_2, a_3 \rangle = a_1\vec{\imath} + a_2\vec{\jmath} + a_3\vec{k}$.

0.5.3 *Dot (scalar) product*

The *dot product* or the *scalar product* of two vectors $\vec{a} = \langle a_1, a_2, a_3 \rangle$ and $\vec{b} = \langle b_1, b_2, b_3 \rangle$ is defined as $\vec{a} \cdot \vec{b} = a_1 b_1 + a_2 b_2 + a_3 b_3$.

Properties of the Dot Product. Let $\vec{a}, \vec{b} \in \mathbb{E}^3$ and $c \in \mathbb{R}$. Then

(a) $\vec{a} \cdot \vec{a} = |\vec{a}|^2$.

(b) $\vec{a} \cdot \vec{b} = \vec{b} \cdot \vec{a}$.

(c) $\vec{a} \cdot (\vec{b} + \vec{c}) = \vec{a} \cdot \vec{b} + \vec{a} \cdot \vec{c}$.

(d) $(c\vec{a}) \cdot \vec{b} = c(\vec{a} \cdot \vec{b}) = \vec{a} \cdot (c\vec{b})$.

(e) $\vec{0} \cdot \vec{a} = 0$.

(f) $\vec{a} \cdot \vec{b} = |\vec{a}||\vec{b}| \cos \angle(\vec{a}, \vec{b})$.

(g) Two nonzero vectors $\vec{a}$ and $\vec{b}$ are mutually orthogonal if and only if $\vec{a} \cdot \vec{b} = 0$.

Direction Angles. The *direction angles* of a nonzero vector $\vec{a} = \langle a, b, c \rangle$ are the angles α, β, and γ that $\vec{a}$ makes, respectively, with vectors $\vec{i}$, $\vec{j}$, and $\vec{k}$. From $\cos \alpha = \frac{\vec{a} \cdot \vec{i}}{|\vec{a}|} = \frac{a}{|\vec{a}|}$, $\cos \beta = \frac{\vec{a} \cdot \vec{j}}{|\vec{a}|} = \frac{b}{|\vec{a}|}$, and $\cos \gamma = \frac{\vec{a} \cdot \vec{k}}{|\vec{a}|} = \frac{c}{|\vec{a}|}$, it follows that $\langle \cos \alpha, \cos \beta, \cos \gamma \rangle$ is the unit vector in the direction of $\vec{a}$.

Projections. The *scalar projection* of $\vec{b}$ onto $\vec{a} \neq \vec{0}$ is defined as $\text{comp}_{\vec{a}}\vec{b} = \frac{\vec{a} \cdot \vec{b}}{|\vec{a}|}$. The *vector projection* of $\vec{b}$ onto $\vec{a}$ is defined as $\text{proj}_{\vec{a}}\vec{b} = \frac{\vec{a} \cdot \vec{b}}{|\vec{a}|^2} \vec{a}$.

0.5.4   *Cross (vector) product*

The *cross product*, or the *vector product*, of two vectors $\vec{a} = \langle a_1, a_2, a_3 \rangle$ and $\vec{b} = \langle b_1, b_2, b_3 \rangle$ is defined by $\vec{a} \times \vec{b} = \langle a_2 b_3 - a_3 b_2, a_3 b_1 - a_1 b_3, a_1 b_2 - a_2 b_1 \rangle = \begin{vmatrix} \vec{i} & \vec{j} & \vec{k} \\ a_1 & a_2 & a_3 \\ b_1 & b_2 & b_3 \end{vmatrix}$.

The vector product of nonzero vectors $\vec{a}$ and $\vec{b}$ has two defining properties:

- $\vec{a} \times \vec{b}$ is the vector that is orthogonal to both $\vec{a}$ and $\vec{b}$ with direction given by the right-hand rule.
- The magnitude of $\vec{a} \times \vec{b}$ is the area of the parallelogram determined by $\vec{a}$ and $\vec{b}$, i.e., $|\vec{a} \times \vec{b}| = |\vec{a}||\vec{b}| \sin \angle(\vec{a}, \vec{b})$.

Properties of the Cross Product. Let $\vec{a}, \vec{b}, \vec{c} \in \mathbb{E}^3$ and $c \in \mathbb{R}$. Then

(a) $\vec{a} \times \vec{b} = -\vec{b} \times \vec{a}$.

(b) $\vec{0} \times \vec{a} = \vec{0}$.

(c) $\vec{a} \times (\vec{b} + \vec{c}) = \vec{a} \times \vec{b} + \vec{a} \times \vec{c}$.

(d) $(c\vec{a}) \times \vec{b} = c(\vec{a} \times \vec{b}) = \vec{a} \times (c\vec{b})$.

(e) $(\vec{a} \times \vec{b}) \cdot \vec{c} = \vec{a} \cdot (\vec{b} \times \vec{c})$.

(f) $\vec{a} \times (\vec{b} \times \vec{c}) = (\vec{a} \cdot \vec{c})\vec{b} - (\vec{a} \cdot \vec{b})\vec{c}$.

(g) Two nonzero vectors $\vec{a}$ and $\vec{b}$ are parallel if and only if $\vec{a} \times \vec{b} = \vec{0}$.

Triple Product. The volume of the parallelepiped determined by the vectors $\vec{a}$, $\vec{b}$, and $\vec{c}$ is the magnitude of their triple product: $V = |(\vec{a} \times \vec{b}) \cdot \vec{c}|$.

0.5.5 *Vector functions*

Let $I \subseteq \mathbb{R}$ and let $\vec{a} : I \to \vec{\mathbb{E}}^3$ be a *vector function*. The function $\vec{a}$ is *continuous* at $t_0 \in I$ if $\lim_{t \to t_0} |\vec{a}(t) - \vec{a}(t_0)| = 0$. The function $\vec{a}$ is *differentiable* at $t_0 \in I$ if there is a vector, denoted by $\vec{a}'(t_0)$, such that $\lim_{h \to 0} \left| \dfrac{\vec{a}(t_0 + h) - \vec{a}(t_0)}{h} - \vec{a}'(t_0) \right| = 0$.

If $Oxyz$ is a Cartesian coordinate system in $\mathbb{E}^3$ then we write $\vec{a} = \langle a_1, a_2, a_3 \rangle$, where a_1, a_2, and a_3, the *scalar components* of the vector function $\vec{a}$, are real valued functions defined in I such that, for each $t \in I$, $\vec{a}(t) = \langle a_1(t), a_2(t), a_3(t) \rangle$. The vector function $\vec{a}$ is continuous (differentiable) at $t \in I$ if and only if each of the functions a_1, a_2, and a_3 is continuous (differentiable) at t. In particular, if $\vec{a}$ is differentiable at $t \in I$, then $\vec{a}'(t) = \langle a_1'(t), a_2'(t), a_3'(t) \rangle$. The vector function $\vec{a}$ is *piecewise continuous* (*piecewise smooth*) in I if and only if each of the functions a_1, a_2, and a_3 is piecewise continuous (piecewise smooth) in I.

If $D \subseteq \mathbb{R}^k$, $k \in \{2, 3\}$, the notions of continuity and partial derivatives of a vector function $\vec{a} : D \to \vec{\mathbb{E}}^3$ are defined in a similar manner.

0.6 Series

Partial Sums and Convergence. Given a series $\sum_{n=0}^{\infty} a_n = a_0 + a_1 + a_2 + \cdots$, let s_m denote its mth partial sum: $s_m = \sum_{n=0}^{m} a_n = a_0 + a_1 + a_2 + \cdots + a_m$. If the sequence $\{s_m\}_{m \in \mathbb{N}_0}$ is convergent and $\lim_{m \to \infty} s_m = s$, then we say that the series is *convergent*, and we write

$\sum_{n=0}^{\infty} a_n = s$. The number s is the *sum* of the series. Otherwise, the series is *divergent*.

A Necessary Condition for Convergence. If a series $\sum_{n=0}^{\infty} a_n$ is convergent, then $\lim_{n\to\infty} a_n = 0$.

The Comparison Test. Suppose that $\sum_{n=0}^{\infty} a_n$ and $\sum_{n=0}^{\infty} b_n$ are series with nonnegative terms.

(a) If $\sum_{n=0}^{\infty} b_n$ is convergent and if there is $N \in \mathbb{N}_0$ such that $a_n \le b_n$ for all $n > N$, then $\sum_{n=0}^{\infty} a_n$ is also convergent.
(b) If $\sum_{n=0}^{\infty} b_n$ is divergent and if there is $N \in \mathbb{N}_0$ such that $a_n \ge b_n$ for all $n > N$, then $\sum_{n=0}^{\infty} a_n$ is also divergent.

Absolute and Conditional Convergence. A series $\sum_{n=0}^{\infty} a_n$ is called *absolutely convergent* if the series of absolute values $\sum_{n=0}^{\infty} |a_n|$ is convergent. A series $\sum_{n=0}^{\infty} a_n$ is called *conditionally convergent* if it is convergent but not absolutely convergent. For example, the series $\sum_{n=1}^{\infty} \frac{(-1)^n}{n^2}$ is absolutely convergent, and the series $\sum_{n=1}^{\infty} \frac{(-1)^n}{n}$ is conditionally convergent.

The Ratio Test.

(a) If $\lim_{n\to\infty} \left| \frac{a_{n+1}}{a_n} \right| = L < 1$, the series $\sum_{n=0}^{\infty} a_n$ is absolutely convergent.

(b) If $\lim_{n\to\infty} \left| \frac{a_{n+1}}{a_n} \right| = L > 1$ or $\lim_{n\to\infty} \left| \frac{a_{n+1}}{a_n} \right| = \infty$, the series $\sum_{n=0}^{\infty} a_n$ is divergent.

(c) If $\lim_{n\to\infty} \left| \frac{a_{n+1}}{a_n} \right| = L = 1$, the ratio test is inconclusive.

The Integral Test. Suppose f is a continuous, positive, decreasing function in $[1, \infty)$ and let $a_n = f(n)$, $n \in \mathbb{N}$. The series $\sum_{n=1}^{\infty} a_n$ is convergent if and only if the improper integral $\int_1^{\infty} f(x)\, dx$ is convergent.

Power Series. A *power series* is a series of the form $\sum_{n=0}^{\infty} c_n (x - a)^n = c_0 + c_1(x - a) + c_2(x - a)^2 + c_3(x - a)^3 + \cdots$, where x is a variable, and $c_0, c_1, c_2, c_3, \ldots$ are constants, called the *coefficients* of the series *centred at a*.

Convergence of a Power Series. For a given power series $\sum_{n=0}^{\infty} c_n(x-a)^n$, there are three possibilities: (a) the series converges only when $x = a$; (b) there is a positive number R, called the *radius of convergence*, such that the series converges if $|x - a| < R$ and diverges if $|x - a| > R$, where $(a - R, a + R)$ is called the *interval of convergence*; or (c) the series converges for all $x \in \mathbb{R}$.

Radius of Convergence. The radius of convergence of the power series $\sum_{n=0}^{\infty} c_n(x-a)^n$ can be calculated by using the ratio test:

$$R = \lim_{n \to \infty} \left| \frac{c_n}{c_{n+1}} \right|.$$

Frequently Used Power Series.

(a) $\frac{1}{1-x} = \sum_{n=0}^{\infty} x^n$, $x \in (-1, 1)$.   (c) $\sin x = \sum_{n=0}^{\infty} (-1)^n \frac{x^{2n+1}}{(2n+1)!}$, $x \in \mathbb{R}$.

(b) $e^x = \sum_{n=0}^{\infty} \frac{x^n}{n!}$, $x \in \mathbb{R}$.   (d) $\cos x = \sum_{n=0}^{\infty} (-1)^n \frac{x^{2n}}{(2n)!}$, $x \in \mathbb{R}$.

(e) $(1+x)^\alpha = 1 + \sum_{n=1}^{\infty} \binom{\alpha}{n} x^n$, $\alpha \in \mathbb{R}$, $x \in (-1, 1)$.

(f) $\ln(1 + x) = \sum_{n=1}^{\infty} (-1)^{n-1} \frac{x^n}{n}$, $x \in (-1, 1]$.

0.7   Theorems

Binomial Formula. Let $x, y \in \mathbb{R}$ and let $n \in \mathbb{N}$. Then,

$$(x + y)^n = \sum_{k=0}^{n} \binom{n}{k} x^{n-k} y^k = \sum_{k=0}^{n} \binom{n}{k} x^k y^{n-k}.$$

L'Hôpital's Rule. Suppose that f and g are differentiable functions and $g'(x) \neq 0$ near a, except possibly at a. Suppose that $\lim_{x \to a} f(x) = \lim_{x \to a} g(x) = 0$ or that $\lim_{x \to a} f(x) = \pm\infty$ and $\lim_{x \to a} g(x) = \pm\infty$. Then,

$\lim_{x \to a} \frac{f(x)}{g(x)} = \lim_{x \to a} \frac{f'(x)}{g'(x)}$ if the limit on the right-hand side exists or is

∞ or $-\infty$ [33].[12]

---

[12] Guillaume François Antoine, Marquis de L'Hôpital, 1661–1704, French mathematician.

The First Derivative Test. Suppose that c is a critical number of a continuous function f and that f is differentiable in $(c-\varepsilon, c)\cup(c, c+\varepsilon)$, for some $\varepsilon > 0$.

(a) If the derivative f' changes from positive to negative at c, then the function f has a local maximum at c.
(b) If the derivative f' changes from negative to positive at c, then the function f has a local minimum at c.
(c) If the derivative f' does not change sign at c, then the function f has no local minimum or maximum at c [33].

The Second Derivative Test. Suppose that f'' is a continuous function near c.

(a) If $f'(c) = 0$ and $f''(c) > 0$, then the function f has a local minimum at c.
(b) If $f'(c) = 0$ and $f''(c) < 0$, then the function f has a local maximum at c [33].

The Fundamental Theorem of Calculus. If f is a continuous function in $[a, b]$, then: (a) the function g, defined by $g(x) = \int_a^x f(t)dt$, $x \in [a, b]$, is continuous in $[a, b]$ and differentiable in (a, b) with $g'(x) = f(x)$; (b) $\int_a^b f(x)dx = F(b) - F(a)$, where F is any *antiderivative* of f, i.e., any continuous function $F : [a, b] \to \mathbb{R}$ such that, for all $x \in (a, b)$, $F'(x) = f(x)$ [33].

Leibniz Rule for Differentiation Under the Integral. Let D be a region in $\mathbb{R}^2$, and let a function $(x, t) \mapsto f(x, t)$ be continuous with a continuous partial derivative $\frac{\partial f}{\partial t}$ in D. Let $a, b, c, d \in \mathbb{R}$ be such that $a < b$, $c < d$, and $[a, b] \times [c, d] \subset D$. Let functions $\alpha, \beta : [c, d] \to [a, b]$ have continuous derivatives in (c, d). Then, for any $t \in (c, d)$,

$$\frac{d}{dt} \int_{\alpha(t)}^{\beta(t)} f(x, t)dx = f(\beta(t), t)\beta'(t) - f(\alpha(t), t)\alpha'(t) + \int_{\alpha(t)}^{\beta(t)} \frac{\partial f}{\partial t}(x, t)\, dx.$$

In other words, under the stated conditions, the order of differentiation and integration can be interchanged [16, 37].[13]

---

[13]Gottfried Wilhelm Leibniz, 1646–1716, a German mathematician, philosopher, scientist, and diplomat.

Fubini's Theorem. Let $f : \mathbb{R}^2 \to \mathbb{R}$ be an absolutely integrable function. Then there exist absolutely integrable functions $F : \mathbb{R} \to \mathbb{R}$ and $G : \mathbb{R} \to \mathbb{R}$ such that for almost every x, $f(x, y)$ is absolutely integrable in y with $F(x) = \int_{\mathbb{R}} f(x, y)\, dy$, and for almost every y, $f(x, y)$ is absolutely integrable in x with $G(y) = \int_{\mathbb{R}} f(x, y)\, dx$. Finally, we have $\int_{\mathbb{R}} F(x)\, dx = \int_{\mathbb{R}^2} f = \int_{\mathbb{R}} G(y)\, dy$ [35].[14]

Poincaré's Lemma. Let $D \subset \mathbb{R}^2$ be a simply connected open set. Continuously differentiable functions, $P = P(x, y)$ and $Q = Q(x, y)$, $(x, y) \in D$, satisfy $\frac{\partial Q}{\partial x}(x, y) = \frac{\partial P}{\partial y}(x, y)$ if and only if there is a twice continuous differentiable function $(x, y) \mapsto F(x, y)$, $(x, y) \in D$, such that $\frac{\partial F}{\partial x}(x, y) = P(x, y)$ and $\frac{\partial F}{\partial y}(x, y) = Q(x, y)$.[15]

Note: This statement is a special case of Poincaré's lemma. Rudin in [30] proves the lemma in the case when $D \subseteq \mathbb{R}^n$, $n \geq 2$, is an open convex set.

Implicit Function Theorem. Let D be an open set in $\mathbb{R}^2$ and let $(a, b) \in D$. Suppose that $F : D \to \mathbb{R}$ has continuous partial derivatives and that $F(a, b) = 0$, and $\frac{\partial F}{\partial y}(a, b) \neq 0$. Then, there are open intervals $I, J \subseteq \mathbb{R}$, such that $(a, b) \in I \times J \subseteq D$ and a differentiable function $f : I \to J$ such that $f(a) = b$ and, for all $x \in I$, $F(x, f(x)) = 0$ [3].

De Moivre's Theorem: For any $\alpha \in \mathbb{R}$ and any $n \in \mathbb{N}$, $(\cos \alpha + i \sin \alpha)^n = \cos n\alpha + i \sin n\alpha$.[16]

---

[14]Guido Fubini, 1879–1943, Italian mathematician.

[15]Jules Henri Poincaré, 1854–1912, French mathematician, theoretical physicist, engineer, and philosopher of science.

[16]Abraham de Moivre, 1667–1754, French mathematician.

Chapter 1

Fourier Analysis

1.1 Introduction

Use the following definitions, techniques, properties, and algorithms to solve the problems contained in this chapter. For more details, see [13, 18, 20, 27, 32].

Fourier Coefficients. Let $f : \mathbb{R} \to \mathbb{R}$ be a periodic function with the fundamental period T. If f is a piecewise continuous function in the interval $[0, T]$, then, for all $k \in \mathbb{N}_0 = \mathbb{N} \cup \{0\}$, $a_k = \frac{2}{T} \int_0^T f(x) \cos \frac{2k\pi x}{T} \, dx$ and, for all $k \in \mathbb{N}$, $b_k = \frac{2}{T} \int_0^T f(x) \sin \frac{2k\pi x}{T} \, dx$ are called the *Fourier coefficients of the function f*.[1]

Fourier Series. Let $\{a_k\}_{k \in \mathbb{N}_0}$ and $\{b_k\}_{k \in \mathbb{N}}$ be the Fourier coefficients of f, a piecewise continuous periodic function with the fundamental period T. The *Fourier series of the function f* is the series $\frac{a_0}{2} + \sum_{k=1}^{\infty} \left(a_k \cos \frac{2k\pi x}{T} + b_k \sin \frac{2k\pi x}{T} \right)$. We write $f(x) \sim \frac{a_0}{2} + \sum_{k=1}^{\infty} \left(a_k \cos \frac{2k\pi x}{T} + b_k \sin \frac{2k\pi x}{T} \right)$.

Dirichlet's Theorem. Let $f : \mathbb{R} \to \mathbb{R}$ be a piecewise continuous periodic function, with the fundamental period T, such that f' is a piecewise continuous function in the interval $[0, T]$.[2]

---

[1] Jean-Baptiste Joseph Fourier, 1768–1830, French mathematician and physicist.

[2] Equivalently, f is a piecewise smooth function.

1

Then, for all $x \in \mathbb{R}$,[3]

$$\frac{1}{2}\left(\lim_{t \to x^-} f(t) + \lim_{t \to x^+} f(t)\right)$$

$$= \frac{a_0}{2} + \sum_{k=1}^{\infty}\left(a_k \cos\frac{2k\pi x}{T} + b_k \sin\frac{2k\pi x}{T}\right) \; [18, 32].$$

Parseval's Identity. Let $f : [-\pi, \pi] \to \mathbb{R}$ be a square-integrable function, and let $\{a_k\}_{k \in \mathbb{N}_0}$ and $\{b_k\}_{k \in \mathbb{N}}$ be the Fourier coefficients of the periodic extension of the function f.[4] Then,

$$\frac{a_0^2}{2} + \sum_{k=1}^{\infty}\left(a_k^2 + b_k^2\right) = \frac{1}{\pi}\int_{-\pi}^{\pi}\left(f(x)\right)^2 dx \; [32].$$

Term-by-Term Differentiation of Fourier Series. Let f be a continuous piecewise smooth function with the fundamental period T. Then, from, for any $x \in \mathbb{R}$, $f(x) = \frac{a_0}{2} + \sum_{k=1}^{\infty}\left(a_k \cos\frac{2k\pi x}{T} + b_k \sin\frac{2k\pi x}{T}\right)$, it follows that $f'(x) \sim \frac{2\pi}{T}\sum_{k=1}^{\infty} k\left(b_k \cos\frac{2k\pi x}{T} - a_k \sin\frac{2k\pi x}{T}\right)$ [32].

Integration of Fourier Series. Let f be a piecewise continuous function with the fundamental period T, let $\int_0^T f(x)\, dx = 0$, and let $f(x) \sim \sum_{k=1}^{\infty}\left(a_k \cos\frac{2k\pi x}{T} + b_k \sin\frac{2k\pi x}{T}\right)$. If, for $x \in \mathbb{R}$, $F(x) = \int_0^x f(t)\, dt$, then, for all $x \in \mathbb{R}$, $F(x) = \frac{A_0}{2} + \frac{T}{2\pi}\sum_{k=1}^{\infty}\left(\frac{a_k}{k}\sin\frac{2k\pi x}{T} - \frac{b_k}{k}\cos\frac{2k\pi x}{T}\right)$, where $A_0 = \frac{2}{T}\int_0^T F(x)\, dx$. The function F is periodic with a period T, continuous, and with the piecewise continuous derivative $F'(x) = f(x)$. The Fourier series of the function F converges uniformly [32].

Fourier Integral. Let $f : \mathbb{R} \to \mathbb{R}$ be a piecewise smooth absolutely integrable function, i.e., let f be such that:

(a) both f and f' are piecewise continuous functions;

---

[3] Johann Peter Gustav Lejeune Dirichlet, 1805–1859, German mathematician.
[4] Marc-Antoine Parseval des Chênes, 1755–1836, French mathematician.

(b) $\int_{-\infty}^{\infty} |f(x)|dx < \infty.$

Then, for any $x \in \mathbb{R}$, the *Fourier integral* of f is

$$\int_0^{\infty} (a(u)\cos ux + b(u)\sin ux)\, du = \frac{1}{2}\left(\lim_{t \to x^-} f(t) + \lim_{t \to x^+} f(t) \right),$$

where

$$a(u) = \frac{1}{\pi}\int_{-\infty}^{\infty} f(t)\cos ut\, dt \text{ and}$$

$$b(u) = \frac{1}{\pi}\int_{-\infty}^{\infty} f(t)\sin ut\, dt \, [3].$$

Fourier Transform. Let $f : \mathbb{R} \to \mathbb{R}$ be an absolutely integrable function over $\mathbb{R}$, i.e., f is such that $\int_{-\infty}^{\infty} |f(t)|\, dt < \infty$. The function $u \mapsto \mathcal{F}[f(t)](u) = \int_{-\infty}^{\infty} f(t)e^{-iut}\, dt,\ u \in \mathbb{R}$, is called the *Fourier transform* of the function f.[5]

Fourier Transformation. The mapping $f \mapsto \mathcal{F}[f(t)]$ is called the *Fourier transformation*. The Fourier transformation is a linear mapping, i.e., if f and g are absolutely integrable functions over $\mathbb{R}$ and if $a, b \in \mathbb{R}$, then $\mathcal{F}[af(t) + bg(t)] = a\mathcal{F}[f(t)] + b\mathcal{F}[g(t)]$.

The Inversion Theorem. If both the function f and its Fourier transform $\mathcal{F}[f(t)]$ are absolutely integrable over and continuous in $\mathbb{R}$, then $f(t) = \frac{1}{2\pi}\int_{-\infty}^{\infty} \mathcal{F}[f(t)](s)e^{its}\, ds$.

Fourier Sine and Cosine Transforms. Let $f : [0, \infty) \to \mathbb{R}$ be an absolutely integrable function over $[0, \infty)$. The function $u \mapsto \mathcal{F}_{\cos}[f(t)](u) = \sqrt{\frac{2}{\pi}}\int_0^{\infty} f(t)\cos ut\, dt,\ u \in [0, \infty)$, is called the *Fourier cosine transform* of the function f.[6]

---

[5]We follow Körner [18]. There are several commonly used definitions of the Fourier transform of a function. The default definition used by Maple is $u \mapsto \mathcal{F}[f(t)](u) = \int_{-\infty}^{\infty} f(t)e^{-iut}\, dt$. As a default, Mathematica uses $u \mapsto \mathcal{F}[f(t)](u) = \frac{1}{\sqrt{2\pi}}\int_{-\infty}^{\infty} f(t)e^{iut}\, dt$. Another commonly used definition is $u \mapsto \mathcal{F}[f(t)](u) = \int_{-\infty}^{\infty} f(t)e^{-2i\pi ut}\, dt$ [32].

[6]This is the default definition of the Fourier cosine transform used by both Maple and Mathematica.

Similarly, the *Fourier sine transform* of f is defined as the function $u \mapsto \mathcal{F}_{\sin}[f(t)](u) = \sqrt{\frac{2}{\pi}} \int_0^\infty f(t) \sin ut \ dt$, $u \in [0, \infty)$.

1.2   Fourier Series

1.2.1   *Functions with the fundamental period $T = 2\pi$*

Problem 1.2.1. Let $T \in (0, \infty)$. Let $f : \mathbb{R} \to \mathbb{R}$ be a continuous, odd, and periodic function, with the fundamental period $2T$, such that for any $x \in (T, 2T)$, $f(x) = (x - T) \ln \frac{x-T}{T}$. Sketch a graph of the function f and evaluate $f\left(\frac{mT}{2}\right)$, $m \in \mathbb{Z}$.

Solution 1.2.1. Since f is an odd function defined in the set of all real numbers $\mathbb{R}$, it follows that $f(0) = 0$. This, together with the fact that f is periodic, implies that $f(2T) = f(0) = 0$. In particular, this means that, for each $x \in (T, 2T]$, $f(x) = (x - T) \ln \frac{x-T}{T}$. From, for each $x \in (-T, 0]$, $x + 2T \in (T, 2T]$, since the function f is periodic with the fundamental period $2T$, it follows that, for all $x \in (-T, 0]$, $f(x) = f(x + 2T) = (x + T) \ln \frac{x+T}{T}$. This implies, since f is an odd function, that for all $x \in [0, T)$, $f(x) = -f(-x) = -(T - x) \ln \frac{T-x}{T}$. Since $\lim\limits_{x \to T^-} f(x) = \lim\limits_{x \to T^-} \left(-(T - x) \ln \dfrac{T - x}{T} \right) = 0$ and since the function is continuous at T, it follows that $f(T) = 0$.

Therefore,

$$
f(x) = \begin{cases} (x + T) \ln \frac{x+T}{T}, & x \in (-T, 0], \\ (x - T) \ln \frac{T-x}{T}, & x \in (0, T), \\ 0, & x \in \{-T, T\}, \end{cases}
$$

and, for any integer k and any real number x, $f(x + 2kT) = f(x)$. To sketch a graph, observe that, for all $x \in (0, T)$, $\frac{T-x}{T} < 1$, which implies $f(x) > 0$. Also, for all $x \in (0, T)$, $f'(x) = \ln \frac{T-x}{T} + 1$ and $f''(x) = \frac{1}{x-T}$. It follows that, for $x \in (0, T)$, $f'(x) = 0$ if and only if $x = T(1 - e^{-1})$. By the second derivative test, the only local extremum of the function f in $(0, T)$ is the local maximum at $(T(1 - e^{-1}), Te^{-1})$. It follows that, for $k \in \mathbb{Z}$, $f\left(2k \cdot \frac{T}{2}\right) = f(kT) = 0$,

$$f\left((4k+1)\tfrac{T}{2}\right) = f\left(2kT + \tfrac{T}{2}\right) = f\left(\tfrac{T}{2}\right) = \tfrac{T\ln 2}{2}, \text{ and } f\left((4k+3)\tfrac{T}{2}\right) =$$
$$f\left(2kT + \tfrac{3T}{2}\right) = f\left(\tfrac{3T}{2}\right) = -\tfrac{T\ln 2}{2}.$$

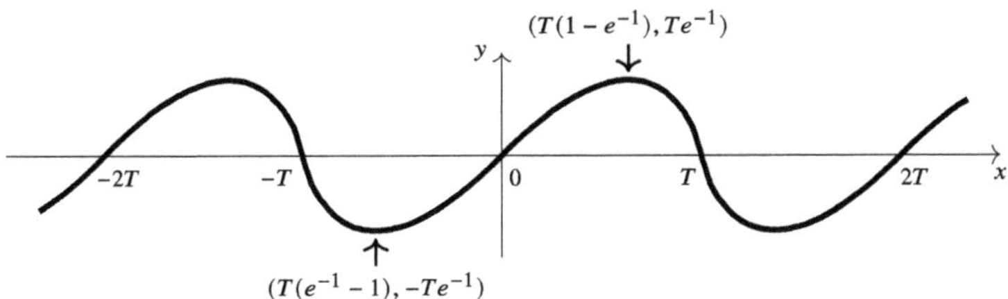

Problem 1.2.2. For $n \in \mathbb{N}_0$, let $f_n : [-\pi, \pi] \to \mathbb{R}$ and $g_n : [-\pi, \pi] \to \mathbb{R}$ be defined by $f_n(x) = \cos nx$ and $g_n(x) = \sin nx$. Establish that the sequence $f_0, f_1, g_1, \ldots, f_n, g_n, \ldots$ is an orthogonal system of functions, i.e., establish that $\int_{-\pi}^{\pi} f_m(x) f_n(x) \, dx = \begin{cases} 0 & \text{if } m \neq n, \\ \pi & \text{if } m = n \neq 0, \\ 2\pi & \text{if } m = n = 0, \end{cases}$

$\int_{-\pi}^{\pi} f_m(x) g_n(x) \, dx = 0$, and $\int_{-\pi}^{\pi} g_m(x) g_n(x) \, dx = \begin{cases} 0 & \text{if } m \neq n, \\ \pi & \text{if } m = n. \end{cases}$

Note: The term *orthogonal* comes from the fact that, in the vector space of all real-valued functions that are square-integrable over the interval $[-\pi, \pi]$, the scalar product can be defined as $\langle f, g \rangle = \int_{-\pi}^{\pi} f(x) g(x) \, dx$. The question is to establish that, for $m, n \in \mathbb{N}_0$ and $p, q \in \mathbb{N}$, if $m \neq n$ and $p \neq q$, then $\langle f_m, f_n \rangle = \langle g_p, g_q \rangle = 0$, and $\langle f_m, g_p \rangle = 0$.

Solution 1.2.2. The claim follows from, for any $m, n \in \mathbb{N}_0$ and $p, q \in \mathbb{N}$,

$$\int_{-\pi}^{\pi} \cos mx \cos nx \, dx = \int_{0}^{\pi} (\cos(m+n)x + \cos(m-n)x) \, dx$$

$$= \begin{cases} 0 & \text{if } m \neq n, \\ \pi & \text{if } m = n \neq 0, \\ 2\pi & \text{if } m = n = 0, \end{cases}$$

$\int_{-\pi}^{\pi} \cos mx \sin px \, dx = 0$, and

$$\int_{-\pi}^{\pi} \sin px \sin qx \, dx = \int_{0}^{\pi} (\cos(p-q)x - \cos(p+q)x) \, dx$$

$$= \begin{cases} 0 & \text{if } p \neq q, \\ \pi & \text{if } p = q. \end{cases}$$

Problem 1.2.3. Let $m \in \mathbb{N}_0$. Let $f : \mathbb{R} \to \mathbb{R}$ be a periodic function with the fundamental period 2π such that its mth derivative $f^{(m)} : \mathbb{R} \to \mathbb{R}$ is a continuous function.[7] Let $\{a_k\}_{k \in \mathbb{N}_0}$ and $\{b_k\}_{k \in \mathbb{N}}$ be the Fourier coefficients of the function f. Use the fact[8] that, for any $a, b \in \mathbb{R}$, if a function $g : [a,b] \to \mathbb{R}$ is integrable, then $\lim\limits_{k \to \infty} \left| \int_{a}^{b} g(x) e^{-ikx} \, dx \right| = 0$ to prove that $\lim\limits_{k \to \infty} k^m |a_k| = \lim\limits_{k \to \infty} k^m |b_k| = 0$.

Solution 1.2.3. For $k \geq 1$, let $c_k = a_k - ib_k = \frac{1}{\pi} \int_{-\pi}^{\pi} f(x) e^{-ikx} \, dx$. Clearly, $\lim\limits_{k \to \infty} \sqrt{a_k^2 + b_k^2} = \lim\limits_{k \to \infty} |c_k| = \frac{1}{\pi} \lim\limits_{k \to \infty} \left| \int_{-\pi}^{\pi} f(x) e^{-ikx} \, dx \right| = 0$, which implies that the claim is true if $m = 0$. Suppose that $m \geq 1$. Then,

$$c_k = \frac{1}{\pi} \int_{-\pi}^{\pi} f(x) e^{-ikx} \, dx = \begin{vmatrix} u = f(x) & du = f'(x) \, dx \\ dv = e^{-ikx} \, dx & v = \frac{i}{k} e^{-ikx} \end{vmatrix}$$

$$= \frac{i}{k\pi} \left(f(x) e^{-ikx} \Big|_{-\pi}^{\pi} - \int_{-\pi}^{\pi} f'(x) e^{-ikx} \, dx \right)$$

$$= \frac{i}{k\pi} \left((f(\pi) - f(-\pi))(-1)^k - \int_{-\pi}^{\pi} f'(x) e^{-ikx} \, dx \right)$$

$$= -\frac{i}{k\pi} \int_{-\pi}^{\pi} f'(x) e^{-ikx} \, dx = \cdots = \frac{(-1)^m i^m}{k^m \pi} \int_{-\pi}^{\pi} f^{(m)}(x) e^{-ikx} \, dx.$$

---

[7] In the case of $m = 0$, the function f is continuous.

[8] This is the claim of the Riemann–Lebesgue lemma [18, 32]. Henri Léon Lebesgue, 1875–1941, was a French mathematician. Georg Friedrich Bernhard Riemann, 1826–1866, was a German mathematician.

It follows that $\lim_{k\to\infty} k^m|c_k| = \frac{1}{\pi}\lim_{k\to\infty}\left|\int_{-\pi}^{\pi} f^{(m)}(x)e^{-ikx}\,dx\right| = 0,$ which implies that the claim is true for $m \in \mathbb{N}_0$.

Note: The fact $\lim_{k\to\infty} k^m|a_k| = \lim_{k\to\infty} k^m|b_k| = 0$ implies that the Fourier coefficients a_k and b_k decay faster than $\frac{1}{k^m}$ when k is increasing. Moreover, if $f : \mathbb{R} \to \mathbb{R}$ is a periodic function with the fundamental period 2π such that, for some $\ell \geq 0$, $f^{(\ell)} : \mathbb{R} \to \mathbb{R}$ is not continuous, but it is a piecewise continuous and a piecewise differentiable[9] function and, in the case of $\ell > 0$, if $f^{(\ell-1)}$ is a continuous function, then $\max\{|a_k|, |b_k|\} \leq |c_k| \leq \frac{M}{k^{\ell+1}} + \sigma_k$, where $M > 0$ and $\lim_{k\to\infty}\sigma_k = 0$. We illustrate this fact by demonstrating a special case when $\ell = 0$. Assume that a periodic piecewise differentiable function f, with the fundamental period $T = 2\pi$, has only one point of discontinuity, α, in the interval $[-\pi, \pi)$, i.e., $f(\alpha^+) = \lim_{x\to\alpha^+} f(x) \neq \lim_{x\to\alpha^-} f(x) = f(\alpha^-)$. Then,

$$c_k = \frac{1}{\pi}\int_{\alpha}^{\alpha+2\pi} f(x)e^{-ikx}\,dx = \begin{vmatrix} u = f(x) & du = f'(x)\,dx \\ dv = e^{-ikx}\,dx & v = \frac{i}{k}e^{-ikx} \end{vmatrix}$$

$$= \frac{i}{k\pi}(f(\alpha^-) - f(\alpha^+))e^{-ik\alpha} - \frac{i}{k\pi}\int_{\alpha}^{\alpha+2\pi} f'(x)e^{-ikx}\,dx.$$

Then, for $M = \frac{1}{\pi}|f(\alpha^-) - f(\alpha^+)| > 0$ and $\sigma_k = \frac{1}{k\pi}\left|\int_{\alpha}^{\alpha+2\pi} f'(x)e^{-ikx}\,dx\right| \to 0$, when $k \to \infty$, we have $|c_k| \leq \frac{M}{k} + \sigma_k$.

Problem 1.2.4. Determine the Fourier series of the following periodic functions with the fundamental period $T = 2\pi$:

(a) $f(x) = x - |x|$, $x \in [-\pi, \pi)$.
(b) $f(x) = x + |x|$, $x \in [-\pi, \pi)$. Evaluate $\sum_{k=0}^{\infty} \frac{1}{(2k+1)^2}$.

---

[9]This means that there is a non-empty finite set $A \subset [-\pi, \pi)$ such the set $\{A + 2n\pi : n \in \mathbb{Z}\}$ is the set of all points of discontinuity of $f^{(\ell)}$ and that at each $a \in A$ both $\lim_{x\to a^-} f^{(\ell)}(x)$ and $\lim_{x\to a^+} f^{(\ell)}(x)$ exist. In addition, there is a finite set $B \subset [-\pi, \pi]$, $A \subseteq B$, such that $\{B + 2n\pi : n \in \mathbb{Z}\}$ is the set of all points of discontinuity of $f^{(\ell+1)}$ and at each $b \in B$ both $\lim_{x\to b^-} f^{(\ell+1)}(x)$ and $\lim_{x\to b^+} f^{(\ell+1)}(x)$ exist. Observe that under these conditions functions $f^{(\ell)}$ and $f^{(\ell+1)}$, because of their periodicity, are integrable over $[\alpha, \alpha + 2\pi]$, for any $\alpha \in \mathbb{R}$.

(c) $f(x) = \sin x + |\sin x|$.
(d) $f(x) = \text{sign}(\cos x + \sin x) + 1$.

Solution 1.2.4.

(a) Observe that

$$f(x) = \begin{cases} 2x, & x \in [-\pi, 0), \\ 0, & x \in [0, \pi). \end{cases}$$

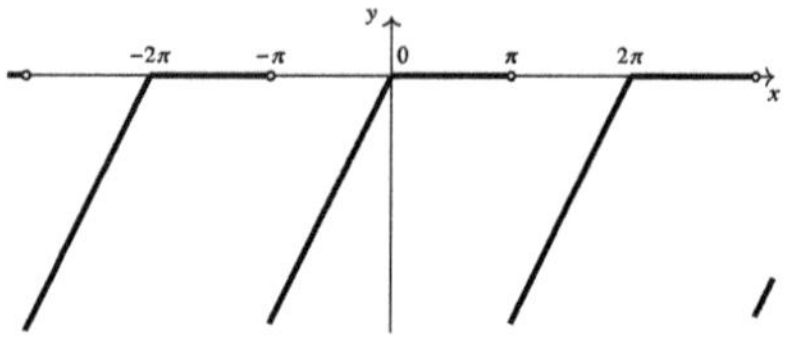

As a continuous and bounded function in the interval $[-\pi, \pi)$, the function f is absolutely integrable over the interval $[-\pi, \pi]$. This implies that the Fourier series of the function f exists. From, for $k \in \mathbb{N}_0$, $a_k = \frac{1}{\pi} \int_{-\pi}^{\pi} f(x) \cos kx \, dx = \frac{2}{\pi} \int_{-\pi}^{0} x \cos kx \, dx$, it follows that $a_0 = \frac{2}{\pi} \int_{-\pi}^{0} x \, dx = -\pi$ and, for $k \in \mathbb{N}$, $a_k = \frac{2}{\pi} \int_{-\pi}^{0} x \cos kx \, dx = \frac{2}{k^2 \pi}(1 - (-1)^k)$. Also, for $k \in \mathbb{N}$, $b_k = \frac{1}{\pi} \int_{-\pi}^{\pi} f(x) \sin kx \, dx = \frac{2}{\pi} \int_{-\pi}^{0} x \sin kx \, dx = \frac{2}{k}(-1)^{k+1}$. Therefore,

$$f(x) \sim -\frac{\pi}{2} + \frac{2}{\pi} \sum_{k=1}^{\infty} \frac{(-1)^k}{k^2}(((-1)^k - 1) \cos kx - k\pi \sin kx).$$

(b) Observe that

$$f(x) = \begin{cases} 0, & x \in [-\pi, 0), \\ 2x, & x \in [0, \pi). \end{cases}$$

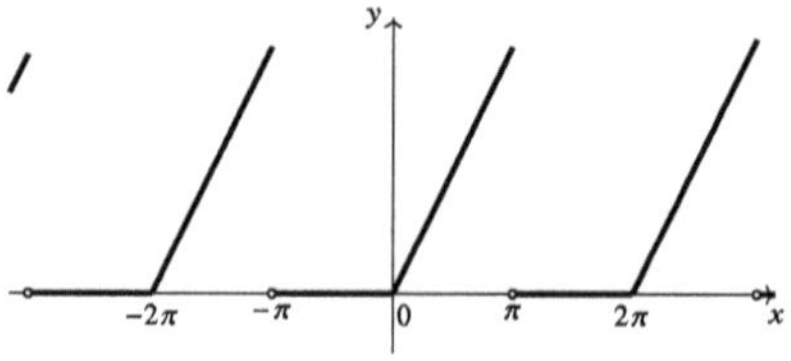

It follows that $f(x) \sim \frac{\pi}{2} + \frac{2}{\pi} \sum_{k=1}^{\infty} \frac{(-1)^k}{k^2}(((1 - (-1)^k) \cos kx - k\pi \sin kx)$. Since the function f is bounded, piecewise monotone, and continuous in the interval $(-\pi, \pi)$, by Dirichlet's theorem,

$$0 = f(0) = \frac{\pi}{2} + \frac{2}{\pi} \sum_{k=1}^{\infty} \frac{1}{k^2}((-1)^k - 1) = \frac{\pi}{2} + \frac{2}{\pi} \sum_{k=0}^{\infty} \frac{-2}{(2k+1)^2}.$$

Therefore, $\sum_{k=0}^{\infty} \frac{1}{(2k+1)^2} = \frac{\pi^2}{8}$.

Note: The Fourier series obtained in Parts (a) and (b) of Problem 1.2.4 are the sum and difference of the Fourier series of periodic extensions of the functions $x \mapsto f_1(x) = x$ and $x \mapsto f_2(x) = |x|$, respectively, in the interval $[-\pi, \pi)$. The Fourier series of the extension of the function f_1 provides the Fourier coefficients $\{b_k\}_{k \in \mathbb{N}}$ and the Fourier series of the extension of the function f_2 provides the coefficients $\{a_k\}_{k \in \mathbb{N}_0}$ of the function f. Since the periodic extension of the function f_1 is not continuous, its Fourier coefficients asymptotically behave like $\frac{1}{k}$. The periodic extension of the function f_2 is an even and continuous function, but it is not differentiable, so its Fourier coefficients asymptotically behave like $\frac{1}{k^2}$.

(c) The function f is continuous and periodic with the fundamental period $T = 2\pi$ and such that $f(x) = \begin{cases} 0 & \text{if } x \in [-\pi, 0], \\ 2\sin x & \text{if } x \in [0, \pi). \end{cases}$

From, for $k \in \mathbb{N}_0$, $a_k = \frac{1}{\pi} \int_0^\pi 2\sin x \cos kx\, dx = \frac{1}{\pi} \int_0^\pi (\sin(1 + k)x + \sin(1 - k)x)\, dx$, it follows that $a_1 = \frac{1}{\pi} \int_0^\pi \sin 2x\, dx = 0$ and, for $k \neq 1$, $a_k = \frac{2(1+(-1)^k)}{\pi(1-k^2)}$. Hence, for $k \in \mathbb{N}_0$, $a_{2k} = \frac{4}{\pi(1-4k^2)}$ and $a_{2k+1} = 0$. From, for $k \in \mathbb{N}$, $b_k = \frac{1}{\pi} \int_0^\pi 2\sin x \sin kx\, dx = \frac{1}{\pi} \int_0^\pi (\cos(1 - k)x - \cos(1 + k)x)\, dx = \frac{1}{\pi} \int_0^\pi \cos(1 - k)x\, dx$, it follows that $b_1 = 1$ and, for $k \in \mathbb{N}\backslash\{1\}$, $b_k = 0$. By Dirichlet's theorem, for all $x \in \mathbb{R}$, $f(x) = \frac{2}{\pi} + \sin x + \frac{4}{\pi} \sum_{k=1}^{\infty} \frac{\cos 2kx}{1-4k^2}$.

Note: The Fourier series obtained in this part of Problem 1.2.4 is the sum of the Fourier series of the function $x \mapsto f_1(x) = \sin x$ and the Fourier series of the function $x \mapsto f_2(x) = |\sin x|$. Thus, $b_k = 0$ for $k \in \mathbb{N}\backslash\{1\}$.

(d) The function f is periodic, with a period 2π. Since, for $x \in \mathbb{R}$, $f(x) = \text{sign}\left(\sin\left(x + \frac{\pi}{4}\right)\right) + 1$, it follows that

$$f(x) = \begin{cases} 2 & \text{if } x \in \left(-\frac{\pi}{4}, \frac{3\pi}{4}\right), \\ 0 & \text{if } x \in \left(\frac{3\pi}{4}, \frac{7\pi}{4}\right), \\ 1 & \text{if } x \in \left\{-\frac{\pi}{4}, \frac{3\pi}{4}\right\}. \end{cases}$$

From, for all $k \in \mathbb{N}_0$, $a_k = \frac{1}{\pi}\int_{-\pi}^{\pi} f(x)\cos kx\, dx = \frac{2}{\pi}\int_{-\frac{\pi}{4}}^{\frac{3\pi}{4}}\cos kx\, dx$, it follows that $a_0 = 2$ and, for $k \in \mathbb{N}$, $a_k = \frac{4}{k\pi}\sin\frac{k\pi}{2}\cos\frac{k\pi}{4}$. Similarly, for $k \in \mathbb{N}$, $b_k = \frac{1}{\pi}\int_{-\pi}^{\pi} f(x)\sin kx\, dx = \frac{4}{k\pi}\sin\frac{k\pi}{2}\sin\frac{k\pi}{4}$. By Dirichlet's theorem, for all $x \in \mathbb{R}$,

$$f(x) = 1 + \frac{4}{\pi}\sum_{k=1}^{\infty}\frac{1}{k}\sin\frac{k\pi}{2}\left(\cos\frac{k\pi}{4}\cos kx + \sin\frac{k\pi}{4}\sin kx\right)$$

$$= 1 + \frac{4}{\pi}\sum_{k=1}^{\infty}\frac{(-1)^{k+1}}{2k-1}\cos(2k-1)\left(x - \frac{\pi}{4}\right).$$

Note: The following argument provides another way to obtain the result. Start with a periodic function $g : \mathbb{R} \to \mathbb{R}$, with the fundamental period $T = 2\pi$, that is determined by $g(-\pi) = 0$ and, for $x \in (-\pi, \pi)$, $g(x) = \operatorname{sign}(x)$. By Dirichlet's theorem, for all $x \in \mathbb{R}$, $g(x) = \frac{4}{\pi}\sum_{k=1}^{\infty}\frac{\sin(2k-1)x}{2k-1}$. Now, for all $x \in \mathbb{R}$,

$$f(x) = 1 + g\left(\sin\left(x + \frac{\pi}{4}\right)\right) = 1 + g\left(x + \frac{\pi}{4}\right)$$

$$= 1 + \frac{4}{\pi}\sum_{k=1}^{\infty}\frac{\sin(2k-1)\left(x + \frac{\pi}{4}\right)}{2k-1}$$

$$= 1 + \frac{4}{\pi}\sum_{k=1}^{\infty}\frac{(-1)^{k+1}}{2k-1}\cos(2k-1)\left(x - \frac{\pi}{4}\right).$$

Problem 1.2.5. Determine the Fourier series of the following periodic functions with the fundamental period $T = 2\pi$:

(a) $f(x) = x|x|$, $x \in [-\pi, \pi)$.

(b) $f(x) = \begin{cases} \pi x - x^2 & \text{if } x \in [0, \pi], \\ x^2 - 3\pi x + 2\pi^2 & \text{if } x \in (\pi, 2\pi). \end{cases}$

    Establish that $\sum_{k=1}^{\infty}\frac{(-1)^{k-1}}{(2k-1)^3} = \frac{\pi^3}{32}$.

Solution 1.2.5.

(a) Since f is an odd function, it follows that, for $k \in \mathbb{N}_0$, $a_k = 0$ and, for all $k \in \mathbb{N}$,

$$b_k = \frac{2}{\pi} \int_0^\pi f(x) \sin kx \; dx = \frac{2}{\pi} \int_0^\pi x^2 \sin kx \; dx$$

$$= \frac{2(-1)^{k+1}}{k^3 \pi} (k^2 \pi^2 + 2((-1)^k - 1)).$$

Therefore, $f(x) \sim \frac{2}{\pi} \sum_{k=1}^\infty \frac{(-1)^{k+1}}{k^3} (k^2 \pi^2 + 2((-1)^k - 1)) \sin kx$.

(b) The function f is continuous in both intervals $(0, \pi)$ and $(\pi, 2\pi)$. From $0 = f(0) = f(\pi) = \lim_{x \to \pi^+} f(x) = \lim_{x \to 2\pi^-} f(x) = f(2\pi)$, it follows that the function f is continuous in the interval $[0, 2\pi]$. Next, note that if $x \in [-\pi, 0]$, then $x + 2\pi \in [\pi, 2\pi]$ and $-x \in [0, \pi]$. Since f is periodic, this implies that, for all $x \in [-\pi, 0]$, $f(x) = f(x + 2\pi) = (x + 2\pi)^2 - 3\pi(x + 2\pi) + 2\pi^2 = \pi x + x^2 = -(\pi(-x) - (-x)^2) = -f(-x)$. The function f is odd. Consequently, for $k \in \mathbb{N}_0$, $a_k = 0$ and, for $k \in \mathbb{N}$, $b_k = \frac{2}{\pi} \int_0^\pi (\pi x - x^2) \sin kx \; dx = \frac{4}{k^3 \pi}(1 - (-1)^k)$. For $k \in \mathbb{N}$, $b_{2k} = 0$ and $b_{2k-1} = \frac{8}{\pi(2k-1)^3}$. By Dirichlet's theorem, for all $x \in \mathbb{R}$, $f(x) = \frac{8}{\pi} \sum_{k=1}^\infty \frac{\sin(2k-1)x}{(2k-1)^3}$. In particular, if $x = \frac{\pi}{2}$, then

$$\frac{\pi^2}{4} = f\left(\frac{\pi}{2}\right) = \frac{8}{\pi} \sum_{k=1}^\infty \frac{\sin \frac{(2k-1)\pi}{2}}{(2k-1)^3} = \frac{8}{\pi} \sum_{k=1}^\infty \frac{(-1)^{k-1}}{(2k-1)^3},$$

or, what is the same, $\sum_{k=1}^\infty \frac{(-1)^{k-1}}{(2k-1)^3} = \frac{\pi^3}{32}$.

Problem 1.2.6. Let $a \in \left(0, \frac{\pi}{2}\right)$ and let $f(x) = \begin{cases} \sin x & \text{if } x \in [0, a] \cup [\pi - a, \pi], \\ \sin a & \text{if } x \in (a, \pi - a). \end{cases}$ Let $F : \mathbb{R} \to \mathbb{R}$ be the odd periodic extension of the function f, with the fundamental period $T = 2\pi$. Determine the Fourier series of the function F.

Solution 1.2.6. The function F is determined by $F(x) = \begin{cases} f(x) & \text{if } x \in [0, \pi], \\ -f(-x) & \text{if } x \in [-\pi, 0), \end{cases}$ and, for $x \in \mathbb{R}$ and $k \in \mathbb{Z}$, $F(x + 2k\pi) = F(x)$.

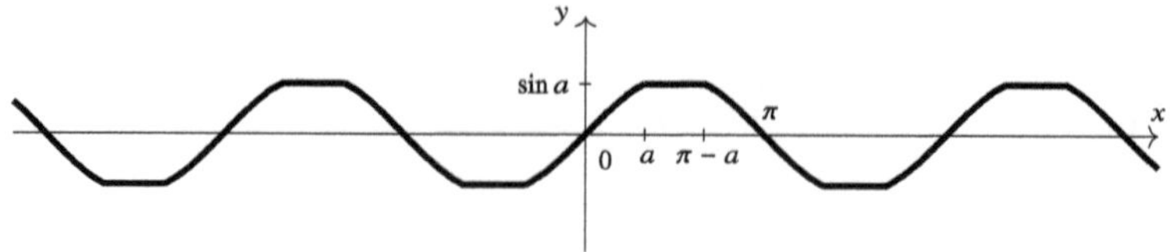

Since F is an odd function, it follows that, for $k \in \mathbb{N}_0$, $a_k = 0$, and, for $k \in \mathbb{N}$, $b_k = \frac{2}{\pi} \int_0^\pi f(x) \sin kx \ dx$. Since, for all $k \in \mathbb{N}$, $\int_{\pi-a}^\pi \sin x \sin kx \ dx = \int_0^a \sin(\pi - x) \sin k(\pi - x) \ dx = (-1)^{k+1} \int_0^a \sin x \sin kx \ dx$ and $\int_{\frac{\pi}{2}}^{\pi-a} \sin a \sin kx \ dx = (-1)^{k+1} \int_a^{\frac{\pi}{2}} \sin a \sin kx \ dx$, it follows that, for all $k \in \mathbb{N}$, $b_k = \frac{2}{\pi}(1 + (-1)^{k+1}) \left(\int_0^a \sin x \sin kx \ dx + \int_a^{\frac{\pi}{2}} \sin a \sin kx \ dx \right)$. Hence, for $k \in \mathbb{N}$, $b_{2k} = 0$ and, for $k \in \mathbb{N}_0$,

$$b_{2k+1} = \frac{2}{\pi} \int_0^a (\cos 2kx - \cos 2(k+1)x) \ dx + \frac{4 \sin a}{(2k+1)\pi} \cos(2k+1)a.$$

Thus, $b_1 = \frac{1}{\pi}(2a + \sin 2a)$ and $b_{2k+1} = \frac{1}{(2k+1)\pi} \left(\frac{\sin 2ka}{k} + \frac{\sin 2(k+1)a}{k+1} \right)$, for $k \in \mathbb{N}$. By Dirichlet's theorem, for all $x \in \mathbb{R}$,

$$F(x) = \frac{2a + \sin 2a}{\pi} \sin x$$

$$+ \frac{1}{\pi} \sum_{k=1}^\infty \frac{\sin(2k+1)x}{2k+1} \left(\frac{\sin 2ka}{k} + \frac{\sin 2(k+1)a}{k+1} \right).$$

Problem 1.2.7. Determine the Fourier series of the following functions:

(a) $f(x) = \arcsin(\sin x)$.   (b) $f(x) = \arccos(\cos x)$.

Solution 1.2.7.

(a) Since the range of the function $x \mapsto \sin x$ is the domain of the function $x \mapsto \arcsin x$, the function f is defined for all real numbers. From, for all $x \in \mathbb{R}$, $f(-x) = \arcsin(\sin(-x)) = \arcsin(-\sin x) = -\arcsin(\sin x) = -f(x)$ and $f(x + 2\pi) = \arcsin(\sin(x + 2\pi)) = \arcsin(\sin x) = f(x)$, it follows that the function f is odd and periodic with a period $T = 2\pi$. Hence, for $k \in \mathbb{N}_0$, $a_k = 0$ and, for $k \in \mathbb{N}$, $b_k = \frac{2}{\pi} \int_0^\pi f(x) \sin kx \ dx$. Since, by definition, for $x \in \left[-\frac{\pi}{2}, \frac{\pi}{2}\right]$, $f(x) = x$, the fact that, for all $x \in \mathbb{R}$, $f(x + \pi) = \arcsin(\sin(x + \pi)) = \arcsin(-\sin x) = -f(x)$, implies that, for $x \in \left(\frac{\pi}{2}, \frac{3\pi}{2}\right]$, $x - \pi \in \left(-\frac{\pi}{2}, \frac{\pi}{2}\right]$ and $f(x) = f(\pi + (x - \pi)) = -f(x - \pi) = \pi - x$. Therefore,

$$f(x) = \begin{cases} x & \text{if } x \in \left[0, \frac{\pi}{2}\right], \\ \pi - x & \text{if } x \in \left(\frac{\pi}{2}, \pi\right]. \end{cases}$$

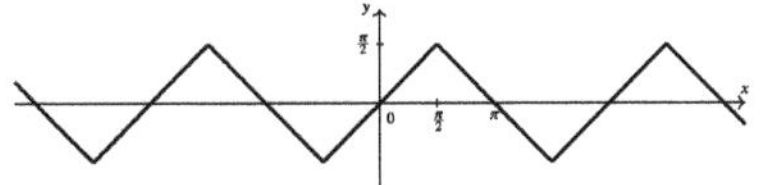

It follows that, for all $k \in \mathbb{N}$, $b_k = \frac{2}{\pi} \int_0^\pi f(x) \sin kx \, dx = \frac{4}{k^2 \pi} \sin \frac{k\pi}{2}$, i.e., for $k \in \mathbb{N}$, $b_{2k} = 0$ and $b_{2k-1} = \frac{4}{\pi} \frac{(-1)^{k+1}}{(2k-1)^2}$. By Dirichlet's theorem, for $x \in \mathbb{R}$, $\arcsin(\sin x) = \frac{4}{\pi} \sum_{k=1}^\infty \frac{(-1)^{k+1} \sin(2k-1)x}{(2k-1)^2}$.

Note: The graph of the function f is called a *triangular wave*.

(b) Since the range of the function $x \mapsto \cos x$ is the domain of the function $x \mapsto \arccos x$, the function f is defined for all real numbers. The function f is even, periodic with a period $T = 2\pi$, and defined in $\mathbb{R}$. Since f is even from, for $x \in [0, \pi]$, $f(x) = x$ it follows that

$$f(x) = \begin{cases} x & \text{if } x \in [0, \pi], \\ -x & \text{if } x \in (-\pi, 0). \end{cases}$$

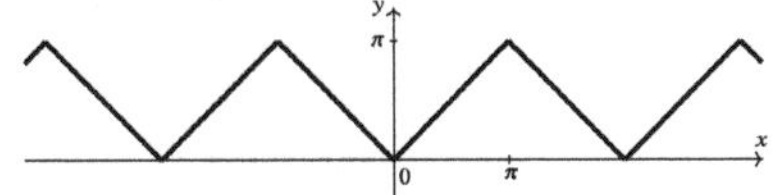

For all $k \in \mathbb{N}$, $b_k = 0$ and, for all $k \in \mathbb{N}_0$, $a_k = \frac{2}{\pi} \int_0^\pi x \cos kx \, dx$. Thus, $a_0 = \pi$ and, for $k \in \mathbb{N}$, $a_{2k} = 0$ and $a_{2k-1} = -\frac{4}{\pi(2k-1)^2}$. By Dirichlet's theorem, for all $x \in \mathbb{R}$, $\arccos(\cos x) = \frac{\pi}{2} - \frac{4}{\pi} \sum_{k=1}^\infty \frac{\cos(2k-1)x}{(2k-1)^2}$.

Problem 1.2.8. Let $A = \left(0, \frac{\pi}{2}\right)$, $B = \left(\pi, \frac{\pi}{2}\right)$, and $C = \left(\frac{\pi}{2}, 0\right)$. Let Γ be the set of all points in the rectangle $[0, \pi] \times \left[0, \frac{\pi}{2}\right]$ that are equally distant from the line segment $\overline{AB}$ and the point C. Determine the function $f : [0, \pi] \to \left[0, \frac{\pi}{4}\right]$ that has Γ as its graph. Let F be the even periodic, with the fundamental period $T = 2\pi$, extension of the function f. Determine the Fourier series of the function F.

Solution 1.2.8. For $x \in [0, \pi]$, let $M = (x, y = f(x)) \in \Gamma$. The distance between the point M and the line segment $\overline{AB}$ is $d(M, \overline{AB}) = \frac{\pi}{2} - y$, and the distance between points M and C is $d(C, M) = \sqrt{\left(x - \frac{\pi}{2}\right)^2 + y^2}$. From $d(M, \overline{AB}) = d(M, C)$, it follows that $\frac{\pi}{2} - y = \sqrt{\left(x - \frac{\pi}{2}\right)^2 + y^2}$, which implies $y = \frac{1}{\pi}\left(\pi x - x^2\right)$. Hence, for $x \in [0, \pi]$, $f(x) = x - \frac{x^2}{\pi}$. Since the function F is even, it follows that $b_k = 0$, for all $k \in \mathbb{N}$, and that, for $k \in \mathbb{N}_0$, $a_k = \frac{2}{\pi} \int_0^\pi \left(x - \frac{x^2}{\pi}\right) \cos kx \, dx$. Thus, $a_0 = \frac{\pi}{3}$ and, for $k \in \mathbb{N}$, $a_{2k-1} = 0$ and $a_{2k} = -\frac{4}{\pi k^2}$. By Dirichlet's theorem, for all $x \in \mathbb{R}$, $F(x) = \frac{\pi}{6} - \frac{4}{\pi} \sum_{k=1}^\infty \frac{\cos 2kx}{k^2}$.

Note: Recall that a parabola is defined as the set of all points in the plane equally distant from the given line and the given point not on the line.

Problem 1.2.9. Determine the Fourier cosine series of the function $f(x) = \sin mx$, $x \in [0, \pi]$, $m \in \mathbb{Z}\backslash\{0\}$.

Solution 1.2.9. This problem is equivalent to the problem of determining the Fourier series of the function F that is the even periodic, with the fundamental period $T = 2\pi$, extension of the function f. It follows that, for all $k \in \mathbb{N}$, $b_k = 0$ and that, for all $k \in \mathbb{N}_0$, $a_k = \frac{2}{\pi} \int_0^\pi \sin mx \cos kx \, dx = \frac{1}{\pi} \int_0^\pi (\sin(m + k)x + \sin(m - k)x) \, dx$. Thus, $a_{|m|} = 0$ and, for $k \in \mathbb{N}_0\backslash\{|m|\}$, $a_k = -\frac{1}{\pi}\left((-1)^{k+m} - 1\right)\frac{2m}{m^2 - k^2}$. By Dirichlet's theorem, for all $x \in [0, \pi]$,

$$\sin mx = \frac{2m}{\pi} \sum_{k \in \mathbb{N}_0\backslash\{|m|\}} \left((-1)^{k+m} - 1\right) \frac{\cos kx}{k^2 - m^2} = \frac{4m}{\pi} \sum_{\substack{k \in \mathbb{N}_0 \\ k+m \notin 2\mathbb{Z}}} \frac{\cos kx}{m^2 - k^2}.$$

Problem 1.2.10. Let $a \in \mathbb{R}\backslash\mathbb{Z}$. Use the Fourier series of the function $f(x) = \cos ax$, $x \in (-\pi, \pi)$, to establish that, for $t \in \mathbb{R}\backslash\{n\pi : n \in \mathbb{Z}\}$,

$$\frac{1}{\sin t} = \frac{1}{t} + \sum_{k=1}^\infty (-1)^k \left(\frac{1}{t + k\pi} + \frac{1}{t - k\pi}\right).$$

Solution 1.2.10. Since f is an even function, it follows that $b_k = 0$, for all $k \in \mathbb{N}$. From $a_0 = \frac{2}{\pi} \int_0^\pi \cos ax \, dx = \frac{2 \sin a\pi}{a\pi}$ and, for all $k \in \mathbb{N}$,

$$a_k = \frac{2}{\pi} \int_0^\pi \cos ax \cos kx \, dx = \frac{1}{\pi} \int_0^\pi (\cos(a+k)x + \cos(a-k)x) \, dx$$

$$= \frac{1}{\pi} \left(\frac{\sin(a+k)\pi}{a+k} + \frac{\sin(a-k)\pi}{a-k} \right) = \frac{2a(-1)^k \sin a\pi}{\pi(a^2 - k^2)},$$

it follows that, for all $x \in (-\pi, \pi)$, $\cos ax = \frac{\sin a\pi}{a\pi} + \frac{2a \sin a\pi}{\pi} \sum_{k=1}^\infty \frac{(-1)^k \cos kx}{a^2 - k^2}$. In particular, for $x = 0$,

$$\frac{1}{\sin a\pi} = \frac{1}{a\pi} + 2 \sum_{k=1}^\infty \frac{(-1)^k a}{\pi(a^2 - k^2)} = \frac{1}{a\pi} + 2 \sum_{k=1}^\infty \frac{(-1)^k a\pi}{(a\pi)^2 - (k\pi)^2}.$$

It follows that, for $t \in \mathbb{R} \backslash \{n\pi : n \in \mathbb{Z}\}$ and $a = \frac{t}{\pi}$,

$$\frac{1}{\sin t} = \frac{1}{t} + 2 \sum_{k=1}^\infty \frac{(-1)^k t}{t^2 - (k\pi)^2} = \frac{1}{t} + \sum_{k=1}^\infty (-1)^k \left(\frac{1}{t + k\pi} + \frac{1}{t - k\pi} \right).$$

Problem 1.2.11. Prove that, for all $x \in (0, \pi)$, $\sum_{k=1}^\infty \frac{\sin(2k-1)x}{2k-1} = \frac{\pi}{4}$ and conclude that $\frac{\pi}{4} = 1 - \frac{1}{3} + \frac{1}{5} - \frac{1}{7} + \cdots$.

Solution 1.2.11. Let f be the odd periodic, with the fundamental period $T = 2\pi$, extension of the function $x \mapsto \frac{\pi}{4}$, $x \in (0, \pi]$. From, for $k \in \mathbb{N}$,

$$b_k = \frac{2}{\pi} \int_0^\pi \frac{\pi}{4} \sin kx \, dx = \frac{1}{2k} \left(1 - (-1)^k \right) = \begin{cases} \frac{1}{k} & \text{if } k \text{ is odd}, \\ 0 & \text{if } k \text{ is even}, \end{cases}$$

it follows that $f(x) \sim \sum_{k=1}^\infty \frac{\sin(2k-1)x}{2k-1}$. In particular, for all $x \in (0, \pi)$, $\frac{\pi}{4} = \sum_{k=0}^\infty \frac{\sin(2k-1)x}{2k-1}$. For $x = \frac{\pi}{2} \in (0, \pi)$, $\frac{\pi}{4} = \sum_{k=1}^\infty \frac{\sin \frac{(2k-1)\pi}{2}}{2k-1} = 1 - \frac{1}{3} + \frac{1}{5} - \frac{1}{7} + \cdots$.

Problem 1.2.12. Prove that, for all $x \in [0, \pi]$,

$$\sum_{k=1}^\infty \frac{\cos(2k-1)x}{(2k-1)^4} = \frac{\pi}{96}(\pi - 2x)(\pi^2 + 2\pi x - 2x^2).$$

Solution 1.2.12. Since, for all $x \in \mathbb{R}$, $f(x) = (\pi - 2x)(\pi^2 + 2\pi x - 2x^2) = \pi^3 - 6\pi x^2 + 4x^3$, the strategy is to first determine the Fourier cosine series of functions $x \mapsto x^2$ and $x \mapsto x^3$ in the interval $[0, \pi]$. From $a_0 = \frac{2}{\pi} \int_0^\pi x^2 \, dx = \frac{2\pi^2}{3}$ and, for $k \in \mathbb{N}$, $a_k = \frac{2}{\pi} \int_0^\pi x^2 \cos kx \, dx = \frac{4(-1)^k}{k^2}$, by Dirichlet's theorem, if follows that, for all $x \in [0, \pi]$,

$$x^2 = \frac{\pi^2}{3} + 4 \sum_{k=1}^{\infty} \frac{(-1)^k}{k^2} \cos kx.$$

Similarly, from $\alpha_0 = \frac{2}{\pi} \int_0^\pi x^3 \, dx = \frac{\pi^3}{2}$ and, for $k \in \mathbb{N}$, $\alpha_k = \frac{2}{\pi} \int_0^\pi x^3 \cos kx \, dx = -\frac{6}{k\pi} \left(\frac{\pi^2 \, (-1)^{k+1}}{k} + \frac{2 \, ((-1)^k - 1)}{k^3} \right)$, by Dirichlet's theorem, for all $x \in [0, \pi]$,

$$x^3 = \frac{\pi^3}{4} + \frac{6}{\pi} \sum_{k=1}^{\infty} \frac{(-1)^k}{k^4} \left(\pi^2 k^2 - 2 \left((-1)^k - 1 \right) \right) \cos kx.$$

It follows that, for all $x \in [0, \pi]$,

$$f(x) = 4x^3 - 6\pi x^2 + \pi^3 = \pi^3 - 2\pi^3 + \pi^3$$

$$+ 24 \sum_{k=1}^{\infty} \left(\frac{\pi \, (-1)^k}{k^2} - \frac{2 \, ((-1)^k - 1)}{k^4 \pi} - \frac{\pi \, (-1)^k}{k^2} \right) \cos kx$$

$$= -\frac{48}{\pi} \sum_{k=1}^{\infty} \frac{(-1)^k - 1}{k^4} \cos kx = \frac{96}{\pi} \sum_{k=1}^{\infty} \frac{\cos(2k - 1)x}{(2k - 1)^4}.$$

Therefore, for all $x \in [0, \pi]$, $\sum_{k=1}^{\infty} \frac{\cos(2k-1)x}{(2k-1)^4} = \frac{\pi}{96}(\pi - 2x) \cdot (\pi^2 + 2\pi x - 2x^2)$.

Problem 1.2.13. Let $f : \mathbb{R} \to \mathbb{R}$ be a piecewise continuous periodic function with a period $T = 2\pi$. Prove the following statements:

(a) If, for all $x \in [0, 2\pi]$, $f(2\pi - x) = f(x)$, i.e., if the graph of f over the interval $[0, 2\pi]$ is symmetric with respect to the line $x = \pi$, then $b_n = 0$, for all $n \in \mathbb{N}$.

(b) If, for all $x \in [0, 2\pi]$, $f(2\pi - x) = -f(x)$, i.e., if the graph of f over the interval $[0, 2\pi]$ is symmetric with respect to the point $(\pi, 0)$, then $a_n = 0$, for all $n \in \mathbb{N}_0$.

(c) If, for all $x \in [-\pi, 0]$, $f(x+\pi) = f(x)$, i.e., if the graph of f over the interval $[0, \pi]$ is a translate of the graph of f over the interval $[-\pi, 0]$, then $a_{2n-1} = b_{2n-1} = 0$, for all $n \in \mathbb{N}$.

(d) If, for all $x \in [-\pi, 0]$, $f(x+\pi) = -f(x)$, i.e., if the graph of f over the interval $[0, \pi]$ is a reflection across the x-axis of the translate of the graph of f over the interval $[-\pi, 0]$, then $a_{2n} = b_{2n} = 0$, for all $n \in \mathbb{N}$.

Solution 1.2.13.

(a) From, for any $k \in \mathbb{N}$, $\int_{\pi}^{2\pi} f(x) \sin kx \; dx = |t = 2\pi - x| = \int_0^{\pi} f(2\pi - t) \sin(2k\pi - kt) \; dt = -\int_0^{\pi} f(t) \sin kt \; dt$, it follows that

$$b_k = \frac{1}{\pi} \int_0^{2\pi} f(x) \sin kx \; dx$$

$$= \frac{1}{\pi} \left(\int_0^{\pi} f(x) \sin kx \; dx + \int_{\pi}^{2\pi} f(x) \sin kx \; dx \right) = 0.$$

Note: Another way to obtain the result is to observe that, under the given condition, the function f is even.

(b) The function f is odd.

(c) The claim follows from, for all $k \in \mathbb{N}$, $\int_0^{\pi} f(x) \cos kx \; dx = \int_{-\pi}^0 f(\pi + t) \cos k(\pi + t) \; dt = (-1)^k \int_{-\pi}^0 f(t) \cos kt \; dt$ and $\int_0^{\pi} f(x) \sin kx \; dx = (-1)^k \int_{-\pi}^0 f(x) \sin kx \; dx$.

(d) The claim follows from, for all $k \in \mathbb{N}_0$, $\int_{-\pi}^0 f(x) \cos kx \; dx = \int_0^{\pi} f(x - \pi) \cos k(x - \pi) \; dx = (-1)^{k+1} \int_0^{\pi} f(x) \cos kx \; dx$ and, for any $k \in \mathbb{N}$, $\int_{-\pi}^0 f(x) \sin kx \; dx = (-1)^{k+1} \int_0^{\pi} f(x) \sin kx \; dx$.

Note: For example, the periodic function with the fundamental period $T = 2\pi$ that is defined by $f(x) = \text{sign}(x)$ if $x \in (-\pi, \pi)$ and $f(\pi) = 0$ satisfies all conditions required in Problem 1.2.13(d).

Problem 1.2.14. Let $f : [0, \pi] \to \mathbb{R}$ be a piecewise continuous function. Prove the following statements:

(a) If, for all $x \in [0, \pi]$, $f(\pi - x) = f(x)$, i.e., if the graph of f over the interval $[0, \pi]$ is symmetric with respect to the line $x = \frac{\pi}{2}$, then $a_{2n-1} = 0$ for the Fourier cosine series of f and $b_{2n} = 0$ for the Fourier sine series of f.

(b) If, for all $x \in [0, \pi]$, $f(\pi - x) = -f(x)$, i.e., if the graph of f over the interval $[0, \pi]$ is symmetric with respect to the point $\left(\frac{\pi}{2}, 0\right)$, then $a_{2n} = 0$ for the Fourier cosine series of f and $b_{2n-1} = 0$ for the Fourier sine series of f.

Solution 1.2.14.

(a) The claim follows from, for any $k \in \mathbb{N}_0$, $\int_{\frac{\pi}{2}}^{\pi} f(x) \cos kx \, dx = -\int_{\frac{\pi}{2}}^{0} f(\pi - x) \cos k(\pi - x) \, dx = (-1)^{k+1} \int_{0}^{\frac{\pi}{2}} f(x) \cos kx \, dx$ and, for any $k \in \mathbb{N}$, $\int_{\frac{\pi}{2}}^{\pi} f(x) \sin kx \, dx = (-1)^k \int_{0}^{\frac{\pi}{2}} f(x) \sin kx \, dx$.

Note: Compare Problem 1.2.14(a) with Problem 1.2.5(b).

Problem 1.2.15. Suppose that f and g are piecewise continuous periodic functions, both with the fundamental period $T = 2\pi$. If f and g are such that, for all $x \in \mathbb{R}$, $f(-x) = g(x)$, i.e., if graphs of f and g are reflections of each other across the y-axis, determine the relationship between their Fourier coefficients.

Solution 1.2.15. Let $\{a_k\}_{k \in \mathbb{N}_0}$ and $\{b_k\}_{k \in \mathbb{N}}$ be the sequences of the Fourier coefficients of the function f, and let $\{\alpha_k\}_{k \in \mathbb{N}_0}$ and $\{\beta_k\}_{k \in \mathbb{N}}$ be the sequences of the Fourier coefficients of the function g. Then, for any $k \in \mathbb{N}_0$,

$$a_k = \frac{1}{\pi} \int_{-\pi}^{\pi} f(x) \cos kx \, dx = \frac{1}{\pi} \int_{-\pi}^{\pi} f(-x) \cos(-kx) \, dx$$

$$= \frac{1}{\pi} \int_{-\pi}^{\pi} g(x) \cos kx \, dx = \alpha_k.$$

Similarly, $b_k = -\beta_k$, for all $k \in \mathbb{N}$.

Problem 1.2.16. Suppose that f is a piecewise continuous periodic function with the fundamental period $T = 2\pi$. If $\{a_k\}_{k \in \mathbb{N}_0}$ and $\{b_k\}_{k \in \mathbb{N}}$ are the sequences of the Fourier coefficients of the function f and if h is a real number, determine the Fourier coefficients of the function $g(x) = f(x + h)$, $x \in \mathbb{R}$.

Solution 1.2.16. The function g is a periodic function with the fundamental period $T = 2\pi$ and piecewise continuous in the interval

$[-\pi, \pi]$. Let $\{\alpha_k\}_{k \in \mathbb{N}_0}$ and $\{\beta_k\}_{k \in \mathbb{N}}$ be the sequences of the Fourier coefficients of the function g. Then,

$$\alpha_0 = \frac{1}{\pi} \int_{-\pi}^{\pi} g(x) \, dx = \frac{1}{\pi} \int_{-\pi+h}^{\pi+h} f(x) \, dx = a_0$$

and, for all $k \in \mathbb{N}$,

$$\alpha_k = \frac{1}{\pi} \int_{-\pi}^{\pi} g(x) \cos kx dx = \frac{1}{\pi} \int_{-\pi}^{\pi} f(x+h) \cos kx \, dx$$

$$= \frac{1}{\pi} \int_{-\pi+h}^{\pi+h} f(x) \cos k(x-h) \, dx$$

$$= \frac{1}{\pi} \int_{-\pi}^{\pi} f(x) \left(\cos kx \cos kh + \sin kx \sin kh \right) \, dx$$

$$= a_k \cos kh + b_k \sin kh.$$

Similarly, for all $k \in \mathbb{N}$, $\beta_k = b_k \cos kh - a_k \sin kh$.

Problem 1.2.17. Suppose that f is an even piecewise continuous periodic function with the fundamental period $T = 2\pi$ such that $f(x) \sim \frac{a_0}{2} + \sum_{k=1}^{\infty} a_k \cos kx$. Let F be a periodic function, with the fundamental period $T = 2\pi$, defined by, for each $x \in [-\pi, \pi]$, $F(x) = \int_0^x f(t) \, dt$. Establish that F is an odd function such that $F(x) \sim \sum_{k=1}^{\infty} \frac{a_k + (-1)^{k+1} a_0}{k} \sin kx$.

Solution 1.2.17. Let $x \in [-\pi, \pi]$. Then,

$$F(-x) = \int_0^{-x} f(t) \, dt = -\int_0^x f(-u) \, du = -\int_0^x f(u) \, du = -F(x),$$

which establishes that F is an odd function. Hence, $F(x) \sim \sum_{n=1}^{\infty} \beta_k \sin kx$ where, for $k \in \mathbb{N}$,

$$\beta_k = \frac{2}{\pi} \int_0^{\pi} \left(\int_0^x f(t) \, dt \right) \sin kx \, dx = \begin{vmatrix} u = \int_0^x f(t) \, dt & du = f(x) \, dx \\ dv = \sin kx \, dx & v = -\frac{\cos kx}{k} \end{vmatrix}$$

$$= \frac{2}{\pi} \left(-\frac{\cos kx}{k} \int_0^x f(t) \, dt \Big|_0^{\pi} + \frac{1}{k} \int_0^{\pi} f(x) \cos kx \, dx \right)$$

$$= \frac{1}{k} \left((-1)^{k+1} a_0 + a_k \right).$$

Problem 1.2.18. Let $f : \left[0, \frac{\pi}{2}\right] \to \mathbb{R}$ be a piecewise continuous function. Determine a periodic extension $F : \mathbb{R} \to \mathbb{R}$ of the function f so that $F(x) \sim \sum_{k=1}^{\infty} a_{2k-1} \cos(2k-1)x$.

Solution 1.2.18. Let $F : \mathbb{R} \to \mathbb{R}$ be *any* even piecewise continuous periodic, with the fundamental period $T = 2\pi$, extension of the function f. Then, $F(x) \sim \frac{a_0}{2} + \sum_{k=1}^{\infty} a_k \cos kx$, where, for all $k \in \mathbb{N}_0$, $a_k = \frac{2}{\pi} \int_0^{\pi} F(x) \cos kx \ dx$. Since, for all $k \in \mathbb{N}_0$, $\int_{\frac{\pi}{2}}^{\pi} F(x) \cos 2kx \ dx = \int_0^{\frac{\pi}{2}} F(\pi - x) \cos 2kx \ dx$, it follows that, for all $k \in \mathbb{N}_0$, $a_{2k} = 0$ if and only if $\int_0^{\frac{\pi}{2}} (f(x) + F(\pi - x)) \cos 2kx \ dx = 0$. Therefore, a periodic function $F : \mathbb{R} \to \mathbb{R}$ such that

$$F(x) = \begin{cases} f(x) & \text{if } x \in \left[0, \frac{\pi}{2}\right], \\ -f(\pi - x) & \text{if } x \in \left(\frac{\pi}{2}, \pi\right), \\ f(-x) & \text{if } x \in \left[-\frac{\pi}{2}, 0\right], \\ -f(\pi + x) & \text{if } x \in \left[-\pi, -\frac{\pi}{2}\right), \end{cases}$$

is a function with the required property.

Note: Compare with Problem 1.2.14(b).

Problem 1.2.19. Let $\alpha \in \left(0, \frac{\pi}{2}\right)$. Use the Fourier series of a periodic extension of the function $f(x) = \begin{cases} 1 - \frac{|x|}{2\alpha} & \text{if } x \in [-2\alpha, 2\alpha], \\ 0 & \text{if } |x| \in (2\alpha, \pi], \end{cases}$ to evaluate $\sum_{k=1}^{\infty} \frac{\sin^2 k\alpha}{k^2}$.

Solution 1.2.19. The function f is even, so $b_k = 0$ for all $k \in \mathbb{N}$. From $a_0 = \frac{2}{\pi} \frac{1 \cdot 2\alpha}{2} = \frac{2\alpha}{\pi}$ and, for $k \in \mathbb{N}$,

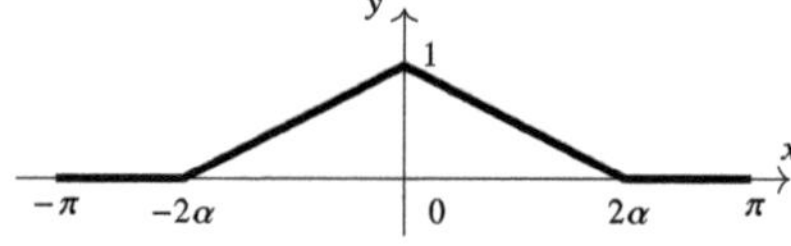

$$a_k = \frac{2}{\pi} \int_0^{2\alpha} \left(1 - \frac{x}{2\alpha}\right) \cos kx \ dx$$

$$= \frac{2}{\pi} \left(\frac{\sin 2k\alpha}{k} - \frac{1}{2\alpha} \left(\frac{2\alpha \sin 2k\alpha}{k} + \frac{\cos 2k\alpha - 1}{k^2}\right)\right) = \frac{2 \sin^2 k\alpha}{\alpha \pi k^2},$$

it follows that, for all $x \in [-\pi, \pi]$, $f(x) = \frac{\alpha}{\pi} + \frac{2}{\alpha\pi} \sum_{k=1}^{\infty} \frac{\sin^2 k\alpha}{k^2} \cos kx$.
For $x = 0$, $1 = \frac{\alpha}{\pi} + \frac{2}{\alpha\pi} \sum_{k=1}^{\infty} \frac{\sin^2 k\alpha}{k^2}$, which implies that $\sum_{k=1}^{\infty} \frac{\sin^2 k\alpha}{k^2} = \frac{\alpha(\pi - \alpha)}{2}$.

Problem 1.2.20. Let $a > 0$.

(a) Determine the Fourier series of the function $f(x) = e^{ax}$, $x \in (-\pi, \pi)$.
(b) Determine the Fourier cosine series and the Fourier sine series of the function $g(x) = e^{ax}$, $x \in (0, \pi)$.
(c) Determine the Fourier sine series of the function $h(x) = \cosh ax$, $x \in (0, \pi)$, and the Fourier cosine series of the function $k(x) = \sin ax$, $x \in (0, \pi)$.

Solution 1.2.20.

(a) From, for $k \in \mathbb{N}_0$,

$$\int_{-\pi}^{\pi} e^{ax} e^{ikx} \, dx = \int_{-\pi}^{\pi} e^{a+ik} x \, dx = \frac{e^{(a+ik)\pi} - e^{-(a+ik)\pi}}{a + ik}$$

$$= \frac{2(-1)^k (a - ik)}{a^2 + k^2} \sinh a\pi,$$

it follows that $a_0 = \frac{2\sinh a\pi}{a\pi}$ and, for $k \in \mathbb{N}$, $a_k = \frac{2(-1)^k a \sinh a\pi}{\pi(a^2 + k^2)}$ and $b_k = \frac{2(-1)^{k+1} k \sinh a\pi}{\pi(a^2 + k^2)}$. By Dirichlet's theorem, for $x \in (-\pi, \pi)$,

$$e^{ax} = \frac{2\sinh a\pi}{\pi} \left(\frac{1}{2a} + \sum_{k=0}^{\infty} \frac{(-1)^k}{a^2 + k^2} (a \cos kx - k \sin kx) \right).$$

(b) From, for $k \in \mathbb{N}_0$,

$$\int_{0}^{\pi} e^{ax} e^{ikx} \, dx = \int_{0}^{\pi} e^{a+ik} x \, dx = \frac{1}{a + ik} \left(e^{(a+ik)\pi} - 1 \right)$$

$$= \frac{a - ik}{a^2 + k^2} ((-1)^k e^{a\pi} - 1),$$

it follows that, for $x \in (0, \pi)$,

$$e^{ax} = \frac{2a}{\pi} \sum_{k=0}^{\infty} \frac{(-1)^k e^{a\pi} - 1}{a^2 + k^2} \cos kx$$

$$= \frac{2}{\pi} \sum_{k=1}^{\infty} \frac{k(1 + (-1)^{k+1} e^{a\pi})}{a^2 + k^2} \sin kx.$$

(c) For $x \in (0, \pi)$,

$$\cosh ax = \frac{e^{ax} + e^{-ax}}{2} = \frac{1}{\pi} \sum_{k=1}^{\infty} \frac{k(2 + (-1)^{k+1}(e^{a\pi} + e^{-a\pi}))}{a^2 + k^2} \sin kx$$

$$= \frac{2}{\pi} \sum_{k=1}^{\infty} \frac{k(1 + (-1)^{k+1} \cosh a\pi)}{a^2 + k^2} \sin kx.$$

If $a > 0$ is not a natural number, since $\cos ax = \frac{e^{axi} + e^{-axi}}{2}$ and $\sin ax = \frac{e^{axi} - e^{-axi}}{2i}$, for $x \in (0, \pi)$,

$$\sin ax = \frac{1}{2i} \frac{2ai}{\pi} \sum_{k=0}^{\infty} \frac{(-1)^k (e^{a\pi i} + e^{-a\pi i}) - 2}{k^2 - a^2} \cos kx$$

$$= \frac{2a}{\pi} \sum_{k=0}^{\infty} \frac{(-1)^k \cos a\pi - 1}{k^2 - a^2} \cos kx.$$

If a is a natural number, then from $\int_0^{\pi} \sin ax \cos ax \, dx = 0$, it follows that, for $x \in (0, \pi)$,

$$\sin ax = \frac{2a}{\pi} \sum_{k \neq a} \frac{(-1)^k \cos a\pi - 1}{k^2 - a^2} \cos kx = \frac{4a}{\pi} \sum_{k \neq a (\mathrm{mod}\ 2)} \frac{\cos kx}{a^2 - k^2}.$$

Note: Compare with Problem 1.2.9.

Problem 1.2.21.

(a) Establish that the function $f(x) = \ln\left(2\cos\frac{x}{2}\right)$, $x \in (-\pi, \pi)$, is absolutely integrable and then determine the Fourier series of its periodic extension.

(b) Determine the Fourier series of the periodic function determined by $g(x) = \cos x \ln\left(2\cos\frac{x}{2}\right)$, $x \in (-\pi, \pi)$.

Solution 1.2.21.

(a) Since $\lim\limits_{x\to\pi^-} f(x) = \infty$, there is no piecewise continuous extension of f. From

$$I = \int_0^\pi \ln\left(2\cos\frac{x}{2}\right)\,dx = \pi\ln 2 + \int_0^\pi \ln\cos\frac{x}{2}\,dx$$

$$= \left|\,\frac{x}{2} = \frac{\pi}{2} - t\,\right| = \pi\ln 2 + 2\int_0^{\frac{\pi}{2}} \ln\sin t\,dt$$

$$= \pi\ln 2 + 2\left(\frac{\pi}{2}\ln 2 + \int_0^{\frac{\pi}{2}}\ln\sin\frac{t}{2}\,dt + \int_0^{\frac{\pi}{2}}\ln\cos\frac{t}{2}\,dt\right)$$

and

$$\int_0^{\frac{\pi}{2}}\ln\sin\frac{t}{2}\,dt = \left|\,\frac{t}{2} = \frac{\pi}{2} - \frac{u}{2}\,\right| = \int_{\frac{\pi}{2}}^\pi \ln\cos\frac{u}{2}\,du,$$

it follows that $I = 2\left(\pi\ln 2 + \int_0^\pi \ln\cos\frac{t}{2}\,dt\right) = 2I$, i.e., $I = \int_0^\pi \ln\left(2\cos\frac{x}{2}\right)\,dx = 0$. Since the function f is even, this establishes that it is absolutely integrable over $(-\pi, \pi)$. Moreover, $a_0 = \frac{2}{\pi}\int_0^\pi \ln\left(2\cos\frac{x}{2}\right)\,dx = 0$. Since, for $k \in \mathbb{N}$ and $t \in \mathbb{R}$,

$$2\left(\frac{1}{2} + \sum_{j=1}^k \cos jt\right)\sin\frac{t}{2} = \sin\frac{t}{2} + \sum_{j=1}^k\left(\sin\frac{(2j+1)t}{2} - \sin\frac{(2j-1)t}{2}\right)$$

$$= \sin\left(k + \frac{1}{2}\right)t,$$

i.e., since for all $t \in (0, \pi]$, $\frac{\sin\left(k+\frac{1}{2}\right)t}{2\sin\frac{t}{2}} = \frac{1}{2} + \sum_{j=1}^k \cos jt$, it follows that

$$a_k = \frac{2}{\pi}\int_0^\pi \ln\left(2\cos\frac{x}{2}\right)\cos kx\,dx$$

$$= \left|\,\begin{array}{ll} u = \ln\left(2\cos\frac{x}{2}\right) & du = -\dfrac{\sin\frac{x}{2}}{2\cos\frac{x}{2}}\,dx \\[2mm] dv = \cos kx & v = \dfrac{\sin kx}{k}\end{array}\,\right|$$

$$= \frac{1}{k\pi} \int_0^\pi \frac{\sin kx \sin \frac{x}{2}}{\cos \frac{x}{2}}\, dx = |x = \pi - t| =$$

$$= \frac{(-1)^{k+1}}{k\pi} \int_0^\pi \frac{\sin kt \cos \frac{t}{2}}{\sin \frac{t}{2}}\, dt$$

$$= \frac{(-1)^{k+1}}{k\pi} \int_0^\pi \frac{\sin\left(k + \frac{1}{2}\right) t + \sin\left(k - \frac{1}{2}\right) t}{2 \sin \frac{t}{2}}\, dt$$

$$= \frac{(-1)^{k+1}}{k\pi} \int_0^\pi \left(2\left(\frac{1}{2} + \sum_{j=1}^{k-1} \cos jt \right) + \cos kt \right)\, dt$$

$$= \frac{(-1)^{k+1}}{k}.$$

Therefore, for $x \in (-\pi, \pi)$, $\ln\left(2 \cos \frac{x}{2}\right) = \sum_{k=1}^{\infty} \frac{(-1)^{k+1}}{k} \cos kx$.

Note 1: For $x \in (0, \pi)$,

$$\ln\left(2 \sin \frac{x}{2}\right) = \ln\left(2 \cos \frac{\pi - x}{2}\right) = \sum_{k=1}^{\infty} \frac{(-1)^{k+1}}{k} \cos k(\pi - x)$$

$$= -\sum_{k=1}^{\infty} \frac{\cos kx}{k}.$$

Note 2: Another way to obtain the Fourier series of the function f is to use the fact that, for $\alpha \in \mathbb{R}\setminus\{(2n+1)\pi \,|\, n \in \mathbb{Z}\}$, $\ln(1+e^{i\alpha}) = \sum_{k=1}^{\infty} \frac{(-1)^{k+1}e^{ik\alpha}}{k}$. For $x \in (0, \pi)$,

$$f(x) = \frac{1}{2} \ln\left(4 \cos^2 \frac{x}{2}\right) = \frac{1}{2} \ln\left(2(1 + \cos x)\right)$$

$$= \frac{1}{2} \ln\left(2 + e^{ix} + e^{-ix}\right) = \frac{1}{2} \ln\left(1 + e^{ix}\right)\left(1 + e^{-ix}\right)$$

$$= \frac{1}{2} \ln\left(1 + e^{ix}\right) + \frac{1}{2} \ln\left(1 + e^{-ix}\right)$$

$$= \frac{1}{2} \sum_{k=1}^{\infty} \frac{(-1)^{k+1}}{k} \left(e^{ikx} + e^{-ikx}\right) = \sum_{k=1}^{\infty} \frac{(-1)^{k+1}}{k} \cos kx.$$

(b) For $x \in (-\pi, \pi)$,

$$\cos x \ln\left(2\cos\frac{x}{2}\right) = \cos x \sum_{n=1}^{\infty} \frac{(-1)^{n+1}}{n} \cos nx$$

$$= \frac{1}{2} \sum_{n=1}^{\infty} \frac{(-1)^{n+1}}{n}(\cos(n+1)x + \cos(n-1)x)$$

$$= \frac{1}{2} \sum_{k=2}^{\infty} \frac{(-1)^k}{k-1} \cos kx + \frac{1}{2} \sum_{k=0}^{\infty} \frac{(-1)^k}{k+1} \cos kx$$

$$= \frac{1}{2} - \frac{\cos x}{4} + \sum_{k=2}^{\infty} \frac{(-1)^k k}{k^2 - 1} \cos kx.$$

Problem 1.2.22. Let $f : \mathbb{R} \to \mathbb{R}$ be a continuous periodic function with the fundamental period $T = 2\pi$, and let $\{a_k\}_{k\in\mathbb{N}_0}$ and $\{b_k\}_{k\in\mathbb{N}}$ be the sequences of the Fourier coefficients of the function f. Let h be a positive real number. Determine the Fourier coefficients of the Steklov[10] function $x \mapsto f_h(x) = \frac{1}{2h} \int_{x-h}^{x+h} f(t)\, dt$.

Solution 1.2.22. The function f_h is periodic with a period $T = 2\pi$. By the fundamental theorem of calculus, the function f_h is continuous and differentiable in the set of real numbers, with, for all $x \in \mathbb{R}$, $f_h'(x) = \frac{1}{2h}(f(x+h) - f(x-h))$. Let $\{A_k\}_{k\in\mathbb{N}_0}$ and $\{B_k\}_{k\in\mathbb{N}}$ be the sequences of the Fourier coefficients of the Steklov function f_h. Then,

$$A_0 = \frac{1}{\pi} \int_{-\pi}^{\pi} f_h(x)\, dx = \frac{1}{2h\pi} \int_{-\pi}^{\pi} \left(\int_{x-h}^{x+h} f(t) dt \right) dx$$

$$= \begin{vmatrix} u = \int_{x-h}^{x+h} f(t)\, dt & du = (f(x+h) - f(x-h))\, dx \\ dv = dx & v = x \end{vmatrix}$$

$$= \frac{x}{\pi} f_h(x) \Big|_{-\pi}^{\pi} - \frac{1}{2h\pi} \int_{-\pi}^{\pi} x(f(x+h) - f(x-h))\, dx$$

$$= f_h(\pi) + f_h(-\pi) + I_h = 2f_h(\pi) + I_h,$$

---

[10]Vladimir Andreevich Steklov, 1864–1926, Russian mathematician and physicist.

where

$$I_h = \frac{1}{2h\pi} \int_{-\pi}^{\pi} x(f(x-h) - f(x+h))\, dx$$

$$= \frac{1}{2h\pi} \left(\int_{-\pi-h}^{\pi-h} (t+h)f(t)\, dt - \int_{-\pi+h}^{\pi+h} (t-h)f(t)\, dt \right)$$

$$= a_0 + \frac{1}{2h\pi} \left(\int_{-\pi-h}^{\pi-h} tf(t)\, dt - \int_{-\pi+h}^{\pi+h} tf(t)\, dt \right)$$

$$= a_0 + \frac{1}{2h\pi} \left(\int_{-\pi-h}^{-\pi+h} tf(t)\, dt - \int_{\pi-h}^{\pi+h} tf(t)\, dt \right).$$

From

$$\int_{-\pi-h}^{-\pi+h} tf(t)\, dt = \begin{vmatrix} s = t + 2\pi,\ ds = dt \\ t = -\pi - h \Rightarrow s = \pi - h \\ t = -\pi + h \Rightarrow s = \pi + h \end{vmatrix} = \int_{\pi-h}^{\pi+h} (s - 2\pi)f(s)\, ds,$$

it follows that $I_h = a_0 - 2f_h(\pi)$ and

$$A_0 = 2f_h(\pi) + (a_0 - 2f_h(\pi)) = a_0.$$

For $k \in \mathbb{N}$,

$$2h\pi A_k = \int_{-\pi}^{\pi} \left(\cos kx \int_{x-h}^{x+h} f(t)\, dt \right) dx$$

$$= \begin{vmatrix} u = \int_{x-h}^{x+h} f(t)\, dt & du = (f(x+h) - f(x-h))\, dx \\ dv = \cos kx & v = \frac{1}{k}\sin kx \end{vmatrix}$$

$$= \frac{\sin kx}{k} \int_{x-h}^{x+h} f(t)\, dt \,\bigg|_{-\pi}^{\pi}$$

$$- \frac{1}{k} \int_{-\pi}^{\pi} (f(x+h) - f(x-h)) \sin kx\, dx$$

$$= \frac{1}{k} \left(\int_{-\pi}^{\pi} f(x-h) \sin kx\, dx - \int_{-\pi}^{\pi} f(x+h) \sin kx\, dx \right)$$

$$= \frac{1}{k} \left(\int_{-\pi-h}^{\pi-h} f(t) \sin k(t+h)\, dt - \int_{-\pi+h}^{\pi+h} f(t) \sin k(t-h)\, dt \right)$$

$$= \frac{1}{k} \left(\cos kh \left(\int_{-\pi-h}^{\pi-h} f(t) \sin kt \, dt - \int_{-\pi+h}^{\pi+h} f(t) \sin kt \, dt \right) \right.$$

$$\left. + \sin kh \left(\int_{-\pi-h}^{\pi-h} f(t) \cos kt \, dt + \int_{-\pi+h}^{\pi+h} f(t) \cos kt \, dt \right) \right).$$

Since both functions $t \mapsto f(t) \cos kt$ and $t \mapsto f(t) \sin kt$ are periodic with a period 2π, it follows that $\int_{-\pi+h}^{\pi+h} f(t) \cos kt \, dt = \int_{-\pi-h}^{\pi-h} f(t) \cos kt \, dt = \pi a_k$ and $\int_{-\pi+h}^{\pi+h} f(t) \sin kt \, dt = \int_{-\pi-h}^{\pi-h} f(t) \sin kt \, dt = \pi b_k$. Therefore, for all $k \in \mathbb{N}$, $A_k = \frac{a_k \sin hk}{hk}$. Similarly, for all $k \in \mathbb{N}$, $B_k = \frac{b_k \sin hk}{hk}$.

Note 1. If f is a periodic continuous function, then it is always possible to integrate its Fourier series term-by-term. The obtained series uniformly converges to f. Therefore, from, for all $t \in \mathbb{R}$, $f(t) = \frac{a_0}{2} + \sum_{k=1}^{\infty}(a_k \cos kt + b_k \sin kt)$ it follows that, for all $h > 0$ and all $x \in \mathbb{R}$,

$$f_h(x) = \frac{1}{2h} \int_{x-h}^{x+h} f(t) \, dt$$

$$= \frac{1}{2h} \int_{x-h}^{x+h} \left(\frac{a_0}{2} + \sum_{k=1}^{\infty}(a_k \cos kt + b_k \sin kt) \right) dt$$

$$= \frac{1}{2h} \int_{x-h}^{x+h} \frac{a_0}{2} \, dt + \sum_{k=1}^{\infty} \left(\frac{1}{2h} \int_{x-h}^{x+h} (a_k \cos kt + b_k \sin kt) \, dt \right)$$

$$= \frac{a_0}{2} + \frac{1}{2h} \sum_{k=1}^{\infty} \left(\frac{a_k}{k} \sin kt - \frac{b_k}{k} \cos kt \, \bigg|_{x-h}^{x+h} \right).$$

Since, for all $x, k, h \in \mathbb{R}$, $\sin k(x+h) - \sin k(x-h) = 2 \cos kx \sin hk$ and $\cos k(x+h) - \cos k(x-h) = -2 \sin kx \sin hk$, it follows that

$$f_h(x) = \frac{a_0}{2} + \frac{1}{h} \sum_{k=1}^{\infty} \left(\frac{a_k \sin hk}{k} \cos kx + \frac{b_k \sin hk}{k} \sin kx \right).$$

Note 2. Another way to determine the Fourier coefficients of the Steklov function is as follows. For any $k \in \mathbb{N}$, $a_k + ib_k = $

$\frac{1}{\pi}\int_{-\pi}^{\pi} f(x)e^{ikx}\,dx$ and $A_k + iB_k = \frac{1}{\pi}\int_{-\pi}^{\pi} f_h(x)e^{ikx}\,dx$. From

$$\int_{-\pi}^{\pi} f_h(x)e^{ikx}\,dx = \frac{1}{2h}\int_{-\pi}^{\pi} e^{ikx}\cdot\left(\int_{x-h}^{x+h} f(t)\,dt\right)dx$$

$$= \left|\begin{array}{l} u = \int_{x-h}^{x+h} f(t)\,dt \\ dv = e^{ikx} \end{array}\right| = \frac{e^{ikx}}{2hk}\int_{x-h}^{x+h} f(t)\,dt\,\Bigg|_{-\pi}^{\pi}$$

$$+\frac{i}{2hk}\int_{-\pi}^{\pi}(f(x+h)-f(x-h))e^{ikx}\,dx$$

$$= \frac{i}{2hk}\left(\int_{-\pi}^{\pi} f(x+h)e^{ikx}\,dx -\int_{-\pi}^{\pi} f(x-h))e^{ikx}\,dx\right),$$

together with

$$\int_{-\pi}^{\pi} f(x+h)e^{ikx}\,dx = \int_{-\pi+h}^{\pi+h} f(t)e^{ik(t-h)}\,dt = e^{-ihk}\int_{-\pi+h}^{\pi+h} f(t)e^{ikt}\,dt$$

$$= \pi e^{-ihk}(a_k + ib_k)$$

and

$$\int_{-\pi}^{\pi} f(x-h)e^{ikx}\,dx = \int_{-\pi-h}^{\pi-h} f(t)e^{ik(t+h)}\,dt = e^{ihk}\int_{-\pi-h}^{\pi-h} f(t)e^{ikt}\,dt$$

$$= \pi e^{ihk}(a_k + ib_k),$$

it follows that, for $k \in \mathbb{N}$,

$$A_k+iB_k = \frac{i}{2hk}\left(e^{-ihk}(a_k + ib_k) - e^{ihk}(a_k + ib_k)\right) = \frac{\sin hk}{hk}(a_k+ib_k).$$

Problem 1.2.23. Let $f : \mathbb{R} \to \mathbb{R}$ be an odd periodic function with the fundamental period $T = 2\pi$. Suppose that there is an integer $m \geq 1$ such that the function $f^{(2m-1)} : \mathbb{R} \to \mathbb{R}$, the $(2m-1)^{\text{st}}$ derivative of the function f, is continuous. Establish that, for all $x \in \mathbb{R}$,

$$f(x) = \frac{2(-1)^{m+1}}{\pi}\sum_{n=1}^{\infty}\frac{\sin nx}{n^{2m-1}}\int_{0}^{\pi} f^{(2m-1)}(x)\cos nx\,dx.$$

Solution 1.2.23. The question is to establish that, under the given conditions, for any $n \in \mathbb{N}$,

$$b_n = \frac{2}{\pi} \int_0^\pi f(x) \sin nx = \frac{2(-1)^{m+1}}{\pi n^{2m-1}} \int_0^\pi f^{(2m-1)}(x) \cos nx \, dx.$$

Observe that, if f is odd, from $f(-x) = -f(x)$, it follows that $-f'(-x) = -f'(x)$, i.e., f' is an even function. By induction, since $2m - 1$ is odd, $f^{(2m-1)}$ is an even function. In addition, from $f(x + 2\pi) = f(x)$, it follows that $f'(x + 2\pi) = f'(x)$, i.e., f' is a periodic function with a period 2π. By induction, for any $\ell \leq 2m-1$, the function $f^{(\ell)}$ is periodic with a period 2π. Moreover, since, for $0 \leq 2\ell < 2m - 1$, $f^{(2\ell)}$ is a continuous odd function with a period 2π, it follows that $f^{(2\ell)}(-\pi) = f^{(2\ell)}(0) = f^{(2\ell)}(\pi) = 0$.

Let M be the set of all natural numbers m such that if $f : \mathbb{R} \to \mathbb{R}$ is an odd periodic function with the fundamental period $T = 2\pi$ such that the function $f^{(2m-1)} : \mathbb{R} \to \mathbb{R}$ is continuous, then, for all $n \in \mathbb{N}$,

$$b_n = \frac{2}{\pi} \int_0^\pi f(x) \sin nx = \frac{2(-1)^{m+1}}{\pi n^{2m-1}} \int_0^\pi f^{(2m-1)}(x) \cos kx \, dx.$$

Suppose that $f : \mathbb{R} \to \mathbb{R}$ is an odd periodic function with the fundamental period $T = 2\pi$ such that f' is a continuous function. Then, for $n \in \mathbb{N}$,

$$b_n = \frac{2}{\pi} \int_0^\pi f(x) \sin nx \, dx = \begin{vmatrix} u = f(x) & du = f'(x) \, du \\ dv = \sin nx \, dx & v = -\frac{\cos nx}{n} \end{vmatrix}$$

$$= -\frac{2f(x) \cos nx}{\pi n} \Big|_0^\pi + \frac{2}{\pi n} \int_0^\pi f'(x) \cos nx \, dx$$

$$= \frac{2}{\pi n} \int_0^\pi f'(x) \cos nx \, dx,$$

which implies that $1 \in M$. Suppose that $m \in M$ and let $f : \mathbb{R} \to \mathbb{R}$ be an odd periodic function with the fundamental period $T = 2\pi$ such that $f^{(2m+1)}$ is a continuous function. Since $m \in M$, for $n \in \mathbb{N}$,

$$b_n = \frac{2(-1)^{m+1}}{\pi n^{2m-1}} \int_0^\pi f^{(2m-1)}(x) \cos nx \, dx$$

$$= \begin{vmatrix} u = f^{(2m-1)}(x) & du = f^{(2m)}(x) \, du \\ dv = \cos nx \, dx & v = \frac{\sin nx}{n} \end{vmatrix}$$

$$= \frac{2(-1)^{m+1}}{\pi n^{2m}} \left(f^{(2m-1)}(x) \sin nx \Big|_0^\pi - \int_0^\pi f^{(2m)}(x) \sin nx \, dx \right)$$

$$= \frac{2(-1)^{m+2}}{\pi n^{2m}} \int_0^\pi f^{(2m)}(x) \sin nx \, dx = \begin{vmatrix} u = f^{(2m)}(x) \\ dv = \sin nx \, dx \end{vmatrix}$$

$$= \frac{2(-1)^{m+2}}{\pi n^{2m+1}} \int_0^\pi f^{(2m+1)}(x) \cos nx \, dx,$$

which establishes that $m + 1 \in M$. By the principle of mathematical induction, $M = \mathbb{N}$, which proves the claim.

Note: For a more general statement, see Solution 1.2.3.

Problem 1.2.24. Let $a > 1$. Determine the Fourier series of the following functions:

(a) $f(x) = \dfrac{\sin x}{a^2 - 2a \cos x + 1}.$ $\qquad$ (b) $g(x) = \ln(a^2 - 2a \cos x + 1).$

Solution 1.2.24.

(a) The function f is odd, periodic with the fundamental period $T = 2\pi$, and continuous in the set of real numbers. From, for all $x \in \mathbb{R}$, $a^2 - 2a \cos x + 1 = (a - e^{ix})(a - e^{-ix})$ and, for all $x \neq k\pi$, $k \in \mathbb{Z}$,

$$\frac{1}{(a - e^{ix})(a - e^{-ix})} = \frac{1}{2i \sin x} \left(\frac{1}{a - e^{ix}} - \frac{1}{a - e^{-ix}} \right)$$

$$= \frac{1}{2ai \sin x} \left(\frac{1}{1 - \frac{e^{ix}}{a}} - \frac{1}{1 - \frac{e^{-ix}}{a}} \right)$$

$$= \frac{1}{2ai \sin x} \sum_{n=1}^{\infty} \frac{e^{inx} - e^{-inx}}{a^n}$$

$$= \frac{1}{a \sin x} \sum_{n=1}^{\infty} \frac{\sin nx}{a^n},$$

it follows that, for all $x \in \mathbb{R}$, $f(x) = \sum_{n=1}^{\infty} \frac{\sin nx}{a^{n+1}}.$

(b) Since $a > 1$, for all real numbers x, $a^2 - 2a\cos x + 1 = (a-\cos x)^2 + \sin^2 x > 0$. It follows that the periodic function g is defined and differentiable for all $x \in \mathbb{R}$. By Part (a), for all $x \in \mathbb{R}$,

$$g'(x) = \frac{2a\sin x}{a^2 - 2a\cos x + 1} = 2af(x) = 2\sum_{n=1}^{\infty} \frac{\sin nx}{a^n}.$$

By integrating this trigonometric series term-by-term, for some $C \in \mathbb{R}$, $g(x) = C - 2\sum_{n=1}^{\infty} \frac{\cos nx}{na^n}$, $x \in \mathbb{R}$. In particular,

$$2\ln(1+a) = g(\pi) = C + 2\sum_{n=1}^{\infty} \frac{(-1)^{n+1}}{n}\frac{1}{a^n} = C + 2\ln\left(1 + \frac{1}{a}\right),$$

implies that $C = 2\ln a$ and $g(x) = 2\ln a - 2\sum_{n=1}^{\infty} \frac{\cos nx}{na^n}$, $x \in \mathbb{R}$.

Problem 1.2.25. Establish that for any $(x, \theta) \in (-1, 1) \times \mathbb{R}$, $F(x, \theta) = \frac{x\sin\theta}{1 - 2x\cos\theta + x^2} = \sum_{n=1}^{\infty} x^n \sin n\theta$. Evaluate $\int_{-\pi}^{\pi} (F(x, \theta))^2\, d\theta$.

Solution 1.2.25. See Problem 1.2.24. By Parseval's identity,

$$\int_{-\pi}^{\pi} (F(x, \theta))^2\, d\theta = \pi \sum_{n=1}^{\infty} x^{2n} = \pi\left(\frac{1}{1 - x^2} - 1\right) = \frac{\pi x^2}{1 - x^2}.$$

Problem 1.2.26. Determine the sums of the following trigonometric series:

(a) $\sum_{n=0}^{\infty} \frac{\cos nx}{n!}$ and $\sum_{n=1}^{\infty} \frac{\sin nx}{n!}$.

(b) $\sum_{n=1}^{\infty} \frac{(-1)^{n+1}\cos(2n-1)x}{(2n-1)!}$ and $\sum_{n=1}^{\infty} \frac{(-1)^{n+1}\sin(2n-1)x}{(2n-1)!}$.

(c) $1 + \sum_{n=1}^{\infty} \frac{(-1)^{n+1}\cos nx}{n(n+1)}$ and $\sum_{n=1}^{\infty} \frac{(-1)^{n+1}\sin nx}{n(n+1)}$.

Solution 1.2.26.

(a) We use the fact that for all $z \in \mathbb{C}$, $e^z = \sum_{n=0}^{\infty} \frac{z^n}{n!}$.

Let $x \mapsto C(x) = \sum_{n=0}^{\infty} \frac{\cos nx}{n!}$ and $x \mapsto S(x) = \sum_{n=1}^{\infty} \frac{\sin nx}{n!}$. From, for any $x \in \mathbb{R}$ and any $n \in \mathbb{N}$, $\left|\frac{\cos nx}{n!}\right| \leq \frac{1}{n!}$ and $\left|\frac{\sin nx}{n!}\right| \leq \frac{1}{n!}$, by the comparison test, the domain of both functions is the set of all real numbers. For all $x \in \mathbb{R}$,

$$C(x) + iS(x) = \sum_{n=0}^{\infty} \frac{\cos nx + i\sin nx}{n!} = \sum_{n=0}^{\infty} \frac{(e^{ix})^n}{n!} = e^{e^{ix}}$$

$$= e^{\cos x + i\sin x} = e^{\cos x}(\cos(\sin x) + i\sin(\sin x)).$$

For $x \in \mathbb{R}$, $C(x) = e^{\cos x}\cos(\sin x)$ and $S(x) = e^{\cos x}\sin(\sin x)$.

(b) We use the fact that for all $z \in \mathbb{C}$, $\sin z = \sum_{n=1}^{\infty} \frac{(-1)^{n+1}}{(2n-1)!} z^{2n-1}$.

Let $x \mapsto C(x) = \sum_{n=1}^{\infty} \frac{(-1)^{n+1}\cos(2n-1)x}{(2n-1)!}$ and $x \mapsto S(x) = \sum_{n=1}^{\infty} \frac{(-1)^{n+1}\sin(2n-1)x}{(2n-1)!}$. By the comparison test, the domain of both functions is the set of all real numbers. For all $x \in \mathbb{R}$, by De Moivre's theorem

$$C(x) + iS(x) = \sum_{n=0}^{\infty} \frac{(-1)^{n+1}(e^{ix})^{2n-1}}{(2n-1)!} = \sin e^{ix}$$

$$= \sin\left(\cos x + i\sin x\right)$$

$$= \sin(\cos x)\cosh(\sin x) + i\cos(\cos x)\sinh(\sin x).$$

Therefore, for $x \in \mathbb{R}$, $C(x) = \sin(\cos x)\cosh(\sin x)$ and $S(x) = \cos(\cos x)\sinh(\sin x)$.

(c) We use the fact that for $x \in (-\pi, \pi)$, $\ln(1 + e^{ix}) = \sum_{n=1}^{\infty} \frac{(-1)^{n+1}e^{inx}}{n}$. Here, $\ln z = \ln|z| + i\mathrm{Arg}(z)$ is the principal branch of the complex logarithm.

Let $x \mapsto C(x) = 1 + \sum_{n=1}^{\infty} \frac{(-1)^{n+1}\cos nx}{n(n+1)}$ and $x \mapsto S(x) = \sum_{n=1}^{\infty} \frac{(-1)^{n+1}\sin nx}{n(n+1)}$. From, for any $x \in \mathbb{R}$ and any $n \in \mathbb{N}$, $\left|\frac{(-1)^{n+1}\cos nx}{n(n+1)}\right| \leq \frac{1}{n(n+1)} < \frac{1}{n^2}$ and $\left|\frac{(-1)^{n+1}\sin nx}{n(n+1)}\right| \leq \frac{1}{n^2}$, by the comparison test, the domain of both functions is the set of all real numbers. For all $x \in (-\pi, \pi)$,

$$C(x) + iS(x) = 1 + \sum_{n=1}^{\infty} \frac{(-1)^{n+1}e^{inx}}{n(n+1)}$$

$$= 1 + \sum_{n=1}^{\infty} (-1)^{n+1}e^{inx}\left(\frac{1}{n} - \frac{1}{n+1}\right)$$

$$= \sum_{n=1}^{\infty} \frac{(-1)^{n+1}e^{inx}}{n} + e^{-ix}\sum_{n=1}^{\infty} \frac{(-1)^{n+1}e^{inx}}{n}$$

$$= (1 + e^{-ix})\ln(1 + e^{ix})$$

$$= (1 + e^{-ix})\ln(1 + \cos x + i\sin x)$$

$$= (1 + e^{-ix})\ln\left(2\cos\frac{x}{2}\left(\cos\frac{x}{2} + i\sin\frac{x}{2}\right)\right)$$

$$= (1 + \cos x - i\sin x)\left(\ln\left(2\cos\frac{x}{2}\right) + i\frac{x}{2}\right).$$

Therefore, for $x \in (-\pi, \pi)$, $C(x) = 2\cos^2\frac{x}{2}\ln\left(2\cos\frac{x}{2}\right) + \frac{x\sin x}{2}$ and $S(x) = x\cos^2\frac{x}{2} - \sin x\ln\left(2\cos\frac{x}{2}\right)$. This, together with $C(\pi) = C(-\pi) = S(\pi) = S(-\pi) = 0$ and the fact that the functions C and C are periodic, determines the sums of the two series for any $x \in \mathbb{R}$.

Problem 1.2.27. Determine the Fourier series of the following functions:

(a) $f(x) = \cos^{2m} x$, $m \in \mathbb{N}$.   (b) $f(x) = \frac{1}{1 - \sin\alpha\cos x}$, $\alpha \in \left(0, \frac{\pi}{2}\right)$.

(c) $f(x) = \frac{1}{a + b\cos x}$, $0 < |b| < a$.

(d) $f(x) = \begin{cases} \sum_{n=1}^{\infty} \frac{\beta^n \sin nx}{\sin x} & \text{if } x \neq k\pi,\ k \in \mathbb{Z}, \\ \frac{1}{(1-\beta)^2} & \text{if } x = 2k\pi,\ k \in \mathbb{Z}, \\ \frac{1}{(1+\beta)^2} & \text{if } x = 2(k+1)\pi,\ k \in \mathbb{Z}, \end{cases}$ $\beta \in (0, 1)$.

Solution 1.2.27.

(a) By the binomial formula, for all $x \in \mathbb{R}$,

$$f(x) = \frac{(e^{ix} + e^{-ix})^{2m}}{2^{2m}} = \frac{1}{4^m}\sum_{n=0}^{2m}\binom{2m}{n}e^{nix}e^{-(2m-n)ix}$$

$$= \frac{1}{4^m}\sum_{n=0}^{2m}\binom{2m}{n}e^{2(n-m)ix}.$$

Since,

$$\sum_{n=m+1}^{2m}\binom{2m}{n}e^{2(n-m)ix} = \sum_{k=0}^{m-1}\binom{2m}{2m-k}e^{-2(k-m)ix}$$

$$= \sum_{k=0}^{m-1}\binom{2m}{k}e^{-2(k-m)ix},$$

it follows that, for all $x \in \mathbb{R}$,

$$f(x) = \frac{1}{4^m}\left(\binom{2m}{m} + \sum_{n=0}^{m-1}\binom{2m}{n}\left(e^{2(n-m)ix} + e^{-2(n-m)ix}\right)\right)$$

$$= \frac{1}{4^m}\left(\binom{2m}{m} + 2\sum_{n=0}^{m-1}\binom{2m}{n}\cos 2(m-n)x\right)$$

$$= \frac{1}{4^m}\binom{2m}{m} + \frac{1}{2^{2m-1}}\sum_{k=1}^{m}\binom{2m}{m-k}\cos 2kx.$$

(b) The domain of the function f is the set of all real numbers. Since, for $x \in \mathbb{R}$, $\cos x = \frac{e^{2ix}+1}{2e^{ix}}$, it follows that $f(x) = -\frac{2e^{ix}}{e^{2ix}\sin\alpha - 2e^{ix} + \sin\alpha}$. From

$$e^{2ix}\sin\alpha - 2e^{ix} + \sin\alpha = \sin\alpha\left(e^{ix} - \frac{1-\cos\alpha}{\sin\alpha}\right)\left(e^{ix} - \frac{1+\cos\alpha}{\sin\alpha}\right)$$

$$= \sin\alpha\left(e^{ix} - \tan\frac{\alpha}{2}\right)\left(e^{ix} - \cot\frac{\alpha}{2}\right),$$

we obtain $\dfrac{1}{e^{2ix}\sin\alpha - 2e^{ix} + \sin\alpha} = -\dfrac{1}{2\cos\alpha}\left(\dfrac{1}{e^{ix} - \tan\frac{\alpha}{2}} - \dfrac{1}{e^{ix} - \cot\frac{\alpha}{2}}\right)$
and

$$f(x) = \frac{e^{ix}}{\cos\alpha}\left(\frac{1}{e^{ix} - \tan\frac{\alpha}{2}} - \frac{1}{e^{ix} - \cot\frac{\alpha}{2}}\right)$$

$$= \frac{1}{\cos\alpha}\left(\frac{1}{1 - e^{-ix}\tan\frac{\alpha}{2}} + \frac{e^{ix}\tan\frac{\alpha}{2}}{1 - e^{ix}\tan\frac{\alpha}{2}}\right).$$

Since $\alpha \in (0, \frac{\pi}{2})$ implies that $\left|e^{ix}\tan\frac{\alpha}{2}\right| = \left|e^{-ix}\tan\frac{\alpha}{2}\right| = \tan\frac{\alpha}{2} < 1$, for all $x \in \mathbb{R}$,

$$f(x) = \frac{1}{\cos\alpha}\left(\sum_{n=0}^{\infty}e^{-inx}\tan^n\frac{\alpha}{2} + e^{ix}\tan\frac{\alpha}{2}\sum_{n=0}^{\infty}e^{inx}\tan^n\frac{\alpha}{2}\right)$$

$$= \frac{1}{\cos\alpha}\left(1 + \sum_{n=1}^{\infty}(e^{inx} + e^{-inx})\tan^n\frac{\alpha}{2}\right)$$

$$= \frac{1}{\cos\alpha}\left(1 + 2\sum_{n=1}^{\infty}\tan^n\frac{\alpha}{2}\cos nx\right).$$

(c) The domain of the function f is the set of all real numbers. From, for $x \in \mathbb{R}$,

$$f(x) = \frac{1}{a + b\cos x} = \frac{2e^{ix}}{be^{2ix} + 2ae^{ix} + b}$$

$$= \frac{2e^{ix}}{b\left(e^{ix} - \frac{-a+\sqrt{a^2-b^2}}{b}\right)\left(e^{ix} - \frac{-a-\sqrt{a^2-b^2}}{b}\right)},$$

it follows that

$$f(x) = \frac{e^{ix}}{\sqrt{a^2 - b^2}}\left(\frac{1}{e^{ix} + \frac{a-\sqrt{a^2-b^2}}{b}} - \frac{1}{e^{ix} + \frac{a+\sqrt{a^2-b^2}}{b}}\right).$$

Since $\left|\frac{a+\sqrt{a^2-b^2}}{b}\right| > 1$ implies $\left|\frac{a-\sqrt{a^2-b^2}}{b}\right| = \left|\frac{b}{a+\sqrt{a^2-b^2}}\right| < 1$, for all $x \in \mathbb{R}$,

$$f(x) = \frac{1}{\sqrt{a^2-b^2}}\left(\frac{1}{1 - e^{-ix}\frac{\sqrt{a^2-b^2}-a}{b}} - \frac{a-\sqrt{a^2-b^2}}{b}\frac{e^{ix}}{1 - e^{ix}\frac{\sqrt{a^2-b^2}-a}{b}}\right)$$

$$= \frac{1}{\sqrt{a^2-b^2}}\left(\sum_{n=0}^{\infty}\left(\frac{\sqrt{a^2-b^2}-a}{b}\right)^n e^{-inx} + \sum_{n=1}^{\infty}\left(\frac{\sqrt{a^2-b^2}-a}{b}\right)^n e^{inx}\right)$$

$$= \frac{1}{\sqrt{a^2-b^2}}\left(1 + 2\sum_{n=1}^{\infty}\left(\frac{\sqrt{a^2-b^2}-a}{b}\right)^n \cos nx\right).$$

(d) From, for all $x \in \mathbb{R}$, $\sum_{n=1}^{\infty}\beta^n e^{inx} = \frac{1}{1-\beta e^{ix}} - 1$ and $\sum_{n=1}^{\infty}\beta^n e^{-inx} = \frac{1}{1-\beta e^{-ix}} - 1$, it follows that, for all $x \in \mathbb{R}$,

$$\sum_{n=1}^{\infty}\beta^n \sin nx = \frac{1}{2i}\left(\frac{1}{1-\beta e^{ix}} - \frac{1}{1-\beta e^{-ix}}\right) = \frac{\beta\sin x}{1 + \beta^2 - 2\beta\cos x}.$$

Hence, for all $x \in \mathbb{R}$, $f(x) = \frac{\beta}{1+\beta^2-2\beta\cos x}$. By Part (c), for all $x \in \mathbb{R}$, $f(x) = \frac{1}{1-\beta^2}\left(1 + 2\sum_{n=1}^{\infty}\beta^n \cos nx\right)$.

Problem 1.2.28. Let f be the periodic extension with the fundamental period $T = 2\pi$ of the function $x \mapsto e^x$, $x \in [-\pi, \pi)$. If $\{a_n\}_{n\in\mathbb{N}_0}$ and $\{b_n\}_{n\in\mathbb{N}}$ are the sequences of the Fourier coefficients of the function f, then, for all $x \in (-\pi, \pi)$, $e^x = \frac{a_0}{2} +$

$\sum_{n=1}^{\infty}(a_n \cos nx + b_n \sin nx)$. One may be tempted to differentiate term-by-term this Fourier series to obtain, for $x \in (-\pi, \pi)$, $e^x = \sum_{n=1}^{\infty}(nb_n \cos nx - na_n \sin nx)$. But this would yield, for all $n \in \mathbb{N}$, $a_n = nb_n$ and $b_n = -na_n$, i.e., $a_n = b_n = 0$ for all $n \in \mathbb{N}$. What went wrong?

Solution 1.2.28. Observe that $\lim_{x \to \pi^-} f(x) = e^\pi \neq e^{-\pi} = f(\pi) = \lim_{x \to \pi^+} f(x)$ implies the function f is not continuous, so it is not guaranteed that differentiation term-by-term would yield the Fourier series of the derivative of f.

Note: In Problem 1.2.20, we have established that for any $x \in (-\pi, \pi)$, $e^x = \frac{2\sinh \pi}{\pi}\left(\frac{1}{2} + \sum_{k=0}^{\infty}\frac{(-1)^n}{1+n^2}(\cos nx - n \sin nx)\right)$. The term-by-term differentiation yields the series $S(x) = -\frac{2\sinh \pi}{\pi}\sum_{k=0}^{\infty}\frac{(-1)^n}{1+n^2}(n^2 \cos nx + n \sin nx)$. But the series $S(0) = -\frac{2\sinh \pi}{\pi}\sum_{k=0}^{\infty}\frac{(-1)^n n^2}{1+n^2}$ is divergent since $\lim_{n \to \infty}\frac{(-1)^n n^2}{1+n^2} \neq 0$.

Problem 1.2.29. Let $\alpha \in (0, \pi]$. Use Parseval's identity for the Fourier coefficients of an extension of the function defined by

$$f(x) = \begin{cases} x & \text{if } x \in (0, \alpha), \\ 0 & \text{if } x \in [\alpha, \pi], \end{cases}$$

to evaluate $\sum_{k=1}^{\infty}\frac{1}{k^2}$ and $\sum_{k=1}^{\infty}\frac{1}{(2k-1)^4}$.

Solution 1.2.29. Consider the function F, an odd periodic, with the fundamental period $T = 2\pi$, extension of the function f. For a graph of the function F in the case of $\alpha = 1$, see the figure below.

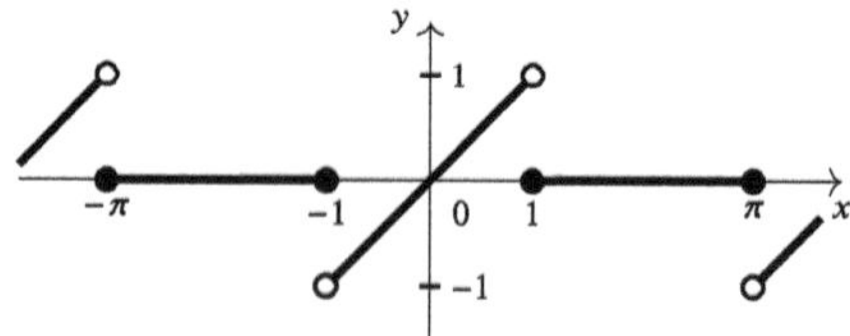

From $\int_{-\pi}^{\pi}(f(x))^2\,dx = 2\int_0^\alpha x^2 dx = \frac{2\alpha^3}{3}$, it follows that f is a square-integrable function. Since f is an odd function, $a_k = 0$ for all $k \in \mathbb{N}_0$, and, for all $k \in \mathbb{N}$, $b_k = \frac{2}{\pi}\int_0^\alpha x \sin kx\,dx = \frac{2}{k^2\pi}(\sin \alpha k - \alpha k \cos \alpha k)$.

By Parseval's identity, for $\alpha \in (0, \pi]$, $\frac{4}{\pi^2} \sum_{k=1}^{\infty} \frac{1}{k^4} (\sin \alpha k - \alpha k \cos \alpha k)^2 = \frac{2\alpha^3}{3\pi}$.

If $\alpha = \pi$, then the identity becomes $\frac{4}{\pi^2} \sum_{k=1}^{\infty} \frac{1}{k^4} \left(-\pi k(-1)^k\right)^2 = \frac{2\pi^3}{3\pi}$, which implies $\sum_{k=1}^{\infty} \frac{1}{k^2} = \frac{\pi^2}{6}$. If $\alpha = \frac{\pi}{2}$, then Parseval's identity becomes

$$\frac{\pi^2}{12} = \frac{4}{\pi^2} \sum_{k=1}^{\infty} \frac{1}{k^4} \left(\sin \frac{\pi k}{2} - \frac{\pi k}{2} \cos \frac{\pi k}{2}\right)^2$$

$$= \frac{4}{\pi^2} \sum_{k=1}^{\infty} \frac{1}{(2k-1)^4} + \sum_{k=1}^{\infty} \frac{1}{4k^2} = \frac{4}{\pi^2} \sum_{k=1}^{\infty} \frac{1}{(2k-1)^4} + \frac{\pi^2}{24},$$

which implies $\sum_{k=1}^{\infty} \frac{1}{(2k-1)^4} = \frac{\pi^2}{4} \left(\frac{\pi^2}{12} - \frac{\pi^2}{24}\right) = \frac{\pi^4}{96}$.

Problem 1.2.30. By integrating term-by-term the Fourier series of the periodic function f determined by $f(x) = x$, $x \in [-\pi, \pi)$, obtain the Fourier series of periodic functions g and h such that $g(x) = x^2$ and $h(x) = x^3$, $x \in [-\pi, \pi)$.

Solution 1.2.30. Note that g is a continuous function and that f and h are piecewise continuous functions. Since the restriction f in $(-\pi, \pi)$ is an odd function, it follows that $a_k = 0$ for all $k \in \mathbb{N}_0$. For all $x \in \mathbb{N}$, $b_k = \frac{2}{\pi} \int_0^{\pi} x \sin kx \, dx = \frac{2(-1)^{k+1}}{k}$. By Dirichlet's theorem, for all $x \in (-\pi, \pi)$, $x = 2 \sum_{k=1}^{\infty} \frac{(-1)^{k+1}}{k} \sin kx$. It follows that, for all $x \in (-\pi, \pi)$,

$$g(x) = x^2 = 2 \int_0^x t \, dt = 4 \int_0^x \sum_{k=0}^{\infty} \frac{(-1)^{k+1}}{k} \sin kt \, dt$$

$$= 4 \sum_{k=1}^{\infty} \frac{(-1)^{k+1}}{k} \left(\int_0^x \sin kt \, dt\right) = 4 \sum_{k=1}^{\infty} \frac{(-1)^{k+1}}{k^2} (1 - \cos kx).$$

This, together with $\sum_{k=1}^{\infty} \frac{(-1)^{k+1}}{k^2} = \frac{\pi^2}{12}$, implies that, for all $x \in \mathbb{R}$, $g(x) = \frac{\pi^2}{3} + 4 \sum_{k=1}^{\infty} \frac{(-1)^k}{k^2} \cos kx$.

The Fourier series of the function g converges uniformly since it is obtained by integrating term-by-term the Fourier series of a piecewise continuous function. For $x \in [-\pi, \pi]$,

$$h(x) = x^3 = 3 \int_0^x t^2 \, dt = 3 \int_0^x \left(\frac{\pi^2}{3} + 4 \sum_{k=1}^{\infty} \frac{(-1)^k}{k^2} \cos kt \right) dt$$

$$= 3 \left(\int_0^x \frac{\pi^2}{3} \, dt + 4 \sum_{k=1}^{\infty} \frac{(-1)^k}{k^2} \left(\int_0^x \cos kt \, dt \right) \right)$$

$$= \pi^2 x + 12 \sum_{k=1}^{\infty} \frac{(-1)^k}{k^3} \sin kx.$$

This, together with, for $x \in (-\pi, \pi)$, $x = 2 \sum_{k=1}^{\infty} \frac{(-1)^{k+1}}{k} \sin kx$, implies that, for $x \in (-\pi, \pi)$,

$$h(x) = 2\pi^2 \sum_{k=1}^{\infty} \frac{(-1)^{k+1}}{k} \sin kx + 12 \sum_{k=1}^{\infty} \frac{(-1)^k}{k^3} \sin kx.$$

Therefore, $h(x) \sim 2 \sum_{k=1}^{\infty} \frac{(-1)^{k+1}}{k^3} (\pi^2 k^2 - 6) \sin kx.$

1.2.2 *Functions with the fundamental period $T \neq 2\pi$*

Problem 1.2.31. Let $f : \mathbb{R} \to \mathbb{R}$ be a continuous piecewise smooth periodic function with a period $T > 0$, and let, for $k \in \mathbb{N}_0$, $a_k = \frac{1}{T} \int_0^T f(x) \cos \frac{2k\pi x}{T} \, dx$, and, for $k \in \mathbb{N}$, $b_k = \frac{1}{T} \int_0^T f(x) \sin \frac{2k\pi x}{T} \, dx$. Let $m > 1$ be a natural number. Establish that, for $n \in \mathbb{N}_0$,

$$A_n = \frac{1}{mT} \int_0^{mT} f(x) \cos \frac{2n\pi x}{mT} \, dx = \begin{cases} a_{\frac{n}{m}} & \text{if } m \mid n, \\ 0 & \text{if } m \nmid n, \end{cases}$$

and, for $n \in \mathbb{N}$,

$$B_n = \frac{1}{mT} \int_0^{mT} f(x) \sin \frac{2n\pi x}{mT} \, dx = \begin{cases} b_{\frac{n}{m}} & \text{if } m \mid n, \\ 0 & \text{if } m \nmid n. \end{cases}$$

Solution 1.2.31. The function f is periodic with the period mT. From, for all $x \in \mathbb{R}$,

$$f(x) = \frac{a_0}{2} + \sum_{k=1}^{\infty} \left(a_k \cos \frac{2k\pi x}{T} + b_k \sin \frac{2k\pi x}{T} \right)$$

$$= \frac{A_0}{2} + \sum_{n=1}^{\infty}\left(A_n \cos \frac{2n\pi x}{mT} + B_n \sin \frac{2n\pi x}{mT}\right),$$

it follows that, for all $x \in \mathbb{R}$,

$$0 = \frac{A_0 - a_0}{2} + \sum_{k=1}^{\infty}\left((A_{km} - a_k)\cos \frac{2k\pi x}{T} + (B_{km} - b_k)\sin \frac{2k\pi x}{T}\right)$$

$$+ \sum_{m \nmid n}\left(A_n \cos \frac{2n\pi x}{mT} + B_n \sin \frac{2n\pi x}{mT}\right).$$

This, together with the fact that

$$\frac{1}{2}, \quad \cos \frac{2\pi x}{mT}, \quad \sin \frac{2\pi x}{mT}, \quad \ldots, \quad \cos \frac{2n\pi x}{mT}, \quad \sin \frac{2n\pi x}{mT}, \quad \ldots$$

is a complete orthogonal system, establishes the required claim.

Note: Problem 1.2.31 confirms the fact that the Fourier series is unique in the sense that replacing the period T by its multiple mT does not change the series.

Problem 1.2.32. Determine the Fourier cosine series of the function

$$f(x) = \begin{cases} 1 & \text{if } x \in \left(\frac{3}{2}, 2\right], \\ 3 - x & \text{if } x \in (2, 3]. \end{cases}$$

Solution 1.2.32. Let F be an even periodic, with the fundamental period $T = 3$, extension of the function f. Then, $b_k = 0$ for all $k \in \mathbb{N}$. From, for all $x \in (0, 1]$, $F(x) = F(-x) = F(3 - x) = f(3 - x) = x$ and, for all $x \in \left[1, \frac{3}{2}\right)$, $F(x) = F(-x) = F(3 - x) = f(3 - x) = 1$, it follows that, for all $k \in \mathbb{N}_0$,

$$a_k = \frac{2}{\frac{3}{2}} \int_0^{\frac{3}{2}} F(x) \cos \frac{k\pi x}{\frac{3}{2}}\, dx$$

$$= \frac{4}{3}\left(\int_0^1 x \cos \frac{2k\pi x}{3}\, dx + \int_1^{\frac{3}{2}} \cos \frac{2k\pi x}{3}\, dx\right).$$

Since F is a continuous piecewise smooth function, by Dirichlet's theorem, for all $x \in \left[\frac{3}{2}, 3\right]$,

$$f(x) = \frac{2}{3} - \frac{6}{\pi^2} \sum_{k=1}^{\infty} \frac{1}{k^2} \sin^2 \frac{k\pi}{3} \cos \frac{2k\pi x}{3}.$$

Problem 1.2.33. Use the Fourier series of the periodic function $f : \mathbb{R} \to \mathbb{R}$ determined by

$$f(x) = \begin{cases} \cos x & \text{if } x \in (0, \pi), \\ 0 & \text{if } x = 0, \\ -\cos x & \text{if } x \in (-\pi, 0), \end{cases}$$

to evaluate $\sum_{n=1}^{\infty} \frac{n^2}{(4n^2-1)^2}$.

Solution 1.2.33. The function f is odd and periodic, with the fundamental period $T = \pi$.

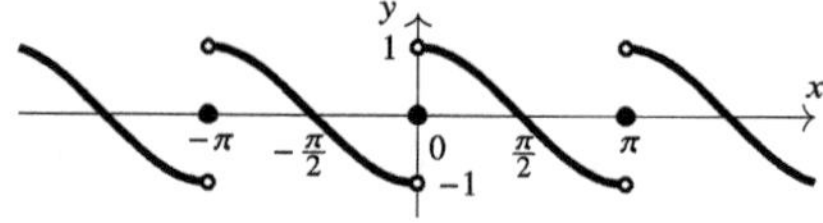

For all $n \in \mathbb{N}_0$, $a_n = 0$ and for $n \in \mathbb{N}$,

$$b_n = \frac{4}{\pi} \int_0^{\frac{\pi}{2}} \cos x \sin 2nx \; dx = \frac{2}{\pi} \int_0^{\frac{\pi}{2}} (\sin(2n+1)x + \sin(2n-1)x) \; dx$$

$$= -\frac{2}{\pi} \left(\frac{\cos(2n+1)x}{2n+1} + \frac{\cos(2n-1)x}{2n-1} \right) \Big|_0^{\frac{\pi}{2}} = \frac{8n}{\pi(4n^2-1)}.$$

Since, for any $k \in \mathbb{Z}$, $0 = f(k\pi) = \frac{1}{2}\left(\lim_{x \to k\pi^-} f(x) + \lim_{x \to k\pi^+} f(x) \right)$, it follows that, for all $x \in \mathbb{R}$, $f(x) = \frac{8}{\pi} \sum_{n=1}^{\infty} \frac{n \sin 2nx}{4n^2-1}$. By Perseval's identity, $\sum_{n=1}^{\infty} \frac{n^2}{(4n^2-1)^2} = \frac{\pi^2}{64} \frac{4}{\pi} \int_0^{\frac{\pi}{2}} \cos^2 x \; dx = \frac{\pi^2}{64}$.

Problem 1.2.34. Determine the Fourier series of the periodic function f, with the fundamental period $T = 8$, such that, for all $x \in [-4, 4]$, $f(x) = \frac{x(4-x)}{3}$.

Solution 1.2.34. Since the fundamental period of the function f is $T = 8$, it follows that, for all $k \in \mathbb{N}_0$,

$$a_k = \frac{1}{4} \int_{-4}^{4} f(x) \cos \frac{k\pi x}{4} \, dx = \frac{1}{4} \int_{-4}^{4} \frac{x(4-x)}{3} \cos \frac{k\pi x}{4} \, dx$$

$$= \frac{1}{12} \left(\int_{-4}^{4} 4x \cos \frac{k\pi x}{4} \, dx - \int_{-4}^{4} x^2 \cos \frac{k\pi x}{4} \, dx \right).$$

The function $x \mapsto x$ is odd and the function $x \mapsto x^2$ is even, which implies $a_k = -\frac{1}{6} \int_{0}^{4} x^2 \cos \frac{k\pi x}{4} \, dx$. It follows that $a_0 = -\frac{32}{9}$ and, for $k \in \mathbb{N}$, $a_k = \frac{64(-1)^{k+1}}{3k^2\pi^2}$. Similarly, for $k \in \mathbb{N}$, $b_k = \frac{32(-1)^{k+1}}{3k\pi}$. Therefore,

$$f(x) \sim -\frac{16}{9} + \frac{32}{3\pi^2} \sum_{k=1}^{\infty} \frac{(-1)^{k+1}}{k^2} \left(2\cos \frac{k\pi x}{4} + k\pi \sin \frac{k\pi x}{4} \right).$$

Problem 1.2.35.

(a) Determine the fundamental period and the Fourier series of the function $f(x) = \sin(x\sqrt{2})$.
(b) Let a be a positive real number, and let $f(x) = ax$ and $g(x) = \cos x$. Determine fundamental periods and Fourier series of the functions $f \circ g$ and $g \circ f$.
(c) Determine the fundamental period and the Fourier series of the function $f(x) = \sin \frac{x}{2} + \sin \frac{x}{3}$.

Solution 1.2.35.

(a) The fundamental period of f is $T = \pi\sqrt{2}$. Since the function f belongs to the complete orthogonal system

$$\frac{1}{2}, \cos\left(x\sqrt{2}\right), \sin\left(x\sqrt{2}\right), \ldots, \cos\left(kx\sqrt{2}\right), \sin\left(kx\sqrt{2}\right), \ldots,$$

it follows that the Fourier series of the function f is the function f itself.
(b) Since $x \mapsto (f \circ g)(x) = a\cos x$ and $x \mapsto (g \circ f)(x) = \cos ax$, the required Fourier series are the functions themselves.
(c) The fundamental period of the function $x \mapsto \sin \frac{x}{2}$ is 4π, and the fundamental period of the function $x \mapsto \sin \frac{x}{3}$ is 6π. It follows that the fundamental period of the function f is $T = 12\pi$.

Since functions $x \mapsto \sin \frac{x}{2}$ and $x \mapsto \sin \frac{x}{3}$ belong to the complete orthogonal system

$$\frac{1}{2}, \cos \frac{x}{6}, \sin \frac{x}{6}, \ldots, \cos \frac{kx}{6}, \sin \frac{kx}{6}, \ldots,$$

it follows that the Fourier series of the function f is the function f itself.

Problem 1.2.36. The function f is periodic with the fundamental period $T = \pi$, and such that, for all $x \in \left[-\frac{\pi}{2}, \frac{\pi}{2}\right]$, $f(x) = x \cos x$. Determine the Fourier series of f.

Solution 1.2.36. The function f is odd, so for all $k \in \mathbb{N}_0$, $a_k = 0$, and for all $k \in \mathbb{N}$,

$$b_k = \frac{4}{\pi} \int_0^{\frac{\pi}{2}} x \cos x \sin \frac{2k\pi x}{\pi} \, dx$$

$$= \frac{2}{\pi} \int_0^{\frac{\pi}{2}} (x \sin(1 + 2k)x - x \sin(1 - 2k)x) \, dx = \frac{16}{\pi} \frac{(-1)^{k+1} k}{(1 - 4k^2)^2}.$$

Since f is a continuous piecewise smooth function, by Dirichlet's theorem, for all $x \in \mathbb{R}$, $f(x) = \frac{16}{\pi} \sum_{k=1}^{\infty} \frac{(-1)^{k+1} k \sin 2kx}{(1-4k^2)^2}$.

Problem 1.2.37. The function $x \mapsto y = f(x)$ is defined by,

$$\text{for all } x \in \mathbb{R}, \ \sin y = \sin 2x \text{ and } y \in \left[-\frac{\pi}{2}, \frac{\pi}{2}\right].$$

(a) Explain why the condition $y \in \left[-\frac{\pi}{2}, \frac{\pi}{2}\right]$ is necessary for the function f to be well-defined.
(b) Draw a graph of the function f, and determine its fundamental period and the Fourier series.

Solution 1.2.37.

(a) If we fix a real number x and consider the expression $\sin y = \sin 2x$ as an equation with the unknown y, then the set of all solutions of that equation is determined by $y = \arcsin(\sin 2x) + k\pi$, $k \in \mathbb{Z}$. The condition $y \in \left[-\frac{\pi}{2}, \frac{\pi}{2}\right]$ implies that $k = 0$, i.e., $y = f(x) = \arcsin(\sin 2x)$, for all $x \in \mathbb{R}$.

(b) The function f is odd and periodic, with the fundamental period $T = \pi$ and such that, for all $x \in \left[-\frac{\pi}{4}, \frac{\pi}{4}\right]$, $f(x) = 2x$ and, for all $x \in \left[\frac{\pi}{4}, \frac{3\pi}{4}\right]$, $f(x) = \pi - 2x$.

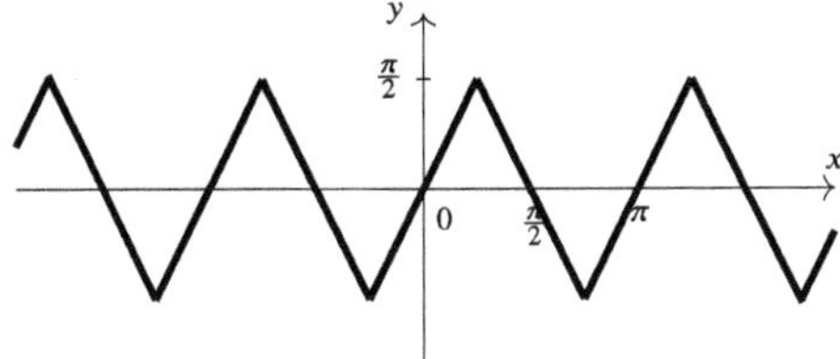

It follows that $a_k = 0$, for all $k \in \mathbb{N}_0$ and for all $k \in \mathbb{N}$,

$$b_k = \frac{4}{\pi} \int_0^{\frac{\pi}{2}} f(x) \sin 2kx \; dx = \frac{8}{\pi} \left(1 + (-1)^{k+1}\right) \int_0^{\frac{\pi}{4}} x \sin 2kx \; dx$$

$$= \frac{1}{k\pi} \left(1 + (-1)^{k+1}\right) \left(-\pi \cos \frac{k\pi}{2} + \frac{2}{k} \sin \frac{k\pi}{2}\right).$$

Hence, for all $k \in \mathbb{N}$, $b_{2k} = 0$ and $b_{2k-1} = \frac{4}{\pi} \frac{(-1)^{k+1}}{(2k-1)^2}$. By Dirichlet's theorem, for all $x \in \mathbb{R}$, $f(x) = \frac{4}{\pi} \sum_{k=1}^{\infty} \frac{(-1)^{k+1}}{(2k-1)^2} \sin 2(2k-1)x$.

Note 1: In Problem 1.2.7, we established that for all $x \in \mathbb{R}$,

$$\arcsin(\sin x) = \frac{4}{\pi} \sum_{k=1}^{\infty} \frac{(-1)^{k+1} \sin(2k - 1)x}{(2k - 1)^2}.$$

It follows that for all $x \in \mathbb{R}$,

$$\arcsin(\sin 2x) = \frac{4}{\pi} \sum_{k=1}^{\infty} \frac{(-1)^{k+1} \sin 2(2k - 1)x}{(2k - 1)^2},$$

which is exactly what we obtained in Problem 1.2.37.

Note 2: Another way to look at the relationship between the Fourier series of the function $x \mapsto \arcsin(\sin x)$ and the Fourier series of the function $x \mapsto \arcsin(\sin 2x)$ is to note that the graph of the latter function is a horizontal compression by coefficient 2 of the graph of the former function. This makes the period of the function $x \mapsto \arcsin(\sin 2x)$ two times smaller than the period of the function $x \mapsto \arcsin(\sin x)$. Hence, instead of "$(2k - 1)x$" in the Fourier series of the function $x \mapsto \arcsin(\sin x)$, we have "$2(2k-1)x$" in the Fourier series of the function $x \mapsto \arcsin(\sin 2x)$.

Problem 1.2.38. The function f is odd and periodic, with the fundamental period $T = 2$, and such that, for all $x \in (1, 2]$, $f(x) = (2-x)^2$. Determine the Fourier series of f.

Solution 1.2.38. The function f is periodic with the fundamental period $T = 2$, which implies that, for all $x \in (-1, 0]$, $f(x) = f(x + 2) = (2 - (x+2))^2 = x^2$ and $f(-1) = f(1)$. Since f is an odd function, for all $x \in [0, 1)$, $f(x) = -f(-x) = -x^2$ and $f(-1) = -f(1)$.

It follows that the function f is determined by

$$f(x) = \begin{cases} -x|x| & \text{if } x \in (-1, 1), \\ 0 & \text{if } x \in \{-1, 1\}, \end{cases}$$

and, for $x \in \mathbb{R}$ and $k \in \mathbb{Z}$, $f(x + 2k) = f(x)$.

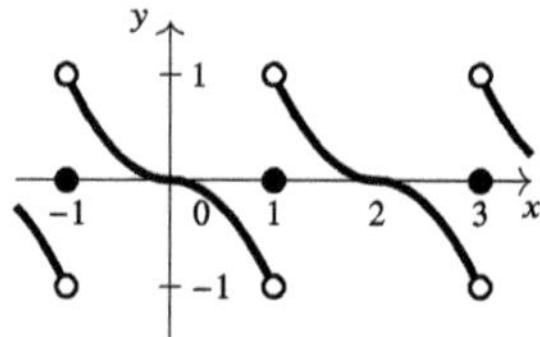

Hence, $a_k = 0$, for all $k \in \mathbb{N}_0$ and, for all $k \in \mathbb{N}$,

$$b_k = 2 \int_0^1 f(x) \sin k\pi x \; dx = -2 \int_0^1 x^2 \sin k\pi x \; dx$$

$$= \frac{2}{k^3 \pi^3} \left((-1)^k k^2 \pi^2 + 2 \left(1 - (-1)^k \right) \right).$$

By Dirichlet's theorem, for all $x \in \mathbb{R}$,

$$f(x) = \frac{2}{\pi^3} \sum_{k=1}^{\infty} \frac{1}{k^3} \left((-1)^k k^2 \pi^2 + 2 \left(1 - (-1)^k \right) \right) \sin k\pi x.$$

Note: We can write the obtained Fourier series in the following form:

$$f(x) = \frac{2}{\pi} \sum_{k=1}^{\infty} \frac{(-1)^k \sin k\pi x}{k} + \frac{8}{\pi^3} \sum_{k=1}^{\infty} \frac{\sin(2k-1)\pi x}{(2k-1)^3}.$$

This corresponds to the fact that, for $x \in [0,1)$, $f(x) = -x + (x - x^2)$. Now, the first sum corresponds to the Fourier sine series of the function $x \mapsto -x$, $x \in [0,1)$. The second sum is the Fourier sine series of the function $x \mapsto x - x^2$, $x \in [0,1)$. We observe that an odd periodic, with the fundamental period 2, extension of the function $x \mapsto x - x^2$, $x \in [0,1)$, has a continuous first derivative and *only* a piecewise continuous second derivative. This is why its Fourier coefficients behave like $\frac{1}{k^3}$. See Problem 1.2.3.

Problem 1.2.39. Let a be a positive number. The function f is periodic, with the fundamental period $T = 2a$, and such that, for all $x \in [-a, a)$, $f(x) = |\sinh ax|$. Determine the Fourier series of f.

Solution 1.2.39. The function f is even, so $b_k = 0$, for all $k \in \mathbb{N}$. For all $k \in \mathbb{N}_0$,

$$a_k = \frac{2}{a} \int_0^a \sinh ax \cos \frac{k\pi x}{a} \, dx = \frac{2a^2 \left((-1)^k \cosh a^2 - 1\right)}{k^2 \pi^2 + a^4}.$$

Since f is a continuous piecewise smooth function, by Dirichlet's theorem, for all $x \in \mathbb{R}$,

$$f(x) = \frac{\cosh a^2 - 1}{a^2} + 2a^2 \sum_{k=1}^{\infty} \frac{(-1)^k \cosh a^2 - 1}{k^2 \pi^2 + a^4} \cos \frac{k\pi x}{a}.$$

Problem 1.2.40. The function f is periodic, with the fundamental period $T = 4$, and such that, for all $x \in (-2, 2]$, $f(x) = 2^{\lfloor |x| \rfloor} \cdot \big| |x| - 1 \big|$. Determine the Fourier series of f.

Solution 1.2.40. The function f is even and such that

$$f(x) = \begin{cases} 1 - x & \text{if } x \in [0, 1), \\ 2(x - 1) & \text{if } x \in [1, 2), \\ 4 & \text{if } x = 2. \end{cases}$$

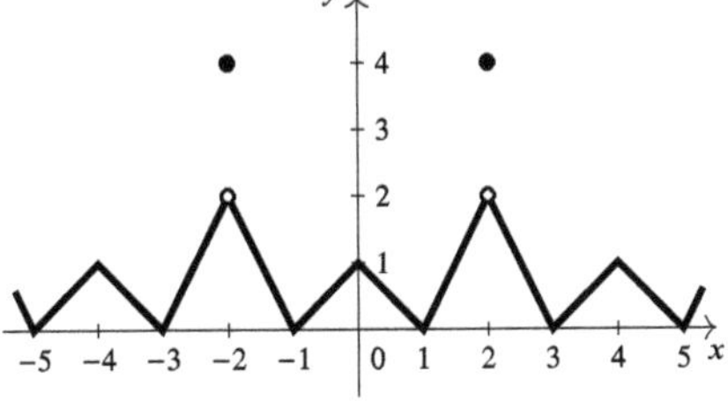

It follows that, for all $k \in \mathbb{N}$, $b_k = 0$ and, for all $k \in \mathbb{N}_0$,

$$a_k = \int_0^2 f(x) \cos \frac{k\pi x}{2}\, dx$$

$$= \int_0^1 (1-x) \cos \frac{k\pi x}{2}\, dx + 2 \int_1^2 (x-1) \cos \frac{k\pi x}{2}\, dx.$$

Hence, $a_0 = \frac{3}{2}$, and for all $k \in \mathbb{N}$, $a_k = \frac{4}{k^2 \pi^2} \left(1 + 2(-1)^k - 3 \cos \frac{k\pi}{2} \right)$. Thus,

$$f(x) \sim \frac{3}{4} + \frac{4}{\pi^2} \sum_{k=1}^{\infty} \frac{1}{k^2} \left(1 + 2(-1)^k - 3\cos \frac{k\pi}{2} \right) \cos \frac{k\pi x}{2}$$

$$= \frac{3}{4} + \frac{2}{\pi^2} \sum_{k=1}^{\infty} \frac{1}{(2k-1)^2} \left(3 \cos(2k-1)\pi x - 2 \cos \frac{(2k-1)\pi x}{2} \right).$$

Note: One rough check of this result is to see what happens if $x = 1$. From $f(1) = 0$ and, for all $k \in \mathbb{N}$, $\cos \frac{(2k-1)\pi}{2} = 0$ and $\cos(2k-1)\pi = -1$, it follows that $0 = \frac{3}{4} - \frac{6}{\pi^2} \sum_{k=1}^{\infty} \frac{1}{(2k-1)^2}$, which is equivalent to $\sum_{k=1}^{\infty} \frac{1}{(2k-1)^2} = \frac{\pi^2}{8}$.

Problem 1.2.41. Let $n \in \mathbb{N} \backslash \{1\}$. Determine the Fourier series of the periodic function f, with the fundamental period $T = 2n$, and such that

$$f(x) = \begin{cases} \frac{n^2(n-|x|)}{n^2-1} & \text{if } |x| \in \left[\frac{1}{n}, n \right], \\ n & \text{if } |x| < \frac{1}{n}. \end{cases}$$

Solution 1.2.41. The image depicts the case of $n = 2$. Since f is an even function, $b_k = 0$ for all $k \in \mathbb{N}$, and, for all $k \in \mathbb{N}_0$,

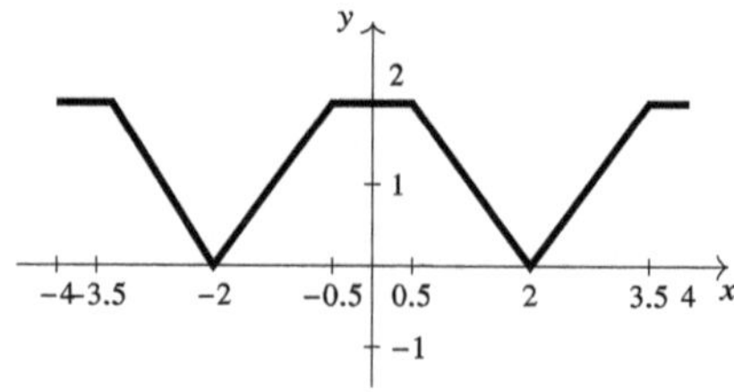

$$a_k = \frac{2}{n} \int_0^n f(x) \cos \frac{k\pi x}{n} \, dx$$

$$= \frac{2}{n} \left(\int_0^{\frac{1}{n}} n \cos \frac{k\pi x}{n} \, dx + \int_{\frac{1}{n}}^n \frac{n^2(n-x)}{n^2-1} \cos \frac{k\pi x}{n} \, dx \right).$$

Hence, $a_0 = \frac{n^2+1}{n}$ and, for $k \in \mathbb{N}$, $a_k = \frac{2n^3}{k^2\pi^2(n^2-1)} \left(\cos \frac{k\pi}{n^2} - (-1)^k \right)$. Since f is a continuous piecewise smooth function, by Dirichlet's theorem, for all $x \in \mathbb{R}$,

$$f(x) = \frac{n^2+1}{2n} + \frac{2n^3}{\pi^2(n^2-1)} \sum_{k=1}^{\infty} \frac{1}{k^2} \left(\cos \frac{k\pi}{n^2} - (-1)^k \right) \cos \frac{k\pi x}{n}.$$

Problem 1.2.42. Determine the Fourier series of the periodic function f, with the fundamental period $T = 2$, and such that
$$f(x) = \begin{cases} x & \text{if } x \in [0,1], \\ 2-x & \text{if } x \in (1,2]. \end{cases}$$

Answer 1.2.42. For all $x \in \mathbb{R}$, $f(x) = \frac{1}{2} - \frac{4}{\pi} \sum_{k=1}^{\infty} \frac{\cos(2k-1)\pi x}{(2k-1)^2}$.

Problem 1.2.43. Determine the Fourier series of the periodic function f, with the fundamental period $T = \pi$, and such that
$$f(x) = \begin{cases} x & \text{if } x \in \left[0, \frac{\pi}{3}\right], \\ \frac{\pi-x}{2} & \text{if } x \in \left(\frac{\pi}{3}, \pi\right). \end{cases}$$

Solution 1.2.43. From, for all $k \in \mathbb{N}_0$,

$$a_k = \frac{2}{\pi} \int_0^\pi f(x) \cos 2kx \, dx$$

$$= \frac{2}{\pi} \left(\int_0^{\frac{\pi}{3}} x \cos 2kx \, dx + \frac{1}{2} \int_{\frac{\pi}{3}}^\pi (\pi - x) \cos 2kx \, dx \right),$$

it follows that $a_0 = \frac{\pi}{3}$ and, for all $k \in \mathbb{N}$, $a_k = -\frac{3}{2\pi k^2} \sin^2 \frac{k\pi}{3}$. Also, for all $k \in \mathbb{N}$,

$$b_k = \frac{2}{\pi} \int_0^\pi f(x) \sin 2kx \, dx = \frac{3}{2\pi k^2} \sin \frac{k\pi}{3} \cos \frac{k\pi}{3}.$$

Since f is a continuous piecewise smooth function, by Dirichlet's theorem, for all $x \in \mathbb{R}$,

$$f(x) = \frac{\pi}{6} - \frac{3}{2\pi} \sum_{k=1}^{\infty} \frac{1}{k^2} \sin \frac{k\pi}{3} \sin k \left(\frac{\pi}{3} - 2x \right).$$

Problem 1.2.44. Determine the Fourier series of the periodic function f, with the fundamental period $T = \pi$, and such that
$$f(x) = \begin{cases} \pi \sin x & \text{if } x \in \left[0, \frac{\pi}{2}\right], \\ \pi \cos x & \text{if } x \in \left(\frac{\pi}{2}, \pi\right). \end{cases} \quad \text{Evaluate } \sum_{k=1}^{\infty} \frac{1}{4(2k-1)^2-1}.$$

Solution 1.2.44. Observe from the graph that $a_0 = \int_{-\frac{\pi}{2}}^{\frac{\pi}{2}} f(x) = 0$
and $f(x) = \begin{cases} \pi \sin x & \text{if } x \in \left[0, \frac{\pi}{2}\right], \\ -\pi \cos x & \text{if } x \in \left(-\frac{\pi}{2}, 0\right). \end{cases}$ From, for $x \in \left(-\frac{\pi}{2}, 0\right)$,

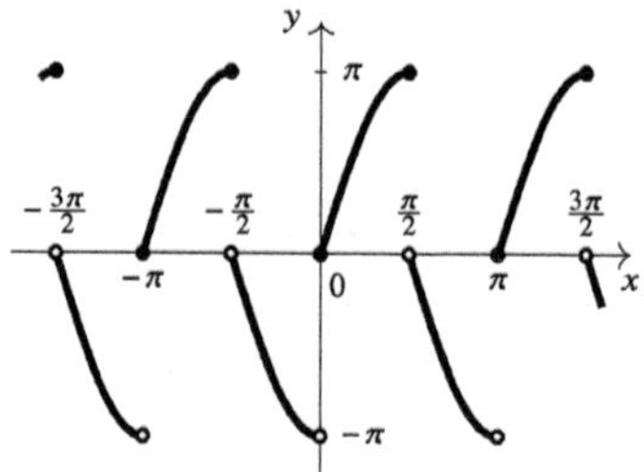

$$f\left(x + \frac{\pi}{2}\right) = \pi \sin\left(x + \frac{\pi}{2}\right) = \pi \cos x = -f(x),$$

it follows[11] that, for $k \in \mathbb{N}$, $a_k = 2((-1)^{k+1} + 1) \int_0^{\frac{\pi}{2}} \sin x \cos 2kx \, dx$
and $b_k = 2((-1)^{k+1} + 1) \int_0^{\frac{\pi}{2}} \sin x \sin 2kx \, dx$. Therefore, for all $k \in \mathbb{N}$, $a_{2k} = b_{2k} = 0$, and

$$a_{2k-1} = 4 \int_0^{\frac{\pi}{2}} \sin x \cos 2(2k - 1)x \, dx = -\frac{4}{(4k - 1)(4k - 3)},$$

$$b_{2k-1} = 4 \int_0^{\frac{\pi}{2}} \sin x \sin 2(2k - 1)x \, dx = \frac{8(2k - 1)}{(4k - 1)(4k - 3)}.$$

---

[11] We used the same argument in Problem 1.2.13(d).

It follows that

$$f(x) \sim 4\sum_{k=1}^{\infty} \frac{2(2k-1)\sin 2(2k-1)x - \cos 2(2k-1)x}{(4k-1)(4k-3)}.$$

By Dirichlet's theorem,

$$\frac{\pi}{2} = \frac{1}{2}\left(\lim_{x\to\frac{\pi}{2}^-} f(x) + \lim_{x\to\frac{\pi}{2}^+} f(x)\right)$$

$$= 4\sum_{k=1}^{\infty} \frac{2(2k-1)\sin 2(2k-1)\frac{\pi}{2} - \cos 2(2k-1)\frac{\pi}{2}}{(4k-1)(4k-3)}$$

$$= 4\sum_{k=1}^{\infty} \frac{1}{4(2k-1)^2 - 1}.$$

Therefore, $\sum_{k=1}^{\infty} \frac{1}{4(2k-1)^2-1} = \frac{\pi}{8}$.

Problem 1.2.45. Determine the Fourier series of the distance-from-the-nearest-integer function $f(x) = (x)$. For example, $f(3) = (3) = 0$, $f(\pi) = \pi - 3$, and $f(3.5) = 0.5$.

Solution 1.2.45. The function f is an even and periodic, with the fundamental period $T = 1$, extension of the function $x \mapsto x$, $x \in \left[0, \frac{1}{2}\right]$. See the figure.

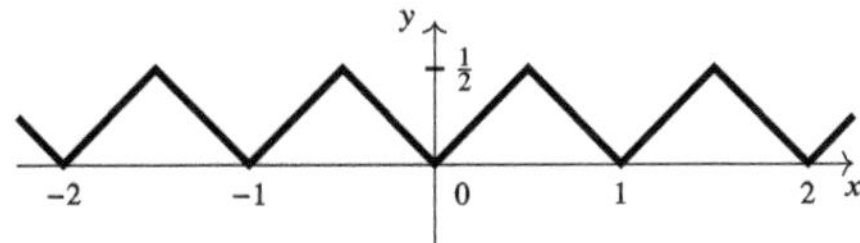

For all $x \in \mathbb{R}$, $(x) = \frac{1}{8} - \frac{2}{\pi^2}\sum_{k=1}^{\infty} \frac{\cos 2(2k-1)\pi x}{(2k-1)^2}$.

Problem 1.2.46.

(a) Establish that $\sin x = \frac{2}{\pi} - \frac{4}{\pi}\sum_{k=1}^{\infty} \frac{\cos 2kx}{4k^2-1}$, for all $x \in (0, \pi)$.
(b) Differentiating and integrating the obtained identity yields two different series expansions of the function $x \mapsto \cos x$, $x \in (0, \pi)$. Demonstrate that it is possible to obtain one from the other.

Solution 1.2.46.

(a) The function $f(x) = |\sin x|$, $x \in \mathbb{R}$, is even, continuous, piecewise smooth, and periodic, with the fundamental period $T = \pi$. From, $b_k = 0$, $k \in \mathbb{N}$, and for $k \in \mathbb{N}_0$,

$$
\begin{aligned}
a_k &= \frac{4}{\pi} \int_0^{\frac{\pi}{2}} \sin x \cos 2kx \; dx \\[2mm]
&= \frac{2}{\pi} \int_0^{\frac{\pi}{2}} (\sin(2k+1k)x - \sin(2k-1)x) \; dx \\[2mm]
&= \frac{2}{\pi} \left(-\frac{\cos(2k+1)x}{2k+1} + \frac{\cos(2k-1)x}{2k-1} \right) \Big|_0^{\frac{\pi}{2}} = -\frac{4}{\pi(4k^2-1)},
\end{aligned}
$$

it follows that $|\sin x| = \frac{2}{\pi} - \frac{4}{\pi} \sum_{k=1}^{\infty} \frac{\cos 2kx}{4k^2-1}$, for all $x \in \mathbb{R}$. In particular, $\sin x = \frac{2}{\pi} - \frac{4}{\pi} \sum_{k=1}^{\infty} \frac{\cos 2kx}{4k^2-1}$, for all $x \in (0, \pi)$.

(b) Since the function f is twice differentiable in the interval $(0, \pi)$, the differentiation of the Fourier series yields the identity

$$
\cos x = \frac{8}{\pi} \sum_{k=1}^{\infty} \frac{k \sin 2kx}{4k^2 - 1}, \quad x \in (0, \pi).
$$

From, for $x \in (0, \pi)$,

$$
\begin{aligned}
\int_0^x \sin t \; dt &= \int_0^x \left(\frac{2}{\pi} - \frac{4}{\pi} \sum_{k=1}^{\infty} \frac{\cos 2kt}{4k^2-1} \right) dt \\[2mm]
&= \frac{2}{\pi} \int_0^x dt - \frac{4}{\pi} \sum_{k=1}^{\infty} \int_0^x \frac{\cos 2kt}{4k^2-1} \; dt,
\end{aligned}
$$

it follows that

$$
\cos x = 1 - \frac{2x}{\pi} + \frac{2}{\pi} \sum_{k=1}^{\infty} \frac{\sin 2kx}{k(4k^2-1)}, \quad x \in (0, \pi).
$$

Indeed, we have obtained two different series expansions of the function $x \mapsto \cos x$, $x \in (0, \pi)$. Observe that the latter expansion is not a trigonometric series.

Let g be an odd periodic, with the fundamental period $T = 2\pi$, extension of the function $x \mapsto 1 - \frac{2x}{\pi}$, $x \in (0, \pi)$, and let $\{A_k\}_{n \in \mathbb{N}_0}$ and $\{B_k\}_{k \in \mathbb{N}}$ be the sequences of the Fourier coefficients of the function g. Then, from $A_k = 0$, $k \in \mathbb{N}_0$, and for $k \in \mathbb{N}$,

$$B_k = \frac{2}{\pi} \int_0^\pi \left(1 - \frac{2x}{\pi}\right) \sin kx \; dx = \frac{2}{k\pi} \left(1 + (-1)^k\right),$$

it follows that, for $x \in (0, \pi)$, $1 - \frac{2x}{\pi} = \frac{2}{\pi} \sum_{k=1}^\infty \frac{\sin 2kx}{k}$. This fact connects the two series expansions of the function $x \mapsto \cos x$, $x \in (0, \pi)$, obtained earlier.

1.3 Fourier Integral

Determine the Fourier integral of the following functions.

Problem 1.3.1. $f(x) = \mathrm{sign}(x - a) - \mathrm{sign}(x - b)$, with $a, b \in \mathbb{R}$, $a < b$.

Solution 1.3.1. The function $f(x) = \begin{cases} 0 & \text{if } x \in (-\infty, a) \cup (b, \infty), \\ 1 & \text{if } x \in \{a, b\}, \\ 2 & \text{if } x \in (a, b), \end{cases}$

is absolutely integrable. For all $u \in (0, \infty)$,

$$a(u) = \frac{1}{\pi} \int_{-\infty}^\infty f(t) \cos ut \; dt = \frac{2}{\pi} \int_a^b \cos ut \; dt$$

$$= \frac{4}{u\pi} \cos \frac{(a+b)u}{2} \sin \frac{(b-a)u}{2}$$

and

$$b(u) = \frac{1}{\pi} \int_{-\infty}^\infty f(t) \sin ut \; dt = \frac{2}{\pi} \int_a^b \sin ut \; dt$$

$$= \frac{4}{u\pi} \sin \frac{(a+b)u}{2} \sin \frac{(b-a)u}{2}.$$

Also, $a(0) = \frac{1}{\pi} \int_{-\infty}^\infty f(t) \; dt = \frac{2(b-a)}{\pi}$. Since the function f is piecewise smooth and since the average of the left-and right-hand side limits of the function f at any $x \in \mathbb{R}$ is $f(x)$, it follows that, for all $x \in \mathbb{R}$,

$$f(x) = \frac{1}{\pi} \int_0^\infty (a(u) \cos ux + b(u) \sin ux) \; du$$

$$= \frac{4}{\pi^2} \int_0^\infty \frac{1}{u} \sin \frac{(b-a)u}{2}$$

$$\cdot \left(\cos \frac{(a+b)u}{2} \cos ux + \sin \frac{(a+b)u}{2} \sin ux \right) du$$

$$= \frac{4}{\pi^2} \int_0^\infty \frac{1}{u} \sin \frac{(b-a)u}{2} \cos \left(\frac{a+b}{2} - x \right) u \, du.$$

Note: Since $\lim\limits_{u \to 0^+} \dfrac{1}{u} \sin \dfrac{(b-a)u}{2} = \dfrac{b-a}{2}$, both functions

$$a(u) = \begin{cases} \frac{4}{u\pi} \cos \frac{(a+b)u}{2} \sin \frac{(b-a)u}{2} & \text{if } u \in (0, \infty) \\ \frac{2(b-a)}{\pi} & \text{if } u = 0 \end{cases}$$

and

$$b(u) = \begin{cases} \frac{4}{u\pi} \sin \frac{(a+b)u}{2} \sin \frac{(b-a)u}{2} & \text{if } u \in (0, \infty) \\ 0 & \text{if } u = 0 \end{cases}$$

are continuous in $[0, \infty)$.

Problem 1.3.2. The function $f : \mathbb{R} \to \mathbb{R}$ is such that, for $k \in \{1, 2\}$, if $|x| \in [k - 1, k)$, then $f(x) = 3 - k + \frac{\text{sign}(x)}{k+1}$ and if $|x| \geq 2$, then $f(x) = 0$.

Solution 1.3.2. Observe that, for $x \in (-1, 1)$, $f(x) = 2 + \frac{\text{sign}(x)}{2}$ and, for $x \in (-2, -1] \cup [1, 2)$, $f(x) = 2 + \frac{\text{sign}(x)}{3}$. See the figure.

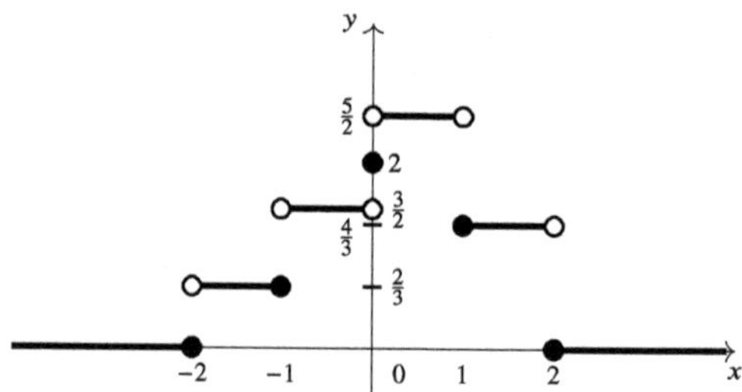

It follows that $a(0) = \frac{1}{\pi} \int_{-\infty}^\infty f(t)dt = \frac{6}{\pi}$ and, for all $u \in (0, \infty)$,

$$a(u) = \frac{1}{\pi} \int_{-\infty}^\infty f(t) \cos ut \, dt = \frac{1}{\pi} \left(\frac{2}{3} \int_{-2}^{-1} \cos ut \, dt + \frac{3}{2} \int_{-1}^0 \cos ut \, dt \right.$$

$$\left. + \frac{5}{2} \int_0^1 \cos ut \, dt + \frac{4}{3} \int_1^2 \cos ut \, dt \right)$$

$$= \frac{2}{\pi u} (\sin 2u + \sin u).$$

Similarly, for all $u \in (0, \infty)$,

$$b(u) = \frac{1}{\pi} \int_{-\infty}^{\infty} f(t) \sin ut\, dt = \frac{1}{3\pi u}(3 - 2\cos 2u - \cos u).$$

Therefore,

$$f(x) \sim \frac{1}{3\pi} \int_0^{\infty} \frac{1}{u}\big(6(\sin 2u + \sin u)\cos ux + (3 - 2\cos 2u - \cos u)\sin ux\big)\, du.$$

Note: Observe that $\lim\limits_{u \to 0^+} a(u) = \dfrac{6}{\pi}$ and $\lim\limits_{u \to 0^+} b(u) = 0$.

Problem 1.3.3. Let $\alpha, \beta > 0$. Use the Fourier integral of the function

$$f(x) = \begin{cases} \beta & \text{if } |x| < \alpha, \\ \frac{\beta}{2} & \text{if } |x| = \alpha, \\ 0 & \text{if } |x| > \alpha, \end{cases} \quad \text{to evaluate } \int_0^{\infty} \frac{\sin \alpha u}{u}\, du.$$

Solution 1.3.3. Since f is an even function, for all $u \in (0, \infty)$, $b(u) = 0$ and

$$a(u) = \frac{2}{\pi} \int_0^{\infty} f(t) \cos ut\, dt = \frac{2\beta}{\pi} \int_0^{\alpha} \cos ut\, dt = \frac{2\beta \sin \alpha u}{\pi u}.$$

It follows that, for all $x \in \mathbb{R}$, $f(x) = \frac{2\beta}{\pi} \int_0^{\infty} \frac{\sin \alpha u}{u} \cos ux\, du$. If $x = 0$ and $\beta = 1$, then $1 = f(0) = \frac{2}{\pi} \int_0^{\infty} \frac{\sin \alpha u}{u}\, du$, which implies that, for $\alpha > 0$, $\int_0^{\infty} \frac{\sin \alpha u}{u}\, du = \frac{\pi}{2}$.[12]

Problem 1.3.4. Let $\lambda > 0$. Use the Fourier integral of the function

$$f(x) = \begin{cases} \cos \lambda x & \text{if } |x| \leq \frac{\pi}{2\lambda}, \\ 0 & \text{if } |x| > \frac{\pi}{2\lambda}, \end{cases} \quad \text{to evaluate } \int_0^{\infty} \frac{\cos u}{\pi^2 - 4u^2}\, du.$$

Solution 1.3.4. Since f is an even function, for all $u \in (0, \infty)$, $b(u) = 0$. For all $u \in [0, \infty)$,

$$a(u) = \frac{2}{\pi} \int_0^{\infty} f(t) \cos ut\, dt = \frac{2}{\pi} \int_0^{\frac{\pi}{2\lambda}} \cos \lambda t \cos ut\, dt$$

---

[12] $\int_0^{\infty} \frac{\sin t}{t}\, dt = \frac{\pi}{2}$ is known as the Dirichlet integral.

$$= \frac{1}{\pi} \int_0^{\frac{\pi}{2\lambda}} \left(\cos(\lambda + u)t + \cos(\lambda - u)t \right) dt$$

$$= \begin{cases} \frac{2\lambda}{\pi(\lambda^2 - u^2)} \cos \frac{\pi u}{2\lambda} & \text{if } u \neq \lambda, \\ \frac{1}{2\lambda} & \text{if } u = \lambda. \end{cases}$$

Since f is a continuous piecewise smooth function in $\mathbb{R}$, for all $x \in \mathbb{R}$, $f(x) = \frac{2\lambda}{\pi} \int_0^\infty \frac{\cos \frac{\pi u}{2\lambda}}{\lambda^2 - \pi^2} \cos ux \, du$. If $x = 0$ and $\lambda = \frac{\pi}{2}$, then $1 = f(0) = \int_0^\infty \frac{\cos u}{\frac{\pi^2}{4} - u^2} \, du$, which implies $\int_0^\infty \frac{\cos u}{\pi^2 - 4u^2} \, du = \frac{1}{4}$.

Note: Observe that $\displaystyle \lim_{u \to \frac{\pi}{2}} \frac{\cos u}{\pi^2 - 4u^2} = \frac{1}{4\pi}$.

Problem 1.3.5. Use the Fourier integral of the function

$$f(x) = \begin{cases} 1 - \frac{|x|}{2} & \text{if } |x| \leq 2, \\ 0 & \text{if } |x| > 2, \end{cases} \quad \text{to evaluate } \int_0^\infty \frac{\sin^2 u}{u^2} \, du.$$

Solution 1.3.5. Since f is an even function, for all $u \in (0, \infty)$, $b(u) = 0$. Also, $a(0) = \frac{2}{\pi} \int_0^2 \left(1 - \frac{t}{2}\right) dt = \frac{2}{\pi}$ and for all $u \in (0, \infty)$,

$$a(u) = \frac{2}{\pi} \int_0^\infty f(t) \cos ut \, dt = \frac{2}{\pi} \int_0^2 \left(1 - \frac{t}{2}\right) \cos ut \, dt = \frac{2}{\pi} \frac{\sin^2 u}{u^2}.$$

Since f is a continuous piecewise smooth function in $\mathbb{R}$, $f(x) = \frac{2}{\pi} \cdot \int_0^\infty \frac{\sin^2 u}{u^2} \cos ux \, du$. If $x = 0$, then $1 = f(0) = \frac{2}{\pi} \int_0^\infty \frac{\sin^2 u}{u^2} du$, which implies $\int_0^\infty \frac{\sin^2 u}{u^2} du = \frac{\pi}{2}$.

Problem 1.3.6. Use the Fourier integral of the function

$$f(x) = \begin{cases} 1 - x^2 & \text{if } |x| < 1, \\ 0 & \text{if } |x| \geq 1, \end{cases} \quad \text{to evaluate } \int_0^\infty \frac{1}{x^3} \left(\sin 2x - 2x \cos 2x \right) \cdot$$

$\cos x \, dx$.

Solution 1.3.6. Since f is an even function, for all $u \in (0, \infty)$, $b(u) = 0$. Also, $a(0) = \frac{2}{\pi} \int_0^1 (1 - t^2) \, dt = \frac{4}{3\pi}$ and for all $u \in (0, \infty)$,

$$a(u) = \frac{2}{\pi} \int_0^1 (1 - t^2) \cos ut \, dt = \frac{4}{\pi} \frac{\sin u - u \cos u}{u^3}.$$

Since f is a continuous piecewise smooth function in $\mathbb{R}$, $f(x) = \frac{4}{\pi} \cdot$ $\int_0^\infty \frac{\sin u - u \cos u}{u^3} \cos ux \, du$. It follows that

$$I = \int_0^\infty \frac{1}{x^3} (\sin 2x - 2x \cos 2x) \cos x \, dx = \begin{vmatrix} 2x = u \\ dx = \frac{du}{2} \end{vmatrix}$$

$$= 4 \int_0^\infty \frac{\sin u - u \cos u}{u^3} \cos \frac{u}{2} \, du = 4 \cdot \frac{\pi}{4} f\left(\frac{1}{2}\right) = \frac{3\pi}{4}.$$

Problem 1.3.7. Let $\alpha > 0$. Use the Fourier integral of the function $f_\alpha(x) = 2^{-\alpha|x|}$ to evaluate $\int_0^\infty \frac{\cos u}{u^2 + 4} \, du$.

Solution 1.3.7. Since f_α is an even function, for all $u \in (0, \infty)$, $b(u) = 0$. Also, for all $u \in [0, \infty)$,

$$a(u) = \frac{2}{\pi} \int_0^\infty 2^{-\alpha t} \cos ut \, dt = \frac{2}{\pi} \int_0^\infty e^{-\alpha t \ln 2} \cos ut \, dt$$

$$= \frac{2}{\pi} \frac{1}{\alpha^2 \ln^2 2 + u^2} \left(2^{-\alpha t}(-\alpha \ln 2 \cos ut + u \sin ut)\right|_0^\infty$$

$$= \frac{2}{\pi} \frac{\alpha \ln 2}{\alpha^2 \ln^2 2 + u^2}.$$

Since f_α is a continuous piecewise smooth function in $\mathbb{R}$, $f_\alpha(x) = \frac{2\alpha \ln 2}{\pi} \int_0^\infty \frac{\cos ux \, du}{\alpha^2 \ln^2 2 + u^2}$. It follows that, for $x = 1$ and $\alpha = \frac{2}{\ln 2}$,

$$\int_0^\infty \frac{\cos u}{u^2 + 4} \, du = \frac{\pi}{2 \cdot \frac{2}{\ln 2} \cdot \ln 2} f_{\frac{2}{\ln 2}}(1) = \frac{\pi}{4e^2}.$$

Problem 1.3.8. Use the Fourier integral of the function

$$f(x) = \begin{cases} e^{-x} \cos x & \text{if } x > 0, \\ 0 & \text{if } x = 0, \\ -e^x \cos x & \text{if } x < 0, \end{cases}$$ to evaluate $\int_0^\infty \frac{u^3 \sin u}{u^4 + 4} \, du$.

Solution 1.3.8. The function f is odd; therefore, for $u \in [0, \infty)$, $a(u) = 0$, and for $u \in (0, \infty)$, $b(u) = \frac{2}{\pi} \int_0^\infty f(t) \sin ut \, dt$. Thus,

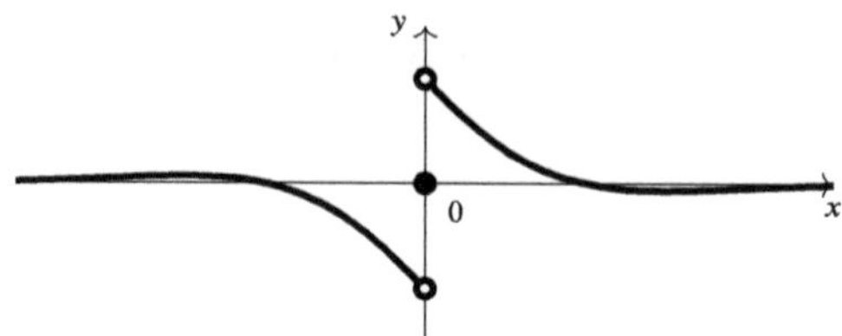

$$b(u) = \frac{2}{\pi} \int_0^\infty e^{-t} \cos t \sin ut \; dt$$

$$= \frac{1}{\pi} \int_0^\infty e^{-t}(\sin(1+u)t - \sin(1-u)t) \; dt = \frac{2}{\pi}\frac{u^3}{u^4+4},$$

and, for all $x \in \mathbb{R}$, $f(x) = \frac{2}{\pi} \int_0^\infty \frac{u^3 \sin xu}{u^4+4} \, du$. In particular, $\int_0^\infty \frac{u^3 \sin u}{u^4+4} \, du = \frac{\pi}{2} f(1) = \frac{\pi \cos 1}{2e}$.

Problem 1.3.9. Use the Fourier integral of the function $f(x) = e^{-a|x|} \sin bx$, $a, b > 0$, to evaluate $\int_0^\infty \frac{u \sin u}{u^4+4} \, du$.

Solution 1.3.9. Since the function f is odd, for $u \in [0, \infty)$, $a(u) = 0$ and

$$b(u) = \frac{2}{\pi} \int_0^\infty e^{-at} \sin bt \sin ut \; dt$$

$$= \frac{1}{\pi} \int_0^\infty e^{-at}(\cos(b-u)t - \cos(b+u)t) \; dt$$

$$= \frac{4abu}{\pi(a^2 + (b-u)^2)(a^2 + (b+u)^2)}.$$

For all $x \in \mathbb{R}$, $f(x) = \frac{4ab}{\pi} \int_0^\infty \frac{u \sin ux \; du}{(a^2+(b-u)^2)(a^2+(b+u)^2)}$. Since for $a = b = x = 1$,

$$e^{-1} \sin 1 = \frac{4}{\pi} \int_0^\infty \frac{u \sin u \; du}{(1+(1-u)^2)(1+(1+u)^2)} = \frac{4}{\pi} \int_0^\infty \frac{u \sin u \; du}{u^4+4},$$

it follows that $\int_0^\infty \frac{u \sin u}{u^4+4} \, du = \frac{\pi \sin 1}{4e}$.

1.4 Fourier Transform

Problem 1.4.1. Let $a > 0$ and let the function $f : \mathbb{R} \to \mathbb{R}$ be absolutely integrable. Establish the following properties of the Fourier transformation:

(a) $\mathcal{F}\left[f(t-a)\right](u) = e^{-iau}\mathcal{F}\left[f(t)\right](u)$.
(b) $\mathcal{F}\left[f(at)\right](u) = \frac{1}{a}\mathcal{F}\left[f(t)\right]\left(\frac{u}{a}\right)$.

Solution 1.4.1. By definition, for any $u \in \mathbb{R}$,

$$\mathcal{F}\left[f(t-a)\right](u) = \int_{-\infty}^{\infty} f(t-a)e^{-iut}\,dt = |s = t - a|$$

$$= \int_{-\infty}^{\infty} f(s)e^{-iu(s+a)}\,ds = e^{-iau}\mathcal{F}\left[f(t)\right](u),$$

and

$$\mathcal{F}\left[f(at)\right](u) = \int_{-\infty}^{\infty} f(at)e^{-iut}\,dt = |s = at|$$

$$= \frac{1}{a}\int_{-\infty}^{\infty} f(s)e^{-i\frac{u}{a}s}\,ds = \frac{1}{a}\mathcal{F}\left[f(t)\right]\left(\frac{u}{a}\right).$$

Problem 1.4.2. Consider functions

$$f(x) = \begin{cases} \frac{1}{\sqrt{x}} & \text{if } x \in (0,1), \\ 0 & \text{otherwise,} \end{cases} \quad \text{and} \quad g(x) = \begin{cases} \frac{1}{\sqrt{1-x}} & \text{if } x \in (0,1), \\ 0 & \text{otherwise.} \end{cases}$$

(a) Establish that f and g are absolutely integrable functions over $\mathbb{R}$.
(b) Establish that there is $x \in \mathbb{R}$ such that the function $t \mapsto f(t)g(x-t)$ is not absolutely integrable over $\mathbb{R}$.

Solution 1.4.2.

(a) From

$$\int_{-\infty}^{\infty} f(x)\,dx = \int_{0}^{1} \frac{dx}{\sqrt{x}} = \lim_{T \to 0^+} \int_{T}^{1} \frac{dx}{\sqrt{x}} = \lim_{T \to 0^+} 2\sqrt{x}\Big|_{T}^{1} = 2$$

and

$$\int_{-\infty}^{\infty} g(x)\,dx = \int_{0}^{1} \frac{dx}{\sqrt{1-x}} = |t = 1 - x| = \int_{-\infty}^{\infty} f(t)\,dt,$$

it follows that both functions are absolutely integrable over $\mathbb{R}$.

(b) If $x = 1$, then

$$F(t) = f(t)g(1-t) = \begin{cases} \frac{1}{\sqrt{t}} \frac{1}{\sqrt{1-(1-t)}} = \frac{1}{t} & \text{if } t \in (0,1), \\ 0 & \text{otherwise.} \end{cases}$$

From

$$\int_{-\infty}^{\infty} F(t)\, dt = \int_0^1 \frac{dt}{t} = \lim_{T \to 0+} \int_T^1 \frac{dt}{t} = \lim_{T \to 0+} \ln T = -\infty,$$

it follows that the function $F(t) = f(t)g(1-t)$ is not absolutely integrable over $\mathbb{R}$.

Note: Problem 1.4.2 establishes the fact that there are functions f and g, both absolutely integrable over $\mathbb{R}$, such that the function $x \mapsto \int_{-\infty}^{\infty} f(t)g(x-t)\, dt$ is not defined for all $x \in \mathbb{R}$.

Problem 1.4.3. Let f and g be absolutely integrable functions over $\mathbb{R}$ such that for every $x \in \mathbb{R}$, the integral $\int_{-\infty}^{\infty} |f(t)g(x-t)|\, dt$ exists. The function $f * g : \mathbb{R} \to \mathbb{R}$ defined by $(f * g)(x) = \int_{-\infty}^{\infty} f(t)g(x - t)\, dt = \int_{-\infty}^{\infty} f(x-t)g(x)\, dt$ is called the *convolution* of f and g. Establish that $\mathcal{F}[(f * g)(t)] = \mathcal{F}[f(t)] \cdot \mathcal{F}[g(t)]$.

Solution 1.4.3. By the definition of the Fourier transform, Fubini's theorem,[13] and Problem 1.4.1(1),

$$\mathcal{F}[(f * g)(t)](u) = \int_{-\infty}^{\infty} \left(\int_{-\infty}^{\infty} f(t)g(s-t)\, dt \right) e^{-ius}\, ds$$

$$= \int_{-\infty}^{\infty} f(t) \left(\int_{-\infty}^{\infty} g(s-t)e^{-ius}\, ds \right) dt = \begin{vmatrix} s = v + t \\ ds = dv \end{vmatrix}$$

$$= \int_{-\infty}^{\infty} f(t)e^{-iut} \left(\int_{-\infty}^{\infty} g(v)e^{-iuv}\, dv \right) dt$$

$$= \mathcal{F}[g(t)](u) \int_{-\infty}^{\infty} f(t)e^{-iut}\, dt = \mathcal{F}[g(t)](u) \cdot \mathcal{F}[f(t)](u).$$

---

[13]Fubini's theorem justifies switching the order of integration.

Problem 1.4.4. Let $f : \mathbb{R} \to \mathbb{R}$ be a differentiable function such that both f and f' are absolutely integrable functions over $\mathbb{R}$. Prove that $\mathcal{F}[f'(t)](u) = iu\mathcal{F}[f(t)](u)$.

Solution 1.4.4. By definition,

$$\mathcal{F}[f'(t)](u) = \int_{-\infty}^{\infty} f'(t)e^{-iut} \, dt = \begin{vmatrix} U = e^{-iut} & dU = -iue^{-iut}dt \\ dV = f'(t) \, dt & V = f(t) \end{vmatrix}$$

$$= f(t)e^{-iut}\Big|_{-\infty}^{\infty} + iu \int_{-\infty}^{\infty} f(t)e^{-iut} \, dt.$$

Since, for all $t, u \in \mathbb{R}$, $\left|f(t)e^{-iut}\right| = |f(t)|$, it follows that $\lim_{t\to\pm\infty} \left|f(t)e^{-iut}\right| = \lim_{t\to\pm\infty} |f(t)| = 0$, which implies the claim.

Problem 1.4.5. Let $a > 0$. Determine the Fourier transforms of each of the following functions:

(a) $f(t) = e^{-a|t|}$.   (b) $f(t) = \frac{1}{t^2+a^2}$.

Solution 1.4.5.

(a) By definition, for all $u \in \mathbb{R}$,

$$\mathcal{F}[f(t)](u) = \int_{-\infty}^{\infty} e^{-a|t|}e^{-iut} \, dt$$

$$= \int_{-\infty}^{0} e^{(a-iu)t} \, dt + \int_{0}^{\infty} e^{-(a+iu)t} \, dt$$

$$= \frac{1}{a-iu} e^{(a-iu)t}\Big|_{-\infty}^{0} - \frac{1}{a+iu} e^{(a+iu)t}\Big|_{0}^{\infty}.$$

Since, $\lim_{t\to-\infty} e^{(a-iu)t} = \lim_{t\to-\infty} e^{at}(\cos ut - i\sin ut) = \lim_{t\to\infty} e^{-(a+iu)t} = 0$, it follows that $\mathcal{F}[f(t)](u) = \frac{1}{a-iu} + \frac{1}{a+iu} = \frac{2a}{a^2+u^2}$.

(b) The function f is even, so its Fourier transform is $\mathcal{F}[f(t)](u) = 2\int_0^\infty \frac{\cos ut \, dt}{t^2+a^2}$. By the Leibniz rule for differentiation in the case of an improper integral,[14] from

$$\frac{d}{du}\mathcal{F}[f(t)](u) = -2\int_0^\infty \frac{(t^2+a^2-a^2)\sin ut\, dt}{t(t^2+a^2)}$$

$$= -2\int_0^\infty \frac{\sin ut}{t}\, dt + 2a^2\int_0^\infty \frac{\sin ut\, dt}{t(t^2+a^2)}$$

and, by Problem 1.3.3, $\int_0^\infty \frac{\sin ut}{t}\, dt = \frac{\pi}{2}\mathrm{sign}(u)$, it follows that

$$\frac{d^2}{du^2}\mathcal{F}[f(t)](u) = 2a^2\int_0^\infty \frac{\cos ut\, dt}{t^2+a^2} = a^2\,\mathcal{F}[f(t)](u).$$

Since $\mathcal{F}[f(t)](0) = 2\int_0^\infty \frac{dt}{t^2+a^2} = \frac{\pi}{a}$ and $\lim_{u\to 0^\pm}\frac{d}{du}\mathcal{F}[f(t)](u) =$ $-\lim_{u\to 0^\pm}\pi\mathrm{sign}(u) = \mp\pi$, it follows that the function $u \mapsto$ $\mathcal{F}[f(t)](u)$ is the solution of the initial value problem $y''(u) - a^2 y(u) = 0$, $y(0) = \frac{\pi}{a}$, $\lim_{u\to 0^\pm} y'(u) = \mp\pi$. Hence, $\mathcal{F}[f(t)](u) =$

$$c_1 e^{au} + c_2 e^{-au} \text{ with } c_1 + c_2 = \frac{\pi}{a} \text{ and } c_1 - c_2 = \begin{cases} -\frac{\pi}{a} & \text{if } u > 0, \\ \frac{\pi}{a} & \text{if } u < 0. \end{cases}$$

It follows that $\mathcal{F}[f(t)](u) = \begin{cases} \frac{\pi}{a}e^{-au} & \text{if } u \geq 0, \\ \frac{\pi}{a}e^{au} & \text{if } u < 0, \end{cases} = \frac{\pi}{a}e^{-a|u|}.$

Note: We may use Problems 1.4.5(a) and (b) to illustrate the inversion theorem for the Fourier transformation: $f(t) = \frac{1}{2\pi}\int_{-\infty}^\infty \mathcal{F}[f(t)](s)e^{its}\, ds$. Recall that, by Problem 1.4.5(a), $\mathcal{F}[e^{-a|t|}](u) = \frac{2a}{a^2+u^2}$. By Problem 1.4.5(b), for any $t \in \mathbb{R}$,

$$\frac{1}{2\pi}\int_{-\infty}^\infty \mathcal{F}\left[e^{-a|t|}\right](s)e^{its}\, ds = \frac{a}{\pi}\int_{-\infty}^\infty \frac{e^{its}\, ds}{a^2+s^2} = \frac{a}{\pi}\mathcal{F}\left[\frac{1}{a^2+t^2}\right](-t)$$

$$= \frac{a}{\pi}\frac{\pi}{a}\,e^{-a|-t|} = e^{-a|t|}.$$

---

[14]For a detailed discussion on the topic of differentiation under the integral, see [18]. Here, we are using [18, Theorem 53.5]: *Let $(x,t) \mapsto g(x,t)$, $(x,t) \in \mathbb{R}^2$, be a continuous function such that $\frac{\partial g}{\partial t}$ exists and is continuous. Suppose $\int_{-\infty}^\infty |g(x,t)|\, dx$ and $\int_{-\infty}^\infty |\frac{\partial g}{\partial t}(x,t)|\, dx$ exist for each t and that $\int_{|x|>R} |\frac{\partial g}{\partial t}(x,t)|\, dx \to 0$ as $R \to \infty$ uniformly in t in each $[a,b]$. Then, $\int_{-\infty}^\infty g(x,t)\, dx$ is differentiable with $\frac{d}{dt}\int_{-\infty}^\infty g(x,t)\, dx = \int_{-\infty}^\infty \frac{\partial g}{\partial t}(x,t)\, dx$.*

Problem 1.4.6. Let $a > 0$. Determine the Fourier transforms of each of the following functions:

(a) $f(t) = e^{-at^2}$.   (b) $f(t) = te^{-at^2}$.

Solution 1.4.6.

(a) The function f is even and hence its Fourier transform is $\mathcal{F}[f(t)](u) = 2\int_0^\infty e^{-at^2}\cos ut\ dt$. Recall that $\mathcal{F}[f(t)](0) = 2\int_0^\infty e^{-at^2}\ dt = \sqrt{\frac{\pi}{a}}$. From[15]

$$\frac{d}{du}\mathcal{F}[f(t)](u) = -2\int_0^\infty te^{-at^2}\sin ut\ dt =$$

$$= \begin{vmatrix} U = \sin ut & dU = u\cos ut\ dt \\ dV = te^{-at^2}\ dt & V = -\frac{1}{2a}e^{-at^2} \end{vmatrix}$$

$$= \frac{1}{a}\ e^{-at^2}\sin ut\ \Big|_0^\infty - \frac{u}{a}\int_0^\infty e^{-at^2}\cos ut\ dt$$

$$= -\frac{u}{2a}\ \mathcal{F}[f(t)](u),$$

i.e., from the initial value problem $y'(u) = -\frac{u}{2a}y(u)$, $y(0) = \sqrt{\frac{\pi}{a}}$, it follows that $\mathcal{F}[f(t)](u) = \sqrt{\frac{\pi}{a}}e^{-\frac{u^2}{4a}}$, $u \in \mathbb{R}$.

(b) By definition, for any $u \in \mathbb{R}$,

$$\mathcal{F}[f(t)](u) = \int_{-\infty}^\infty te^{-at^2}e^{-itu}\ dt = -\frac{1}{i}\frac{d}{du}\left(\int_{-\infty}^\infty e^{-at^2}e^{-itu}\ dt\right)$$

$$= -\frac{1}{i}\frac{d}{du}\mathcal{F}\left[e^{-ax^2}\right](u) = -\frac{1}{i}\frac{d}{du}\left(\sqrt{\frac{\pi}{a}}\ e^{-\frac{u^2}{4a}}\right)$$

$$= -\frac{ui}{2a}\sqrt{\frac{\pi}{a}}\ e^{-\frac{u^2}{4a}}.$$

Problem 1.4.7. Let $a > 0$. Determine the Fourier transform of each of the following functions:

(a) $f(t) = te^{-a|t|}$.   (b) $f(t) = \frac{t}{t^2+a^2}$.

---

[15]In both parts of this problem, as well as in Problem 1.4.7, we use the Leibniz rule for differentiation under an improper integral. See Footnote 14.

Solution 1.4.7. We use the Fourier transforms obtained in Problem 1.4.5.

By definition, for any $u \in \mathbb{R}$:

(a) $\mathcal{F}[f(t)](u) = \displaystyle\int_{-\infty}^{\infty} t e^{-a|t|} e^{-itu}\, dt = -\frac{1}{i}\frac{d}{du}\left(\int_{-\infty}^{\infty} e^{-a|t|} e^{-itu}\, dt \right)$

$\qquad = -\dfrac{1}{i}\dfrac{d}{du} \mathcal{F}\left[e^{-a|x|} \right](u) = 2ai\, \dfrac{d}{du}\left(\dfrac{1}{a^2 + u^2} \right)$

$\qquad = -\dfrac{4aui}{(a^2 + u^2)^2},\ \text{and}$

(b) $\mathcal{F}[f(t)](u) = \displaystyle\int_{-\infty}^{\infty} \frac{t}{t^2 + a^2} e^{-itu}\, dt = -\frac{1}{i}\frac{d}{du}\left(\int_{-\infty}^{\infty} \frac{e^{-itu}\, dt}{t^2 + a^2} \right)$

$\qquad = i\, \dfrac{d}{du} \mathcal{F}\left[\dfrac{1}{x^2 + a^2} \right](u)$

$\qquad = i\, \dfrac{d}{du}\left(\dfrac{\pi}{a}\, e^{-a|u|} \right) = \begin{cases} -i\pi e^{-au} & \text{if } u > 0, \\ i\pi e^{au} & \text{if } u < 0, \end{cases}$

$\qquad = i\pi \left(H(-u) e^{au} - H(u) e^{-au} \right),$

where $u \mapsto H(u)$ is the Heaviside step function.

Problem 1.4.8. Determine the Fourier transform of the function

$$f(x) = \begin{cases} 1 & \text{if } |x| \in \left[0, \tfrac{1}{2}\right), \\ 1 - |x| & \text{if } |x| \in \left[\tfrac{1}{2}, 1\right], \\ 0 & \text{if } |x| \in (1, \infty). \end{cases}$$

Answer 1.4.8. $\mathcal{F}[f(x)](u) = \begin{cases} \frac{1}{u} \sin \frac{u}{2} + \frac{2}{u^2}\left(\cos \frac{u}{2} - \cos u \right) & \text{if } u \neq 0, \\ \frac{5}{4} & \text{if } u = 0. \end{cases}$

Problem 1.4.9. Sketch a graph of the function $f : (0, \infty) \to \mathbb{R}$ such that, for any $n \in \mathbb{N}$, if $x \in (n - 1, n]$, then $f(x) = \int_{0}^{\pi e^{nx}} \cos\left((n-1) e^{-nx} t \right) dt$. Determine the Fourier sine transform of the function f.

Solution 1.4.9. For $x \in (0,1]$, $f(x) = \int_0^{\pi e^x} dt = \pi e^x$ and, for $x \in (1,\infty)$, $f(x) = \int_0^{\pi e^{x\lceil x\rceil}} \cos\left((\lceil x\rceil - 1)\, e^{-x\lceil x\rceil}t\right)\, dt = 0$. It follows that, for $u \in [0,\infty)$, $\mathcal{F}_{\sin}[f(x)](u) = \sqrt{\frac{2}{\pi}}\int_0^\infty f(x)\sin ux\, dx = \sqrt{\frac{2}{\pi}}\int_0^1 \pi e^x \sin ux\, dx = \frac{\sqrt{2\pi}}{1+u^2}\left(e(\sin u - u\cos u) + u\right)$.

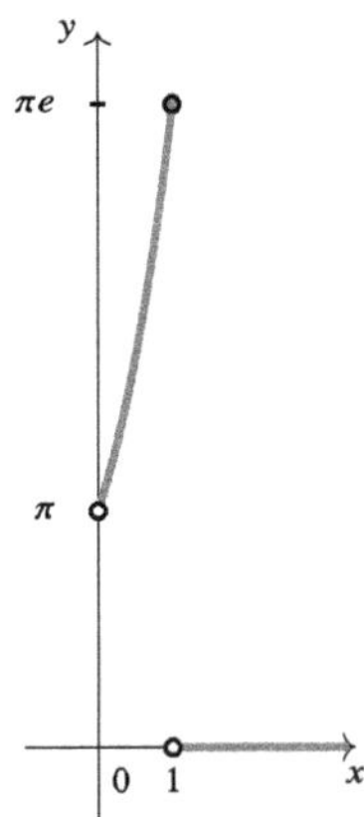

Problem 1.4.10. Let F_c be the Fourier cosine transform of the function $f : [0,\infty) \to \mathbb{R}$, where f is a continuous absolutely integrable function such that, for any $n \in \mathbb{N}$, $t \mapsto t^n f(t)$ is absolutely integrable. Prove that, for any $n \in \mathbb{N}$, the Fourier sine transform of the function $\varphi_n(t) = t^{2n-1}f(t)$ is $\Phi_n(u) = (-1)^n \frac{d^{2n-1}F_c(u)}{du^{2n-1}}$. [16]

Solution 1.4.10. Let $M = \left\{n \in \mathbb{N} : \Phi_n(u) = \sqrt{\frac{2}{\pi}}\int_0^\infty \varphi_n(t)\cdot \sin utdt\right\}$. If $n = 1$, then

$$\Phi_1(u) = -\frac{d\,F_c(u)}{du} = -\frac{d}{du}\left(\sqrt{\frac{2}{\pi}}\int_0^\infty f(t)\cos ut\, dt\right)$$

$$= -\sqrt{\frac{2}{\pi}}\int_0^\infty \frac{\partial}{\partial u}\left(f(t)\cos tu\right)dt = \sqrt{\frac{2}{\pi}}\int_0^\infty tf(t)\sin ut\, dt$$

$$= \sqrt{\frac{2}{\pi}}\int_0^\infty \varphi_1(t)\sin ut\, dt = \mathcal{F}_{\sin}[\varphi_1(t)](u).$$

---

[16] The stated conditions guarantee that Φ_n is differentiable and that, when finding its derivative, we can apply the Leibniz rule for differentiation [18, Lemma 53.6].

Therefore, $1 \in M$. Let $n \in M$. Then, for all $u \in (0, \infty)$,

$$\Phi_{n+1}(u) = (-1)^{n+1} \frac{d^{2n+1} F_c(u)}{du^{2n+1}} = -\frac{d^2 \Phi_n(u)}{du^2}$$

$$= -\frac{d^2}{du^2} \left(\sqrt{\frac{2}{\pi}} \int_0^\infty \varphi_n(t) \sin ut \, dt \right)$$

$$= -\sqrt{\frac{2}{\pi}} \int_0^\infty \frac{\partial^2}{\partial u^2} \left(\varphi_n(t) \sin ut \right) \, dt$$

$$= \sqrt{\frac{2}{\pi}} \int_0^\infty t^2 \varphi_n(t) \sin ut \, dt = \sqrt{\frac{2}{\pi}} \int_0^\infty \varphi_{n+1}(t) \sin ut \, dt$$

$$= \mathcal{F}_{\sin}[\varphi_{n+1}(t)](u),$$

which means that $n + 1 \in M$. By the principle of mathematical induction, $M = \mathbb{N}$.

Problem 1.4.11. Use the fact[17] $\int_0^\infty \frac{\sin t}{\sqrt{t}} \, dt = \int_0^\infty \frac{\cos t}{\sqrt{t}} \, dt = \sqrt{\frac{\pi}{2}}$ to establish that the function $f(t) = \frac{1}{\sqrt{t}}$, $t \in (0, \infty)$, is its own cosine and sine Fourier transform.

Solution 1.4.11. By definition, for $u > 0$,

$$\mathcal{F}_{\sin}\left[\frac{1}{\sqrt{t}}\right](u) = \sqrt{\frac{2}{\pi}} \int_0^\infty \frac{\sin ut}{\sqrt{t}} \, dt = \left| \begin{array}{c} s = ut \Rightarrow dt = \frac{1}{u} \, ds \\ t = 0 \Rightarrow s = 0 \\ t \to \infty \Rightarrow s \to \infty \end{array} \right.$$

$$= \sqrt{\frac{2}{\pi}} \int_0^\infty \frac{\sin s}{\sqrt{\frac{s}{u}}} \frac{ds}{u} = \sqrt{\frac{2}{\pi}} \frac{1}{\sqrt{u}} \int_0^\infty \frac{\sin s}{\sqrt{s}} \, ds = \frac{1}{\sqrt{u}}.$$

Problem 1.4.12. Solve the integral equation

$$\int_0^\infty f(t) \cos tu \, dt = \begin{cases} \frac{\pi}{2} \sin u & \text{if } u \in [0, \pi], \\ 0 & \text{if } u \in (\pi, \infty). \end{cases}$$

---

[17]These are the well-known Fresnel's integrals. Augustin-Jean Fresnel, 1788–1827, was a French civil engineer and physicist.

Solution 1.4.12. Let $F : \mathbb{R} \to \mathbb{R}$ be an even extension of the (still unknown) function f. Then, for all $u \in \mathbb{R}$,

$$\mathcal{F}[F(t)](u) = 2 \int_0^\infty f(t) \cos tu \; dt = \begin{cases} \pi \, |\sin u| & \text{if } |u| \le \pi, \\ 0 & \text{if } |u| \ge \pi. \end{cases}$$

By the inversion theorem for the Fourier transformation,

$$F(t) = \frac{1}{2\pi} \int_{-\infty}^\infty \mathcal{F}[F(t)](u) e^{itu} \; du = \frac{1}{2\pi} \int_{-\pi}^\pi \pi \, |\sin u| e^{itu} \; du$$

$$= \int_0^\pi \sin u \cos tu \; du = \frac{1}{2} \int_0^\pi (\sin(1+t)u + \sin(1-t)u) \; du$$

$$= \frac{\cos \pi t + 1}{1 - t^2}, \; t \ne \pm 1.$$

Hence, $f(t) = \begin{cases} \frac{\cos \pi t + 1}{1 - t^2} & \text{if } t \in [0, 1) \cup (1, \infty), \\ 0 & \text{if } t = 1. \end{cases}$

Problem 1.4.13. Solve the integral equation

$$\int_0^\infty f(t) \sin tu \; dt = \begin{cases} \frac{\pi}{2} \cos u & \text{if } u \in [0, \pi), \\ -\frac{\pi}{4} & \text{if } u = \pi, \\ 0 & \text{if } u \in (\pi, \infty). \end{cases}$$

Solution 1.4.13. Let $F : \mathbb{R} \to \mathbb{R}$ be an odd extension of the (still unknown) function f. Then, for all $u \in \mathbb{R}$,

$$\mathcal{F}[F(t)](u) = -2i \int_0^\infty f(t) \sin tu \; dt$$

$$= \begin{cases} -i\pi \, \text{sign}(u) \cdot \cos u & \text{if } |u| < \pi, \\ \mp \frac{i\pi}{2} & \text{if } u = \pm\pi, \\ 0 & \text{if } |u| > \pi. \end{cases}$$

By the inversion theorem for the Fourier transformation,

$$F(t) = \frac{1}{2\pi} \int_{-\infty}^{\infty} \mathcal{F}[F(t)](u)e^{itu} \; du = -\frac{i}{2} \int_{-\pi}^{\pi} \operatorname{sign}(u) \; \cos u \; e^{itu} \; du$$

$$= \int_{0}^{\pi} \cos u \; \sin tu \; du = \frac{1}{2} \int_{0}^{\pi} (\sin(t+1)u + \sin(t-1)u) \; du$$

$$= \frac{t(\cos \pi t + 1)}{t^2 - 1}, \; t \neq \pm 1.$$

Hence, $f(t) = \begin{cases} \frac{t(\cos \pi t + 1)}{t^2 - 1} & \text{if } t \in [0,1) \cup (1, \infty), \\ 0 & \text{if } t = 1. \end{cases}$

Chapter 2

Ordinary Differential Equations

2.1 Introduction

Use the following definitions, techniques, properties, and algorithms to solve the problems contained in this chapter. For more details, see [1, 4, 8, 9, 20, 27, 28, 34, 37].

(a) **Ordinary Differential Equation.** An *ordinary differential equation* is an equation stated in terms of one or more derivatives of an unknown function $y = y(x)$. For a natural number n, a region $D \subseteq \mathbb{R}^{n+2}$, and a function $F : D \to \mathbb{R}$, the function $y = f(x)$, $x \in I \subseteq \mathbb{R}$, is a *solution* to the equation $F(x, y, y', \ldots, y^{(n)}) = 0$, an nth-order ordinary differential equation, if, for all $x \in I$, $F(x, f(x), f'(x), \ldots, f^{(n)}(x)) = 0$.

(b) **Initial Value Problem.** Let $n \in \mathbb{N}$, $D \subseteq \mathbb{R}^{n+2}$, $F : \mathbb{R}^{n+2} \to \mathbb{R}$, and $(a, \alpha_0, \alpha_1, \ldots, \alpha_{n-1}) \in \mathbb{R}^{n+1}$ be given. To solve the *initial value problem*

$$F(x, y, y', \ldots, y^{(n)}) = 0, \quad y(a) = \alpha_0,$$

$$y'(a) = \alpha_1, \ldots, y^{(n-1)}(a) = \alpha_{n-1},$$

means to determine all functions $y = f(x)$, $x \in I$, such that $a \in I$, $f(a) = \alpha_0$, $f'(a) = \alpha_1, \ldots, f^{(n-1)}(a) = \alpha_{n-1}$, and, for all $x \in I$, $F(x, f(x), f'(x), \ldots, f^{(n)}(x)) = 0$. In other words, a solution to the given initial value problem is any solution to the differential equation that satisfies the *initial conditions* $(a, \alpha_0, \alpha_1, \ldots, \alpha_{n-1})$ [16, 28].

67

(c) **Picard's Existence and Uniqueness Theorem.** Consider the initial value problem $y' = F(x, y)$, $y(x_0) = y_0$. Suppose that $(x, y) \mapsto F(x, y)$ and $(x, y) \mapsto \frac{\partial F}{\partial y}(x, y)$ are continuous functions in some open rectangle $R = (a, b) \times (c, d) = \{(x, y) : a < x < b, c < y < d\}$ that contains the point (x_0, y_0). Then, there exists $\delta > 0$ such that the initial value problem has a unique solution defined in the interval $[x_0 - \delta, x_0 + \delta]$ [8].[1]

(d) **Singular Solution.** A solution $y = y(x)$, $x \in I$, to the differential equation $F(x, y, y') = 0$ is called a *singular solution* if for each point $(x_0, y_0 = y(x_0))$, $x_0 \in I$, there is another solution $z = z(x)$ to the equation with the property that $z(x_0) = y(x_0)$. If $y = y(x)$ is a singular solution to the differential equation $y' = f(x, y)$, then for any other solution $z = z(x)$ such that $z(x_0) = y(x_0)$, $z'(x_0) = y'(x_0)$ also holds. Geometrically, this means that another integral curve with the common tangent line passes through each point on the graph of a singular solution to the equation $y' = f(x, y)$.

(e) **p-discriminant Curve.** Consider a differential equation $F(x, y, y') = 0$ and suppose that the function $(x, y, p) \mapsto F(x, y, p)$ and its partial derivative $\frac{\partial F}{\partial p}$ are continuous in the domain $D \subseteq \mathbb{R}^3$ of the function F. The set

$$\Gamma = \left\{ (x, y) \in \mathbb{R}^2 : (\exists p \in \mathbb{R}) \; F(x, y, p) = 0 \text{ and } \frac{\partial F}{\partial p}(x, y, p) = 0 \right\}$$

is called the *p-discriminant curve* of the differential equation $F(x, y, y') = 0$. The p-discriminant curve may or may not be a graph of a solution to the equation.

(f) **First-Order Differential Equations.**

    (i) If I and J are intervals and if $f : I \to \mathbb{R}$ and $g : J \to \mathbb{R}$ are continuous functions with $g(y) \neq 0$, $y \in J$, then the differential equation $y' = \frac{f(x)}{g(y)}$ is called a *separable differential equation*. See Problems 2.2.2–2.2.4.

    (ii) If p and q are continuous functions in an interval I, then the differential equation $y' + p(x)y = q(x)$ is called a *first-order linear differential equation*. If $q(x) = 0$, $x \in I$, we say that the equation is *homogeneous*; otherwise, it is *non-homogeneous*. See Problems 2.2.5–2.2.7.

---

[1]Charles Émile Picard, 1856–1941, French mathematician.

(iii) Let $n \in \mathbb{R}\backslash\{0,1\}$. If p and q are continuous functions in an interval I, then the differential equation $y' + p(x)y + q(x)y^n = 0$ is called a *Bernoulli differential equation*.[2] See Problems 2.2.8–2.2.11.

(iv) If p, q, and r are continuous functions in an interval I, where r is not the zero function, then the differential equation $y' = p(x) + q(x)y + r(x)y^2$ is called a *Riccati differential equation*.[3] See Problems 2.2.12–2.2.16.

(v) If f is a twice differentiable function in an interval I with $f''(t) \neq 0$, $t \in I$, then the differential equation $y = xy' + f(y')$ is called *Clairaut's differential equation*.[4] See Problems 2.2.17–2.2.19.

(vi) If f and g are continuously differentiable functions in an interval I, then the differential equation $y = xf(y') + g(y')$ is called *Lagrange's differential equation*.[5] See Problems 2.2.20–2.2.21.

(vii) If $(x,y) \mapsto F(x,y)$ is a homogeneous function with the degree of homogeneity zero in a region $D \subseteq \mathbb{R}^2$, then the differential equation $y' = F(x,y)$ is called a *homogeneous differential equation*. See Problems 2.2.22–2.2.28.

(viii) If P and Q are functions with continuous partial derivatives in a simply connected region $D \subseteq \mathbb{R}^2$ such that $\frac{\partial P}{\partial y} = \frac{\partial Q}{\partial x}$ and $Q(x,y) \neq 0$, $(x,y) \in D$, then the differential equation $y' = -\frac{P(x,y)}{Q(x,y)}$ is called an *exact differential equation*. It is common to write an exact differential equation as $P(x,y)dx + Q(x,y)dy = 0$ and to request that $(P(x,y))^2 + (Q(x,y))^2 \neq 0$, $(x,y) \in D$.[6] See Problems 2.2.29–2.2.39.

---

[2] Jacob Bernoulli, 1655–1705, Swiss mathematician.

[3] Jacopo Francesco Riccati, 1676–1754, Venetian mathematician.

[4] Alexis Claude Clairaut, 1713–1765, French mathematician, astronomer, and geophysicist.

[5] Joseph-Louis Lagrange, 1736–1813, Italian French mathematician, physicist, and astronomer.

[6] This underlines the fact that the variables x and y are treated as *equal* in the sense that y may be taken as the independent variable and x as the dependent variable. So, if $P(x,y) \neq 0$, $(x,y) \in D$, then the problem may be formulated as finding a solution $x = x(y)$ to the exact differential equation $\frac{dx}{dy} = -\frac{Q(x,y)}{P(x,y)}$.

(g) **Wronskian.** For n functions $f_1, f_2, \ldots, f_n$ that are $n - 1$ times differentiable in an interval I, the Wronskian[7] $W(f_1, f_2, \ldots, f_n)$ is a function defined by, for $x \in I$,

$$W(f_1, f_2, \ldots, f_n)(x) = \begin{vmatrix} f_1(x) & f_2(x) & \cdots & f_n(x) \\ f_1'(x) & f_2'(x) & \cdots & f_n'(x) \\ \vdots & \vdots & \ddots & \vdots \\ f_1^{(n-1)}(x) & f_2^{(n-1)}(x) & \cdots & f_n^{(n-1)}(x) \end{vmatrix}.$$

See Problems 2.3.3–2.3.11.

(h) **Second-Order Linear Differential Equations.**

    (i) If p, q, and r are continuous functions in an interval I, then the differential equation $y'' + p(x)y' + q(x)y = r(x)$ is called a *second-order linear differential equation*. If $r(x) = 0$, $x \in I$, the equation is *homogeneous*; otherwise, it is *non-homogeneous*.

    (ii) If $p(x) = a$ and $q(x) = b$ are constant functions, we say that the equation $y'' + ay' + by = r(x)$ is a *second-order linear differential equation with constant coefficients*.

    (iii) If $a, b \in \mathbb{R}$, then the differential equation $x^2 y'' + axy' + by = 0$ is called the *Cauchy–Euler differential equation*.[8]

(i) **Existence and Uniqueness Theorem for Second-Order Linear Differential Equations.** Suppose that p, q, and r are continuous functions in an interval I. For any $(a, b, c) \in I \times \mathbb{R} \times \mathbb{R}$, there is a unique solution to the initial value problem $y'' + p(x)y' + q(x)y = r(x)$, $y(a) = b$, $y'(a) = c$, defined for $x \in I$ [4, 8].

(j) **General Solution to Second-Order Linear Differential Equation.** Suppose that p, q, and r are continuous functions in an interval I, where r is not the zero function. Suppose that y_1 and y_2 are two linearly independent solutions to the homogeneous equation $y'' + p(x)y' + q(x)y = 0$ and that y_p is one solution

---

[7] Józef Maria Hoene-Wroński, 1776–1853, Polish philosopher, mathematician, physicist, inventor, and economist.

[8] Augustin-Louis Cauchy, 1789–1857, French mathematician, engineer, and physicist. Leonhard Euler, 1707–1783, Swiss mathematician, physicist, astronomer, geographer, logician, and engineer.

to the non-homogeneous equation $y'' + p(x)y' + q(x)y = r(x)$. Then, we have the following:

- Any solution to the homogeneous equation is a linear combination of y_1 and y_2. The expression $y = c_1 y_1 + c_2 y_2$, $c_1, c_2 \in \mathbb{R}$, is called the *general solution* to the homogeneous second-order linear differential equation $y'' + p(x)y' + q(x)y = 0$. See Problem 2.3.12.

- Any solution to the non-homogeneous equation is the sum of y_p and a solution to the corresponding homogeneous equation. The expression $y = y_p + c_1 y_1 + c_2 y_2$, $c_1, c_2 \in \mathbb{R}$, is called the *general solution* to the non-homogeneous second-order linear differential equation $y'' + p(x)y' + q(x)y = r(x)$. See Problem 2.3.13.

Any solution to a linear second-order differential equation obtained from its general solution for a particular choice of real numbers c_1 and c_2 is called a *particular solution*.

(k) **General Solution to Second-Order Linear Differential Equation with Constant Coefficients.** Let a and b be real numbers and let $r_1, r_2 \in \mathbb{C}$ be roots of the quadratic equation $t^2 + at + b = 0$. The general solution $y = y(x)$ to the equation $y'' + ay' + by = 0$ is determined as follows:

- If $r_1, r_2 \in \mathbb{R}$ with $r_1 \neq r_2$, then $y = c_1 e^{r_1 x} + c_2 e^{r_2 x}$, $c_1, c_2 \in \mathbb{R}$.
- If $r_1 = r_2 = r \in \mathbb{R}$, then $y = (c_1 + c_2 x)e^{rx}$, $c_1, c_2 \in \mathbb{R}$.
- If $r_1, r_2 \in \mathbb{C} \backslash \mathbb{R}$ with $r_{1,2} = e^{\alpha}(\cos \beta \pm i \sin \beta)$, then $y = (c_1 \cos \beta x + c_2 \sin \beta x)e^{\alpha x}$, $c_1, c_2 \in \mathbb{R}$.

See Problem 2.3.20.

(l) **Principle of Superposition.** If a function f is a solution to the equation $y'' + p(x)y' + q(x)y = r(x)$, $x \in I$, and if a function g is a solution to the equation $y'' + p(x)y' + q(x)y = s(x)$, $x \in I$, then the function $f + g$ is a solution to the equation $y'' + p(x)y' + q(x)y = r(x) + s(x)$, $x \in I$. See Problem 2.3.9.

(m) **Lagrange Method of Variation of Parameters.** Let p, q, and r be continuous functions in an interval I, where r is not the zero function, and let $y_1 = y_1(x)$ and $y_2 = y_2(x)$ be two linearly independent solutions to the equation $y'' + p(x)y' + q(x)y = 0$. One solution to the equation $y'' + p(x)y' + q(x)y = r(x)$ is determined by $y = u(x)y_1 + v(x)y_2$, where the functions u and v

satisfy the system of equations $u'(x)y_1(x) + v'(x)y_2(x) = 0$ and $u'(x)y_1'(x) + v'(x)y_2'(x) = r(x)$. See Problems 2.3.16–2.3.17.

(n) **Method of Undetermined Coefficients.** In several special cases, determining a particular solution $y_p = y_p(x)$ to a non-homogeneous second-order differential equation with constant coefficients $y'' + \alpha y' + \beta y = r(x)$ reduces to determining the coefficients of a polynomial:

$r(x)$	$y_p(x)$
$\sum_{k=0}^{n} a_k x^{n-k}$	$x^s \sum_{k=0}^{n} c_k x^{n-k}$
$e^{ax} \sum_{k=0}^{n} a_k x^{n-k}$	$x^s e^{ax} \sum_{k=0}^{n} c_k x^{n-k}$
$e^{ax} \cos bx \sum_{k=0}^{n} a_k x^{n-k}$ or $e^{ax} \sin bx \sum_{k=0}^{n} a_k x^{n-k}$	$x^s e^{ax} \left(\sin bx \sum_{k=0}^{n} c_k x^{n-k} + \cos bx \sum_{k=0}^{n} d_k x^{n-k} \right)$

where s is the smallest nonnegative integer, $s = 0$, 1, or 2, chosen so that y_p has no term that is a solution to the corresponding homogeneous differential equation. See Problems 2.3.25–2.3.30.

2.2   First-Order Differential Equations

Problem 2.2.1. Consider the first-order differential equation

$$\left(e^{y'} - 1 \right) \left(\sqrt{-\ln y} - \sqrt{x} \right) = 0.$$

(a) Determine a function $y = f(x)$ such that its graph is the p-discriminant curve of the differential equation.

(b) Determine if the function f is a singular solution to the equation.

(c) Determine the Fourier cosine transform of the function f.

Solution 2.2.1. The domain of the function $(x, y, p) \mapsto F(x, y, p) = (e^p - 1) \left(\sqrt{-\ln y} - \sqrt{x} \right)$ is the set $D = \{(x, y, p) : x \in [0, \infty) \text{ and } y \in (0, 1] \text{ and } p \in \mathbb{R}\} = [0, \infty) \times (0, 1] \times \mathbb{R}$.

(a) From $\frac{\partial F}{\partial p} = \left(\sqrt{-\ln y} - \sqrt{x} \right) e^p$, it follows that, for all $(x, y, p) \in D$, $\frac{\partial F}{\partial p} = 0$ if and only if $\sqrt{-\ln y} = \sqrt{x}$, or, what is the same,

$y = f(x) = e^{-x}$, $x \in [0, \infty)$. From

$$\Gamma = \left\{ (x, y) : (\exists p \in \mathbb{R}) \ F(x, y, p) = 0 \text{ and } \frac{\partial F}{\partial p}(x, y, p) = 0 \right\}$$

$$= \left\{ (x, y) : (\exists p \in \mathbb{R}) \ (e^p - 1)\left(\sqrt{-\ln y} - \sqrt{x} \right) = 0 \text{ and } \right.$$

$$\left. \left(\sqrt{-\ln y} - \sqrt{x} \right) e^p = 0 \right\} = \{ (x, e^{-x}) : x \in [0, \infty) \},$$

it follows that the p-discriminant curve of the differential equation $F(x, y, y') = 0$ is the graph of the function $f(x) = e^{-x}$, $x \in [0, \infty)$.

(b) For all[9] $x \in [0, \infty)$, $F(x, e^{-x}, -e^{-x}) = \left(e^{-e^{-x}} - 1 \right)\left(\sqrt{x} - \sqrt{x} \right) = 0$, which implies that the function $f(x) = e^{-x}$, $x \in [0, \infty)$, is a solution to the equation. All other solutions to the equation are determined by $e^{y'} - 1 = 0$, $x \in [0, \infty)$, and $y \in (0, 1]$. The family of those solutions is, for $c \in (0, 1]$, $y = c$, $x \in [0, \infty)$. Since, for any $x_0 \in [0, \infty)$, the initial value problem $\left(e^{y'} - 1 \right)\left(\sqrt{-\ln y} - \sqrt{x} \right) = 0$, $y(x_0) = e^{-x_0}$, has two solutions, $y = f(x) = e^{-x}$, $x \in [0, \infty)$, and $y = e^{-x_0}$, $x \in [0, \infty)$, the function $y = f(x) = e^{-x}$, $x \in [0, \infty)$ is a singular solution to the equation.

Note: Solutions $y = e^{-x}$, $x \in [0, \infty)$, and $y = e^{-x_0}$, $x \in [0, \infty)$, to the initial value problem $\left(e^{y'} - 1 \right)\left(\sqrt{-\ln y} - \sqrt{x} \right) = 0$, $y(x_0) = e^{-x_0}$, have different first derivatives at $x_0 \in [0, \infty)$.

(c) Since the function f is absolutely integrable, its Fourier cosine transform exists, and it is determined by, for all $u \in \mathbb{R}$,

$$\mathcal{F}_{\cos}[f(x)](u) = \sqrt{\frac{2}{\pi}} \int_0^\infty e^{-x} \cos ux \, dx = \sqrt{\frac{2}{\pi}} - u^2 \, \mathcal{F}_{\cos}[f(x)](u).$$

It follows, $\mathcal{F}_{\cos}[f(x)](u) = \sqrt{\frac{2}{\pi}} \frac{1}{1+u^2}$, $u \in [0, \infty)$.

Problem 2.2.2. Solve the following separable differential equations and initial value problems:

---

[9]In what follows, for a function $y = y(x)$, $x \in [0, \infty)$, we take $y'(0) = \lim_{x \to 0^+} y'(x)$.

(a) $(x^2 + 2x + 2)y' - y^2 = 1$.
(b) $e^{ay}(y' - 1) = 1$, $a \in \mathbb{R}^+$.
(c) $y - xy' = 2(1 + x^2 y')$, $y(1) = 1$.
(d) $(4 + y^2)\, dx + 2x\sqrt{2x - x^2}\, dy = 0$, $\lim\limits_{x \to 2^-} y(x) = 0$.

Solution 2.2.2.

(a) Since, for any $x \in \mathbb{R}$, $x^2 + 2x + 2 = (x+1)^2 + 1 \neq 0$, and, for any $y \in \mathbb{R}$, $y^2 + 1 \neq 0$, the equation is equivalent to the separable differential equation $\frac{dy}{1+y^2} = \frac{dx}{1+(x+1)^2}$. From $\int \frac{dy}{1+y^2} = \int \frac{dx}{1+(x+1)^2}$, it follows that all solutions to the equation are implicitly determined by $\arctan y = \arctan(1 + x) + c$, $c \in D \subseteq \mathbb{R}$, where D is a set that needs to be determined. Since the range of the function $t \mapsto \arctan t$ is the interval $\left(-\frac{\pi}{2}, \frac{\pi}{2}\right)$, it follows that $c \in D = (-\pi, \pi)$. If $c = \frac{\pi}{2}$, then from $\arctan y = \arctan(1+x) + \frac{\pi}{2}$, it follows that $\arctan(1 + x) < 0$, i.e., $x < -1$, and

$$y = \tan\left(\arctan(1 + x) + \frac{\pi}{2}\right) = -\frac{1}{\tan(\arctan(1 + x))}$$

$$= -\frac{1}{1 + x}, \quad x < -1.$$

If $c = -\frac{\pi}{2}$, then from $\arctan y = \arctan(1 + x) - \frac{\pi}{2}$, it follows that $\arctan(1 + x) > 0$, i.e., $x > -1$, and

$$y = \tan\left(\arctan(1 + x) - \frac{\pi}{2}\right) = -\frac{1}{1 + x}, \quad x > -1.$$

If $c = 0$, then $y = 1 + x$, $x \in \mathbb{R}$.
If $c \in (0, \pi)$, $c \neq \frac{\pi}{2}$, then from $-\frac{\pi}{2} - c < \arctan(1+x) < \frac{\pi}{2} - c < \frac{\pi}{2}$, it follows that $1 + x < \tan\left(\frac{\pi}{2} - c\right) = \cot c$, which, together with $\arctan y = \arctan(1 + x) + c$, implies that

$$y = \frac{1 + x + \tan c}{1 - (1 + x)\tan c}, \quad x \in (-\infty, \cot c - 1).$$

Finally, if $c \in (-\pi, 0)$, $c \neq -\frac{\pi}{2}$, then from $-\frac{\pi}{2} < -\frac{\pi}{2} - c < \arctan(1+x) < \frac{\pi}{2} - c$, it follows that $1+x > \tan\left(-\frac{\pi}{2} - c\right) = \cot c$, which, together with $\arctan y = \arctan(1 + x) + c$, implies that

$$y = \frac{1 + x + \tan c}{1 - (1 + x)\tan c}, \quad x \in (\cot c - 1, \infty).$$

(b) The equation is equivalent to the separable differential equation $\frac{e^{ay}}{1+e^{ay}}\,dy = dx$. From $\int \frac{e^{ay}}{1+e^{ay}}\,dy = \int dx$, it follows that all solutions to the equation are implicitly determined by $\ln(1+e^{ay}) = ax+c$, $c \in \mathbb{R}$. From $e^{ay} = e^{ax+c}-1$, it follows that $y = \frac{1}{a}\ln(e^{ax+c}-1)$, $x > -\frac{c}{a}$, $c \in \mathbb{R}$.

(c) In a neighbourhood of the point $(1,1)$, the equation is equivalent to the separable differential equation $\frac{dy}{y-2} = \frac{dx}{x(1+2x)}$. The solution to the initial value problem in a neighbourhood of the point $(1,1)$ is determined by $\int_1^y \frac{dt}{y-2} = \int_1^x \frac{dt}{t(1+2t)}$, i.e., by $\ln(2-y) = \ln\frac{3x}{1+2x}$. Thus, $y = \frac{2+x}{1+2x}$, $x \in (0,\infty)$.

(d) The domain of the equation is the set $D = \{(x,y) : x \in (0,2) \text{ and } y \in \mathbb{R}\}$. For $x \in (0,2)$, the equation is equivalent to the separable differential equation $\frac{dy}{4+y^2} = -\frac{dx}{2x\sqrt{2x-x^2}}$. From $\int \frac{dy}{4+y^2} = -\int \frac{dx}{2x\sqrt{2x-x^2}}$, it follows that all solutions to the equation are implicitly determined by $\arctan\frac{y}{2} = c + \sqrt{\frac{2}{x}-1}$, $c \in \mathbb{R}$. From $\lim_{x \to 2^-} y(x) = 0$, it follows that $c = 0$, i.e., the solution to the initial value problem is implicitly determined by $\arctan\frac{y}{2} = \sqrt{\frac{2}{x}-1}$. Since the range of the function $t \mapsto \arctan t$ is the interval $\left(-\frac{\pi}{2}, \frac{\pi}{2}\right)$, it follows that the solution to the initial value problem is defined for real numbers x such that $0 < \sqrt{\frac{2}{x}-1} < \frac{\pi}{2}$, i.e., for $x \in \left(\frac{8}{4+\pi^2}, 2\right)$. Therefore, $y = 2\tan\sqrt{\frac{2}{x}-1}$, $x \in \left(\frac{8}{4+\pi^2}, 2\right)$.

Problem 2.2.3. The graph Γ of a function $x \mapsto y = f(x)$ has the property that every point $T \in \Gamma$ is the midpoint of the segment of the normal line to Γ at T cut off by the coordinate axes. Determine the function f so that $f(-1) = -2$.

Solution 2.2.3. Let, for $x \in \mathrm{Dom}(f)$, $T = (x, y = f(x)) \in \Gamma$. An equation of the normal line to Γ at T, in the coordinate system OXY, is $Y - y = -\frac{1}{y'}(X - x)$, with $y' = f'(x) \neq 0$. The X- and Y-intercepts of this line are $M = (x + yy', 0)$ and $N = \left(0, y + \frac{x}{y'}\right)$. Since T is the midpoint of the line segment $\overline{MN}$, it follows that $x = \frac{1}{2}(x + yy')$ and $y = \frac{1}{2}\left(y + \frac{x}{y'}\right)$, or, what is the same, $yy' = x$.

This is a separable differential equation that can be rewritten as $y\,dy = x\,dx$. The function $y = f(x)$ is determined by

$$\int_{-2}^{y} t\,dt = \int_{-1}^{x} t\,dt,$$

i.e., by $y^2 = x^2 + 3$. Since $f(-1) = -2 < 0$, it follows that $f(x) = -\sqrt{x^2 + 3}$, $x \in (-\infty, 0)$.

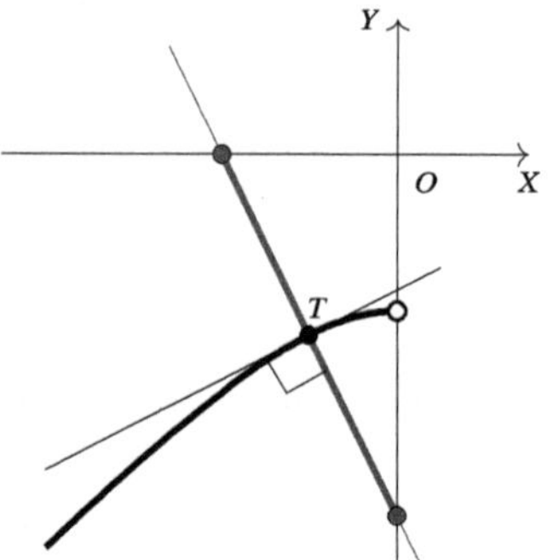

Problem 2.2.4. Consider the differential equation $(y')^2 = y$.

(a) Establish that for any $x_0 \in \mathbb{R}$ and any $y_0 \in \mathbb{R}^+$, the initial value problem $(y')^2 = y$, $y(x_0) = y_0$, has no unique solution.[10]

(b) Determine a family of quadratic functions with the property that every member of the family is a solution to the equation.

(c) Establish that the p-discriminant curve is a graph of a singular solution to the equation.

(d) Establish that the function $\eta(x) = \begin{cases} \frac{x^2}{4} & \text{if } x \in [0, \infty), \\ 0 & \text{if } x \in (-\infty, 0), \end{cases}$ is a solution to the equation. Is this a singular solution?

Solution 2.2.4. From, for $y \geq 0$, $0 = (y')^2 - y = (y' - \sqrt{y})(y' + \sqrt{y})$, it follows that if a differentiable function $y = y(x)$, $x \in I$, is a solution to the separable differential equation $y' = \sqrt{y}$ or a solution to the separable differential equation $y' = -\sqrt{y}$, then it is a solution to the equation $(y')^2 = y$.

---

[10] This means that there is $\varepsilon > 0$ such that there are two solutions φ and ψ, $\varphi \neq \psi$, to the initial value problem defined in $(x_0 - \varepsilon, x_0 + \varepsilon)$.

(a) Consider the equation $y' = \sqrt{y}$. Since both functions $f(x,y) = \sqrt{y}$ and $\frac{\partial f}{\partial y} = \frac{1}{2\sqrt{y}}$ are continuous in the region $\mathbb{R} \times \mathbb{R}^+$, by Picard's theorem, for any $x_0 \in \mathbb{R}$ and any $y_0 \in \mathbb{R}^+$, the initial value problem $y' = \sqrt{y}$, $y(x_0) = y_0$, has a unique solution. Since, for $y > 0$ and $x \in \mathbb{R}$,

$$\int_{y_0}^{y} \frac{dt}{\sqrt{t}} = 2(\sqrt{y} - \sqrt{y_0}) \quad \text{and} \quad \int_{x_0}^{x} dx = x - x_0,$$

the solution to the initial value problem $y' = \sqrt{y}$, $y(x_0) = y_0$ is $2(\sqrt{y} - \sqrt{y_0}) = x - x_0$, $x > x_0 - 2\sqrt{y_0}$, or, explicitly, the function $y = \frac{1}{4}(x - x_0 + 2\sqrt{y_0})^2$, $x > x_0 - 2\sqrt{y_0}$.

Similarly, by Picard's theorem, for $x_0 \in \mathbb{R}$ and any $y_0 \in \mathbb{R}^+$, the initial value problem $y' = -\sqrt{y}$, $y(x_0) = y_0$, has a unique solution: $y = \frac{1}{4}(x - x_0 - 2\sqrt{y_0})^2$, $x < x_0 + 2\sqrt{y_0}$.

For any $x_0 \in \mathbb{R}$ and any $y_0 \in \mathbb{R}^+$, the initial value problem $(y')^2 = y$, $y(x_0) = y_0$ has solutions $\varphi(x) = \frac{1}{4}(x - x_0 + 2\sqrt{y_0})^2$, $x > x_0 - 2\sqrt{y_0}$, and $\psi(x) = \frac{1}{4}(x - x_0 - 2\sqrt{y_0})^2$, $x < x_0 + 2\sqrt{y_0}$. From $\varphi'(x_0) = \sqrt{y_0}$ and $\psi'(x_0) = -\sqrt{y_0}$, it follows that the graphs of φ and ψ intersect at the point (x_0, y_0) but do not share the tangent line there.

(b) For $c \in \mathbb{R}$, let $y = \frac{1}{4}(x - c)^2$. From $y' = \frac{1}{2}(x - c)$, it follows that every member of the family of quadratic functions $\mathcal{S} = \{y = \frac{1}{4}(x - c)^2 : c \in \mathbb{R}\}$ is a solution to the equation. By Part (a), the function $\xi \in \mathcal{S}$ is never a unique solution to the initial value problem $(y')^2 = y$, $y(x_0) = \xi(x_0) > 0$, so it is a singular solution to the equation.

Note: Graphs of $y = \frac{1}{4}(x - a)^2$ and $y = \frac{1}{4}(x - b)^2$, $a \neq b$, intersect at the point $\left(\frac{a+b}{2}, \left(\frac{b-a}{4}\right)^2\right)$ but do not share the tangent line at it.

(c) Let $(x, y, p) \mapsto F(x, y, p) = p^2 - y$. The domain of the function F is the set $\mathbb{R}^3$. The p-discriminant curve is determined by $\Gamma = \{(x, y) \in \mathbb{R}^2 : (\exists p \in \mathbb{R})\ F(x, y, p) = \frac{\partial F(x,y,p)}{\partial p} = 0\} = \{(x, y) \in \mathbb{R}^2 : (\exists p \in \mathbb{R})\ p^2 = y \text{ and } p = 0\} = \{(x, 0) : x \in \mathbb{R}\}$. It follows that the p-discriminant curve is the x-axis, i.e., the graph of the function $x \mapsto \tau(x) = 0$. Since, for any $x \in \mathbb{R}$, $(\tau'(x))^2 = \tau(x)$, the function τ is a singular solution to the given equation. Indeed, at any point $(a, 0)$ on the graph of τ, the initial value problem $(y')^2 = y$, $y(a) = 0$, has another solution:

$\sigma(x) = \frac{1}{4}(x - a)^2$. Since $\sigma'(a) = 0$, the solutions τ and σ share the tangent line at the point $(a, 0)$.

(d) The fact $\eta'(x) = \begin{cases} \frac{x}{2} & \text{if } x \in (0, \infty), \\ 0 & \text{if } x \in (-\infty, 0), \end{cases}$ together with $\displaystyle\lim_{x \to 0+}$ $\dfrac{\frac{x^2}{4}}{x} = 0$ and $\displaystyle\lim_{x \to 0-} = \dfrac{0}{x} = 0$, establishes that the function η is differentiable at any $x \in \mathbb{R}$. For any $x \in \mathbb{R}$, $(\eta'(x))^2 = \eta(x)$, i.e., the function η is a solution to the given equation. By Part (a), the function η is a singular solution to the equation $(y')^2 = y$.

Problem 2.2.5. Let $I \subseteq \mathbb{R}$ be an open interval, and let $p : I \to \mathbb{R}$ and $q : I \to \mathbb{R}$ be two continuous functions.

(a) Determine the general solution to the first-order linear equation $y' + p(x)\, y = q(x)$.[11]
(b) For $a \in I$ and $b \in \mathbb{R}$, solve the initial value problem $y' + p(x)\, y = q(x)$, $y(a) = b$.

Solution 2.2.5.

(a) Let $a \in I$ and let, for $x \in I$, $\mu(x) = e^{\int_a^x p(t)\, dt}$. By the chain rule and the fundamental theorem of calculus, $\mu'(x) = \mu(x)p(x)$. It follows that, for all $x \in I$,

$$\mu(x)q(x) = \mu(x)y' + \mu(x)p(x)y = (\mu(x)y)',$$

i.e., for some $c \in \mathbb{R}$, $\mu(x)y = c + \int_a^x \mu(t)q(t)\, dt$. The general solution to the equation is

$$y = e^{-\int_a^x p(t)\, dt}\left(c + \int_a^x q(t)\, e^{\int_a^t p(u)\, du}\, dt\right), \quad x \in I,\ c \in \mathbb{R}.$$

(b) The solution to the initial value problem is

$$y = e^{-\int_a^x p(t)\, dt}\left(b + \int_a^x q(t)\, e^{\int_a^t p(u)\, du}\, dt\right).$$

Problem 2.2.6. Use Picard's theorem to establish that the initial value problem $xy' + x^2 = y$, $y(x_0) = y_0$, has a unique solution if $x_0 < 0$. Draw a graph of the solution $y = f(x)$ for which $f(-1) = 0$.

---

[11]The general solution to a first-order linear differential equation is a family of functions $x \mapsto y = F(x, c)$, $c \in \mathbb{R}$, with the property that $y = f(x)$, $x \in I$, is a solution to the equation if and only if there is $c \in \mathbb{R}$ such that, for each $x \in I$, $f(x) = F(x, c)$.

Solution 2.2.6. For any $x_0 < 0$, the initial value problem $xy' + x^2 = y$, $y(x_0) = y_0$, is equivalent to the initial value problem $y' = \frac{y}{x} - x$, $y(x_0) = y_0$. Since the function $F(x,y) = \frac{y}{x} - x$, $(x,y) \in (-\infty, 0) \times \mathbb{R}$, and its partial derivative $\frac{\partial F}{\partial y} = \frac{1}{x}$ are continuous, by Picard's theorem, the initial value problem $y' = \frac{y}{x} - x$, $y(x_0) = y_0$, $x_0 < 0$, has a unique solution.

The equation $y' - \frac{y}{x} = -x$, $x < 0$, is a non-homogeneous first-order linear differential equation, and by Problem 2.2.5, its solution f such that $f(-1) = 0$ is determined by $f(x) = e^{\int_{-1}^{x} \frac{dt}{t}} \left(0 + \int_{-1}^{x} (-t) e^{-\int_{-1}^{t} \frac{du}{u}} \, dt \right) = -x(x+1)$, $x < 0$.

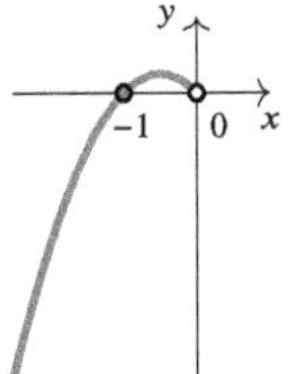

Problem 2.2.7. Determine a function $x \mapsto f(x)$ with the graph Γ such that $(1, -1) \in \Gamma$ and that, for any $T \in \Gamma$, the area of the trapezoid determined by the coordinate axes, the ordinate of the point T, and the tangent line to Γ at T is equal 3.

Solution 2.2.7. Let f be a function with the required property and let $x \in \mathrm{Dom}(f)$. Consider the coordinate system OXY, with the origin $O = (0,0)$. Let $T = (x, y = f(x))$ and $P = (x, 0)$. The equation of the tangent line to Γ at T in the coordinate system OXY is $Y - y = y'(X - x)$, where $y' = f'(x)$. The intersection of the tangent line and the Y-axis is the point $S = (0, y - xy')$. The points O, P, T, and S are the vertices of a trapezoid if and only if $x \neq 0$ and $\mathrm{sign}(y) = \mathrm{sign}(y - xy') \neq 0$. Since $f(1) = -1 < 0$, it follows that $y < 0$ and that the area of the trapezoid is $3 = \frac{1}{2} |\overline{OP}| \left(|\overline{OS}| + |\overline{PT}| \right) = -\frac{x}{2}(y + y - xy')$.

Since $y' - \frac{2}{x} y = \frac{6}{x^2}$, $x > 0$ and $y < 0$, is a non-homogeneous first-order linear differential equation, its solution f such that $f(1) = -1$ is

$$f(x) = e^{2 \int_1^x \frac{dt}{t}} \left(-1 + 6 \int_1^x \frac{1}{t^2} e^{-2 \int_1^t \frac{du}{u}} \, dt \right) = x^2 - \frac{2}{x}, \ x \in \left(0, \sqrt[3]{2} \right).$$

Problem 2.2.8. Let $I \subseteq \mathbb{R}$ be an open interval, and let $p : I \to \mathbb{R}$ and $q : I \to \mathbb{R}$ be continuous functions. Demonstrate that the Bernoulli equation $y' + p(x)y + q(x)y^n = 0$, $n \in \mathbb{R}\backslash\{0, 1\}$, transforms into a first-order linear equation via substitution $z = \frac{1}{y^{n-1}}$.[12] Solve the Bernoulli equation.

Solution 2.2.8. From $z = \frac{1}{y^{n-1}}$, it follows that $y = y^n z$ and $y' = -\frac{y^n z'}{n-1}$. The Bernoulli equation becomes $-\frac{y^n z'}{n-1} + p(x)y^n z + q(x)y^n = 0$, or, what is the same, $z' - (n-1)p(x)z = (n-1)q(x)$. It follows that, for some fixed $a \in I$ and for all $x \in I$,

$$z = e^{(n-1)\int_a^x p(t)\,dt}\left(c + (n-1)\int_a^x q(t)e^{-(n-1)\int_a^t p(u)\,du}\,dt\right), \quad c \in \mathbb{R}.$$

All solutions to the Bernoulli equation are determined by, for some fixed $a \in I$ and for all $x \in I$,

$$y = e^{-\int_a^x p(t)\,dt}\left(c + (n-1)\int_a^x q(t)e^{-(n-1)\int_a^t p(u)\,du}\,dt\right)^{\frac{1}{1-n}},$$

with $c \in J \subseteq \mathbb{R}$, where the set J depends of n and the functions p and q.

Note: If $n > 0$, $n \neq 1$, then the function $x \mapsto y = 0$, $x \in I$, is also a solution to the equation. If $n > 1$, this solution is obtained when $c \to \infty$. If $n \in (0, 1)$, the function $x \mapsto y = 0$, $x \in I$, is a singular solution to the equation.

Problem 2.2.9. Let n be a nonnegative integer. Solve the initial value problem $y^{n-1}(y' + y) = x$, $y(1) = 1$.

Solution 2.2.9. If $n = 0$, this is a homogeneous first-order linear differential equation $y' + (1 - x)y = 0$. The solution to the initial value problem is $y = e^{\frac{x^2+1}{2}-x}$, $x \in \mathbb{R}$.

    If $n = 1$, this is a non-homogeneous first-order linear differential equation $y' + y = x$. The solution to the initial value problem is $y = e^{1-x} + x - 1$, $x \in \mathbb{R}$.

    If $n \geq 2$, this is the Bernoulli equation $y' + y = xy^{1-n}$. We use the substitution $z = y^n$ to obtain $z' + nz = nx$ and

---

[12] Assume that the function $y = y(x)$ is nowhere zero, if necessary.

$z = ce^{-nx} + x - \frac{1}{n}$, $c \in \mathbb{R}$. The solution to the initial value problem is $y = \left(ce^{-nx} + x - \frac{1}{n}\right)^{\frac{1}{n}}$, where c is a real number such that $y(1) = 1$. Thus $c = \frac{e^n}{n}$ and $y = \left(\frac{1}{n}e^{n(1-x)} + x - \frac{1}{n}\right)^{\frac{1}{n}}$, $x \in \mathbb{R}$.

Note: Since $\lim\limits_{x \to \pm\infty}\left(\frac{1}{n}e^{n(1-x)} + x - \frac{1}{n}\right) = \infty$ and $\frac{d}{dx}\left(\frac{1}{n}e^{n(1-x)} + x - \frac{1}{n}\right) = -e^{n(1-x)} + 1$, by the first derivative test, the function $x \mapsto \frac{1}{n}e^{n(1-x)} + x - \frac{1}{n}$ has the number 1 as its absolute minimum value. This justifies the statement that the solution to the initial value problem is defined for all real numbers, regardless to the parity of n.

Problem 2.2.10. Let $I \subset \mathbb{R}$ be an interval, and let $f : I \to \mathbb{R}^+$ be a continuously differentiable function. Solve the equation $y' = \frac{yf'(x) - y^2}{f(x)}$.

Solution 2.2.10. This is a Bernoulli equation, $y' - \frac{f'(x)}{f(x)}y = -\frac{y^2}{f(x)}$. The function $x \mapsto y = 0$, $x \in I$, is one solution to the equation. Suppose that $y = y(x)$, $x \in I$, is a nowhere-zero solution to the equation. The substitution $z = \frac{1}{y}$ yields a first-order linear equation $z' + \frac{f'(x)}{f(x)}z = \frac{1}{f(x)}$. From

$$z = e^{-\int \frac{f'(x)}{f(x)}\,dx}\left(c + \int \frac{1}{f(x)}e^{\int \frac{f'(x)}{f(x)}dx}dx\right) = \frac{c + x}{f(x)}, \quad c \in \mathbb{R},$$

it follows that, for $x \in I$, $y = \frac{f(x)}{x-c}$, where $c \in \mathbb{R}\backslash I$.

Note: The equation is equivalent to the equation $\left(\frac{y}{f(x)}\right)' = -\left(\frac{y}{f(x)}\right)^2$. The substitution $u = \frac{y}{f(x)}$ transforms this equation into a separable differential equation $u' = -u^2$.

Problem 2.2.11. Solve the equation $y' = 1 + e^{x+2y}$.

Solution 2.2.11. The substitution $z = e^{2y}$ transforms the equation into the Bernoulli equation $z' - 2z = 2e^x z^2$. It follows that, referring to Problem 2.2.8, for $c \in \mathbb{R}^+$ and $x \in \mathbb{R}$ such that $x < \frac{1}{3}\ln\frac{3c}{2}$, $z = e^{2x}\left(c - \frac{2}{3}e^{3x}\right)^{-1}$. Therefore, $y = x - \frac{1}{2}\ln\left(c - \frac{2}{3}e^{3x}\right)$, $c \in \mathbb{R}^+$, $x < \frac{1}{3}\ln\frac{3c}{2}$.

Problem 2.2.12. Let $I \subseteq \mathbb{R}$ be an interval, and let p, q, and r be functions continuous in I and such that, for any $x \in I$, $r(x) \neq 0$. Consider the Riccati differential equation $y' = p(x) + q(x)y + r(x)y^2$.

(a) Prove that the Riccati equation has no singular solution.
(b) Prove that if $y_1 = y_1(x)$ is a solution to the Riccati equation, then the substitution $y = y_1 + \frac{1}{z}$ transforms the Riccati equation into a non-homogeneous first-order linear equation.
(c) Prove that if $y_1 = y_1(x)$ is a solution to the Riccati equation, then the substitution $y = z + y_1$ transforms the Riccati equation into a Bernoulli equation.

Solution 2.2.12.

(a) Both functions $(x, y) \mapsto F(x, y) = p(x) + q(x)y + r(x)y^2$ and $(x, y) \mapsto \frac{\partial F}{\partial y}(x, y) = q(x) + 2r(x)y$, $(x, y) \in I \times \mathbb{R}$, are continuous; therefore, by Picard's existence and uniqueness theorem, for any $(x_0, y_0) \in I \times \mathbb{R}$, the initial value problem $y' = p(x) + q(x)y + r(x)y^2$, $y(x_0) = y_0$, has a unique solution. Therefore, the equation has no singular solution.
(b) Since $y = y_1 + \frac{1}{z}$ yields $y' = y_1' - \frac{z'}{z^2}$, the Riccati equation becomes the equation

$$y_1' - \frac{z'}{z^2} = p(x) + q(x)y_1 + \frac{q(x)}{z} + \left(y_1^2 + \frac{2y_1}{z} + \frac{1}{z^2} \right) r(x),$$

i.e., a non-homogeneous first-order linear differential equation $z' + (q(x) + 2r(x)y_1)z = -r(x)$, $z(x) \neq 0$, $x \in I$.
(c) From $y = z + y_1$, it follows that $y' = z' + y_1'$. The Riccati equation becomes $z' + y_1' = p(x) + q(x)(z + y_1) + r(x)(z + y_1)^2$, i.e., the Bernoulli equation $z' = (q(x) + 2r(x)y_1)z + r(x)z^2$.

Note: Every Bernoulli equation with $n = 2$ is a Riccati equation.

Problem 2.2.13. For $a, b \in \mathbb{R}^+$, solve the Riccati equation $y' = a + \frac{y}{2x} + \frac{by^2}{x}$, $x > 0$.

Solution 2.2.13. Suppose that $y = y(x)$ is a solution to the Riccati equation, and let $z = \frac{y}{\sqrt{x}}$. Then, $y' = z'\sqrt{x} + \frac{z}{2\sqrt{x}}$ and z is a solution to the equation

$$z'\sqrt{x} + \frac{z}{2\sqrt{x}} = a + \frac{z}{2\sqrt{x}} + bz^2,$$

i.e., z is a solution to the separable differential equation $z'\sqrt{x} = a + bz^2$. It follows that $\arctan\left(z\sqrt{\frac{b}{a}}\right) = 2\sqrt{ab}x + c$, $c \in \mathbb{R}$, $x \in J(c)$, where $J(c) \subset (0, \infty)$ is an interval that depends on c. This implies that all solutions to the Riccati equation are determined by $\arctan\left(y\sqrt{\frac{b}{ax}}\right) = 2\sqrt{ab}x + c$, $c \in \left(-\infty, \frac{\pi}{2}\right)$, $x \in J(c)$.

Note: From $\arctan\left(z\sqrt{\frac{b}{a}}\right) = 2\sqrt{ab}x + c$, it follows that $-\frac{\pi}{2} < 2\sqrt{ab}x + c < \frac{\pi}{2}$. This implies that $-\frac{\pi+2c}{4\sqrt{ab}} < \sqrt{x} < \frac{\pi-2c}{4\sqrt{ab}}$. In particular, the constant c must be less than $\frac{\pi}{2}$. For such a c and for $\alpha = \alpha(c) = \max\left\{0, -\frac{\pi+2c}{4\sqrt{ab}}\right\}$, $J(c) = \left(\alpha^2, \frac{(\pi-2c)^2}{16ab}\right)$.

Problem 2.2.14. Let $I \subseteq \mathbb{R}$ be an interval, let p and q be functions continuous in I, and let r be a function differentiable in I and such that, for any $x \in I$, $r(x) \neq 0$. Consider the Riccati differential equation $y' = p(x) + q(x)y + r(x)y^2$.

(a) Prove that the substitution $y = -\frac{z'}{zr(x)}$, where z is an unknown function, transforms the Riccati equation into a homogeneous second-order linear differential equation.

(b) Prove that any second-order homogeneous equation, $y'' + \alpha(x)y' + \beta(x)y = 0$, reduces to a Riccati equation via the substitution $y = e^{\int u(x)dx}$, where $u = u(x)$ is another unknown function.

Solution 2.2.14.

(a) Since $y = -\frac{z'}{zr(x)}$ yields

$$y' = -\frac{z''zr(x) - z'(z'r(x) + zr'(x))}{z^2(r(x))^2} = -\frac{z''}{zr(x)} + \frac{(z')^2}{z^2r(x)} + \frac{z'r'(x)}{z(r(x))^2},$$

the Riccati equation becomes the equation

$$-\frac{z''}{zr(x)} + \frac{(z')^2}{z^2r(x)} + \frac{z'r'(x)}{z(r(x))^2} = p(x) - \frac{z'q(x)}{zr(x)} + r(x) \cdot \frac{(z')^2}{z^2r(x))^2},$$

i.e., the homogeneous second-order linear differential equation

$$r(x)z'' - (q(x)r(x) + r'(x))z' + p(x)r^2(x)z = 0.$$

(b) From $y = e^{\int u(x)dx}$, it follows that $y' = u(x)e^{\int u(x)dx}$ and $y'' = (u'(x) + (u(x))^2)\, e^{\int u(x)dx}$. The second-order linear equation transforms into $u'(x)e^{\int u(x)dx} + (u(x))^2 e^{\int u(x)dx} + \alpha(x)u(x)e^{\int u(x)dx} + \beta(x)e^{\int u(x)dx} = 0$, i.e., into the Riccati equation $u'(x) + (u(x))^2 + \alpha(x)u(x) + \beta(x) = 0$.

Problem 2.2.15. Solve the initial value problem $xy' - (2x+1)y + y^2 + x^2 = 0$, $y(1) = 2$.

Solution 2.2.15. For $x > 0$, the given equation is a Riccati equation: $y' - \frac{2x+1}{x}y + \frac{y^2}{x} = -x$. Moreover, if $x > 0$, the equation is equivalent to the equation $\left(\frac{y}{x}\right)' + \left(1 - \frac{y}{x}\right)^2 = 0$. It follows that the function $\varphi(x) = x$, $x > 0$, is a solution to the Riccati equation. Next, since $(1,2) \in D = \{(x,y) : 0 < x < y\}$, we restrict our attention to finding the solution whose graph is contained in D. We follow Problem 2.2.12(b) and, via the substitution $y = x + \frac{1}{z}$ since $y' = 1 - \frac{z'}{z^2}$, obtain a non-homogeneous first-order linear differential equation: $z' + \frac{z}{x} = \frac{1}{x}$, $x > 0$, $z > 0$. The solution to the initial value problem $z' + \frac{z}{x} = \frac{1}{x}$, $z(1) = 1$, is

$$z = e^{-\int_1^x \frac{dt}{t}}\left(1 + \int_1^x \frac{1}{t}e^{\int_1^t \frac{du}{u}}\,dt\right) = 1,$$

which implies that the solution to the original initial value problem is $y = x + 1$, $x > 0$.

Note: Through the substitution $u = 1 - \frac{y}{x}$, the equation $\left(\frac{y}{x}\right)' + \left(1 - \frac{y}{x}\right)^2 = 0$ becomes a separable differential equation $u' = -u^2$. This implies that $1 - \frac{y}{x} = -\frac{1}{x+c}$, for some $c \in \mathbb{R}$. The initial condition $y(1) = 2$ yields $y = x + 1$, $x > 0$.

Problem 2.2.16. Consider the Riccati equation $y' = \frac{2x}{x^3-1} + \frac{x^2 y}{x^3-1} - \frac{y^2}{x^3-1}$, $x \in (1, \infty)$.

(a) Determine $a \in \mathbb{R}$ so that the function $y_1 = ax^2$, $x \in (1, \infty)$, is a solution to the equation, and then solve the equation.

(b) Determine all $a, b \in \mathbb{C}$ so that the function $y_1 = ax + b$, $x \in (1, \infty)$, is a solution to the equation, and then solve the equation.

Solution 2.2.16.

(a) Suppose that $a \in \mathbb{R}$ is such that the function $y_1 = ax^2$, $x \in (1, \infty)$, is a solution to the equation. Since $y_1' = 2ax$, it follows that the number a has the property that, for all $x \in (1, \infty)$,

$$2ax(x^3 - 1) = 2x + ax^4 - a^2 x^4 = 2x + a(1 - a)x^4,$$

i.e., such that $-2a = 2$ and $2a = a(1 - a)$. Hence, $a = -1$ and $y_1 = -x^2$. By Problem 2.2.12(b), the substitution $y = y_1 + \frac{1}{z} = -x^2 + \frac{1}{z}$ transforms the Riccati equation into the first-order linear equation $(x^3 - 1)z' + 3x^2 z = 1$, or, what is the same, the equation $((x^3 - 1)z)' = 1$. From $z = \frac{x+c}{x^3-1}$, $x \in (1, \infty)$, $c \in \mathbb{R}$, it follows that the solution to the Riccati equation is

$$y = -x^2 + \frac{x^3 - 1}{x + c} = -\frac{cx^2 + 1}{x + c}, \quad x \in J, \ c \in \mathbb{R},$$

where $J \subseteq (1, \infty)$ is an interval such that $-c \notin J$.

Note: For $c > 0$, let $y = y^{(c)}(x) = -\frac{cx^2+1}{x+c}$, $x \in (1, \infty)$, i.e., let $y^{(c)}$ be the solution to the equation that corresponds to c. Then, for $x \in (1, \infty)$, $y_1 = -x^2 = \lim_{c \to \infty} y^{(c)}(x)$.

(b) Suppose that $a, b \in \mathbb{C}$ are such that $y = ax + b$ is a solution to the Riccati equation. Then, $y' = a$. From $(x^3 - 1)a = 2x + x^2(ax + b) - (ax + b)^2$, it follows that a and b satisfy the system of equations $a = b^2$, $a^2 = b$, and $ab = 1$. This system has three solutions, $(a, b) = (1, 1)$, $(a, b) = \left(-\frac{1}{2} + i\frac{\sqrt{3}}{2}, -\frac{1}{2} - i\frac{\sqrt{3}}{2}\right)$, and $(a, b) = \left(-\frac{1}{2} - i\frac{\sqrt{3}}{2}, -\frac{1}{2} + i\frac{\sqrt{3}}{2}\right)$, which correspond to three solutions to the equation,

$$y_1 = x + 1 \text{ and } y_{2,3} = \left(-\frac{1}{2} \pm i\frac{\sqrt{3}}{2}\right)x - \frac{1}{2} \mp i\frac{\sqrt{3}}{2}.$$

The substitution $y = x + 1 + \frac{1}{z}$ yields the equation $(x^3 - 1)z' + (x^2 - 2(x + 1))z = 1$, $x \in (1, \infty)$. This equation is equivalent to the equation $((x^3 - 1)z)' - \frac{2}{x-1}(x^3 - 1)z = 1$, which implies that $z = \frac{cx-c-1}{x^2+x+1}$, $c \in \mathbb{R}$. All real-valued solutions to the equation are determined by $y = x + 1 + \frac{x^2+x+1}{cx-c-1} = \frac{(c+1)x^2-c}{cx-c-1}$, $c \in \mathbb{R}$, $x \in J$, where $J \subseteq (1, \infty)$ is an interval such that $1 - \frac{1}{c} \notin J$.

Note 1: For $c > 0$, let $y = y^{(c)}(x) = \frac{(c+1)x^2-c}{cx-c-1}$, $x \in \left(1 + \frac{1}{c}, \infty\right)$, i.e., let $y^{(c)}$ be a solution to the equation that corresponds to c. Then, for $x \in (1, \infty)$, $y_1 = x + 1 = \lim_{c \to \infty} y^{(c)}(x)$.

Note 2: For $c \in \mathbb{R}\backslash\{0\}$ and $d = -\frac{c+1}{c}$, $y = \frac{(c+1)x^2-c}{cx-c-1} = \frac{\frac{c+1}{c} \cdot x^2 - 1}{x - \frac{c+1}{c}} = -\frac{dx^2+1}{x+d}$, i.e., for $c \neq 0$, any function determined by $y = \frac{(c+1)x^2-c}{cx-c-1}$, the expression obtained in Part (b), is also determined by $y = -\frac{dx^2+1}{x+d}$, $d = \frac{c-1}{c}$, the expression obtained in Part (a). The solution that corresponds to $c = 0$ is $y = -x^2$, the particular solution used to obtain the family of solutions in Part (a).

Note 3: From $\frac{cx^2+1}{x+c} = cx - c^2 + \frac{c^3+1}{x+c}$, it follows that $y = -\frac{cx^2+1}{x+c}$ is a linear function, i.e., $y = -cx+c^2$, if and only if $c^3 = -1$. This is another way to obtain the three linear solutions established in Part (b).

Problem 2.2.17. Let $I \subseteq \mathbb{R}$ be an interval, and let $f : I \to \mathbb{R}$ be a twice differentiable function such that $f''(p) \neq 0$, for all $p \in I$. Solve Clairaut's equation $y = xy' + f(y')$.

Solution 2.2.17. The assumption $f''(p) \neq 0$, $p \in I$, guarantees that f is not a constant function or a linear function, i.e., the equation $y = xy' + f(y')$ is not a linear differential equation.

The domain of the function $(x, y, p) \mapsto F(x, y, p) = xp + f(p) - y$ is the set $D = \{(x, y, p) : x, y \in \mathbb{R} \text{ and } p \in I\}$. Since $\frac{\partial F}{\partial p} = x + f'(p)$, it follows that the p-discriminant curve of Clairaut's equation is $\Gamma = \{(x, y) \in \mathbb{R}^2 : (\exists\, p \in I)\ x = -f'(p) \text{ and } y = f(p) - cf'(p)\} = \{(-f'(p), f(p) - pf'(p)) : p \in I\}$.

Let $y = y(x)$ be a function determined by the parametric equations $x = -f'(p)$, $y = f(p) - pf'(p)$, $p \in J \subseteq I$. Then, for $p \in J$,

$$y' = \frac{dy}{dx} = \frac{\frac{dy}{dp}}{\frac{dx}{dp}} = \frac{f'(p) - f'(p) - pf''(p)}{-f''(p)} = p.$$

It follows that the nonlinear[13] function $x \mapsto y$ determined by the parametric equations $x = -f'(p)$, $y = f(p) - pf'(p)$, $p \in J \subseteq I$, is

---

[13] The derivative $y' = p$ is not a constant.

a solution to Clairaut's equation, $y = f(p) - pf'(p) = f(y') + xy'$. Let, for $c \in I$, $P_c = (-f'(c), f(c) - cf'(c)) \in \Gamma$. The equation of the tangent line to the p-discriminant curve Γ at the point P_c is $y - (f(c) - cf'(c)) = c(x - (-f'(c)))$, i.e., $y = cx + f(c)$. Since the function $x \mapsto y = cx + f(c)$ is such that $y'(x) = c$, for all $x \in \mathbb{R}$, it follows that for any $c \in I$, the linear function $y = cx + f(c)$ is a solution to Clairaut's equation. Hence, there is a failure of uniqueness of a solution to the initial value problem at every *smooth* point on the curve Γ. Any function that determines a smooth piece of the curve Γ is a singular solution to the equation.

Note: For a Clairaut's equation $y = xy' + f(y')$, the family of linear functions $y = cx + f(c)$, $c \in \mathrm{Dom}(f)$, is called the *general solution* to the equation.

Problem 2.2.18. Determine the p-discriminant curve Γ and the general solution to the differential equation $y = xy' + (y')^3$. Draw the graph of the curve Γ as well as several lines determined by the general solution. Determine the singular solution(s) to the equation.

Solution 2.2.18. This is Clairaut's equation with $f(p) = p^3$, $p \in \mathbb{R}$. From $f'(p) = 3p^2$, it follows that the p-discriminant curve is $\Gamma = \{(-3p^2, -2p^3) : p \in \mathbb{R}\}$. The general solution is $y = cx + c^3$, $c \in \mathbb{R}$. From $x = -3p^2$ and $y = -2p^3$, $p \neq 0$, it follows that $p = \frac{3y}{2x}$, which implies that $x = -\frac{27y^2}{4x^2}$. Hence, all singular solutions are implicitly determined by $4x^3 + 27y^2 = 0$. Observe that the curve Γ contains the graphs of two singular solutions: $y = \frac{2x\sqrt{-x}}{3\sqrt{3}}$, $x \in (-\infty, 0)$, and $y = -\frac{2x\sqrt{-x}}{3\sqrt{3}}$, $x \in (-\infty, 0)$. The six lines $y = cx + c^3$ in the graph correspond to $c = \pm 1.2, \ \pm 1, \ \pm 0.8$.

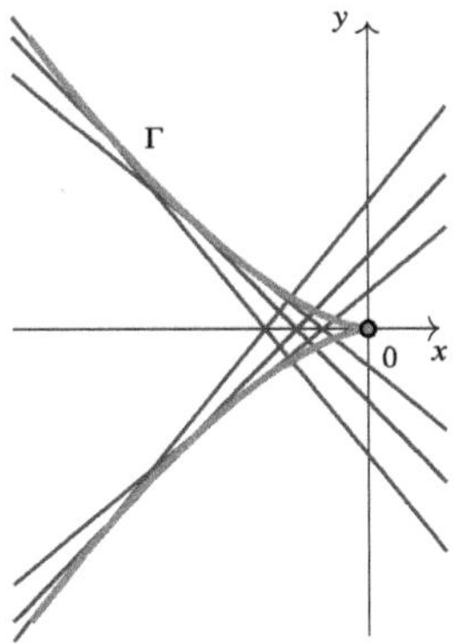

Problem 2.2.19. Solve the equation $(y')^2 - 2x\sqrt{y}\, y' + 4y\sqrt{y} = 0$.

Solution 2.2.19. The domain of the function $(x, y, p) \mapsto F(x, y, p) = p^2 - 2xp\sqrt{y} + 4y\sqrt{y}$ is the set $D = \{(x, y, p) : x, p \in \mathbb{R} \text{ and } y \in [0, \infty)\}$. The p-discriminant curve of the equation is determined by

$$\Gamma = \{(x, y) \in \mathbb{R} \times [0, \infty) : (\exists p \in \mathbb{R})\ \ p^2 - 2xp\sqrt{y} + 4y\sqrt{y} = 0 \quad \text{and}$$
$$2p - 2x\sqrt{y} = 0\} = \{(x, y) \in \mathbb{R} \times [0, \infty) : y(4\sqrt{y} - x^2) = 0\}.$$

The p-discriminant Γ is the union of the graphs of the functions $\alpha(x) = 0$, $x \in \mathbb{R}$, and $\beta(x) = \frac{x^4}{16}$, $x \in \mathbb{R}$. Both of these functions are solutions to the equation. Moreover, from $0 = (y')^2 - 2x\sqrt{y}\, y' + 4y\sqrt{y} = (y' - x\sqrt{y})^2 - y(x^2 - 4\sqrt{y})$, it follows that if $y = y(x)$ is a solution to the equation, then its graph lies in the set $E = \left\{(x, y) : 0 \le y \le \frac{x^4}{16}\right\}$, i.e., in the region bounded by the quartic parabola $y = \frac{x^4}{16}$ and the x-axis.

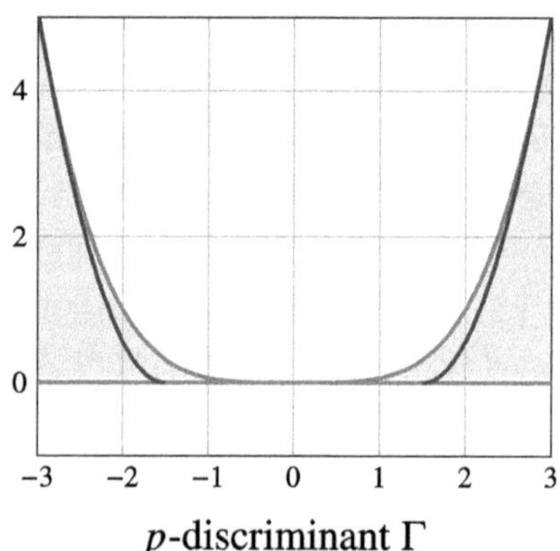

p-discriminant Γ

The substitution $u = \sqrt{y} > 0$ transforms the equation into the equation $4u^2((u')^2 - xu' + 2u) = 0$, i.e., into Clairaut's equation $u = xu' - (u')^2$, $u > 0$. It follows that for $c > 0$, $u = cx - c^2$, $x \in (c, \infty)$, and for $c < 0$, $u = cx - c^2$, $x \in (-\infty, c)$. Consequently, a family of solutions to the equation $(y')^2 - 2x\sqrt{y}\, y' + 4y\sqrt{y} = 0$ is determined by:

- if $c > 0$, $y = (cx - c^2)^2$, $x \in (c, \infty)$;
- if $c < 0$, $y = (cx - c^2)^2$, $x \in (-\infty, c)$.

If we consider the initial value problem $(y')^2 - 2x\sqrt{y}\, y' + 4y\sqrt{y} = 0$, $y(a) = 0$, $a \in \mathbb{R}$, then the point $(a, 0)$ belongs to the graph of the solution $\alpha(x) = 0$, $x \in \mathbb{R}$. If $a > 0$, another solution to this initial value problem is the function $f_a(x) = \begin{cases} 0 & \text{if } x \in (-\infty, a], \\ (ax - a^2)^2 & \text{if } x \in (a, \infty). \end{cases}$ If $a < 0$, the function $g_a(x) = \begin{cases} (ax - a^2)^2 & \text{if } x \in (-\infty, a), \\ 0 & \text{if } x \in [a, -\infty), \end{cases}$ is another solution to the initial value problem. If $a = 0$, then $\beta(x) = \frac{x^4}{16}$ is a solution to the initial value problem. Hence, the function α is a singular solution to the equation.

If we consider the initial value problem $(y')^2 - 2x\sqrt{y}\, y' + 4y\sqrt{y} = 0$, $y(2a) = a^4$, $a \in \mathbb{R}$, then the point $(2a, a^4)$ belongs to the graph of the solution $\beta = \beta(x) = \frac{x^4}{16}$. If $a > 0$, another solution to this initial value problem is the function $h_a(x) = \begin{cases} \frac{x^4}{16} & \text{if } x \in (-\infty, 2a], \\ (ax - a^2)^2 & \text{if } x \in (2a, \infty). \end{cases}$ If $a < 0$, the function $k_a(x) = \begin{cases} (ax - a^2)^2 & \text{if } x \in (-\infty, 2a), \\ \frac{x^4}{16} & \text{if } x \in [2a, -\infty), \end{cases}$ is a solution to the initial value problem. The function $\beta(x) = \frac{x^4}{16}$ is a singular solution to the equation.

Note: Let $x > c > 0$. Then, $\frac{x^4}{16} - (cx - c^2)^2 = \left(\frac{x^2}{4} - cx + c^2\right)\left(\frac{x^2}{4} + cx - c^2\right) = \frac{1}{16}(x - 2c)^2\left((x + 2c)^2 - 8c^2\right) \geq 0$, which establishes that the graph of the solution $y = (cx - c^2)^2$, $x \in (c, \infty)$, $c > 0$, lies in the region bounded by the quartic parabola and the x-axis.

Problem 2.2.20. Let $I \subseteq \mathbb{R}$ be an interval, and let $f : I \to \mathbb{R}$ and $g : I \to \mathbb{R}$ be two continuously differentiable functions such that $f(t) \neq t$, for all $t \in I$. Solve Lagrange's equation, $y = xf(y') + g(y')$.

Solution 2.2.20. Let $y = y(x)$ be a solution to Lagrange's equation, and let $u = u(x) = y'(x)$. By differentiating the equation with respect to x, we obtain $u = f(u) + xf'(u)\frac{du}{dx} + g'(u)\frac{du}{dx}$. We rewrite this equation in the form $(u - f(u))\, dx - (xf'(u) + g'(u))\, du = 0$. This yields a first-order linear equation, $\frac{dx}{du} - \frac{f'(u)}{u - f(u)}x = \frac{g'(u)}{u - f(u)}$, with an unknown function $x = x(u)$. It follows that

$$x(u) = e^{\int \frac{f'(u)\,du}{u - f(u)}} \left(c + \int \frac{g'(u)}{u - f(u)} e^{-\int \frac{f'(u)\,du}{u - f(u)}}\,du \right) = F(u, c), \ c \in \mathbb{R}.$$

The solution $y = y(x)$ is defined parametrically by $x = F(u, c)$, $y = F(u, c)f(u) + g(u)$, $u \in J \subseteq I$, for some $c \in \mathbb{R}$.

Note 1: Clairaut's equation is a special case of Lagrange's equation.

Note 2: If $\alpha \in I$ is such that $f(\alpha) = \alpha$, then the linear function $y = \alpha x + g(\alpha)$ is a solution to Lagrange's equation $y = xf(y') + g(y')$.

Problem 2.2.21. Consider Lagrange's equation $y = 3xy' - 7(y')^3$.

(a) Solve the equation.
(b) Determine the p-discriminant of the equation. Establish if the p-discriminant determines a singular solution to the equation.
(c) Establish if the p-discriminant is a locus of cusps of the integral curves of the equation.

Solution 2.2.21.

(a) We follow the procedure established in Problem 2.2.20.
Let $y = y(x)$ be a solution to the equation, and let $u = y'$ with $u(x) \neq 0$ in the domain of u. Then, after differentiating the equation with respect to x, $u = 3u + 3x\frac{du}{dx} - 21u^2\frac{du}{dx}$. It follows that $\frac{dx}{du} + \frac{3}{2u}x = \frac{21u}{2}$ and

$$x = e^{-\int \frac{3}{2u}\,du} \left(c + \frac{21}{2} \int u e^{\int \frac{3}{2u}\,du}\,du \right) = c|u|^{-\frac{3}{2}} + 3u^2, \ u \neq 0, \ c \in \mathbb{R}.$$

A family of parametrically defined solutions $y = y(x)$ is, for $c \in \mathbb{R}$,

$$x(u) = c|u|^{-\frac{3}{2}} + 3u^2, \ y(u) = 3cu|u|^{-\frac{3}{2}} + 2u^3, \ u \in J_c,$$

where $J_c \subset \mathbb{R}\backslash\{0\}$ is an interval that depends on the chosen constant c.

The function $y = y(x) = 0$, $x \in \mathbb{R}$, is a solution to the equation that does not belong to the above family of solutions.

(b) Let $F(x, y, p) = 3px - 7p^3 - y$. The p-discriminant of the equation is determined by the system of equations $F(x, y, p) = 3px - 7p^3 - y = 0$ and $\frac{\partial F(x,y,p)}{\partial p} = 3x - 21p^2 = 0$. The p-discriminant is defined parametrically by $x = 7p^2$ and $y = 3px - 7p^3 = 21p^3 - 7p^3 = 14p^3$, $p \in \mathbb{R}$, or implicitly by $4x^3 = 7y^2$.

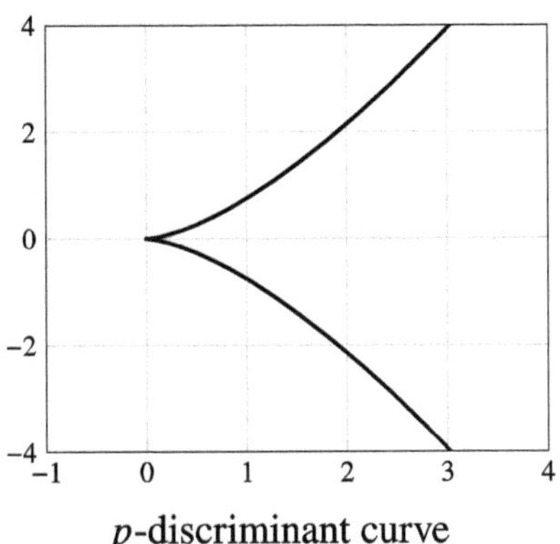

p-discriminant curve

From $4x^3 = 7y^2$, it follows that $12x^2 = 14yy'$, i.e., for $y \neq 0$, $y' = \frac{6x^2}{7y}$. Since

$$3xy' - 7(y')^3 = \frac{6x^2}{7y}\left(3x - 7\left(\frac{6x^2}{7y}\right)^2\right) = -9y \neq y,$$

no solution to the equation is determined implicitly by $4x^3 = 7y^2$. In other words, neither of the two smooth branches of the p-discriminant is the graph of a solution to the equation.

(c) For $c > 0$, consider the curve $\mathcal{C}_c$, determined by the parametric equations $x(u) = cu^{-\frac{3}{2}} + 3u^2$, $y(u) = 3cu^{-\frac{1}{2}} + 2u^3$, $u \in J_c \subseteq (0, \infty)$. Then, $\frac{dx}{du} = -\frac{3}{2}u^{-\frac{5}{2}}(c - 4u^{\frac{7}{2}})$ and $\frac{dy}{du} = -\frac{3}{2}u^{-\frac{3}{2}}(c - 4u^{\frac{7}{2}})$. It follows that if $u = \left(\frac{c}{4}\right)^{\frac{2}{7}}$, then $\frac{dx}{du} = \frac{dy}{du} = 0$. Consequently, the point

$$T_c = \left(x\left(\left(\frac{c}{4}\right)^{\frac{2}{7}}\right), y\left(\left(\frac{c}{4}\right)^{\frac{2}{7}}\right)\right) = \left(7\left(\frac{c}{4}\right)^{\frac{4}{7}}, 14\left(\frac{c}{4}\right)^{\frac{6}{7}}\right)$$

is a cusp of the curve $\mathcal{C}_c$. By removing the cusp, the curve $\mathcal{C}_c$ splits into two integral curves of the equation. From

$$4\left(7\left(\frac{c}{4}\right)^{\frac{4}{7}}\right)^{3} = 4\cdot 7^3\left(\frac{c}{4}\right)^{\frac{12}{7}} = 7\left(14\left(\frac{c}{4}\right)^{\frac{6}{7}}\right)^{2},$$

it follows that the cusp T_c lies on the p-discriminant curve of the equation. See the images below.

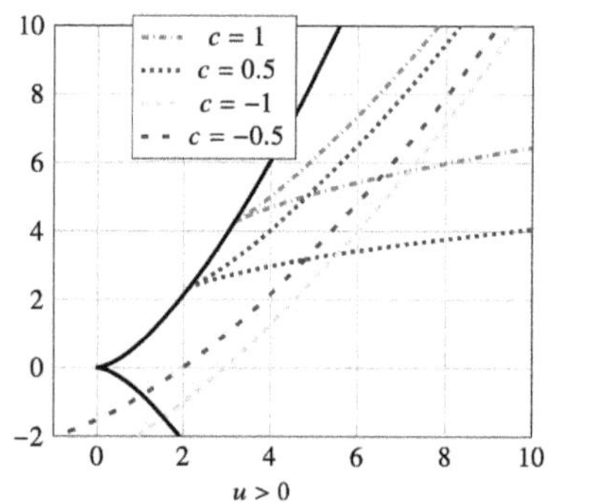

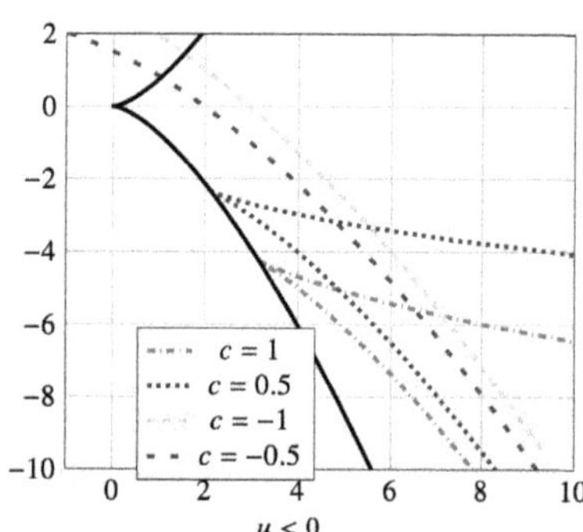

The p-discriminant is a locus of cusps of the integral curves of the equation.

Problem 2.2.22. Suppose that $(x, y) \mapsto F(x, y)$ is a homogeneous function with the degree of homogeneity zero, i.e., suppose that, for any $(x, y) \in \mathrm{Dom}(F)$ and any $t \in \mathbb{R}\backslash\{0\}$ such that $(tx, ty) \in \mathrm{Dom}(F)$, $F(tx, ty) = F(x, y)$. Establish that the homogeneous differential equation $y' = F(x, y)$ reduces to a separable differential equation via the substitution $y = xz$, where $z = z(x)$ is an unknown function.

Solution 2.2.22. From $y' = z + xz'$ and $F(x, xz) = F(1, z)$ it follows that, via the substitution $y = xz$, the homogeneous equation becomes a separable differential equation $z + xz' = F(1, z)$.

Problem 2.2.23. Let a, b, $c \in \mathbb{R}$, with $c \neq 0$. Determine a substitution that transforms the Riccati equation $y' = a + \frac{by}{x} + \frac{cy^2}{x^2}$, $x > 0$, into a separable differential equation.

Solution 2.2.23. From, for any $t > 0$, $F(x, y) = a + \frac{by}{x} + \frac{cy^2}{x^2} = F(tx, ty)$, it follows that this is a homogeneous differential equation.

The substitution $y = xz$, transforms the Riccati equation into the separable differential equation $xz' = a + (b-1)z + cz^2$, $x > 0$.

Problem 2.2.24. Solve the following initial value problems:

(a) $y' = \frac{x}{2y} + \frac{3y}{2x}$, $y(1) = 1$.

(b) $y' = \frac{x-y}{x+y}$, $y(1) = 1$.

Solution 2.2.24.

(a) Let $F(x,y) = \frac{x}{2y} + \frac{3y}{2x}$. Then, for any $t, x, y \in \mathbb{R}\backslash\{0\}$, $F(tx, ty) = \frac{tx}{2ty} + \frac{3ty}{2tx} = F(x,y)$. The equation is a homogeneous differential equation and via the substitution $y = xz$, the initial value problem becomes the initial value problem $z + xz' = \frac{1}{2z} + \frac{3z}{2}$, $z(1) = 1$, i.e., the problem $\frac{2z}{z^2+1}\,dz = \frac{dx}{x}$, $z(1) = 1$. In a neighbourhood of the point $(1,1)$ in the xz-plane, the solution is determined by $\int_1^z \frac{2t\,dt}{t^2+1} = \int_1^x \frac{dt}{t}$, i.e., by $1 + z^2 = 2x$, $x > 0$. The solution to the original initial value problem is determined by $1 + \frac{y^2}{x^2} = 2x$ in a neighbourhood of the point $(1,1)$ in the xy-plane. It follows that $y = x\sqrt{2x - 1}$, $x \in \left(\frac{1}{2}, \infty\right)$.

(b) Let $F(x,y) = \frac{x-y}{x+y}$. Then, for any $t, x, y \in \mathbb{R}$ such that $t(x+y) \neq 0$, $F(tx, ty) = \frac{tx-ty}{tx+ty} = F(x,y)$. The equation is a homogeneous differential equation and via the substitution $y = xz$, the given initial value problem becomes the initial value problem $z + xz' = \frac{1-z}{1+z}$, $z(1) = 1$, i.e., the problem $\frac{1+z}{1-2z-z^2}\,dz = \frac{dx}{x}$, $z(1) = 1$. In a neighbourhood of the point $(1,1)$ in the xz-plane, the solution is determined by $z^2 + 2z - 1 = \frac{2}{x^2}$, $x > 0$. The solution to the initial value problem is $y = -x + \sqrt{2(x^2 + 1)}$, $x \in (0, \infty)$.

Problem 2.2.25. Solve the initial value problem $y' = \frac{x-y+1}{x+y-3}$, $y(2) = 3$.

Solution 2.2.25. The strategy is to transform this equation into a homogeneous differential equation via the substitution $x = X + \alpha$, $y = Y + \beta$, where $\alpha, \beta \in \mathbb{R}$, are such that $\alpha - \beta + 1 = 0$ and $\alpha + \beta - 3 = 0$. Thus, $x = X + 1$ and $y = Y + 2$, and the given initial value problem becomes the problem $\frac{dY}{dX} = \frac{X-Y}{X+Y}$, $Y(1) = 1$.

Proceed like in Problem 2.2.24(b). The solution to the initial value problem is $y = 3 - x + \sqrt{2(x^2 - 2x + 2)}$, $x \in \mathbb{R}$.

Problem 2.2.26. Let $a \in (0,1)$. Use the substitution $u = u(x) = \frac{1}{y}$ to transform the equation $\frac{dy}{dx} = \frac{y^2(xy+a)}{axy+4}$ into a homogeneous differential equation. Determine all solutions $x \mapsto y = f(x)$ such that $f(2a) = -a^{-2}$.

Solution 2.2.26. The domain of the function $F(x,y) = \frac{y^2(xy+a)}{axy+4}$ is the set $D = \left\{(x,y) : xy \neq -\frac{4}{a}\right\}$. Since F is a rational function, both F and its partial derivative $(x,y) \mapsto \frac{\partial F}{\partial y}(x,y)$ are continuous functions in D. By Picard's Theorem, since $P_0 = (2a, -a^{-2}) \in D$, there is only one solution $x \mapsto f(x)$ such that $f(2a) = -a^{-2}$. Let $D_0 = \left\{(x,y) : x > 0 > y > -\frac{4}{ax}\right\}$. Then $P_0 \in D_0 \subset D$. The substitution $u = \frac{1}{y}$ maps the set D_0 onto the set $E_0 = \{(x,u) : x > 0 \text{ and } ax + 4u < 0\}$ and the point P_0 onto the point $Q_0 = (2a, -a^2)$. Since $u = \frac{1}{y}$ implies that $\frac{dy}{dx} = -\frac{1}{u^2}\frac{du}{dx}$, the task becomes the question of solving the initial value problem $\frac{du}{dx} = -\frac{au+x}{4u+ax}$, $u(2a) = -a^2$.

The homogeneous differential equation $\frac{du}{dx} = -\frac{au+x}{4u+ax}$, $(x,u) \in E_0$, via substitution $u = vx$, transforms into a separable differential equation $v + x\frac{dv}{dx} = -\frac{axv+x}{4xv+ax}$. The substitution $u = vx$ maps the set E_0 onto the set $F_0 = \{(x,v) : x > 0 \text{ and } a + 4v < 0\}$ and the point Q_0 onto the point $R_0 = \left(2a, -\frac{a}{2}\right)$. The solution to the initial value problem $\frac{4v+a}{4v^2+2av+1}dv = -\frac{dx}{x}$, $v(2a) = -\frac{a}{2}$, is determined by

$$\int_{-\frac{a}{2}}^{v} \frac{4t + a}{4t^2 + 2at + 1}\, dt = -\int_{2a}^{x} \frac{dt}{t}, \quad (x,v) \in F_0.$$

Since $a \in (0,1)$, it follows that $4v^2 + 2av + 1 = \frac{4a^2}{x^2}$, $(x,v) \in F_0$, which yields $\frac{4}{x^2y^2} + \frac{2a}{xy} + 1 = \frac{4a^2}{x^2}$, $(x,y) \in D_0$. The required solution $x \mapsto y = f(x)$, $f(2a) = -a^{-2}$, is implicitly determined by $(x^2 - 4a^2)y^2 + 2axy + 4 = 0$, $(x,y) \in D_0$.

Problem 2.2.27. Let $a, b, m \in \mathbb{R}\setminus\{0\}$. Determine $p, q \in \mathbb{R}$ so that, via the substitution $y = z^m$, the equation $y' = ax^p + by^q$ transforms into a homogeneous differential equation.

Solution 2.2.27. Since $y = z^m$ and $y' = mz^{m-1}z'$, by the suggested substitution, the equation becomes $mz^{m-1}z' = ax^p + bz^{mq}$, i.e., the equation $z' = \frac{1}{m}\left(ax^p z^{1-m} + bz^{mq+1-m}\right)$. This is a homogeneous differential equation if and only if, for the function $(x,z) \mapsto F(x,z) = ax^p z^{1-m} + bz^{mq+1-m}$, there exist intervals I_1 and I_2 such that, for any $(x,u) \in I_1 \times I_2$, $F(x,xu) = F(1,u)$. Hence, $p, q \in \mathbb{R}$ must be such that, for any $(x, xu) \in I_1 \times I_2$,

$$\left(ax^{p+1-m} + bx^{mq+1-m}u^{mq}\right)u^{1-m} = (a + bu^{mq})u^{1-m}.$$

This is satisfied if and only if $p + 1 - m = mq + 1 - m = 0$, i.e., if $p = m - 1$ and $q = 1 - \frac{1}{m}$.

Problem 2.2.28. Determine all solutions to the initial value problem $xy' = \frac{y+\ln x}{y-\ln x}$, $y(1) = 1$, by using the fact that the substitution $T(x,y) = (e^{u+v}, u - v)$ transforms the equation into a homogeneous differential equation.

Solution 2.2.28. In the set $D = \{(x,y) : x > 0 \text{ and } y > \ln x\}$, the initial value problem is equivalent to the initial value problem $y' = \frac{y+\ln x}{x(y-\ln x)}$, $y(1) = 1$. Since the function $F(x,y) = \frac{y+\ln x}{x(y-\ln x)}$, $(x,y) \in D$, is a ratio of continuously differentiable functions, by Picard's theorem, the initial value problem has a unique solution. Let $P_0 = (1,1)$, and let $D_0 = \{(x,y) : x > 0 \text{ and } y + \ln x > 0 \text{ and } y - \ln x > 0\} \subset D$. From $x = e^{u+v}$, $y = u - v$, it follows that

$$\frac{dy}{dx} = \frac{\frac{dy}{du}}{\frac{dx}{du}} = \frac{1 - \frac{dv}{du}}{\left(1 + \frac{dv}{du}\right)e^{u+v}} = \frac{1 - \frac{dv}{du}}{1 + \frac{dv}{du}}e^{-(u+v)}.$$

The substitution T transforms the equation into the equation

$$\frac{1 - \frac{dv}{du}}{1 + \frac{dv}{du}}e^{-(u+v)} = -\frac{u}{v}e^{-(u+v)},$$

i.e., into the homogeneous differential equation $\frac{dv}{du} = \frac{v+u}{v-u}$. In addition, $T(P_0) = Q_0 = \left(\frac{1}{2}, -\frac{1}{2}\right)$ and $T(D_0) = E_0 = \{(u,v) : u > 0 \text{ and } v < 0\}$. It follows that,[14] $v^2 - 2uv - u^2 = \frac{1}{2}$, $(u,v) \in \{(u,v) : u > 0 \text{ and } v < 1 - \sqrt{2}u\} \subset E_0$.

Since $u = \frac{y+\ln x}{2}$ and $v = \frac{y-\ln x}{2}$, the solution to the given initial value problem is determined by $y^2 - 2y\ln x - \ln^2 x = 1$,

---

[14]Proceed like in Problem 2.2.24(b).

 Advanced Engineering Calculus

$(x, y) \in \{(x, y) : x > 0 \text{ and } y > \max\{|\ln x|, (1 + \sqrt{2}) \ln x\}$, i.e., by $y = \ln x + \sqrt{1 + 2\ln^2 x}$, $x > 0$.

Problem 2.2.29. Let P and Q be functions with continuous partial derivatives in a simply connected region $D \subset \mathbb{R}^2$ and such that $\frac{\partial P}{\partial y} = \frac{\partial Q}{\partial x}$ and $(P(x, y))^2 + (Q(x, y))^2 \neq 0$, $(x, y) \in D$. Solve the exact differential equation $P(x, y)dx + Q(x, y)dy = 0$ in a neighbourhood of a point $(x_0, y_0) \in D$.

Solution 2.2.29. The strategy is to determine a function, $(x, y) \mapsto F(x, y)$, $(x, y) \in D_0 \subseteq D$, such that $dF = P(x, y)dx + Q(x, y)dy$, i.e., such that $\frac{\partial F}{\partial x}(x, y) = P(x, y)$ and $\frac{\partial F}{\partial y}(x, y) = Q(x, y)$, $(x, y) \in D_0$. This will imply that each solution to the equation is implicitly defined by $F(x, y) = c$, $(x, y) \in D_0$, for some $c \in \mathbb{R}$. Since D is a simply connected region and since $(x_0, y_0) \in D$, there are $\varepsilon > 0$ and $\eta > 0$ such that the rectangle $D_0 = [x_0 - \varepsilon, x_0 + \varepsilon] \times [y_0 - \eta, y_0 + \eta]$ is contained in D. Let, for $(x, y) \in D_0$,

$$F(x, y) = \int_{x_0}^{x} P(t, y) \, dt + \int_{y_0}^{y} Q(x_0, t) \, dt.$$

By the fundamental theorem of calculus and the Leibniz rule for differentiation under the integral, for $(x, y) \in D_0$,

$$\frac{\partial F}{\partial x}(x, y) = P(x, y),$$

$$\frac{\partial F}{\partial y}(x, y) = \int_{x_0}^{x} \frac{\partial P}{\partial y}(t, y) \, dt + Q(x_0, y) = \int_{x_0}^{x} \frac{\partial Q}{\partial t}(t, y) \, dt + Q(x_0, y)$$

$$= Q(x, y) - Q(x_0, y) + Q(x_0, y) = Q(x, y).$$

Therefore, in a neighbourhood of a point $(x_0, y_0) \in D$, each solution to the equation is implicitly defined by $\int_{x_0}^{x} P(t, y) \, dt + \int_{y_0}^{y} Q(x_0, t) \, dt = c$, for some $c \in \mathbb{R}$.

Note: Here is a common way to determine the function F. From $\frac{\partial F}{\partial x}(x, y) = P(x, y)$, it follows that $F(x, y) = \int P(x, y) \, dx + \varphi(y)$. This, together with $\frac{\partial F}{\partial y}(x, y) = Q(x, y)$, implies that the function φ is determined by $\varphi'(y) = Q(x, y) - \frac{\partial}{\partial y} \left(\int P(x, y) \, dx \right)$.

Problem 2.2.30. Determine all real numbers a for which the equation $axyy' + ax + y^2 = 0$, $a \neq 0$, is an exact differential equation, and for the obtained value(s) of a, solve the equation.

Solution 2.2.30. We rewrite the equation in the form of $(ax + y^2)\, dx + axy\, dy = 0$. From, for any $(x, y) \in \mathbb{R}^2$, $P(x, y) = ax + y^2$ and $Q(x, y) = axy$ and $\frac{\partial P}{\partial y} = 2y$ and $\frac{\partial Q}{\partial x} = ay$, it follows that the given equation is exact if $a = 2$. By Poincaré's lemma, there is a function $(x, y) \mapsto F(x, y)$ such that $dF = (2x + y^2)\, dx + 2xy\, dy$, i.e., such that $\frac{\partial F}{\partial x} = 2x + y^2$ and $\frac{\partial F}{\partial y} = 2xy$. From $F(x, y) = \int (2x + y^2)\, dx = x^2 + xy^2 + \varphi(y)$, it follows that $\frac{\partial F}{\partial y} = 2xy + \varphi'(y)$. This, together with $\frac{\partial F}{\partial y} = 2xy$, implies that $\varphi'(y) = 0$, i.e., $\varphi(y) = c$, for some $c \in \mathbb{R}$. Hence, $F(x, y) = x^2 + xy^2 + c$, and each solution to the exact equation is implicitly determined by $x^2 + xy^2 = c$, for some $c \in \mathbb{R}$.

Note 1: Since the functions $P(x, y) = 2x + y^2$ and $Q(x, y) = 2xy$ have continuous partial derivatives in $\mathbb{R}^2$, we can take $F(x, y) = \int_0^x P(t, y)\, dt + \int_0^y Q(0, t)\, dt = \int_0^x (2t + y^2)\, dt = x^2 + xy^2$.

Note 2: Let $G(x, y) = x^2 + xy^2 + c$, $c \in \mathbb{R}$, and let $(\alpha, \beta) \in \mathbb{R}^2$ be such that $G(\alpha, \beta) = 0$. For the existence of a function $y = g(x)$ such that $g(\alpha) = \beta$ and $G(x, g(x)) = 0$, $x \in \text{Dom}(g)$, the implicit function theorem requires that $\frac{\partial G}{\partial y}(\alpha, \beta) = 2\alpha\beta \neq 0$. Note that this is the same as asking whether the point (α, β) belongs to the domain of the initial differential equation $axyy' + ax + y^2 = 0$.

Problem 2.2.31. Solve the initial value problem

$$\left(x + e^{\frac{x}{y}}\right) dx + e^{\frac{x}{y}}\left(1 - \frac{x}{y}\right) dy = 0, \ y(0) = 2.$$

Solution 2.2.31. The domain of the functions $P(x, y) = x + e^{\frac{x}{y}}$ and $Q(x, y) = e^{\frac{x}{y}}\left(1 - \frac{x}{y}\right)$ is the set $D = \{(x, y) : y \neq 0\} \subset \mathbb{R}^2$. For every $(x, y) \in D$, $\frac{\partial P}{\partial y} = \frac{\partial Q}{\partial x} = -\frac{x}{y^2}\, e^{\frac{x}{y}}$. Let $D_0 = \{(x, y) : y > 0\} \subset D$. Since $(0, 2) \in D_0$ and since D_0 is a simply connected region in the plane, the initial value problem has a unique solution in a neighbourhood of the point $(0, 2)$, determined by, for $(x, y) \in D_0$,

$$\int_0^x P(t, y)\, dt + \int_2^y Q(0, t)\, dt = \int_0^x \left(t + e^{\frac{t}{y}}\right) dt + \int_2^y dt = 0.$$

The solution to the initial value problem $y = y(x)$ is implicitly determined by $x^2 + 2ye^{\frac{x}{y}} = 4$, with $x \in (-\alpha, \beta)$, for some $\alpha, \beta > 0$ and $y = y(x) > 0$.

Note 1: Since, for $F(x,y) = x^2 + 2ye^{\frac{x}{y}} - 4$, $\frac{\partial F}{\partial y}(x,y) = 2\left(1 - \frac{x}{y}\right)e^{\frac{x}{y}}$, it follows that $F(0,2) = 0$ and $\frac{\partial F}{\partial y}(0,2) = 2$. By the implicit function theorem, there is a function $y = y(x)$, $x \in (-\alpha, \beta)$, for some $\alpha, \beta > 0$, such that $y(0) = 2$ and $F(x, y(x)) = 0$, $x \in (-\alpha, \beta)$.

Note 2: The figure depicts the set $\Gamma = \{(x,y) \in \mathbb{R}^2 : x^2 + 2ye^{\frac{x}{y}} = 4\}$. The visual inspection of the part of Γ that is above the x-axis suggests that this curve is not a graph of a function. Indeed, technology establishes that the required solution to the initial value problem is defined in the interval $(-2, \beta)$, where $\beta \approx 0.656$.

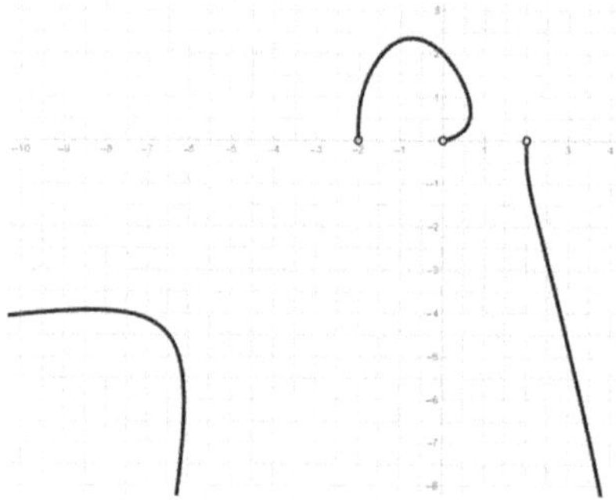

Problem 2.2.32. Solve the initial value problem $(e^x + 2e^x y + 1) \cdot y' = y + 1$, $y(0) = 0$, by using the fact that the substitution $x = -\ln t$ transforms the given equation into an exact differential equation.

Solution 2.2.32. Observe that, for $x, y \in \mathbb{R}$, $e^x + 2e^x y + 1 \neq 0$ if and only if $y \neq -\frac{1}{2}(1 + e^{-x})$. Let $D = \{(x,y) : x \in \mathbb{R} \text{ and } y > -\frac{1}{2}(1 + e^{-x})\}$. The set D is a simply connected region in the xy-plane such that $P_0 = (0,0) \in D$. The substitution $x = -\ln t$, i.e., the mapping $(x,y) \mapsto (t = e^{-x}, y)$, maps D onto the simply connected region $E = \{(t,y) : t > 0 \text{ and } y > -\frac{1}{2}(1 + t)\}$ in the ty-plane and the point P_0 onto the point $T_0 = (1,0)$. Since $x = -\ln t$ implies that $y' = \frac{dy}{dx} = \frac{dy}{dt}\frac{1}{\frac{dx}{dt}} = -t\frac{dy}{dt}$, this substitution transforms the equation $(e^x + 2e^x y + 1) y' = y + 1$, $(x,y) \in D$ into the equation $\left(\frac{1}{t} + \frac{2y}{t} + 1\right)\left(-t\frac{dy}{dt}\right) = y + 1$, i.e., into the equation $(y+1)dt + (t + 2y + 1)dy = 0$, $(t,y) \in E$. Since $\frac{\partial}{\partial y}(y+1) = \frac{\partial}{\partial t}(t + 2y + 1) = 1$, this is an exact differential equation. Its solution $y = y(t)$, $y(1) = 0$, is determined by

$$0 = \int_1^t (y+1)\,du + \int_0^y (2u+2)\,du = y^2 + (1+t)y + t - 1, \quad (t,y) \in E.$$

The solution to the given initial value problem is implicitly determined by $y^2 + (1 + e^{-x})y + e^{-x} - 1 = 0$, $(x,y) \in D$, or explicitly by
$$y = \tfrac{1}{2}\left(\sqrt{4 + (1 - e^{-x})^2} - 1 - e^{-x} \right), \quad x \in \mathbb{R}.$$

Problem 2.2.33. Determine continuously differentiable functions $r \mapsto f(r)$ and $\theta \mapsto g(\theta)$ so that, for $x = r\cos\theta$ and $y = r\sin\theta$, there is a function $(r,\theta) \mapsto F(r,\theta)$ such that $dF = f(r)\,dx - g(\theta)\,dy$.

Solution 2.2.33. Recall that $dx = \cos\theta\,dr - r\sin\theta\,d\theta$ and $dy = \sin\theta\,dr + r\cos\theta\,d\theta$. It follows that

$$f(r)\,dx - g(\theta)\,dy = (f(r)\cos\theta - g(\theta)\sin\theta)\,dr - r(f(r)\sin\theta + g(\theta)\cos\theta)\,d\theta.$$

This is the total differential of a function, $(r,\theta) \mapsto F(r,\theta)$, in a simply connected region, D, in the $r\theta$-plane if and only if

$$\frac{\partial}{\partial\theta}(f(r)\cos\theta - g(\theta)\sin\theta) = -\frac{\partial}{\partial r}(r(f(r)\sin\theta + g(\theta)\cos\theta)), \quad (r,\theta) \in D,$$

or, what is the same, $rf'(r) = g'(\theta)$, $(r,\theta) \in D$. Since in this identity in the region D, the left-hand side is a function of r only and the right-hand side of the identity is a function of θ only, and since r and θ are independent variables, it follows that there is $a \in \mathbb{R}$ such that, for any $(r,\theta) \in D$, $rf'(r) = a$ and $g'(\theta) = a$. Therefore, for some real numbers a, b, and c, $f(r) = a\ln r + b$, $r > 0$, and $g(\theta) = a\theta + c$, $\theta \in \mathbb{R}$.

Problem 2.2.34.

(a) Determine a function $\lambda = \lambda(t)$ such that the function $(x,y) \mapsto \lambda(x+y)$ is an integrating factor of the equation $(x+2y)\,dx + y\,dy = 0$.

(b) Solve the initial value problem $(x+2y)\,dx + y\,dy = 0$, $y(0) = 1$.

Solution 2.2.34.

(a) Suppose that the function $\lambda = \lambda(t)$ is such that the equation $(x+2y)\lambda(x+y)\,dx + y\lambda(x+y)\,dy = 0$ is an exact differential equation in a simply connected region D. Then, for any $(x, y) \in D$, $\frac{\partial}{\partial y}((x+2y)\lambda(x+y)) = \frac{\partial}{\partial x}(y\lambda(x+y))$. It follows that, for any $(x, y) \in D$, $(x+y)\lambda'(x+y) = -2\lambda(x+y)$. The function $\lambda = \lambda(t)$ is a solution to the differential equation $tu'(t) = -2u(t)$, i.e., the equation $(t^2 u(t))' = 0$. We take $\lambda(t) = \frac{1}{t^2}$, $t > 0$.

(b) The set $D = \{(x, y) : x + y > 0 \text{ and } y > 0\} \subset \mathbb{R}^2$ is a simply connected region in the xy-plane such that $(0, 1) \in D$. The equation

$$\frac{x + 2y}{(x + y)^2}\, dx + \frac{y}{(x + y)^2}\, dy = 0, \quad (x, y) \in D,$$

is an exact differential equation. The solution to the initial value problem is implicitly determined by

$$0 = \int_0^x \frac{t + 2y}{(t + y)^2}\, dt + \int_1^y \frac{dt}{t} = \ln(x + y) + \frac{x}{x + y}, \quad (x, y) \in D.$$

Note: For $F(x, y) = \ln(x+y)+\frac{x}{x+y}$, $\frac{\partial F}{\partial y}(x, y) = \frac{y}{(x+y)^2}$. Hence, the implicit function theorem guarantees the existence of the solution to the initial value problem. Observe that the condition $y \neq 0$, required by the implicit function theorem, determines the domain of the initial differential equation rewritten as a homogeneous differential equation $\frac{dy}{dx} = -\frac{x+2y}{y}$.

Problem 2.2.35.

(a) Determine a function $\lambda = \lambda(t)$ such that the function $(x, y) \mapsto \lambda(x^2 + y^2)$ is an integrating factor of the equation $(y^2 - x^2)\, dx - 2xy\, dy = 0$.

(b) Solve the initial value problem $(y^2 - x^2)\, dx - 2xy\, dy = 0$, $y\left(\frac{1}{2}\right) = \frac{1}{2}$.

Solution 2.2.35.

(a) Suppose that the function $\lambda = \lambda(t)$ is such that the equation $(y^2 - x^2)\lambda(x^2 + y^2)\, dx - 2xy\lambda(x^2 + y^2)\, dy = 0$ is an exact differential equation in a simply connected region D. Then, for any $(x, y) \in D$, $\frac{\partial}{\partial y}((y^2 - x^2)\lambda(x^2 + y^2)) = -2\frac{\partial}{\partial x}(xy\lambda(x^2 + y^2))$. It follows that, for any $(x, y) \in D$, $2\lambda(x^2+y^2)+(x^2+y^2)\,\lambda'(x^2+y^2) = 0$. The function $\lambda = \lambda(t)$ is a solution to the differential equation $2u(t) + tu'(t) = 0$, $t > 0$, i.e., the equation $(t^2 u(t))' = 0$, $t > 0$. We take $\lambda(t) = \frac{1}{t^2}$, $t > 0$.

(b) The set $D = \{(x, y) : x > 0 \text{ and } y > 0\} \subset \mathbb{R}^2$ is a simply connected region in the xy-plane such that $\left(\frac{1}{2}, \frac{1}{2}\right) \in D$. The equation

$$\frac{y^2 - x^2}{(x^2 + y^2)^2} \, dx - \frac{2xy}{(x^2 + y^2)^2} \, dy = 0, \quad (x, y) \in D$$

is an exact differential equation, and the solution to the initial value problem is implicitly determined by

$$0 = \int_{\frac{1}{2}}^{x} \frac{y^2 - t^2}{(t^2 + y^2)^2} \, dt - \int_{\frac{1}{2}}^{y} \frac{t \, dt}{\left(\frac{1}{4} + t^2\right)^2} = \frac{x}{x^2 + y^2} - 1, \quad (x, y) \in D.$$

The function $y = \sqrt{x - x^2}$, $x \in (0, 1)$, is the required solution.

Problem 2.2.36. Let $(x, y) \mapsto P(x, y)$ and $(x, y) \mapsto Q(x, y)$ be two continuously differentiable functions in a simply connected region $D \subseteq \mathbb{R}^2$ such that $P(x, y) \neq 0$ and $Q(x, y) \neq 0$ in D. Suppose that $\frac{\partial P}{\partial y} \neq \frac{\partial Q}{\partial x}$, i.e., suppose that the equation $P(x, y) \, dx + Q(x, y) \, dy = 0$ is not exact.

(a) Prove that if there is a continuous function $t \mapsto f(t)$ such that, for all $(x, y) \in D$, $\dfrac{\frac{\partial P(x,y)}{\partial y} - \frac{\partial Q(x,y)}{\partial x}}{Q(x,y)} = f(x)$, then the function $\lambda(x) = e^{\int f(x) \, dx}$ is an integrating factor of the given equation.

(b) Prove that if there is a continuous function $t \mapsto f(t)$ such that, for all $(x, y) \in D$, $\dfrac{\frac{\partial Q(x,y)}{\partial x} - \frac{\partial P(x,y)}{\partial y}}{P(x,y)} = f(y)$, then the function $\lambda(y) = e^{\int f(y) \, dy}$ is an integrating factor of the given equation.

Solution 2.2.36. Suppose that for all $(x, y) \in D$, $\dfrac{\frac{\partial P(x,y)}{\partial y} - \frac{\partial Q(x,y)}{\partial x}}{Q(x,y)} = f(x)$, for some continuous function f. For $\lambda(x) = e^{\int f(x) \, dx}$, consider the equation $\lambda(x)P(x, y) \, dx + \lambda(x)Q(x, y) \, dy = 0$. From, for all $(x, y) \in D$,

$$\frac{\partial}{\partial x} (\lambda(x)Q(x, y)) = \lambda'(x)Q(x, y) + \lambda(x)\frac{\partial Q(x, y)}{\partial x}$$

$$= e^{\int f(x) \, dx} \left(\frac{\frac{\partial P(x,y)}{\partial y} - \frac{\partial Q(x,y)}{\partial x}}{Q(x, y)} Q(x, y) + \frac{\partial Q(x, y)}{\partial x} \right)$$

$$= e^{\int f(x) \, dx} \cdot \frac{\partial P(x, y)}{\partial y} = \frac{\partial}{\partial y} (\lambda(x)P(x, y)),$$

it follows that the obtained equation is exact, i.e., $\lambda(x) = e^{\int f(x)\,dx}$ is an integrating factor of the equation $P(x,y)\,dx + Q(x,y)\,dy = 0$.

Problem 2.2.37. The function $f : \mathbb{R}^+ \to \mathbb{R}^+$ is the solution to the initial value problem $\left(1 + \ln\frac{y}{x}\right)\,dx - \frac{x}{y}\,dy = 0$, $y(1) = \frac{1}{e}$. Determine the Fourier sine transform of the function $F(x) = \begin{cases} f(x) & \text{if } x \in \mathbb{R}^+, \\ 0 & \text{if } x = 0. \end{cases}$

Solution 2.2.37. The set $D = \mathbb{R}^+ \times \mathbb{R}^+$ is a simply connected region in the xy-plane such that $\left(1, \frac{1}{e}\right) \in D$. Let $P(x,y) = 1 + \ln\frac{y}{x}$ and $Q(x,y) = -\frac{x}{y}$. From, for all $(x,y) \in D$,

$$\frac{\frac{\partial P}{\partial y} - \frac{\partial Q}{\partial x}}{Q(x,y)} = -\frac{2}{x},$$

by Problem 2.2.36, the equation has an integrating factor $\lambda = \lambda(x) = e^{-\int \frac{2}{x}\,dx} = \frac{1}{x^2}$. The equation $\frac{1}{x^2}\left(1 + \ln\frac{y}{x}\right)\,dx - \frac{1}{xy}\,dy = 0$, $(x,y) \in D$, is an exact differential equation, and the function $y = f(x)$ is determined by

$$0 = \int_1^x \frac{1}{t^2}\left(1 + \ln\frac{y}{t}\right)\,dt - \int_{\frac{1}{e}}^y \frac{dt}{t} = -\frac{1}{x}\ln\frac{y}{x} - 1, \quad (x,y) \in D.$$

It follows that $f(x) = xe^{-x}$, $x \in \mathbb{R}^+$.

The Fourier sine transform of the function F is, for all $u \in [0, \infty)$,[15]

$$\mathcal{F}_{\sin}[F(x)](u) = \sqrt{\frac{2}{\pi}}\int_0^\infty xe^{-x}\sin ux\,dx = -\sqrt{\frac{2}{\pi}}\frac{d}{du}\left(\int_0^\infty e^{-x}\cos ux\,dx\right)$$

$$= -\sqrt{\frac{2}{\pi}}\frac{d}{du}\left(\frac{1}{1+u^2}\right) = \sqrt{\frac{2}{\pi}}\frac{2u}{(1+u^2)^2}.$$

Problem 2.2.38. Let D be a simply connected region in the xy-plane, and let $F : D \to \mathbb{R}$ and $\frac{\partial F}{\partial y} : D \to \mathbb{R}$ be continuous functions. Prove that $y' = F(x,y)$ is a linear differential equation if and only if it has an integrating factor that is a function of x only.

---

[15]See Problems 1.4.5 and 1.4.7.

Solution 2.2.38. Suppose that a continuously differentiable function $x \mapsto f(x)$, $f(x) \neq 0$, $x \in \mathrm{Dom}(f)$, is an integrating factor of the equation $F(x,y)dx - dy = 0$, i.e., suppose that $f(x)F(x,y)dx - f(x)dy = 0$ is an exact differential equation. Then, for any $(x,y) \in D$, $\frac{\partial}{\partial y}(f(x)F(x,y)) = \frac{\partial}{\partial x}(-f(x))$. It follows that $\frac{\partial F(x,y)}{\partial y} = -\frac{f'(x)}{f(x)}$, which means that there is a continuous function $x \mapsto g(x)$ such that, $F(x,y) = -\frac{f'(x)}{f(x)} \, y + g(x)$, $(x,y) \in D$. Hence, $y' = F(x,y) = -\frac{f'(x)}{f(x)} \, y + g(x)$, $(x,y) \in D$, is a first-order linear differential equation.

Now, suppose that $y' = F(x,y)$ is a linear differential equation, i.e., suppose that $F(x,y) = p(x)y + q(x)$, $(x,y) \in D$, for some continuous functions $p(x)$ and $q(x)$. From

$$\frac{1}{-1}\left(\frac{\partial}{\partial y}(p(x)y + q(x)) - \frac{\partial}{\partial x}(-1)\right) = -p(x),$$

by Problem 2.2.36, it follows that $\lambda(x) = e^{-\int p(x)\,dx}$ is an integrating factor of the equation $(p(x)y + q(x))dx - dy = 0$.

Problem 2.2.39. Solve the initial value problem $xdx + (x^2 + y^2 + y)dy = 0$, $y(-1) = 0$.

Solution 2.2.39. Let $P(x,y) = x$ and $Q(x,y) = x^2 + y^2 + y$. Then, for any $(x,y) \in D = (-\infty, 0) \times \mathbb{R}$,

$$\frac{\frac{\partial Q(x,y)}{\partial x} - \frac{\partial P(x,y)}{\partial y}}{P(x,y)} = \frac{2x - 0}{x} = 2,$$

which, by Problem 2.2.36, implies that the function $y \mapsto \lambda(y) = e^{2\int dy} = e^{2y}$ is an integrating factor of the equation. The equation $xe^{2y}dx + (x^2 + y^2 + y)e^{2y}dy = 0$ is an exact differential equation in the simply connected region D. Since $(-1, 0) \in D$, the solution to the initial value problem is implicitly determined by

$$0 = \int_{-1}^{x} te^{2y}\,dt + \int_{0}^{y}(1 + t^2 + t)e^{2t}\,dt = \frac{1}{2}e^{2y}\left(x^2 + y^2\right) - \frac{1}{2}, \quad (x,y) \in D.$$

Note: The solution to the initial value problem can be written as $x = f(y) = -\sqrt{e^{-2y} - y^2}$. From $e^{-2y} - y^2 = (e^{-y} - y)(e^{-y} + y)$ and the fact $e^{-y} + y > 0$ for all $y \in \mathbb{R}$, it follows that the domain of the function f is the interval $(-\infty, W(1))$, where $W(1) \approx 0.5671432904$ is the value of the Lambert function W at 1.[16]

2.3   Higher-Order Differential Equations

Problem 2.3.1. Solve the initial value problem $y''(1+y^2) = 2y(y')^2$, $y(0) = y'(0) = 1$.

Solution 2.3.1. The problem is equivalent to the initial value problem

$$0 = \frac{y''(1+y^2) - 2y(y')^2}{(1+y^2)^2} = \left(\frac{y'}{1+y^2}\right)', \quad y(0) = y'(0) = 1.$$

The question is to find a function $y = f(x)$, $f(0) = f'(0) = 1$ such that in its domain, for a certain $c \in \mathbb{R}$, $\frac{f'(x)}{1+(f(x))^2} = c$. From $c = \frac{f'(0)}{1+(f(0))^2} = \frac{1}{2}$, it follows that the function f is the *unique* solution to the initial value problem $\frac{y'}{1+y^2} = \frac{1}{2}$, $y(0) = 1$. This is a separable differential equation, and the solution to the initial value problem in a neighbourhood of the point $(0,1)$ is determined by $\int_1^y \frac{dt}{1+t^2} = \frac{1}{2}\int_0^x dt$, i.e., by $\arctan y - \frac{\pi}{4} = \frac{x}{2}$. Therefore, $f(x) = \tan\left(\frac{x}{2} + \frac{\pi}{4}\right)$, $x \in \left(-\frac{3\pi}{2}, \frac{\pi}{2}\right)$.

Note: For $y' \neq 0$, the equation $y''(1+y^2) = 2y(y')^2$ is equivalent to the equation $\frac{y''}{y'} = \frac{2yy'}{1+y^2}$, which implies that $\ln|y'| = \ln c(1+y^2)$, $c > 0$, i.e., $y' = c(1+y^2)$, $c \in \mathbb{R}\setminus\{0\}$.

Problem 2.3.2. Solve the initial value problem $y'y''' = 2(y'')^2$, $y(0) = y'(0) = 1$, and $y''(0) = -1$.

Solution 2.3.2. In a neighbourhood of the point $(0, 1, 1, -1)$, the given initial value problem is equivalent to the problem

$$\left(\frac{y''}{y'}\right)^2 = \frac{y'y''' - (y'')^2}{(y')^2} = \left(\frac{y''}{y'}\right)', \quad y(0) = y'(0) = 1, \quad \text{and } y''(0) = -1.$$

---

[16] Johann Heinrich Lambert, 1728–1777, Swiss polymath.

The substitution $u = \frac{y''}{y'}$ yields the initial value problem $u' = u^2$, $u(0) = \frac{y''(0)}{y'(0)} = -1$. This is a separable differential equation. The solution to the initial value problem is $\int_{-1}^{u} \frac{dt}{t^2} = \int_{0}^{x} dt$, which implies that $u = u(x) = -\frac{1}{x+1}$. It follows that $\frac{y''}{y'} = -\frac{1}{x+1}$, i.e., $((x+1)y')' = 0$. Since $y(0) = y'(0) = 1$, the required solution is $y = \ln(x+1) + 1$, $x \in (-1, \infty)$.

Problem 2.3.3. Let $I \subseteq \mathbb{R}$ be an interval, and let $f : I \to \mathbb{R}$ and $g : I \to \mathbb{R}$ be differentiable functions such that their Wronskian $W(f, g)(x)$, $x \in I$, is not the zero function.[17] Establish that the functions f and g are linearly independent.

Solution 2.3.3. Let c_1, $c_2 \in \mathbb{R}$ be such that, for all $x \in I$, $c_1 f(x) + c_2 g(x) = 0$. Then, for all $x \in I$, $c_1 f'(x) + c_2 g'(x) = 0$. For $x_0 \in I$ such that $W(f, g)(x_0) \neq 0$, the homogeneous system of linear equations with the unknowns c_1 and c_2,

$$c_1 f(x_0) + c_2 g(x_0) = 0 \quad \text{and} \quad c_1 f'(x_0) + c_2 g'(x_0) = 0,$$

has only a trivial solution: $c_1 = c_2 = 0$. Therefore, the functions f and g are linearly independent.

Problem 2.3.4. Establish whether or not the functions $f(x) = \arctan x$, $g(x) = \text{arccot}x$, and $h(x) = 1$, $x \in \mathbb{R}$ are linearly independent.

Solution 2.3.4. Since

$$W(f, g, h)(x) = \begin{vmatrix} \arctan x & \text{arccot}x & 1 \\ \frac{1}{1+x^2} & -\frac{1}{1+x^2} & 0 \\ -\frac{2x}{(1+x^2)^2} & \frac{2x}{(1+x^2)^2} & 0 \end{vmatrix} = 0,$$

for any $x \in \mathbb{R}$, the value of the Wronskian does not provide the answer to the question. Since, for any $x \in \mathbb{R}$, $\text{arccot}x = \frac{\pi}{2} - \arctan x$,

---

[17]This assumption implies that neither f nor g is the zero function.

it follows that $f(x) + g(x) - \frac{\pi}{2}h(x) = 0$, i.e., the functions f, g, and h are linearly dependent.

Problem 2.3.5. Establish whether or not the functions

$$f(x) = \begin{cases} -x^3 & \text{if } x \leq 0, \\ 0 & \text{if } x > 0, \end{cases} \quad \text{and} \quad g(x) = \begin{cases} 0 & \text{if } x \leq 0, \\ x^3 & \text{if } x > 0, \end{cases}$$

are linearly independent.

Solution 2.3.5. From

$$f'(x) = \begin{cases} -3x^2 & \text{if } x \leq 0, \\ 0 & \text{if } x > 0, \end{cases} \quad \text{and} \quad g'(x) = \begin{cases} 0 & \text{if } x \leq 0, \\ 3x^2 & \text{if } x > 0, \end{cases}$$

it follows that both functions f and g are continuously differentiable in $\mathbb{R}$. Their Wronskian is

$$W(f, g)(x) = \begin{vmatrix} f(x) & g(x) \\ f'(x) & g'(x) \end{vmatrix} = f(x)g'(x) - f'(x)g(x) = 0,$$

for all $x \in \mathbb{R}$. If c_1 and c_2 are two real numbers such that $c_1 f(x) + c_2 g(x) = 0$, for all $x \in \mathbb{R}$, then $0 = c_1 f(1) + c_2 g(1) = c_2$ and $0 = c_1 f(-1) + c_2 g(-1) = -c_1$. Hence, $c_1 = c_2 = 0$, which means that the functions f and g are linearly independent.

Problem 2.3.6. Suppose that p and q are continuous functions in the interval I and that y_1 and y_2 are two solutions to the equation $y'' + p(x)y' + q(x)y = 0$, $x \in I$. Use the existence and uniqueness theorem for second-order linear differential equations to prove that y_1 and y_2 are linearly independent solutions if and only if, for all $x \in I$, $W(y_1, y_2)(x) \neq 0$.

Solution 2.3.6. By Problem 2.3.3, if, for all $x \in I$, $W(y_1, y_2)(x) \neq 0$, then y_1 and y_2 are linearly independent solutions to the equation.

Now, suppose that y_1 and y_2 are two linearly independent solutions[18] and that there is $x_0 \in I$ such that $W(y_1, y_2)(x_0) = 0$. This implies that the system of linear equations with unknowns c_1 and c_2,

$$c_1 y_1(x_0) + c_2 y_2(x_0) = 0 \quad \text{and} \quad c_1 y_1'(x_0) + c_2 y_2'(x_0) = 0,$$

has a non-trivial solution, i.e., there are $a, b \in \mathbb{R}$ such that $a^2 + b^2 \neq 0$, and $a y_1(x_0) + b y_2(x_0) = a y_1'(x_0) + b y_2'(x_0) = 0$. It follows that the

---

[18]This assumption implies that neither y_1 nor y_2 is the zero function.

function $f(x) = ay_1(x) + by_2(x)$, $x \in I$, is a solution to the initial value problem $y'' + p(x)y' + q(x)y = 0$, $y(x_0) = y'(x_0) = 0$. However, since the zero function is a solution to this initial value problem, by the existence and uniqueness theorem for second-order linear differential equations, this means that, for all $x \in I$, $f(x) = ay_1(x) + by_2(x) = 0$. Since $a^2 + b^2 \neq 0$, this contradicts the assumption that y_1 and y_2 are linearly independent solutions. Hence, $W(y_1, y_2)(x) \neq 0$, for all $x \in I$.

Problem 2.3.7. Suppose that p and q are continuous functions in the interval I and that y_1 and y_2 are two solutions to the equation $y'' + p(x)y' + q(x)y = 0$, $x \in I$. Prove that, for a fixed $a \in I$, there is $C \in \mathbb{R}$ such that $W(y_1, y_2)(x) = Ce^{-\int_a^x p(t)\, dt}$, for all $x \in I$.[19]

Solution 2.3.7. From $W(y_1, y_2)(x) = y_1 y_2' - y_1' y_2$, it follows that

$$
\begin{aligned}
W'(y_1, y_2)(x) &= y_1' y_2' + y_1 y_2'' - y_1'' y_2 - y_1' y_2' \\
&= y_1 y_2'' - y_1'' y_2 \\
&= y_1(-p(x)y_2' - q(x)y_2) - y_2(-p(x)y_1' - q(x)y_1) \\
&= -p(x)W(y_1, y_2)(x).
\end{aligned}
$$

If y_1 and y_2 are linearly dependent solutions, then $W(y_1, y_2)(x) = 0$ and $C = 0$. If y_1 and y_2 are linearly independent solutions, then, by Problem 2.3.6, $W(y_1, y_2)(x)$ is a nowhere-zero solution to the separable differential equation $z' = -p(x)z$. Hence, for a fixed $a \in I$, there is $C \in \mathbb{R}\backslash\{0\}$ such that $W(y_1, y_2)(x) = Ce^{-\int_a^x p(t)\, dt}$, for all $x \in I$.

Problem 2.3.8. Suppose that p and q are continuous functions in the interval I and that $y_1 = y_1(x) > 0$, $x \in I$, is a solution to the equation $y'' + p(x)y' + q(x)y = 0$. Determine $y_2 = y_2(x)$, $x \in I$, a solution to the equation that is linearly independent from y_1.

Solution 2.3.8. We take $C = 1$ in Abel's formula, referring to Problem 2.3.7, to determine y_2 for which $y_1(x)y_2'(x) - y_1'(x) \cdot$

---

[19]This fact, a special case of the Liouville–Ostrogradski formula, is known as Abel's formula. Niels Henrik Abel, 1802–1829, was a Norwegian mathematician; Joseph Liouville, 1809–1882, was a French mathematician and engineer; and Mikhail Vasilyevich Ostrogradsky, 1801–1862, was an Ukrainian mathematician and physicist.

$y_2(x) = e^{-\int_a^x p(t)\,dt}$, $a \in I$. Since $y_1(x) \neq 0$, $x \in I$, it follows that y_2 is a solution to the first-order linear differential equation

$$y' - \frac{y_1'(x)}{y_1(x)}\,y = \frac{1}{y_1(x)}e^{-\int_a^x p(t)\,dt},$$

where a is a fixed number in I. By Problem 2.2.5, the general solution to this equation is

$$y = e^{\int_a^x \frac{y_1'(t)}{y_1(t)}\,dt}\left(c + \int_a^x \frac{1}{y_1(u)}e^{-\int_a^u p(t)\,dt}e^{-\int_a^u \frac{y_1'(t)}{y_1(t)}\,dt}\,du\right)$$

$$= \frac{y_1(x)}{y_1(a)}\left(c + y_1(a)\int_a^x \frac{1}{y_1^2(u)}e^{-\int_a^u p(t)\,dt}\,du\right)$$

$$= y_1(x)\left(\frac{c}{y_1(a)} + \int_a^x \frac{1}{y_1^2(u)}e^{-\int_a^u p(t)\,dt}\,du\right).$$

For $c = 0$, we obtain $y_2(x) = y_1(x)\int_a^x \frac{1}{y_1^2(u)}e^{-\int_a^u p(t)\,dt}\,du$, a solution that is linearly independent from y_1.

Note: It is common to write $y_2 = y_1 \int \frac{e^{-\int p(x)\,dx}}{y_1^2}\,dx$. We refer to this expression as *Liouville's formula*.

Problem 2.3.9. Suppose that p, q, r, and s are continuous functions in the interval I. Establish the *principle of superposition*: if a function f is a solution to the equation $y'' + p(x)y' + q(x)y = r(x)$, $x \in I$, and if a function g is a solution to the equation $y'' + p(x)y' + q(x)y = s(x)$, $x \in I$, then the function $f + g$ is a solution to the equation $y'' + p(x)y' + q(x)y = r(x) + s(x)$, $x \in I$.

Solution 2.3.9. From $f''(x) + p(x)f'(x) + q(x)f(x) = r(x)$ and $g''(x) + p(x)g'(x) + q(x)g(x) = s(x)$, it follows that, for all $x \in I$,

$$r(x) + s(x) = f''(x) + g''(x) + p(x)(f'(x) + g'(x)) + q(x)(f(x) + g(x)),$$

i.e., it follows that $f + g$ is a solution to the equation $y'' + p(x)y' + q(x)y = r(x) + s(x)$, $x \in I$.

Problem 2.3.10. Suppose that p and q are continuous functions in the interval I. Let y_1 and y_2 be two solutions to the equation

$y'' + p(x)y' + q(x)y = 0$ with the property that they attain a local extremum at the same point $c \in I$. Are the functions y_1 and y_2 linearly independent or not?

Solution 2.3.10. The Wronskian of the functions y_1 and y_2 is, for any $x \in I$, $W(y_1, y_2)(x) = y_1 y_2' - y_1' y_2$. Since both y_1 and y_2 attain a local extremum at the point c, it follows that $y_1'(c) = y_2'(c) = 0$, i.e., $W(y_1, y_2)(c) = 0$ and, consequently, referring to Problem 2.3.7, $W(y_1, y_2)(x) = 0$, for all $x \in \mathbb{R}$. The functions y_1 and y_2 are linearly dependent solutions to the homogeneous differential equation.

Problem 2.3.11. Consider the equation $y'' + p(x)y' + q(x)y = 0$, $x \in I$, where p and q are continuous functions. Prove the following statements:

(a) The quotient of any two linearly independent solutions is a function with no local extremum.
(b) If I is an open interval such that $q(x) < 0$ for all $x \in I$, then no solution has a positive local maximum in I.

Solution 2.3.11.

(a) Let y_1 and y_2 be two linearly independent solutions to the equation. This implies that neither y_1 nor y_2 is the zero function and that $W(y_1, y_2)(x) \neq 0$, for all $x \in I$. Let $J \subset I$ be such that $y_2(x) \neq 0$, for all $x \in J$. From, for all $x \in J$,

$$\left(\frac{y_1(x)}{y_2(x)} \right)' = \frac{y_1(x)y_2'(x) - y_1'(x)y_2(x)}{(y_2(x))^2} = \frac{W(y_1, y_2)(x)}{(y_2(x))^2} \neq 0,$$

it follows that the function $\frac{y_1}{y_2}$ has no local extremum.
(b) Let $y = f(x)$, $x \in I$, be a solution to the equation. Suppose that the function f attains a local maximum at $c \in I$. Then, $f'(c) = 0$ and $f''(c) \leq 0$. From $f''(c) + p(c)f'(c) + q(c)f(c) = 0$, it follows that $f(c) = -\frac{f''(c)}{q(c)} \leq 0$.

Problem 2.3.12. Suppose that p and q are continuous functions in an interval I. Suppose that y_1 and y_2 are two linearly independent solutions to the homogeneous equation $y'' + p(x)y' + q(x)y = 0$. Use the existence and uniqueness theorem for second-order linear differential equations to prove that any solution $y = f(x)$ to the homogeneous equation is a linear combination of y_1 and y_2, i.e., there are c_1, $c_2 \in \mathbb{R}$ such that $f(x) = c_1 y_1(x) + c_2 y_2(x)$.

Solution 2.3.12. We first demonstrate that any linear combination of y_1 and y_2 is a solution to the homogeneous equation. Let c_1, $c_2 \in \mathbb{R}$, and let $f(x) = c_1 y_1(x) + c_2 y_2(x)$, $x \in I$. Then, for any $x \in I$,

$$f''(x) + p(x)f'(x) + q(x)f(x) = (c_1 y_1''(x) + c_2 y_2''(x))$$
$$+ p(x)(c_1 y_1'(x) + c_2 y_2'(x)) + q(x)(c_1 y_1(x) + c_2 y_2(x))$$
$$= c_1(y_1'' + p(x)y_1' + q(x)y_1) + c_2(y_2'' + p(x)y_2' + q(x)y_2) = 0,$$

which establishes that $f(x) = c_1 y_1(x) + c_2 y_2(x)$ is a solution to the homogeneous equation.

Now, let $y = f(x)$, $x \in I$, be a nonzero solution to the equation. Let $\alpha \in I$ be such that $f(\alpha) \neq 0$. Since $W(y_1, y_2)(\alpha) \neq 0$, the non-homogeneous system of linear equations with the unknowns a and b,

$$ay_1(\alpha) + by_2(\alpha) = f(\alpha) \quad \text{and} \quad ay_1'(\alpha) + by_2'(\alpha) = f'(\alpha),$$

has a unique solution: $a = c_1$ and $b = c_2$, $c_1^2 + c_2^2 \neq 0$. This implies that the function $g(x) = c_1 y_1(x) + c_2 y_2(x)$, $x \in I$, is a solution to the initial value problem $y'' + p(x)y' + q(x)y = 0$, $y(\alpha) = f(\alpha)$, $y'(\alpha) = f'(\alpha)$. By the existence and uniqueness theorem for second-order linear differential equations, $f(x) = g(x) = c_1 y_1(x) + c_2 y_2(x)$, $x \in I$, which completes the proof.

Problem 2.3.13. Suppose that p, q, and r are continuous functions in an interval I, where r is not the zero function. Suppose that y_1 and y_2 are two linearly independent solutions to the homogeneous equation and that y_p is one solution to the non-homogeneous equation $y'' + p(x)y' + q(x)y = r(x)$. Prove that any solution $y = F(x)$, $x \in I$, to the non-homogeneous equation is the sum of y_p and a solution to the corresponding homogeneous equation, i.e., there are c_1, $c_2 \in \mathbb{R}$ such that $F(x) = y_p(x) + c_1 y_1(x) + c_2 y_2(x)$, $x \in I$.

Solution 2.3.13. Let $y = F(x)$, $x \in I$, be a solution to the non-homogeneous equation equation. Let $G(x) = F(x) - y_p(x)$, $x \in I$.

Then, for all $x \in I$,

$$G''(x) + p(x)G'(x) + q(x)G(x) = F''(x) + p(x)F'(x) + q(x)F(x)$$
$$- (y_p''(x) + p(x)y_p'(x) + q(x)y_p(x)) = r(x) - r(x) = 0.$$

Since the function G is a solution to the homogeneous equation, by Problem 2.3.12, there are c_1, $c_2 \in \mathbb{R}$ such that $G(x) = c_1 y_1(x) + c_2 y_2(x)$, $x \in I$. Therefore, $F(x) = y_p + c_1 y_1(x) + c_2 y_2(x)$, $x \in I$.

Problem 2.3.14. Solve the initial value problem $x^2(\ln x - 1)y'' - xy' + y = 0$, $y(e^2) = 3e^2$, $y'(e^2) = 2$ by using the fact that the function $y_1(x) = x$ is one solution to this equation.

Solution 2.3.14. Let $D = (e, \infty) \times \mathbb{R}^+ \times \mathbb{R}$. Then, $T_0 = (e^2, 3e^2, 2) \in D$. In the simply connected region D, the problem can be rewritten as

$$y'' - \frac{1}{x(\ln x - 1)}y' + \frac{1}{x^2(\ln x - 1)}y = 0, \quad y(e^2) = 3e^2, \ y'(e^2) = 2.$$

By Liouville's formula, a solution to this homogeneous second-order linear equation that is linearly independent of the solution $y_1(x) = x$ is determined by

$$y_2(x) = x \int \frac{e^{\int \frac{dx}{x(\ln x - 1)}}}{x^2} \, dx = - \ln x, \quad x \in (e, \infty).$$

It remains to determine the constants c_1 and c_2 so that the function $f(x) = c_1 x + c_2 \ln x$ is such that $3e^2 = f(e^2) = e^2 c_1 + 2c_2$ and $2 = f'(e^2) = c_1 + c_2 e^{-2}$. It follows that $c_1 = 1$, $c_2 = e^2$, and $f(x) = x + e^2 \ln x$, $x \in \mathbb{R}^+$.

Note: Observe that Louiville's formula yields a solution with the domain (e, ∞).

Problem 2.3.15. Let $m \in \mathbb{R}\setminus\{0\}$. Determine real numbers a, b, and c so that the function $y_1 = e^{mx}$ is a solution to the equation $(1 + x^2)y'' - 2xy' + (ax^2 + bx + c)y = 0$. For those values of a, b, and c, find another solution to the equation that is linearly independent of the solution y_1.

Answer 2.3.15. $a = c = -m^2$, $b = 2m$, and $y_2 = (2m^2 x^2 + 2mx + 2m^2 + 1)e^{-mx}$.

Problem 2.3.16. Let $I \subseteq \mathbb{R}$ be an interval, and let p, q, and r be functions continuous in I. Consider a non-homogeneous second-order linear differential equation $y'' + p(x)y' + q(x)y = r(x)$, and suppose that $y_h = c_1 y_1 + c_2 y_2$, $c_1, c_2 \in \mathbb{R}$, is the general solution to the corresponding homogeneous equation. Prove that there is a particular solution to the non-homogeneous equation in the form $y_p = \varphi(x)y_1 + \psi(x)y_2$, with the functions $x \mapsto \varphi(x)$ and $x \mapsto \psi(x)$ determined by the following system of functional equations:

$$\varphi'(x)y_1(x)+\psi'(x)y_2(x) = 0 \text{ and } \varphi'(x)y_1'(x)+\psi'(x)y_2'(x) = r(x), \ x \in I.$$

Solution 2.3.16. From $y_p = \varphi(x)y_1 + \psi(x)y_2$, it follows that

$$y_p' = \varphi'(x)y_1 + \varphi(x)y_1' + \psi'(x)y_2 + \psi(x)y_2'.$$

We restrict our search to those functions φ and ψ for which $\varphi'(x)y_1(x) + \psi'(x)y_2(x) = 0$, for all $x \in I$, i.e., we look for a solution $y_p = \varphi(x)y_1 + \psi(x)y_2$ to the non-homogeneous equation so that $y_p' = \varphi(x)y_1' + \psi(x)y_2'$ and $y_p'' = \varphi'(x)y_1' + \varphi(x)y_1'' + \psi'(x)y_2' + \psi(x)y_2''$. The solution y_p must be such that, for all $x \in I$,

$$\begin{aligned}
r(x) = y_p'' + p(x)y_p' + q(x)y_p &= \varphi'(x)y_1' + \varphi(x)y_1'' + \psi'(x)y_2' + \psi(x)y_2'' \\
&\quad + p(x)\left(\varphi(x)y_1' + \psi(x)y_2'\right) + q(x)(\varphi(x)y_1 + \psi(x)y_2) \\
&= \varphi(x)(y_1'' + p(x)y_1' + q(x)y_1) + \psi(x)(y_2'' + p(x)y_2' + q(x)y_2) \\
&\quad + \varphi'(x)y_1' + \psi'(x)y_2' = \varphi(x)\cdot 0 + \psi(x)\cdot 0 + \varphi'(x)y_1' + \psi'(x)y_2' \\
&= \varphi'(x)y_1' + \psi'(x)y_2'.
\end{aligned}$$

Therefore, if the derivatives of the functions φ and ψ satisfy the system of functional equations

$$\varphi'(x)y_1(x) + \psi'(x)y_2(x) = 0 \text{ and } \varphi'(x)y_1'(x) + \psi'(x)y_2'(x) = r(x), \ x \in I,$$

then the function $y_p = \varphi(x)y_1 + \psi(x)y_2$ is a solution to the non-homogeneous equation.

Note 1: Since the determinant of the non-homogeneous linear system is the Wronskian $W(y_1, y_2)(x) \neq 0$, $x \in I$, the system has a unique solution.

Note 2: Problem 2.3.16 establishes the Lagrange method of variation of parameters.

Problem 2.3.17. Solve the initial value problem $(2+x)^2 y'' - 3(2+x)y' + 4y = (2+x)^2$, $y(-3) = 1$, $y'(-3) = -2$, by using the fact that there is $r \in \mathbb{R}$ such that the function $y_1 = (2+x)^r$ is a solution to the corresponding homogeneous equation.

Solution 2.3.17. The point $T_0 = (-3, 1, -2)$ lies in the simply connected region $D = \{(x, y, y') : (x, y, y') \in (-\infty, -2) \times \mathbb{R} \times \mathbb{R}\}$. In the region D, the given equation is equivalent to the non-homogeneous second-order linear equation $y'' - \frac{3}{2+x}y' + \frac{4}{(2+x)^2}y = 1$. The corresponding homogeneous equation is $y'' - \frac{3}{2+x}y' + \frac{4}{(2+x)^2}y = 0$. The function $y_1 = (2+x)^r$ is a solution to this equation if and only if, for all $x \in (-\infty, -2)$,

$$r(r-1)(2+x)^{r-2} - \frac{3}{2+x}\, r(2+x)^{r-1} + \frac{4}{(2+x)^2}(2+x)^r = 0,$$

i.e., if $r = 2$.

By Liouville's formula, another solution to the homogeneous equation, linearly independent of y_1, is the function $y_2 = (2+x)^2 \ln|2+x|$, $x \in (-\infty, -2)$.

To determine the differentiable functions $x \mapsto \varphi(x)$ and $x \mapsto \psi(x)$ so that the function $y_p = \varphi(x)y_1 + \psi(x)y_2$ is a solution to the non-homogeneous equation, by the Lagrange method of variation of parameters, we solve the following system of functional equations:

$$\varphi'(x)y_1(x) + \psi'(x)y_2(x) = 0 \text{ and } \varphi'(x)y_1'(x) + \psi'(x)y_2'(x) = 1.$$

From, for all $x \in (-\infty, -2)$,

$$\varphi'(x)(2+x)^2 + \psi'(x)(2+x)^2 \ln|2+x| = 0$$

$$2\varphi'(x)(2+x) + \psi'(x)(2(2+x)\ln|2+x| + 2+x) = 1,$$

it follows that $\varphi(x) = -\frac{1}{2}\ln^2|2+x|$ and $\psi(x) = \ln|2+x|$, $x \in (-\infty, -2)$. A particular solution to the non-homogeneous equation is $y_p = \frac{1}{2}(2+x)^2 \ln^2|2+x|$, $x \in (-\infty, -2)$. There are $c_1, c_2 \in \mathbb{R}$ such that the solution to the initial value problem is a function,

$$f(x) = c_1 y_1(x) + c_2 y_2(x) + y_p(x)$$

$$= \left(c_1 - c_2 \ln|2+x| + \frac{1}{2}\ln^2|2+x|\right)(2+x)^2,$$

with $f(-3) = 1$ and $f'(-3) = -2$. It follows that $c_1 = 1$ and $c_2 = 0$. The solution to the initial value problem is $f(x) = \left(1 + \frac{1}{2}\ln^2|2+x|\right)(2+x)^2$, $x \in (-\infty, -2)$.

Problem 2.3.18. Solve the initial value problem $x^2 y'' - xy' - 3y = 4x^2 \ln x$, $y(1) = -1$, $y'(1) = -3$, by using the fact that the corresponding homogeneous equation has a polynomial as a solution.

Solution 2.3.18. Let $D = \{(x, y, y') : (x, y, y') \in \mathbb{R}^+ \times \mathbb{R} \times \mathbb{R}\}$. In the region D, the equation is equivalent to the non-homogeneous second-order linear equation $y'' - \frac{1}{x}y' - \frac{3}{x^2}y = 4 \ln x$. Suppose that the polynomial $y_1 = \sum_{k=0}^{n} a_k x^k$, $a_n = 1$, is a solution to the corresponding homogeneous equation $y'' - \frac{1}{x}y' - \frac{3}{x^2}y = 0$. Then, for all $x \in \mathbb{R}^+$,

$$0 = \sum_{k=2}^{n} k(k-1)a_k x^{k-2} - \frac{1}{x}\sum_{k=1}^{n} k a_k x^{k-1} - \frac{3}{x^2}\sum_{k=0}^{n} a_k x^k$$

$$= \frac{1}{x^2}\sum_{k=0}^{n}(k+1)(k-3)a_k x^k,$$

which implies that, for all $k \in \{0, 1, 2, \ldots, n\}$, $(k+1)(k-3)a_k = 0$. In particular, since $(n+1)(n-3) = 0$, the degree of the polynomial is $n = 3$. It follows that $a_0 = a_1 = a_2 = 0$ and $y_1 = x^3$. By Liouville's formula, $y_2 = -\frac{1}{4x}$. By the Lagrange method of variation of parameters, a particular solution to the non-homogeneous equation is $y_p = -x^2(\ln x + 1) - \frac{x^2}{9}(3 \ln x - 1) = -\frac{4x^2}{9}(3 \ln x + 2)$. There are $c_1, c_2 \in \mathbb{R}$ such that the solution to the initial value problem is $f(x) = c_1 x^3 + \frac{c_2}{x} - \frac{4x^2}{9}(3 \ln x + 2)$, $x \in \mathbb{R}^+$. From the initial conditions, it follows that $c_1 = 0$, $c_2 = -\frac{1}{9}$, and $f(x) = -\frac{1}{9x} - \frac{4x^2}{9}(3 \ln x + 2)$, $x > 0$.

Problem 2.3.19. Solve the initial value problem $x^2 y'' - xy' - 3y = 5x$, $y(-1) = 2$, $y'(-1) = 0$, by using the fact that the function $y_1 = \frac{1}{x}$ is a solution to the corresponding homogeneous equation.

Answer 2.3.19. $y = \frac{x^3}{8} - \frac{7}{8x} - \frac{5x}{4}$, $x \in (-\infty, 0)$.

Problem 2.3.20. Let $a, b \in \mathbb{R}$. Solve $y'' + ay' + by = 0$, a homogeneous second-order linear differential equation with constant coefficients.

Solution 2.3.20. Let r be a root, real or complex, of the quadratic equation[20] $t^2 + at + b = 0$, and let $f(x) = e^{rx}$, $x \in \mathbb{R}$. Then, for all $x \in \mathbb{R}$,

$$f''(x) + af'(x) + bf(x) = e^{rx}(r^2 + ar + b) = 0,$$

which implies that f is a solution to the differential equation. We distinguish the following three cases:

(a) If $a^2 - 4b > 0$, then the quadratic equation has two distinct real roots: r_1 and r_2. Both $y_1 = y_1(x) = e^{r_1 x}$ and $y_2 = y_2(x) = e^{r_2 x}$ are solutions to the differential equation. Since, for all $x \in \mathbb{R}$,

$$W(y_1, y_2)(x) = \begin{vmatrix} e^{r_1 x} & e^{r_2 x} \\ r_1 e^{r_1 x} & r_2 e^{r_2 x} \end{vmatrix} = (r_2 - r_1)e^{(r_1 + r_2)x} \neq 0,$$

solutions y_1 and y_2 are linearly independent. The general solution to the equation is $y = c_1 e^{r_1 x} + c_2 e^{r_2 x}$, $c_1, c_2 \in \mathbb{R}$.

(b) If $a^2 - 4b = 0$, then the quadratic equation has a single real root $r = -\frac{a}{2}$. The function $y_1 = e^{-\frac{ax}{2}}$, $x \in \mathbb{R}$, is a solution to the differential equation. By Liouville's formula, another solution, linearly independent from y_1, is determined by

$$y_2 = e^{-\frac{ax}{2}} \int e^{-ax} e^{ax} \, dx = xe^{-\frac{ax}{2}}, \quad x \in \mathbb{R}.$$

The general solution to the equation is $y = (c_1 + c_2 x)e^{-\frac{ax}{2}}$, $c_1, c_2 \in \mathbb{R}$.

(c) If $a^2 - 4b < 0$, i.e., $0 \leq a^2 < 4b$, then the quadratic equation has two complex roots, $r_1 = \alpha + i\beta$ and $r_2 = \alpha - i\beta$, where $\alpha = \mathrm{Re}(r_1) = -\frac{a}{2}$ and $\beta = \mathrm{Im}(r_1) = \sqrt{b - \frac{a^2}{4}}$. As observed earlier, two linearly independent solutions to the equation are

$$y_1 = e^{r_1 x} = e^{\alpha x}(\cos \beta x + i \sin \beta x), \quad x \in \mathbb{R},$$

and

$$y_2 = e^{r_2 x} = e^{\alpha x}(\cos \beta x - i \sin \beta x), \quad x \in \mathbb{R}.$$

---

[20] This is the *characteristic equation* of the differential equation $y'' + ay' + by = 0$.

Two linearly independent real-valued solutions are obtained in the following way:

$$Y_1 = \frac{y_1 + y_2}{2} = e^{\alpha x} \cos \beta x \quad \text{and} \quad Y_2 = \frac{y_1 - y_2}{2i} = e^{\alpha x} \sin \beta x, \ x \in \mathbb{R}.$$

The general solution to the equation is $y = (c_1 \cos \beta x + c_2 \sin \beta x) \cdot e^{\alpha x}$, $c_1, c_2 \in \mathbb{R}$.

Problem 2.3.21. Let $\lambda \in \mathbb{R}^+$. Solve the initial value problem $(1 - x^2)y'' - xy' + \lambda^2 y = 0$, $y(0) = 0$, $y'(0) = \lambda$, by using the substitution $x = \cos t$, $t \in (0, \pi)$.

Solution 2.3.21. From, for $t \in (0, \pi)$,

$$y' = \frac{dy}{dx} = \frac{dy}{dt}\frac{1}{\frac{dx}{dt}} = -\frac{1}{\sin t}\frac{dy}{dt} \quad \text{and} \quad y'' = -\frac{1}{\sin^2 t}\left(\frac{dy}{dt}\cot t - \frac{d^2y}{dt^2}\right),$$

it follows that the substitution $x = \cos t$, $t \in (0, \pi)$, transforms the initial value problem into the initial value problem $-\frac{1}{\sin^2 t}\left(\frac{dy}{dt}\cot t - \frac{d^2y}{dt^2}\right)(1 - \cos^2 t) - \left(-\frac{1}{\sin t}\frac{dy}{dt}\right)\cos t + \lambda^2 y = 0$, $y\left(\frac{\pi}{2}\right) = 0$, $\frac{dy}{dt}\Big|_{t=\frac{\pi}{2}} = -\lambda$, i.e., into the initial value problem $\frac{d^2y}{dt^2} + \lambda^2 y = 0$, $y\left(\frac{\pi}{2}\right) = 0$, $\frac{dy}{dt}\Big|_{t=\frac{\pi}{2}} = -\lambda$.

The general solution to the homogeneous second-order linear equation with constant coefficients $\frac{d^2y}{dt^2} + \lambda^2 y = 0$ is $y = c_1 \cos \lambda t + c_2 \sin \lambda t$, $c_1, c_2 \in \mathbb{R}$. From $y\left(\frac{\pi}{2}\right) = 0$ and $\frac{dy}{dt}\Big|_{t=\frac{\pi}{2}} = -\lambda$, it follows that the required solution to the initial value problem is determined by the constants c_1 and c_2 that satisfy the system of equations

$$c_1 \cos\frac{\lambda\pi}{2} + c_2 \sin\frac{\lambda\pi}{2} = 0 \quad \text{and} \quad \lambda\left(-c_1 \sin\frac{\lambda\pi}{2} + c_2 \cos\frac{\lambda\pi}{2}\right) = -\lambda.$$

Hence, $c_1 = \sin\frac{\lambda\pi}{2}$ and $c_2 = -\cos\frac{\lambda\pi}{2}$ and the solution to the initial value problem is the function $y = \sin\frac{\lambda(\pi - 2t)}{2}$, $t \in (0, \pi)$. The solution to the given initial value problem is the function $y = \sin\frac{\lambda(\pi - 2\arccos x)}{2}$, $x \in (-1, 1)$.

Problem 2.3.22. Determine a necessary and sufficient condition so that every solution to the homogeneous second-order linear differential equation with constant coefficients $y'' + ay' + by = 0$, $a, b \in \mathbb{R}$, has an infinite number of zeros.

Solution 2.3.22. If the characteristic equation $r^2 + ar + b = 0$ has a real root, λ, then the function $y = e^{\lambda x}$ is a solution to the equation with no zeros. Hence, for every solution to $y'' + ay' + by = 0$ to have an infinite number of zeros, it is necessary that $a^2 - 4b < 0$. Suppose that real numbers a and b are such that $a^2 - 4b < 0$. Then, there are $\alpha, \beta \in \mathbb{R}$, $\beta \neq 0$, such that $\lambda = \alpha \pm i\beta$ are roots of the characteristic equation, i.e., the general solution to the differential equation is $y = (c_1 \cos \beta x + c_2 \sin \beta x)\, e^{\alpha x}$, $c_1, c_2 \in \mathbb{R}$. Suppose that c_1 and c_2 are such that $c_1^2 + c_2^2 \neq 0$. Let $\varphi \in \left(-\frac{\pi}{2}, \frac{\pi}{2}\right)$ be such that $\sin \varphi = \frac{c_1}{\sqrt{c_1^2 + c_2^2}}$ and $\cos \varphi = \frac{c_2}{\sqrt{c_1^2 + c_2^2}}$. The set of all zeros of the function $y = (c_1 \cos \beta x + c_2 \sin \beta x)\, e^{\alpha x} = \sqrt{c_1^2 + c_2^2}\, \sin(\varphi + \beta x)\, e^{\alpha x}$ is $M = \left\{ \frac{k\pi - \varphi}{\beta} : k \in \mathbb{Z} \right\}$. Since the set M has an infinite number of elements, the condition $a^2 - 4b < 0$ is also sufficient.

Problem 2.3.23. Let the function $g : [0, \infty) \to \mathbb{R}$ be the restriction of the solution $y = f(x)$ to the initial value problem $y^{(4)} + 4y = 0$, $y(0) = -y'(0) = 1$, that has a horizontal asymptote when $x \to \infty$. Determine the Fourier sine transform of the function g.

Solution 2.3.23. The set of roots of the corresponding characteristic equation $r^4 + 4 = 0$ is $S = \left\{ \sqrt{2}\, e^{\frac{(2k+1)\pi}{4} i} : k \in \{0, 1, 2, 3\} \right\} = \{1 + i, 1 - i, -1 + i, -1 - i\}$. Hence, $f(x) = (c_1 \cos x + c_2 \sin x)e^x + (c_3 \cos x + c_4 \sin x)e^{-x}$, for some $c_1, c_2, c_3, c_4 \in \mathbb{R}$. Since the functions $x \mapsto c_1 a \cos x + c_2 \sin x$ and $x \mapsto c_3 \cos x + c_4 \sin x$ are bounded and since $\lim\limits_{x \to \infty} e^x = \infty$ and $\lim\limits_{x \to \infty} e^{-x} = 0$, it follows that $\lim\limits_{x \to \infty} f(x)$ exists if and only if $c_1 = c_2 = 0$. Hence, $f(x) = (c_3 \cos x + c_4 \sin x)e^{-x}$. From $f(0) = -f'(0) = 1$, it follows that $f(x) = e^{-x} \cos x$. The Fourier sine transform of the function $g(x) = e^{-x} \cos x$, $x \in [0, \infty)$ is determined by, for $u \in [0, \infty)$,

$$\mathcal{F}_{\sin}[g(x)](u) = \sqrt{\frac{2}{\pi}} \int_0^\infty e^{-x} \cos x \sin ux\, dx$$

$$= \frac{1}{\sqrt{2\pi}} \left(\frac{u+1}{(u+1)^2 + 1} + \frac{u-1}{(u-1)^2 + 1} \right).$$

Problem 2.3.24. Determine the solution $y = f(x)$ to the differential equation $y'' - 4y = 4e^{-2x}$ such that the graph of the function f has a horizontal asymptote and that its tangent at the y-intercept is parallel to the line $y = x$.

Solution 2.3.24. Since the roots of the characteristic equation are ± 2, the general solution to the corresponding homogeneous equation is $y_h = c_1 e^{2x} + c_2 e^{-2x}$, $c_1, c_2 \in \mathbb{R}$. The Lagrange method of variation of parameters yields a particular solution to the non-homogeneous equation $y_p = -xe^{-2x}$. Hence, $f(x) = c_1 e^{2x} + (c_2 - x)e^{-2x}$, for some constants $c_1, c_2 \in \mathbb{R}$. Since the graph of f has a horizontal asymptote, at least one of $\lim_{x \to \infty} f(x)$ and $\lim_{x \to -\infty} f(x)$ exists. Since, for any $c_1, c_2 \in \mathbb{R}$, $\lim_{x \to -\infty} (c_1 e^{2x} + (c_2 - x)e^{-2x}) = \infty$ and since for any $a, b \in \mathbb{R}$, $a \neq 0$, $\lim_{x \to \infty} |ae^{2x} + (b-x)e^{-2x}| = \infty$, it follows that $c_1 = 0$. Indeed, if $c_1 = 0$, then $f(x) = (c_2 - x)e^{-2x}$, $c_2 \in \mathbb{R}$, and $\lim_{x \to \infty} f(x) = \lim_{x \to \infty} (c_2 - x)e^{-2x} = 0$, i.e., the x-axis is the horizontal asymptote to the graph of the function f when $x \to \infty$. The fact that the tangent to the graph of f at the y-intercept is parallel to the line $y = x$ means that $f'(0) = 1$, i.e., $c_2 = -1$. Therefore, $f(x) = -(1 + x)e^{-2x}$.

Problem 2.3.25. Justify the method of undetermined coefficients for finding a particular solution to the equation $y'' + ay' + by = \sum_{k=0}^{n} a_k x^k$, $a_n \neq 0$.

Solution 2.3.25. We are looking for a polynomial $y_p = \sum_{k=0}^{m} c_k x^k$, $c_m \neq 0$, that is a particular solution to the equation. From $y_p' = \sum_{k=1}^{m} kc_k x^{k-1}$ and $y_p'' = \sum_{k=2}^{m} k(k-1)c_k x^{k-2}$, it follows that, at this moment, the undetermined coefficients, $c_0, c_1, \ldots, c_m$, must be such that

$$\sum_{k=2}^{m} k(k-1)c_k x^{k-2} + a \sum_{k=1}^{m} kc_k x^{k-1} + b \sum_{k=0}^{m} c_k x^k = \sum_{k=0}^{n} a_k x^k.$$

Since the polynomials on the left- and right-hand sides of this expression must be of the same degree, we conclude that

$$m = \begin{cases} n & \text{if } b \neq 0, \\ n+1 & \text{if } b = 0 \text{ and } a \neq 0, \\ n+2 & \text{if } a = b = 0. \end{cases}$$

If $b \neq 0$, then the coefficients $c_0, c_1, \ldots, c_n$ can be determined by solving the system of equations

$$bc_n = a_n$$

$$anc_n + bc_{n-1} = a_{n-1}$$

$$n(n-1)c_n + a(n-1)c_{n-1} + bc_{n-2} = a_{n-2}$$

$$\vdots$$

$$2 \cdot 1 \cdot c_2 + a \cdot 1 \cdot c_1 + bc_0 = a_0.$$

If $b = 0$ and $a \neq 0$, then

$$\sum_{k=0}^{n} a_k x^k = \sum_{k=2}^{n+1} k(k-1)c_k x^{k-2} + a \sum_{k=1}^{n+1} k c_k x^{k-1}$$

$$= a(n+1)c_{n+1}x^n + \sum_{k=0}^{n-1}(k+1)((k+2)c_{k+2} + ac_{k+1})x^k.$$

It follows that the coefficients $c_1, c_2, \ldots, c_{n+1}$ can be determined by solving the system of equations

$$a(n+1)c_{n+1} = a_n$$

$$n((n+1)c_{n+1} + ac_n) = a_{n-1}$$

$$\vdots$$

$$2c_2 + a \cdot c_1 = a_0.$$

We can take $c_0 = 0$.

If $a = b = 0$ then, from $\sum_{k=2}^{n+2} k(k-1)c_k x^{k-2} = \sum_{k=0}^{n} a_k x^k$, it follows that, for $k \in \{2, \ldots, n+2\}$, $c_k = \frac{a_{k-2}}{k(k-1)}$. We take $c_0 = c_1 = 0$.

Note: Observe that in the case of $b = 0$ and $a \neq 0$, one of the roots of the characteristic equation $r^2 + ar = 0$ is $r = 0$. This implies that $y = c$, $c \in \mathbb{R}$, is a solution to the homogeneous equation. By taking $c_0 = 0$ we ensure that the polynomial obtained as a particular solution in the case of $b = 0$ does not have a term that is a solution to the homogeneous equation.

Problem 2.3.26. Let $a, b \in \mathbb{R}$ be such that $a^2 \geq 4b$, and let $\omega \in \mathbb{R} \backslash \{0\}$. Determine the amplitude of all periodic solutions to the equation $y'' + ay' + by = \sin \omega x$.

Solution 2.3.26. Since $a^2 - 4b \geq 0$, the characteristic equation $r^2 + ar + b = 0$ has real roots. The non-homogeneous part of the differential equation is the function $x \mapsto \sin \omega x$, so one particular solution may be determined by the method of undetermined coefficients: $y_p(t) = x^s e^{\alpha x}(P_m(x) \cos \beta x + Q_m(x) \sin \alpha x)$, where $\alpha = 0$, $\beta = \omega$, and $m = 0$. Since $\alpha + i\beta = i\omega$ is not a solution to the characteristic equation, it follows that $s = 0$. Hence, $y_p = c \cos \omega x + d \sin \omega x$, for some $c, d \in \mathbb{R}$. From, for any $x \in \mathbb{R}$,

$$\sin \omega x = y_p'' + a y_p' + b y_p$$

$$= (c(b - \omega^2) + ad\omega) \cos \omega x + (-ac\omega + d(b - \omega^2)) \sin \omega x,$$

since the functions $x \mapsto \cos \omega x$ and $x \mapsto \sin \omega x$ are linearly independent, it follows that $c(b - \omega^2) + ad\omega = 0$ and $-ac\omega + d(b - \omega^2) = 1$. Hence, $c = -\frac{a\omega}{a^2\omega^2 + (b - \omega^2)^2}$, $d = \frac{b - \omega^2}{a^2\omega^2 + (b - \omega^2)^2}$, and

$$y_p = \frac{-a\omega \cos \omega x + (b - \omega^2) \sin \omega x}{a^2\omega^2 + (b - \omega^2)^2}.$$

The fact that the characteristic equation has real roots implies that the only periodic solution to the given equation is the function y_p. Its amplitude is $A = (a^2\omega^2 + (b - \omega^2)^2)^{-\frac{1}{2}}$.

Problem 2.3.27. Let $n \in \mathbb{N}$. Determine the Fourier series of a periodic function $y = f(x)$, with the fundamental period $T = 2\pi$ and such that $f(0) = 0$, $f'(0) = 1$, and, for $x \in [-\pi, \pi)$, $f''(x) + n^2 f(x) = -2n \sin nx$.

Solution 2.3.27. The characteristic equation is $r^2 + n^2 = 0$, and its roots are $r = \pm in$. One particular solution to the equation is, by the method of undetermined coefficients, $y_p = x \cos nx$. Hence, for $x \in [-\pi, \pi)$, $f(x) = c_1 \cos nx + c_2 \sin nx + x \cos nx$, for some $c_1, c_2 \in \mathbb{R}$. From the initial conditions, it follows that $c_1 = c_2 = 0$ and $f(x) = x \cos nx$, $x \in [-\pi, \pi)$. Since f is an odd function, $a_k = 0$, for all $k \in \mathbb{N}_0$ and for all $k \in \mathbb{N}$,

$$b_k = \frac{2}{\pi} \int_0^\pi x \cos nx \sin kx \, dx = \frac{1}{\pi} \int_0^\pi x(\sin(n+k)x - \sin(n-k)x) dx.$$

It follows that, for $k \in \mathbb{N} \setminus \{n\}$, $b_k = \frac{(-1)^{k+n} 2k}{n^2 - k^2}$ and $b_n = -\frac{1}{2n}$. Therefore, $f(x) \sim -\frac{1}{2n} \sin nx + 2(-1)^n \sum_{k \in \mathbb{N} \setminus \{n\}} \frac{(-1)^k k}{n^2 - k^2} \sin kx$.

Note: One can be tempted to start with, for $x \in (-\pi, \pi)$, $f(x) = \frac{a_0}{2} + \sum_{k=1}^{\infty}(a_k \cos kx + b_k \sin kx)$, differentiate this Fourier series term-by-term twice, and then determine the Fourier coefficients by using the given differential equation. Would this work? See Problem 1.2.28.

Problem 2.3.28. Solve the equation $y'' - 4y' + 8y = e^{2x} + \sin 2x$.

Solution 2.3.28. The characteristic equation is $r^2 - 4r + 8 = 0$, and its roots are $r = 2(1 \pm i)$. By the principle of superposition, a particular solution to the equation y_p is the sum of y_{p_1}, a particular solution to the equation $y'' - 4y' + 8y = e^{2x}$, and y_{p_2}, a particular solution to the equation $y'' - 4y' + 8y = \sin 2x$. By the method of undetermined coefficients, $y_{p_1} = \frac{1}{4}e^{2x}$ and $y_{p_2} = \frac{1}{20}(2\cos 2x + \sin 2x)$. Hence, $y_p = \frac{1}{4}e^{2x} + \frac{1}{20}(2\cos 2x + \sin 2x)$, and the general solution to the equation is

$$y = \left(c_1 \cos 2x + c_2 \sin 2x + \frac{1}{4} \right) e^{2x} + \frac{1}{20}(2\cos 2x + \sin 2x), \ c_1, c_2 \in \mathbb{R}.$$

Problem 2.3.29. Solve the equations:

(a) $y''' - 5y'' + 6y' = xe^{2x}$.   (b) $y''' - y'' - y' + y = 2\sinh x$.

Solution 2.3.29.

(a) The characteristic equation is $r^3 - 5r^2 + 6r = 0$, and its roots are $r = 0$, $r = 2$, and $r = 3$. Since the non-homogeneous part of the differential equation is the function $x \mapsto xe^{2x}$, one particular solution may be determined by the method of undetermined coefficients: $y_p(x) = x^s e^{\alpha x} P_m(x)$, where $\alpha = 2$ and $m = 1$. Since $\alpha = 2$ is a root of the characteristic equation of multiplicity one, it follows that $s = 1$. Hence, $y_p = (ax + b)xe^{2x}$, for some $a, b \in \mathbb{R}$ that need to be determined under the assumption that y_p is a solution to the equation, i.e., under the condition that $y_p''' - 5y_p'' + 6y_p' = xe^{2x}$. Thus, a and b are real numbers such that for all $x \in \mathbb{R}$,

$$x = 8ax^2 + 8(3a + b)x + 12(a + b) - 5(4ax^2 + 4(2a + b)x$$
$$+ 2(a + 2b)) + 6(2ax^2 + 2(a + b)x + b) = -4ax + 2(a - b).$$

It follows that $a = b = -\frac{1}{4}$ and $y_p = -\frac{1}{4}(x+1)x\, e^{2x}$. The general solution to the equation is $y = c_1 + c_2 e^{2x} + c_3 e^{3x} - \frac{1}{4}(x+1)xe^{2x}$, $c_1, c_2, c_3 \in \mathbb{R}$.

(b) By the principle of superposition and using the method of undetermined coefficients, a particular solution to the equation is $y_p = \frac{x}{4}\left(xe^x - e^{-x}\right)$. The general solution is $y = (c_1 + c_2 x)e^x + c_3 e^{-x} + \frac{x}{4}\left(xe^x - e^{-x}\right)$, $c_1, c_2, c_3 \in \mathbb{R}$.

Problem 2.3.30. Solve the equations:

(a) $y^{(4)} - 2y''' + y'' = e^x$.   (b) $y^{(4)} + 5y'' + 4y = \sin x \cos 2x$.

Solution 2.3.30.

(a) The general solution is $y = c_1 + c_2 x + \left(c_3 + c_4 x + \frac{1}{2}\, x^2\right)e^x$, c_1, c_2, c_3, $c_4 \in \mathbb{R}$.

(b) Observe that, for all $x \in \mathbb{R}$, $\sin x \cos 2x = \frac{1}{2}(\sin 3x - \sin x)$. By the principle of superposition and using the method of undetermined coefficients, one particular solution to the given equation is $y_p = \frac{1}{80}\sin 3x + \frac{x}{12}\cos x$. The general solution is $y = c_1 \cos 2x + c_2 \sin 2x + c_3 \cos x + c_4 \sin x + \frac{1}{80}\sin 3x + \frac{x}{12}\cos x$, $c_1, c_2, c_3, c_4 \in \mathbb{R}$.

Problem 2.3.31. Let $a, b, c \in \mathbb{R}$, $a \neq 0$. Solve the Cauchy–Euler equation $ax^2 y'' + bxy' + cy = 0$, $x > 0$.

Solution 2.3.31. The substitution $x = e^t$, $t \in \mathbb{R}$, yields

$$\frac{dy}{dx} = e^{-t}\frac{dy}{dt} \quad \text{and} \quad \frac{d^2 y}{dx^2} = e^{-t}\frac{d}{dt}\left(e^{-t}\frac{dy}{dt}\right) = e^{-2t}\left(\frac{d^2 y}{dt^2} - \frac{dy}{dt}\right)$$

and transforms the Cauchy–Euler equation into a second-order homogeneous linear differential equation with constant coefficients: $a\frac{d^2 y}{dt^2} + (b - a)\frac{dy}{dt} + cy = 0$.

If $(a-b)^2 - 4ac > 0$, the characteristic equation $ar^2 + (b-a)r + c = 0$ has two real roots: r_1 and r_2. The general solution to the linear equation with constant coefficients is $y = c_1 e^{r_1 t} + c_2 e^{r_2 t}$, $c_1, c_2 \in \mathbb{R}$, which implies that the general solution to the Cauchy–Euler equation is $y = c_1 x^{r_1} + c_2 x^{r_2}$, $x > 0$, $c_1, c_2 \in \mathbb{R}$.

If $(a-b)^2 - 4ac = 0$, the characteristic equation $ar^2 + (b-a)r + c = 0$ has only one root: $r_0 \in \mathbb{R}$. The general solution to the linear equation with constant coefficients is $y = (c_1 + c_2 t)e^{r_0 t}$, $c_1, c_2 \in \mathbb{R}$, which implies that the general solution to the Cauchy–Euler equation in this case is $y = (c_1 + c_2 \ln x)x^{r_0}$, $x > 0$, $c_1, c_2 \in \mathbb{R}$.

If $(a-b)^2 - 4ac < 0$, the characteristic equation $ar^2 + (b-a)r + c = 0$ has two complex roots: $r_1 = \alpha + i\beta$, and $r_2 = \alpha - i\beta$, $\alpha, \beta \in \mathbb{R}$, $\beta \neq 0$. The general solution to the linear equation with constant coefficients is $y = (c_1 \cos \beta t + c_2 \sin \beta t)e^{\alpha t}$, $c_1, c_2 \in \mathbb{R}$, which implies that the general solution to the Cauchy–Euler equation is $y = (c_1 \cos(\beta \ln x) + c_2 \sin(\beta \ln x))x^\alpha$, $x > 0$, $c_1, c_2 \in \mathbb{R}$.

Problem 2.3.32. Solve the following initial value problems:

(a) $x^2 y'' + 2xy' + y = 0$, $y(1) = 0$, $y'(1) = 1$.
(b) $x^2 y'' + 3xy' + y = 0$, $y(1) = y'(1) = 1$.
(c) $x^2 y'' + 3xy' - 8y = 0$, $y(1) = 1$, $y'(1) = 4$.

Solution 2.3.32.

(a) This is a Cauchy–Euler equation with $a = c = 1$ and $b = 2$. Since the equation $r^2 + (2-1)r + 1 = 0$ has complex roots, $r_{1,2} = \frac{-1 \pm i\sqrt{3}}{2}$, by Problem 2.3.31, the general solution to the differential equation is $y = \frac{1}{\sqrt{x}}\left(c_1 \cos \frac{\sqrt{3}\ln x}{2} + c_2 \sin \frac{\sqrt{3}\ln x}{2}\right)$, $x > 0$, $c_1, c_2 \in \mathbb{R}$. Since $0 = y(1) = c_1$, it follows that $y = \frac{c_2}{\sqrt{x}} \sin \frac{\sqrt{3}\ln x}{2}$. From $y' = -\frac{c_2}{2x\sqrt{x}} \sin \frac{\sqrt{3}\ln x}{2} + \frac{c_2}{\sqrt{x}} \frac{\sqrt{3}}{2x} \cos \frac{\sqrt{3}\ln x}{2}$ and $y'(1) = 1$, it follows that $c_2 = \frac{2}{\sqrt{3}}$. The solution to the initial value problem is $y = \frac{2}{\sqrt{3x}} \sin \frac{\sqrt{3}\ln x}{2}$, $x > 0$.

(b) This is a Cauchy–Euler equation with $a = c = 1$ and $b = 3$. Since the equation $r^2 + (3-1)r + 1 = 0$ has only one real root, $r_0 = -1$, the general solution to the differential equation is $y = \frac{c_1 + c_2 \ln x}{x}$, $x > 0$, $c_1, c_2 \in \mathbb{R}$. Since $1 = y(1) = c_1$, it follows that $y = \frac{1 + c_2 \ln x}{x}$. From $y' = \frac{c_2 - (1 + c_2 \ln x)}{x^2}$ and $y'(1) = 1$, it follows that $c_2 = 2$. The solution to the initial value problem is $y = \frac{1 + 2\ln x}{x}$, $x > 0$.

(c) This is a Cauchy–Euler equation with $a = 1$, $b = 3$, and $c = -8$. Since the equation $r^2 + 2r - 8 = 0$ has two real roots, $r_1 = 2$ and $r_2 = -4$, the general solution to the differential equation is $y = c_1 x^2 + \frac{c_2}{x^4}$, $x > 0$, $c_1, c_2 \in \mathbb{R}$. Since $y(1) = 1$, it follows that $c_1 + c_2 = 1$. From $y' = 2c_1 x - \frac{4c_2}{x^5}$ and $y'(1) = 4$, it follows that $2c_1 - 4c_2 = 4$. Hence, $c_1 = \frac{4}{3}$ and $c_2 = -\frac{1}{3}$. The solution to the initial value problem is $y = \frac{4x^2}{3} - \frac{1}{3x^4}$, $x > 0$.

In Problems 2.3.33–2.3.39, use power series to solve the given differential equations and initial value problems.

Problem 2.3.33. Solve $(1 - x^2)y'' + 2y = 0$, $x \in (-1, 1)$.

Solution 2.3.33. Let, for $x \in (-1, 1)$, $y = \sum_{k=0}^{\infty} a_k x^k$ be a solution to the equation. Then, $y' = \sum_{k=1}^{\infty} k a_k x^{k-1}$ and $y'' = \sum_{k=2}^{\infty} k(k-1)a_k x^{k-2}$. It follows that, for all $x \in (-1, 1)$,

$$0 = (1 - x^2) \sum_{k=2}^{\infty} k(k-1)a_k x^{k-2} + 2 \sum_{k=0}^{\infty} a_k x^k$$

$$= \sum_{k=0}^{\infty} (k+1)((k+2)a_{k+2} - (k-2)a_k)x^k,$$

which implies that, for all $k \in \mathbb{N}_0$, $a_{k+2} = \frac{k-2}{k+2} a_k$. Hence, $a_2 = -a_0$ and, for all $k \in \mathbb{N}_0$, $a_{2k+4} = 0$ and $a_{2k+1} = -\frac{1}{(2k-1)(2k+1)} a_1$. For any $a_0, a_1 \in \mathbb{R}$, the function

$$y = a_0(1 - x^2) - a_1 \sum_{k=0}^{\infty} \frac{x^{2k+1}}{(2k-1)(2k+1)}, \quad x \in (-1, 1)$$

is a solution to the equation. Since, for any $x \in (-1, 1)$,

$$\sum_{k=0}^{\infty} \frac{x^{2k+1}}{(2k-1)(2k+1)} = \frac{1}{2} \sum_{k=0}^{\infty} \frac{x^{2k+1}}{2k-1} - \frac{1}{2} \sum_{k=0}^{\infty} \frac{x^{2k+1}}{2k+1}$$

$$= -\frac{x}{2} + \frac{x^2 - 1}{2} \sum_{k=0}^{\infty} \frac{x^{2k+1}}{2k+1}$$

and

$$\ln \frac{1+x}{1-x} = \ln(1+x) - \ln(1-x) = \sum_{k=1}^{\infty} \frac{(-1)^{k+1} + 1}{k} x^k = 2 \sum_{k=0}^{\infty} \frac{x^{2k+1}}{2k+1},$$

it follows that the general solution to the equation is

$$y = a_0(1 - x^2) + a_1 \left(\frac{x}{2} + \frac{1 - x^2}{4} \ln \frac{1+x}{1-x} \right), \quad x \in (-1, 1), \ a_0, a_1 \in \mathbb{R}.$$

Problem 2.3.34. Solve the initial value problem $y'' + xy = 0$, $y(0) = 1$, $y'(0) = 0$.

Solution 2.3.34. Since the function $x \mapsto x$ has a power series representation in the neighbourhood of $x = 0$, with $\mathbb{R}$ as its interval of convergence, the solution to the initial problem $y = f(x)$ also has a power series representation. Let a sequence of real numbers, $\{a_k\}_{k \in \mathbb{N}_0}$, be such that $y = f(x) = \sum_{k=0}^{\infty} a_k x^k$, $x \in \mathbb{R}$, with $f(0) = a_0 = 1$ and $f'(0) = a_1 = 0$. Since the function f is the solution to the initial value problem, from, for all $x \in \mathbb{R}$,

$$0 = \sum_{k=2}^{\infty} k(k-1)a_k x^{k-2} + x \sum_{k=0}^{\infty} a_k x^k$$

$$= 2a_2 + \sum_{k=1}^{\infty}((k+2)(k+1)a_{k+2} + a_{k-1})x^k,$$

it follows that $a_2 = 0$ and, for all $k \in \mathbb{N}$, $a_{k+2} = -\frac{1}{(k+2)(k+1)} a_{k-1}$. For any $k \in \mathbb{N}$, $a_{3k-1} = a_{3k-2} = 0$ and $a_{3k} = \frac{(-1)^k}{(3k)!} \prod_{m=1}^{k}(3m-2)$. The solution to the initial value problem is the function, for $x \in \mathbb{R}$,

$$f(x) = 1 + \sum_{k=1}^{\infty}\left(\frac{(-1)^k}{(3k)!}\prod_{m=1}^{k}(3m-2)\right)x^{3k}$$

$$= 1 - \frac{1}{3!}x^3 + \frac{1\cdot 4}{6!}x^6 - \frac{1\cdot 4\cdot 7}{9!}x^9 + \cdots.$$

Problem 2.3.35. Solve the equation $y'' + (a - x^2)y = 0$, $a \in \mathbb{R}$, by first using the substitution $y = ze^{-\frac{x^2}{2}}$.

Solution 2.3.35. Since, for $y = ze^{-\frac{x^2}{2}}$, $y' = (z' - xz)e^{-\frac{x^2}{2}}$ and $y'' = (z'' - 2xz' + (x^2-1)z)e^{-\frac{x^2}{2}}$, the substitution $y = ze^{-\frac{x^2}{2}}$ transforms the given equation into the equation $z'' - 2xz' + (a-1)z = 0$. Each of the functions $x \mapsto -2x$ and $x \mapsto a - 1$ has a power series representation with $\mathbb{R}$ as its interval of convergence, so any solution to the equation also has a power series representation. Let $\{a_k\}_{k \in \mathbb{N}_0}$ be the sequence of real numbers such that $z = \sum_{k=0}^{\infty} a_k x^k$, $x \in \mathbb{R}$, is a solution to the equation. Then, for every $x \in \mathbb{R}$,

$$\sum_{k=0}^{\infty} \left((k+2)(k+1)a_{k+2} - (2k+1-a)a_k\right) x^k = 0.$$

Hence, for every $k \in \mathbb{N}_0$, $a_{k+2} = \frac{2k+1-a}{(k+2)(k+1)} a_k$, i.e., for $k \in \mathbb{N}$,

$$a_{2k} = \frac{a_0}{(2k)!} \prod_{m=1}^{k} (4m-3-a) \text{ and } a_{2k+1} = \frac{a_1}{(2k+1)!} \prod_{m=1}^{k} (4m-1-a).$$

The general solution to the given equation is

$$y = a_0 \left(1 + \sum_{k=1}^{\infty} \frac{\prod_{m=1}^{k}(4m-3-a)}{(2k)!} x^{2k}\right) e^{-\frac{x^2}{2}}$$

$$+ a_1 \left(x + \sum_{k=1}^{\infty} \frac{\prod_{m=1}^{k}(4m-1-a)}{(2k+1)!} x^{2k+1}\right) e^{-\frac{x^2}{2}},$$

where $x \in \mathbb{R}$ and $a_0, a_1 \in \mathbb{R}$.

Problem 2.3.36. Represent the solution to the initial value problem $y'' + 2xy' + 2y = 0$, $y(0) = 1$, $y'(0) = 0$, by its Fourier integral.

Solution 2.3.36. Since each of the functions $x \mapsto 2x$ and $x \mapsto 2$ has a power series representation in the neighbourhood of $x = 0$ with $\mathbb{R}$ as its interval of convergence, the solution to the initial value problem, the function $y = f(x)$, also has a power series representation. Let $y = f(x) = \sum_{k=0}^{\infty} a_k x^k$, $x \in \mathbb{R}$, with $f(0) = a_0 = 1$ and $f'(0) = a_1 = 0$. From, for all $x \in \mathbb{R}$,

$$\sum_{k=0}^{\infty} (k+1)((k+2)a_{k+2} + 2a_k)x^k = 0,$$

it follows that, for all $k \in \mathbb{N}_0$, $a_{k+2} = -\frac{2}{k+2}a_k$, which, together with $a_0 = 1$ and $a_1 = 0$, means that, for $k \in \mathbb{N}$, $a_{2k} = \frac{(-1)^k}{k!}$ and $a_{2k-1} = 0$. The solution to the initial value problem is $f(x) = \sum_{k=0}^{\infty} \frac{(-1)^k}{k!} x^{2k} = e^{-x^2}$, $x \in \mathbb{R}$.

Recall that $\int_{-\infty}^{\infty} f(x)\, dx = \sqrt{\pi}$. Since f is an even function, for all $u \in [0, \infty)$, $b(u) = \frac{1}{\pi} \int_{-\infty}^{\infty} f(x)\, \sin ux\, dx = 0$. From, for all $u, x \in \mathbb{R}$, $\cos ux = \sum_{k=0}^{\infty} \frac{(-1)^k}{(2k)!}\, (ux)^{2k}$, since this is a uniformly convergent series and since, for any $k \in \mathbb{N}$, the function $x \mapsto x^{2k} e^{-x^2}$ is integrable over $[0, \infty)$, it follows that, for all $u \in [0, \infty)$,

$$a(u) = \frac{2}{\pi} \sum_{k=0}^{\infty} \frac{(-1)^k u^{2k}}{(2k)!} \int_0^{\infty} x^{2k} e^{-x^2}\, dx.$$

Let, for every $k \in \mathbb{N}_0$, $I_k = \int_0^{\infty} x^{2k} e^{-x^2}\, dx$. Then, $I_0 = \frac{\sqrt{\pi}}{2}$ and, for $k \in \mathbb{N}$, $I_k = \frac{2k-1}{2}\, I_{k-1} = \frac{(2k-1)!!}{2^{k+1}} \sqrt{\pi}$. It follows that, for all $u \in [0, \infty)$,

$$\begin{aligned}
a(u) &= \frac{2}{\pi} \left(\frac{\sqrt{\pi}}{2} + \sum_{k=1}^{\infty} \frac{(-1)^k u^{2k}}{(2k)!}\, \frac{(2k-1)!!}{2^{k+1}} \sqrt{\pi} \right) \\
&= \frac{1}{\sqrt{\pi}} \left(1 + \sum_{k=1}^{\infty} \frac{(-1)^k u^{2k}}{2^k (2k)!!} \right) \\
&= \frac{1}{\sqrt{\pi}} \sum_{k=0}^{\infty} \frac{(-1)^k}{k!} \left(\frac{u}{2} \right)^{2k} = \frac{1}{\sqrt{\pi}}\, e^{-\frac{u^2}{4}}.
\end{aligned}$$

Therefore, for all $x \in \mathbb{R}$, $e^{-x^2} = \frac{1}{\sqrt{\pi}} \int_0^{\infty} e^{-\frac{u^2}{4}} \cos ux\, du$.

Note 1: From $0 = y'' + 2xy' + 2y = (y' + 2xy)'$, it follows that $y' + 2xy = c$, for some $c \in \mathbb{R}$. From the initial conditions, $c = 0$, which yields a separable differential equation: $y' = -2xy$.

Note 2: See Problem 1.4.6.

Problem 2.3.37. Solve the initial value problem $x^2 y'' - 2xy' + (x^2 + 2)y = 0$, $y(\pi) = -\pi$, $y'(\pi) = -(\pi + 1)$.

Solution 2.3.37. Let, in some interval I, $y = f(x) = \sum_{k=0}^{\infty} a_k x^k$, $x \in I$, be the solution to the initial value problem. Then, for all $x \in I$, $2a_0 + \sum_{k=2}^{\infty} ((k-2)(k-1)a_k + a_{k-2})x^k = 0$, which implies that $a_0 = 0$ and, for all $k \in \mathbb{N}\setminus\{1\}$, $(k-2)(k-1)a_k + a_{k-2} = 0$. The condition, obtained for $k = 2$, $(2-2)(2-1)a_2 + a_0 = 0$, is satisfied for any real number a_2. From, for any $k \in \mathbb{N}\backslash\{1,2\}$, $a_k = -\frac{1}{(k-2)(k-1)}a_{k-2}$, it follows that, for any $k \in \mathbb{N}$,

$$a_{2k} = \frac{(-1)^{k-1}}{(2k-1)!}\, a_2 \quad \text{and} \quad a_{2k-1} = \frac{(-1)^{k-1}}{(2k-2)!}a_1.$$

Therefore, for $x \in I$, and $a_1, a_2 \in \mathbb{R}$,

$$f(x) = a_1 \sum_{k=0}^{\infty} \frac{(-1)^k}{(2k)!}\, x^{2k+1} + a_2 \sum_{k=1}^{\infty} \frac{(-1)^{k-1}}{(2k-1)!}\, x^{2k}$$

$$= x(a_1 \cos x + a_2 \sin x).$$

In particular, $I = \mathbb{R}$. From $f(\pi) = -\pi$ and $f'(\pi) = -(\pi+1)$, it follows that $a_1 = a_2 = 1$ and $f(x) = x(\cos x + \sin x)$, $x \in \mathbb{R}$.

Note: The substitution $y = xz$ yields the equation $z'' + z = 0$. The general solution to this homogeneous second-order linear equation with constant coefficients is $z = a_1 \cos x + a_2 \sin x$, $a_1, a_2 \in \mathbb{R}$. The general solution to the original equation is $y = x(a_1 \cos x + a_2 \sin x)$, $a_1, a_2 \in \mathbb{R}$.

Problem 2.3.38. Use the solution to the initial value problem $xy'' + 2y' + xy = 0$, $y(0) = 1$, $y'(0) = 0$, to determine a function $y = g(x)$ that satisfies the equation and for which $g\left(\frac{\pi}{2}\right) = \frac{2}{\pi}$ and $g'\left(\frac{\pi}{2}\right) = -\frac{2}{\pi}\left(\frac{2}{\pi}+1\right)$.

Solution 2.3.38. Let $y = f(x) = \sum_{k=0}^{\infty} a_k x^k$, $x \in \mathbb{R}$, with $f(0) = a_0 = 1$ and $f'(0) = a_1 = 0$, be the solution to the initial value problem. From, for all $x \in \mathbb{R}$, $2a_1 + \sum_{k=1}^{\infty}((k+1)(k+2)a_{k+1} + a_{k-1})x^k = 0$, it follows that, for all $k \in \mathbb{N}$, $a_{k+1} = -\frac{1}{(k+1)(k+2)}a_{k-1}$. This implies that, since $a_0 = 1$ and $a_1 = 0$, for all $k \in \mathbb{N}$, $a_{2k} = \frac{(-1)^k}{(2k+1)!}$ and $a_{2k-1} = 0$. Thus,

$$f(x) = 1 + \sum_{k=1}^{\infty} \frac{(-1)^k}{(2k+1)!} x^{2k} = \operatorname{sinc}(x) = \begin{cases} \frac{\sin x}{x} & \text{if } x \neq 0, \\ 1 & \text{if } x = 0. \end{cases}$$

Let $D = \{(x,y,y') : (x,y,y') \in (0,\pi) \times \mathbb{R} \times \mathbb{R}\}$. Then, $\left(\frac{\pi}{2}, \frac{2}{\pi}, -\frac{2}{\pi}\left(\frac{2}{\pi}+1\right)\right) \in D$, and in the set D, the problem of finding the function $y = g(x)$ becomes the problem of solving the initial value problem $y'' + \frac{2}{x} y' + y = 0$, $y\left(\frac{\pi}{2}\right) = \frac{2}{\pi}$, $y'\left(\frac{\pi}{2}\right) = -\frac{2}{\pi}\left(\frac{2}{\pi} + 1\right)$. Since the function $y_1 = \frac{\sin x}{x}$, $x \in (0, \pi)$ is a solution to the homogeneous second-order linear equation $y'' + \frac{2}{x} y' + y = 0$, by Liouville's formula, the function

$$y_2 = \frac{\sin x}{x} \int \frac{e^{-\int \frac{2\,dx}{x}}}{\frac{\sin^2 x}{x^2}}\,dx = \frac{\sin x}{x} \int \frac{\frac{1}{x^2}}{\frac{\sin^2 x}{x^2}}\,dx = -\frac{\cos x}{x}$$

is another solution to the equation, linearly independent from y_1. There are $c_1, c_2 \in \mathbb{R}$ such that $g(x) = \frac{1}{x}\left(c_1 \cos x + c_2 \sin x\right)$, $x \in (0, \pi)$. The initial conditions imply that $c_1 = c_2 = 1$ and $g(x) = \frac{1}{x}\left(\cos x + \sin x\right)$, $x \in (0, \pi)$.

Note: The substitution $z = xy$ transforms the equation $xy'' + 2y' + xy = 0$ into the equation $z'' + z = 0$. Hence, for $x > 0$, there are real numbers a and b such that $g(x) = \frac{1}{x}(a \sin x + b \cos x)$.

Problem 2.3.39. Solve the initial value problem $(x^2 - x + 1)y'' + 2(2x - 1)y' + 2y = 0$, $y\left(\frac{1}{2}\right) = 2$, $y'\left(\frac{1}{2}\right) = \frac{4}{3}$.

Solution 2.3.39. Since $x^2 - x + 1 \neq 0$ for all $x \in \mathbb{R}$, the initial value problem is equivalent to the initial value problem

$$y'' + \frac{2(2x-1)}{x^2 - x + 1}\,y' + \frac{2}{x^2 - x + 1}\,y = 0, \quad y\left(\frac{1}{2}\right) = 2, \quad y'\left(\frac{1}{2}\right) = \frac{4}{3}.$$

Since, for any $x \in I = \left(\frac{1}{2} - \frac{\sqrt{3}}{2}, \frac{1}{2} + \frac{\sqrt{3}}{2}\right)$, $\left|\frac{4}{3}\left(x - \frac{1}{2}\right)^2\right| < 1$, it follows that, for any $x \in I$,

$$\frac{2(2x-1)}{x^2 - x + 1} = \frac{4}{3}\,\frac{4\left(x - \frac{1}{2}\right)}{1 + \frac{4}{3}\left(x - \frac{1}{2}\right)^2} = \frac{16}{3}\sum_{k=0}^{\infty}(-1)^k \left(\frac{4}{3}\right)^k \left(x - \frac{1}{2}\right)^{2k+1}$$

and

$$\frac{1}{x^2 - x + 1} = \frac{4}{3}\sum_{k=0}^{\infty}(-1)^k \left(\frac{4}{3}\right)^k \left(x - \frac{1}{2}\right)^{2k}.$$

The function $y = f(x)$, the solution to the initial value problem, has a power series representation in the interval I. Let the sequence of real numbers $\{a_k\}_{k\in\mathbb{N}_0}$ be such that $f(x) = \sum_{k=0}^{\infty} a_k \left(x - \frac{1}{2}\right)^k$, $x \in I$, with $a_0 = 2$ and $a_1 = \frac{4}{3}$. It follows that, for $x \in I$,

$$0 = (x^2 - x + 1)f''(x) + 2(2x - 1)f'(x) + 2f(x)$$

$$= \left(\left(x - \frac{1}{2}\right)^2 + \frac{3}{4}\right) \sum_{k=2}^{\infty} k(k-1)a_k \left(x - \frac{1}{2}\right)^{k-2}$$

$$+ 4\left(x - \frac{1}{2}\right) \sum_{k=1}^{\infty} ka_k \left(x - \frac{1}{2}\right)^{k-1} + 2\sum_{k=0}^{\infty} a_k \left(x - \frac{1}{2}\right)^k$$

$$= \sum_{k=0}^{\infty} (k+2)(k+1)\left(\frac{3}{4}a_{k+2} + a_k\right)\left(x - \frac{1}{2}\right)^k.$$

Hence, for all $k \in \mathbb{N}_0$, $a_{k+2} = -\frac{4}{3}a_k$. This, together with $a_0 = 2$ and $a_1 = \frac{4}{3}$, implies that, for all $k \in \mathbb{N}_0$, $a_{2k} = 2(-1)^k \left(\frac{4}{3}\right)^k$ and $a_{2k+1} = (-1)^k \left(\frac{4}{3}\right)^{k+1}$. The solution to the initial value problem is, for $x \in I$,

$$f(x) = 2\sum_{k=0}^{\infty}(-1)^k \left(\frac{4}{3}\right)^k \left(x - \frac{1}{2}\right)^{2k} + \sum_{k=0}^{\infty}(-1)^k \left(\frac{4}{3}\right)^{k+1} \left(x - \frac{1}{2}\right)^{2k+1}$$

$$= \frac{2}{1 + \frac{4}{3}\left(x - \frac{1}{2}\right)^2} + \frac{4}{3}\left(x - \frac{1}{2}\right)\frac{1}{1 + \frac{4}{3}\left(x - \frac{1}{2}\right)^2} = \frac{x + 1}{x^2 - x + 1}.$$

Note: The given equation is equivalent to the equation $\left((x^2 - x + 1)\, y\right)'' = 0$.

Problem 2.3.40. Use trigonometric series and the fact that every solution is a periodic function with a period $T = 2\pi$ to solve the equation $y'' + y = |\sin x|$.

Solution 2.3.40. By Dirichlet's theorem, for all $x \in \mathbb{R}$, $|\sin x| = \frac{1}{2}a_0 + \sum_{k=1}^{\infty} a_k \cos 2kx$, where, for any $k \in \mathbb{N}_0$,

$$a_k = \frac{4}{\pi} \int_0^{\frac{\pi}{2}} \sin x \, \cos 2kx \, dx = \frac{4}{\pi(1 - 4k^2)}.$$

See Problem 1.2.46.

Let $y = f(x)$ be a solution to the given non-homogeneous second-order linear differential equation. Since the solution f is a periodic function, with a period $T = 2\pi$, that is twice differentiable, the functions f, f', and f'' are continuous in $\mathbb{R}$. By differentiating the Fourier series of the function f term-by-term twice, the obtained series uniformly converge to functions f' and f''. Therefore, for all $x \in \mathbb{R}$, $f(x) = \frac{1}{2}A_0 + \sum_{k=1}^{\infty}(A_k \cos kx + B_k \sin kx)$, $f'(x) = \sum_{k=1}^{\infty} k(-A_k \sin kx + B_k \cos kx)$, and $f''(x) = -\sum_{k=1}^{\infty} k^2(A_k \cos kx + B_k \sin kx)$, where $\{A_k\}_{k\in\mathbb{N}_0}$ and $\{B_k\}_{k\in\mathbb{N}}$ are sequences of the Fourier coefficients of the function f. Since, by our assumption, f is a solution to the equation, it follows that, for all $x \in \mathbb{R}$,

$$\frac{2}{\pi} + \frac{4}{\pi}\sum_{k=1}^{\infty}\frac{\cos 2kx}{1 - 4k^2} = |\sin x| = y'' + y$$

$$= \frac{1}{2}A_0 + \sum_{k=1}^{\infty}(1 - k^2)(A_k \cos kx + B_k \sin kx).$$

Hence, $A_0 = \frac{4}{\pi}$, and for all $k \in \mathbb{N}$, $(1 - 4k^2)A_{2k} = \frac{4}{\pi(1-4k^2)}$ and $(1 - (2k - 1)^2)A_{2k-1} = (1 - k^2)B_k = 0$. Therefore, for any $k \in \mathbb{N}_0$, $A_{2k} = \frac{4}{\pi(1-4k^2)^2}$ and $A_{2k+3} = B_{k+2} = 0$. Since $(1 - (2 - 1)^2)^2 A_1 = (1 - 1)B_1 = 0$ for any $A_1, B_1 \in \mathbb{R}$, the general solution to the equation is

$$y = A_1 \cos x + B_1 \sin x + \frac{2}{\pi} + \frac{4}{\pi}\sum_{k=1}^{\infty}\frac{\cos 2kx}{(4k^2 - 1)^2}, \quad x \in \mathbb{R},\ A_1, B_1 \in \mathbb{R}.$$

Problem 2.3.41. Use trigonometric series to determine all periodic solutions to the equation $y'' + y' + y = |\sin x|$.

Solution 2.3.41. Let $y = f(x)$ be a periodic solution with a period $T > 0$. From, for any $x \in \mathbb{R}$,

$$|\sin(x + T)| = f''(x + T) + f'(x + T) + f(x + T)$$

$$= f''(x) + f'(x) + f(x) = |\sin x|,$$

it follows that T is a period of the function $x \mapsto |\sin x|$. Hence, referring to Problem 1.2.31, we can take $T = \pi$. Let

$$f(x) = \frac{a_0}{2} + \sum_{k=1}^{\infty}(a_k \cos 2kx + b_k \sin 2kx), \quad x \in \mathbb{R}.$$

From, for all $x \in \mathbb{R}$, $|\sin x| = \frac{2}{\pi} + \frac{4}{\pi}\sum_{k=1}^{\infty}\frac{\cos 2kx}{1-4k^2}$, and $f'(x) = 2\sum_{k=1}^{\infty}k(-a_k\sin 2kx + b_k\cos 2kx)$, and $f''(x) = -4\sum_{k=1}^{\infty}k^2(a_k\cos 2kx + b_k\sin 2kx)$, it follows that, for $k \in \mathbb{N}$,

$$(1 - 4k^2)a_k + 2kb_k = \frac{4}{\pi(1-4k^2)} \quad \text{and} \quad -2ka_k + (1-4k^2)b_k = 0.$$

The given equation has only one periodic solution:

$$f(x) = \frac{2}{\pi} + \frac{4}{\pi}\sum_{k=1}^{\infty}\frac{1}{1-4k^2+16k^4}$$
$$\cdot\left(\cos 2kx + \frac{2k}{1-4k^2}\sin 2kx\right), \ x \in \mathbb{R}.$$

Problem 2.3.42. Let $p \in \mathbb{R}^+$, and let a function $f : \mathbb{R} \to \mathbb{R}$ be continuous, piecewise smooth, and periodic, with the fundamental period $T = 2\pi$. Use Fourier series to determine all periodic, with a period 2π, solutions to the equation $y' + py = f(x)$.

Solution 2.3.42. Let $f(x) = \frac{a_0}{2} + \sum_{k=1}^{\infty}(a_k\cos kx + b_k\sin kx)$, $x \in \mathbb{R}$. Suppose that $y = y(x)$ is a periodic, with a period 2π, solution to the equation, and let $\{A_k\}_{k\in\mathbb{N}_0}$ and $\{B_k\}_{k\in\mathbb{N}}$ be the sequences of its Fourier coefficients. Since the derivative of the function $y = y(x)$ is a continuous function, it follows that by differentiating term-by-term the Fourier series of the function $y = y(x)$, we obtain the Fourier series of $y' = y'(x)$. Therefore, for $x \in \mathbb{R}$, $y(x) = \frac{A_0}{2} + \sum_{k=1}^{\infty}(A_k\cos kx + B_k\sin kx)$ and $y'(x) = \sum_{k=1}^{\infty}k(B_k\cos kx - A_k\sin kx)$. From this and $y' + py = f(x)$, it follows that $A_0 = \frac{a_0}{p}$ and, for all $k \in \mathbb{N}$,

$$pA_k + kB_k = a_k \quad \text{and} \quad -kA_k + pB_k = b_k.$$

Therefore, there is only one periodic, with a period 2π, solution to the equation:

$$y = \frac{a_0}{2p} + \sum_{k=1}^{\infty}\frac{1}{k^2+p^2}\left((pa_k - kb_k)\cos kx + (ka_k + pb_k)\sin kx\right), \ x \in \mathbb{R}.$$

Problem 2.3.43. Let $p \in \mathbb{R}^+$, and let a function $f : \mathbb{R} \to \mathbb{R}$ be continuous, piecewise smooth, and periodic, with the fundamental period $T = 2\pi$. Use Fourier series to determine all periodic, with a period 2π, solutions to the equation $y'' + py = f(x)$.

Solution 2.3.43. Let $f(x) = \frac{a_0}{2} + \sum_{k=1}^{\infty}(a_k \cos kx + b_k \sin kx)$, $x \in \mathbb{R}$. Suppose that $y = y(x)$ is a periodic, with a period 2π, solution to the equation. Let $y(x) = \frac{A_0}{2} + \sum_{k=1}^{\infty}(A_k \cos kx + B_k \sin kx)$. Since the second derivative of the function $y = y(x)$ is a continuous function, we obtain the Fourier series of $y' = y'(x)$ and $y'' = y''(x)$ by differentiating term-by-term the Fourier series of the functions $y = y(x)$ twice. Therefore, for $x \in \mathbb{R}$, $y'(x) = \sum_{k=1}^{\infty} k(B_k \cos kx - A_k \sin kx)$ and $y''(x) = -\sum_{k=1}^{\infty} k^2(A_k \cos kx + B_k \sin kx)$.
From this and $y'' + py = f(x)$, it follows that $A_0 = \frac{a_0}{p}$ and, for all $k \in \mathbb{N}$,

$$(p - k^2)A_k = a_k \quad \text{and} \quad (p - k^2)B_k = b_k.$$

If $p \notin \mathbb{N}^2$, then there is only one periodic, with a period 2π, solution to the equation:

$$y = \frac{a_0}{2p} + \sum_{k=1}^{\infty} \frac{1}{p - k^2}(a_k \cos kx + b_k \sin kx), \quad x \in \mathbb{R}.$$

If $p = m^2 \in \mathbb{N}^2$ and $a_m = b_m = 0$, then there is an infinite number of periodic, with a period 2π, solutions to the equation. For any $A, B \in \mathbb{R}$, the periodic function

$$y = \frac{a_0}{2\sqrt{m}} + A \cos mx + B \sin mx$$

$$+ \sum_{k \neq m} \frac{1}{m^2 - k^2}(a_k \cos kx + b_k \sin kx), \quad x \in \mathbb{R},$$

is a solution to the equation.

If $p = m^2 \in \mathbb{N}^2$ and $a_m^2 + b_m^2 \neq 0$, then the equation has no solution that is periodic with a period 2π.

2.4 Systems of Differential Equations

Solve the following systems of differential equations and the initial value problems. Also, see Section 4.4.2.

Problem 2.4.1. $x'' - x + 2y'' - 2y = 0$ and $x' - x + y' + y = 0$.

Solution 2.4.1. This is a homogeneous linear system of differential equations. It is possible to solve this system by reducing it to a single differential equation first. From

$$x'' - x = 2(y - y'') \quad \text{and} \quad x' - x = -(y' + y),$$

by subtracting the right-hand equation from the left-hand one, we obtain the equation $x'' - x' = -2y'' + y' + 3y$, i.e., the equation $(x' - x)' = -2y'' + y' + 3y$. This, together with the system's right-hand equation, yields the equation $-(y'+y)' = -2y''+y'+3y$, i.e., the equation $y'' - 2y' - 3y = 0$. The general solution of this homogeneous second-order linear differential equation with constant coefficients is $y = c_1 e^{-t} + c_2 e^{3t}$, $c_1, c_2 \in \mathbb{R}$. We determine the unknown function $x = x(t)$ from the fact that $x' - x = -(y' + y) = -4c_2 e^{3t}$. Thus, $x = c_3 e^t - 2c_2 e^{3t}$, for some $c_3 \in \mathbb{R}$. The general solution to the system is

$$x = c_3 e^t - 2c_2 e^{3t}, \ y = c_1 e^{-t} + c_2 e^{3t}, \ c_1, \ c_2, \ c_3 \in \mathbb{R}.$$

Note: Another way to solve this system is to observe that the system can be rewritten as

$$(x' - x)' + x' - x + 2(y' + y)' - 2(y' + y) = 0 \text{ and } x' - x + y' + y = 0,$$

and then substitute $u = x' - x$ and $v = y' + y$.

Problem 2.4.2. $x' = 5x - 3y + 2e^{3t}$ and $y' = x + y + 5e^{-t}$.

Solution 2.4.2. This is a non-homogeneous linear system of differential equations. It is possible to solve this system by reducing it to a single differential equation first. It follows from the right-hand equation that $x = y' - y - 5e^{-t}$, and after differentiation, $x' = y'' - y' + 5e^{-t}$. This, together with the left-hand equation, yields the equation $y'' - 6y' + 8y = 2e^{3t} - 30e^{-t}$. By using the principle of superposition and the method of undetermined coefficients, one particular solution is $y_p = -2(e^{-t} + e^{3t})$. Hence, $y = c_1 e^{2t} + c_2 e^{4t} - 2(e^{-t} + e^{3t})$, $c_1, c_2 \in \mathbb{R}$. It follows that $x = c_1 e^{2t} + 3c_2 e^{4t} - e^{-t} - 4 e^{3t}$.

Problem 2.4.3. $x' = 2x + y$ and $y' = 4y - x$, with $x(1) = 1$ and $y(1) = 2$.

Solution 2.4.3. The matrix of the system is $A = \begin{bmatrix} 2 & 1 \\ -1 & 4 \end{bmatrix}$. Let $E = \begin{bmatrix} 1 & 0 \\ 0 & 1 \end{bmatrix}$. From, for all $\lambda \in \mathbb{R}$,

$$\det(A - \lambda\, E) = \begin{vmatrix} 2 - \lambda & 1 \\ -1 & 4 - \lambda \end{vmatrix} = (3 - \lambda)^2,$$

it follows that $\lambda = 3$ is an eigenvalue of the matrix A of the algebraic multiplicity 2. The general solution to the system is $t \mapsto Y(t) = \begin{bmatrix} x(t) \\ y(t) \end{bmatrix} = (A_1 + A_2\, t)e^{3t}$, where $A_1 = \begin{bmatrix} \alpha_1 \\ \alpha_2 \end{bmatrix}$ and $A_2 = \begin{bmatrix} \beta_1 \\ \beta_2 \end{bmatrix}$, $\alpha_1, \alpha_2, \beta_1, \beta_2 \in \mathbb{R}$, are such that $Y'(t) = A\, Y(t)$, for all $t \in \mathbb{R}$. It follows that, for all $t \in \mathbb{R}$,

$$(3A_1 + A_2 + 3A_2 t)e^{3t} = A(A_1 + A_2\, t)\, e^{3t},$$

which implies $(A - 3E)A_2 = 0$ and $(A - 3E)A_1 = A_2$.

From, for $\beta_1 = \beta_2 = \beta$, $\beta \begin{bmatrix} 1 \\ 1 \end{bmatrix} = A_2 = (A - 3E)A_1 = \begin{bmatrix} -1 & 1 \\ -1 & 1 \end{bmatrix} \begin{bmatrix} \alpha_1 \\ \alpha_2 \end{bmatrix}$, it follows that $\alpha_2 = \alpha_1 + \beta$.

The general solution to the system is $Y(t) = \left(\begin{bmatrix} \alpha \\ \alpha + \beta \end{bmatrix} + \beta \begin{bmatrix} 1 \\ 1 \end{bmatrix} t \right) e^{3t}$, $t \in \mathbb{R}$, $\alpha, \beta \in \mathbb{R}$, i.e., $x(t) = (\alpha + \beta\, t)e^{3t}$ and $y(t) = (\alpha + \beta\,(1 + t))\, e^{3t}$.

The initial conditions $x(1) = 1$ and $y(1) = 2$ yield the system $\alpha + \beta = e^{-3}$ and $\alpha + 2\beta = 2e^{-3}$. Hence, $\alpha = 0$ and $\beta = e^{-3}$, and the solution to the initial value problem is $x = t\, e^{3(t-1)}$, $y = (1+t)\, e^{3(t-1)}$, $t \in \mathbb{R}$.

Problem 2.4.4. $x' = x - y + z$, $y' = x + y - z$, and $z' = 2z - y$.

Solution 2.4.4. The matrix of the system is $A = \begin{bmatrix} 1 & -1 & 1 \\ 1 & 1 & -1 \\ 0 & -1 & 2 \end{bmatrix}$. Let $E = \begin{bmatrix} 1 & 0 & 0 \\ 0 & 1 & 0 \\ 0 & 0 & 1 \end{bmatrix}$. From, for all $\lambda \in \mathbb{R}$, $\det(A - \lambda E) = (2 - \lambda)(\lambda - 1)^2$, it follows that the matrix A has two eigenvalues: $\lambda_1 = 2$ of the algebraic multiplicity 1 and $\lambda_2 = 1$ of the algebraic multiplicity 2. A solution corresponding to $\lambda_1 = 2$ is of the form $Y_1(t) = A_1 e^{2t}$, where A_1 is an eigenvector corresponding to λ_1. Let $A_1 = \begin{bmatrix} a_1 \\ a_2 \\ a_3 \end{bmatrix}$, for some $a_1, a_2, a_3 \in \mathbb{R}$. The numbers a_1, a_2, and a_3 are determined by the equation $(A - 2E)A_1 = 0$, i.e., by the homogeneous system of linear equations $-a_1 - a_2 + a_3 = 0$, $a_1 - a_2 - a_3 = 0$, and $-a_2 = 0$. It follows that $a_1 = a_3$ and $a_2 = 0$. Hence, $A_1 = a \begin{bmatrix} 1 \\ 0 \\ 1 \end{bmatrix}$, $a \in \mathbb{R}$.

Since the rank of the matrix $A - \lambda_2 E = A - E$ is 2 and since the difference between the order of the system and the multiplicity of the eigenvalue λ_2 is $3 - 2 = 1$, all solutions to the system that correspond to the eigenvalue λ_2 are of the form $Y_2(t) = (B_1 + B_2 t)e^t$. Since $Y_2'(t) = (B_1 + B_2 + B_2 t)e^t$, the vector-columns $B_1 = \begin{bmatrix} b_1 \\ b_2 \\ b_3 \end{bmatrix}$ and $B_2 = \begin{bmatrix} c_1 \\ c_2 \\ c_3 \end{bmatrix}$ must be such that, for all $t \in \mathbb{R}$,

$$(B_1 + B_2 + B_2 t)e^t = A(B_1 + B_2 t)e^t.$$

It follows that B_1 and B_2 are solutions to the system of the two matrix equations $(A - E)B_1 = B_2$ and $(A - E)B_2 = 0$, i.e., $b_1, b_2, b_3, c_1, c_2, c_3 \in \mathbb{R}$ are solutions to the homogeneous system of linear equations $-b_2 + b_3 = c_1$, $b_1 - b_3 = c_2$, $-b_2 + b_3 = c_3$, $-c_2 + c_3 = 0$, $c_1 - c_3 = 0$, and $-c_2 + c_3 = 0$. It follows that $c_1 = c_2 = c_3 = c$, $b_1 = b_3 + c$, $b_2 = b_3 - c$. For $a, b, c \in \mathbb{R}$, the general solution to the system of differential equations is

$$Y(t) = Y_1(t) + Y_2(t) = a \begin{bmatrix} 1 \\ 0 \\ 1 \end{bmatrix} e^{2t} + \left(\begin{bmatrix} b + c \\ b - c \\ b \end{bmatrix} + c \begin{bmatrix} 1 \\ 1 \\ 1 \end{bmatrix} t \right) e^t, \quad t \in \mathbb{R},$$

or, what is the same, $x = e^t(ae^t + b + c + ct)$, $y = e^t(b - c + ct)$, and $z = e^t(ae^t + b + ct)$, $t \in \mathbb{R}$.

Problem 2.4.5. $x' = 4x - y$, $y' = 3x + y - z$, and $z' = x + z$.

Answer 2.4.5. The general solution to the system is, for $t \in \mathbb{R}$ and $a, b, c \in \mathbb{R}$,

$$x = (a + bt + ct^2)\, e^{2t},$$

$$y = (2a - b + 2(b - c)t + 2ct^2)\, e^{2t}, \text{ and}$$

$$z = (a - b + 2c + (b - 2c)t + ct^2)\, e^{2t}.$$

Chapter 3

Difference Equations

3.1 Introduction

Use the following definitions, techniques, properties, and algorithms to solve the problems contained in this chapter. For more details, see $[5, 12, 14, 17, 24, 27, 31]$.

Difference Equation. Let $k \in \mathbb{N}$ and let $F : \mathbb{C}^{k+2} \to \mathbb{C}$. A kth-*order difference equation* is a system with countably many unknowns $x_0, x_1, x_2, \ldots$ and countably many equations $F(n, x_n, \ldots, x_{n+k-1}, x_{n+k}) = 0$, $n \in \mathbb{N}_0$. A solution to this difference equation is any sequence $\{a_n\}_{n \in \mathbb{N}_0}$ such that $F(n, a_n, \ldots, a_{n+k}) = 0$, for all $n \in \mathbb{N}_0$. It is common to write a kth-order difference equation as $x_{n+k} = f(n, x_n, \ldots, x_{n+k-1})$, keeping in mind that n ranges over the set of all nonnegative integers.

Initial Value Problem. Let $k \in \mathbb{N}$, $F : \mathbb{C}^{k+2} \to \mathbb{C}$, and $(\alpha_0, \alpha_1, \ldots, \alpha_{k-1}) \in \mathbb{C}^k$ be given. To solve the *initial value problem*

$$F(n, x_n, \ldots, x_{n+k-1}, x_{n+k}) = 0, \ x_0 = \alpha_0,$$

$$x_1 = \alpha_1, \ \ldots, \ x_{k-1} = \alpha_{k-1},$$

means to determine all sequences $\{a_n\}_{n \in \mathbb{N}_0}$ such that

$$F(n, a_n, \ldots, a_{n+k-1}, a_{n+k}) = 0, \quad \text{for all } n \in \mathbb{N}_0,$$

with $(a_0, a_1, \ldots, a_{k-1}) = (\alpha_0, \alpha_1, \ldots, \alpha_{k-1})$. An initial value problem has at most one solution. See Problem 3.3.3.

139

General and Particular Solutions. Let a kth-order difference equation be given. If a function $(t_1, t_2, \ldots, t_{k+1}) \mapsto f(t_1, t_2, \ldots, t_{k+1})$ has the property that, for each fixed $(c_1, c_2, \ldots, c_k) \in \mathbb{C}^k$ such that $(n, c_1, c_2, \ldots, c_k) \in \text{Dom}(f)$ for all $n \in \mathbb{N}_0$, the sequence $\{f(n, c_1, c_2, \ldots, c_k)\}_{n \in \mathbb{N}_0}$ is a solution to the difference equation, it is common to say and write that $x_n = f(n, c_1, c_2, \ldots, c_k)$ is the *general solution* to the equation. Note that the expression that determines the general solution contains the variable n, which ranges over the set of nonnegative integers, and k complex-valued parameters, $c_1, c_2, \ldots, c_k$. A solution obtained from the general solution for a particular choice of parameters $c_1, c_2, \ldots, c_k$ is called a *particular solution.*

Linear Difference Equations. Let $k \in \mathbb{N}$ and let $k + 1$ sequences of complex numbers be given: $\{\alpha_n^{(m)}\}_{n \in \mathbb{N}_0}$, $m = 1, 2, \ldots, k$, and $\{\beta_n\}_{n \in \mathbb{N}_0}$. We say that $x_{n+k} + \alpha_n^{(1)} x_{n+k-1} + \alpha_n^{(2)} x_{n+k-2} + \cdots + \alpha_n^{(k)} x_n = \beta_n$ is a kth-*order linear difference equation.* If $\beta_n = 0$ for all $n \in \mathbb{N}_0$, we say that the linear difference equation is *homogeneous.* Otherwise, the equation is *non-homogeneous.*

If $\{\{a_n^{(1)}\}_{n \in \mathbb{N}_0}, \{a_n^{(2)}\}_{n \in \mathbb{N}_0}, \ldots, \{a_n^{(k)}\}_{n \in \mathbb{N}_0}\}$ is a *fundamental set of solutions*, i.e., a set of linearly independent solutions to a homogeneous k^{th}-order linear difference equation, then the general solution to the equation is $x_n = c_1 a_n^{(1)} + c_2 a_n^{(2)} + \cdots + c_k a_n^{(k)}$, where $c_i \in \mathbb{C}$ for $i = 1, \ldots, k$. See Problem 3.3.4.

If $\{b_n\}_{n \in \mathbb{N}_0}$ is one solution to a non-homogeneous k^{th}-order linear difference equation and if $\{\{a_n^{(1)}\}_{n \in \mathbb{N}_0}, \{a_n^{(2)}\}_{n \in \mathbb{N}_0}, \ldots, \{a_n^{(k)}\}_{n \in \mathbb{N}_0}\}$ is a set of linearly independent solutions to the corresponding homogeneous equation, then the general solution to the non-homogeneous equation is determined by $x_n = b_n + c_1 a_n^{(1)} + c_2 a_n^{(2)} + \cdots + c_k a_n^{(k)}$, where $c_i \in \mathbb{C}$ for $i = 1, \ldots, k$.

Principle of Superposition. If the sequence $\{a_n\}_{n \in \mathbb{N}_0}$ is a solution to the linear difference equation $x_{n+k} + \alpha_n^{(1)} x_{n+k-1} + \alpha_n^{(2)} x_{n+k-2} + \cdots + \alpha_n^{(k)} x_n = \beta_n$ and if the sequence $\{b_n\}_{n \in \mathbb{N}_0}$ is a solution to the linear difference equation $x_{n+k} + \alpha_n^{(1)} x_{n+k-1} + \alpha_n^{(2)} x_{n+k-2} + \cdots + \alpha_n^{(k)} x_n = \gamma_n$, then the sequence

$\{a_n + b_n\}_{n\in\mathbb{N}_0}$ is a solution to the linear difference equation $x_{n+k} + \alpha_n^{(1)} x_{n+k-1} + \alpha_n^{(2)} x_{n+k-2} + \cdots + \alpha_n^{(k)} x_n = \beta_n + \gamma_n$.

Linear Dependence and Casoratian. The *Casoratian* of the set of sequences $\{\{a_n^{(1)}\}_{n\in\mathbb{N}_0}, \{a_n^{(2)}\}_{n\in\mathbb{N}_0}, \ldots, \{a_n^{(k)}\}_{n\in\mathbb{N}_0}\}$ is the sequence $\{C_n\}_{n\in\mathbb{N}_0}$, where

$$C_n(a_n^{(1)}, a_n^{(2)}, \ldots, a_n^{(k)}) = \begin{vmatrix} a_n^{(1)} & a_n^{(2)} & \cdots & a_n^{(k)} \\ a_{n+1}^{(1)} & a_{n+1}^{(2)} & \cdots & a_{n+1}^{(k)} \\ \vdots & \vdots & & \vdots \\ a_{n+k-1}^{(1)} & a_{n+k-1}^{(2)} & \cdots & a_{n+k-1}^{(k)} \end{vmatrix}.$$

The set $\{\{a_n^{(1)}\}_{n\in\mathbb{N}_0}, \{a_n^{(2)}\}_{n\in\mathbb{N}_0}, \ldots, \{a_n^{(k)}\}_{n\in\mathbb{N}_0}\}$ is a set of linearly dependent sequences if and only if its Casoratian is the zero sequence.[1]

Method of Variation of Parameters. If $\{x_n^{(1)}\}_{n\in\mathbb{N}_0}$ and $\{x_n^{(2)}\}_{n\in\mathbb{N}_0}$ are two linearly independent solutions to the homogeneous second-order difference equation $x_{n+2} + \alpha_n x_{n+1} + \beta_n x_n = 0$, then one particular solution to the non-homogeneous equation $x_{n+2} + \alpha_n x_{n+1} + \beta_n x_n = \gamma_n$ is in the form $\{x_n^{(p)}\}_{n\in\mathbb{N}_0} = \{c_n^{(1)} x_n^{(1)} + c_n^{(2)} x_n^{(2)}\}_{n\in\mathbb{N}_0}$ with $\{c_n^{(1)}\}_{n\in\mathbb{N}_0}$ and $\{c_n^{(2)}\}_{n\in\mathbb{N}_0}$, determined by the system of difference equations:

$$\left(c_{n+1}^{(1)} - c_n^{(1)}\right) x_{n+1}^{(1)} + \left(c_{n+1}^{(2)} - c_n^{(2)}\right) x_{n+1}^{(2)} = 0$$

$$\left(c_{n+1}^{(1)} - c_n^{(1)}\right) x_{n+2}^{(1)} + \left(c_{n+1}^{(2)} - c_n^{(2)}\right) x_{n+2}^{(2)} = \gamma_n.$$

See Problem 3.3.30.

Second-Order Linear Difference Equations with Constant Coefficients. Let $a, b \in \mathbb{R}$, $b \neq 0$, and let $r_1, r_2 \in \mathbb{C}$ be solutions to the quadratic equation $t^2 + at + b = 0$. The general

---

[1] Felice Casorati, 1835–1890, Italian mathematician.

solution $\{x_n\}_{n\in\mathbb{N}_0}$ to the second-order linear difference equations with constant coefficients $x_{n+2} + ax_{n+1} + bx_n = 0$ is determined as follows:

- If $r_1, r_2 \in \mathbb{R}$ with $r_1 \neq r_2$, then $x_n = c_1 r_1^n + c_2 r_2^n$, $c_1, c_2 \in \mathbb{C}$.
- If $r_1 = r_2 = r \in \mathbb{R}$, then $x_n = (c_1 + c_2 n)r^n$, $c_1, c_2 \in \mathbb{C}$.
- If $r_1, r_2 \in \mathbb{C}$ with $r_{1,2} = e^\alpha(\cos\beta \pm i\sin\beta)$, then $x_n = (c_1 \cos n\beta + c_2 \sin n\beta)e^{\alpha n}$, $c_1, c_2 \in \mathbb{C}$.

See Problem 3.3.5.

Method of Undetermined Coefficients. In several special cases, determining a particular solution $\{x_n^{(p)}\}_{n\in\mathbb{N}_0}$ of a non-homogeneous second-order difference equation $x_{n+2}+ax_{n+1}+bx_n = a_n$ reduces to determining coefficients of a polynomial:

a_n	$x_n^{(p)}$
$\sum_{k=0}^m b_k n^{m-k}$	$n^s \sum_{k=0}^m c_k n^{m-k}$
$\alpha^n \sum_{k=0}^m b_k n^{m-k}$	$n^s \alpha^n \sum_{k=0}^m c_k n^{m-k}$
$\alpha^n \cos bn \sum_{k=0}^m b_k n^{m-k}$ or $\alpha^n \sin bx \sum_{k=0}^m b_k n^{m-k}$	$n^s \alpha^n \left(\sin bn \sum_{k=0}^m c_k n^{m-k} + \cos bn \sum_{k=0}^m d_k n^{m-k} \right)$,

where s is the smallest nonnegative integer, $s = 0$, 1, or 2, chosen so that none of the sequences $\{n^s c_k n^{m-k}\}_{n\in\mathbb{N}_0}$, $k = 0, 1, \ldots, m$, is a solution to the corresponding homogeneous difference equation. See Problem 3.3.19 and Problem 3.6.14.

Generating Function. We say that $z \mapsto \sum_{n=0}^\infty a_n z^n$ is the *generating function associated with the sequence* of complex numbers $\{a_n\}_{n\in\mathbb{N}_0}$.

z-Transform. The *z-transform* of a nonzero sequence of complex numbers $\{a_n\}_{n\in\mathbb{N}_0}$ is the function $\mathcal{Z}[a_n](z) = \sum_{n=0}^\infty \frac{a_n}{z^n}$, $|z| > R$, where $R > 0$ depends on the sequence $\{a_n\}_{n\in\mathbb{N}_0}$. By definition, the z-transform of the zero sequence is the zero function. The z-transformation, $\{a_n\}_{n\in\mathbb{N}_0} \mapsto \mathcal{Z}[a_n]$, is a linear mapping, i.e., if $\{a_n\}_{n\in\mathbb{N}_0}$ and $\{b_n\}_{n\in\mathbb{N}_0}$ are sequences and if $\alpha, \beta \in \mathbb{C}$, then $\mathcal{Z}[\alpha a_n + \beta b_n] = \alpha\mathcal{Z}[a_n] + \beta\mathcal{Z}[b_n]$.

Note: In the remainder of this chapter, we will often refer to $\{a_n\}_{n\in\mathbb{N}_0}$ as *the sequence a_n.*

3.2 First-Order Linear Difference Equations

Solve the following first-order difference equations and initial value problems.

Problem 3.2.1. Let $\alpha, \beta, a \in \mathbb{C}\backslash\{0\}$, with $a \neq 1$, and let $\{\beta_n\}_{n\in\mathbb{N}_0}$ be a sequence of complex numbers. Solve:

(a) $x_{n+1} = \alpha x_n$.

(b) $x_{n+1} = \alpha x_n + \beta$.

(c) $x_{n+1} = \alpha x_n + \beta a^n$.

(d) $x_{n+1} = \alpha x_n + \beta_n$.

Solution 3.2.1. Let $\{a_n\}_{n\in\mathbb{N}_0}$ be a solution to the difference equation in the question.

(a) From $a_1 = \alpha a_0$, $a_2 = \alpha a_1 = \alpha^2 a_0$, and $a_3 = \alpha a_2 = \alpha^3 a_0$, by the principle of mathematical induction, it follows that $a_n = \alpha^n a_0$, $n \in \mathbb{N}_0$. The general solution is $x_n = c\alpha^n$, $c \in \mathbb{C}$. For $c \neq 0$, this is a geometric progression.

(b) From $a_1 = \alpha a_0 + \beta$, $a_2 = \alpha a_1 + \beta = \alpha^2 a_0 + (\alpha + 1)\beta$, and $a_3 = \alpha a_2 + \beta = \alpha^3 a_0 + (\alpha^2 + \alpha + 1)\beta$, by the principle of mathematical induction, $a_n = \alpha^n a_0 + (\alpha^{n-1} + \alpha^{n-2} + \cdots + \alpha + 1)\beta$, $a_0 \in \mathbb{C}$, $n \in \mathbb{N}$. From, for $\alpha \neq 1$, $\alpha^{n-1} + \alpha^{n-2} + \cdots + \alpha + 1 = \frac{\alpha^n - 1}{\alpha - 1}$, it follows that $a_n = \alpha^n a_0 + \frac{\alpha^n - 1}{\alpha - 1}\beta = \left(a_0 + \frac{\beta}{\alpha - 1}\right)\alpha^n + \frac{\beta}{1 - \alpha}$, $n \in \mathbb{N}_0$.

Hence, in the case $\alpha \neq 1$, for $c = a_0 + \frac{\beta}{\alpha - 1} \in \mathbb{C}$, the general solution is $x_n = c\alpha^n + \frac{\beta}{1-\alpha}$, $c \in \mathbb{C}$. If $\alpha = 1$, the general solution is $x_n = c + n\beta$, $c \in \mathbb{C}$.

(c) From $a_1 = \alpha a_0 + \beta$, $a_2 = \alpha a_1 + \beta a = \alpha^2 a_0 + \beta(\alpha + a) = \alpha^2 a_0 + \alpha\beta\left(1 + \frac{a}{\alpha}\right)$, and $a_3 = \alpha^3 a_0 + \alpha^2\beta\left(1 + \frac{a}{\alpha} + \left(\frac{a}{\alpha}\right)^2\right)$, by the principle of mathematical induction, $a_n = \alpha^n a_0 + \alpha^{n-1}\beta\left(1 + \frac{a}{\alpha} + \cdots + \left(\frac{a}{\alpha}\right)^{n-1}\right)$, $n \in \mathbb{N}$.

Since, for $a \neq \alpha$, $1 + \frac{a}{\alpha} + \cdots + \left(\frac{a}{\alpha}\right)^{n-1} = \frac{\left(\frac{a}{\alpha}\right)^n - 1}{\frac{a}{\alpha} - 1}$, it follows that, for $a \neq \alpha$, $a_n = \alpha^n a_0 + \alpha^{n-1}\beta\frac{\left(\frac{a}{\alpha}\right)^n - 1}{\frac{a}{\alpha} - 1} = \alpha^n a_0 + \beta\frac{a^n - \alpha^n}{a - \alpha} = \left(a_0 + \frac{\beta}{\alpha - a}\right)\alpha^n + \frac{\beta a^n}{a - \alpha}$, $n \in \mathbb{N}_0$. Hence, in the case $\alpha \neq a$, for

$c = a_0 + \frac{\beta}{a-\alpha} \in \mathbb{C}$, the general solution is $x_n = c\alpha^n + \frac{\beta a^n}{a-\alpha}$, $c \in \mathbb{C}$. If $a = \alpha$, the general solution is $x_n = (c\alpha + n\beta)\alpha^{n-1}$, $c \in \mathbb{C}$.

(d) From $a_1 = \alpha a_0 + \beta_0$, $a_2 = \alpha a_1 + \beta_1 = \alpha^2 a_0 + \alpha\beta_0 + \beta_1$, and $a_3 = \alpha a_2 + \beta_2 = \alpha_3 a_0 + \alpha^2 \beta_0 + \alpha\beta_1 + \beta_2$, by the principle of mathematical induction, the general solution is $x_n = c\alpha^n + \sum_{k=0}^{n-1} \alpha^{n-k-1}\beta_k$, $n \geq 1$, $x_0 = c \in \mathbb{C}$.

Problem 3.2.2. Let $\beta \in \mathbb{C}\backslash\{0\}$, and let $\{\alpha_n\}_{n\in\mathbb{N}_0}$ and $\{\beta_n\}_{n\in\mathbb{N}_0}$ be sequences of complex numbers. Solve the following difference equations:

(a) $x_{n+1} = \alpha_n x_n$.   (b) $x_{n+1} = \alpha_n x_n + \beta$.   (c) $x_{n+1} = \alpha_n x_n + \beta_n$.

Solution 3.2.2. Let $\{a_n\}_{n\in\mathbb{N}_0}$ be a solution to the difference equation in the question.

(a) From $a_1 = \alpha_0 a_0$, $a_2 = \alpha_1 a_1 = \alpha_1 \alpha_0 a_0$, and $a_3 = \alpha_2 a_2 = \alpha_2 \alpha_1 \alpha_0 a_0$, by the principle of mathematical induction, the general solution is $x_n = c\prod_{i=0}^{n-1} \alpha_i$, $n \geq 1$, $x_0 = c \in \mathbb{C}$.

(b) From $a_1 = \alpha_0 a_0 + \beta$, $a_2 = \alpha_1 a_1 + \beta = \alpha_1 \alpha_0 a_0 + \beta(\alpha_1 + 1)$, and $a_3 = \alpha_2 a_2 + \beta = \alpha_2 \alpha_1 \alpha_0 a_0 + \beta(\alpha_2 \alpha_1 + \alpha_2 + 1)$, by the principle of mathematical induction, the general solution is $x_n = c\prod_{k=0}^{n-1} \alpha_k + \beta(1 + \sum_{k=1}^{n-1} \prod_{m=k}^{n-1} \alpha_m)$, $n \geq 2$, $x_1 = c\alpha_0 + \beta$, $x_0 = c \in \mathbb{C}$.

(c) From $a_1 = \alpha_0 a_0 + \beta_0$, $a_2 = \alpha_1 a_1 + \beta_1 = \alpha_1 \alpha_0 a_0 + \alpha_1 \beta_0 + \beta_1$, and $a_3 = \alpha_2 a_2 + \beta_2 = \alpha_2 \alpha_1 \alpha_0 a_0 + \alpha_2 \alpha_1 \beta_0 + \alpha_2 \beta_1 + \beta_2$, by the principle of mathematical induction, the general solution is $x_n = c\prod_{k=0}^{n-1} \alpha_k + \sum_{k=0}^{n-2} \beta_k \prod_{i=k+1}^{n-1} \alpha_i + \beta_{n-1}$, $n \geq 2$, $x_1 = c\alpha_0 + \beta_0$, $x_0 = c \in \mathbb{C}$.

Problem 3.2.3. Determine an initial value problem with the solution:

(a) $20, 15, 10, 5, \ldots$ (b) $20, 10, 5, 2.5, \ldots$ (c) $\{2^n - 5^{n+2}\}_{n\in\mathbb{N}_0}$.

Solution 3.2.3.

(a) $x_{n+1} = x_n - 5$, $x_0 = 20$.   (b) $x_{n+1} = 0.5x_n$, $x_0 = 20$.

(c) Observe that $a_0 = -24$. From $a_n = 2^n - 5^{n+2}$ and $a_{n+1} = 2^{n+1} - 5^{n+3}$, it follows that $a_{n+1} = 2(a_n + 5^{n+2}) - 5 \cdot 5^{n+2} = 2a_n - 3 \cdot 5^{n+2}$. The sequence is the solution to the initial value problem $x_{n+1} = 2x_n - 3 \cdot 5^{n+2}$, $x_0 = -24$. The same sequence is the solution to the initial value problem $x_{n+1} = 5x_n - 3 \cdot 2^n$, $x_0 = -24$.

Problem 3.2.4. Determine a first-order difference equation with the general solution:
(a) $x_n = 2^n c + 3^{n-1} n$, $c \in \mathbb{C}$.   (b) $x_n = 2^{n-1}(2c + 3n)$, $c \in \mathbb{C}$.

Solution 3.2.4.

(a) Eliminate the parameter c from $x_n = 2^n c + 3^{n-1} n$ and $x_{n+1} = 2^{n+1} c + 3^n (n+1)$ to obtain the difference equation $x_{n+1} = 2x_n + 3^{n-1}(n+3)$.
(b) Eliminate the parameter c from $x_n = 2^{n-1}(2c + 3n)$ and $x_{n+1} = 2^n(2c + 3n + 3)$ to obtain the difference equation $x_{n+1} = 2x_n + 3 \cdot 2^n$.

Problem 3.2.5. Solve the following initial value problems:

(a) $x_{n+1} = 2x_n$, $x_0 = 5$.   (c) $5x_{n+1} - 3x_n = 2x_{n+1} - x_n$, $x_0 = 100$.
(b) $x_{n+1} = \frac{3}{4}x_n$, $x_0 = 100$.

Answer 3.2.5. See Problem 3.2.1(a).

(a) $a_n = 5 \cdot 2^n$.   (b) $a_n = 100 \left(\frac{3}{4}\right)^n$.   (c) $a_n = 100 \left(\frac{2}{3}\right)^n$.

Problem 3.2.6. Solve the following homogeneous first-order difference equations and initial value problems:

(a) $x_{n+1} = (n+1)x_n$, $x_0 = 3$.   (b) $x_{n+1} - 3(n+1)x_n = 0$, $x_0 = \alpha$.
(c) $x_{n+1} - \frac{n+1}{n+2}x_n = 0$, $x_0 = \alpha$.   (d) $x_{n+1} - (n+1)e^{2n}x_n = 0$, $x_0 = \alpha$.

Answer 3.2.6. See Problem 3.2.2(a).

(a) $a_n = 3n!$.   (b) $a_n = \alpha 3^n n!$.

(c) $a_n = \alpha \prod_{k=0}^{n-1} \frac{k+1}{k+2} = \frac{\alpha}{n+1}$.
(d) $a_n = \alpha \Pi_{k=0}^{n-1}(k+1)e^{2k} = \alpha n! e^{2\sum_{k=0}^{n-1} k} = \alpha n! e^{n(n-1)}$.

Problem 3.2.7. Solve the following non-homogeneous first-order difference equations and initial value problems:

(a) $x_{n+1} - \frac{1}{2}x_n = 2$, $x_0 = \alpha$.   (c) $x_{n+1} = 4x_n - 3$.

(b) $2x_{n+1} - x_n = 6$, $x_0 = 6$.   (d) $x_{n+1} - 2x_n = 3^n n^2$.

Answer 3.2.7. See Problem 3.2.2(b) and (c).

(a) $a_n = \frac{\alpha}{2^n} + \frac{\frac{1}{2^n}-1}{\frac{1}{2}-1} \cdot 2 = \frac{\alpha-4}{2^n} + 4$.   (b) $a_n = \left(6 + \frac{3}{\frac{1}{2}-1}\right)\frac{1}{2^n} + \frac{3}{1-\frac{1}{2}} = 6$.

(c) $a_n = (a_0 - 1)4^n + 1$, $a_0 \in \mathbb{C}$.

(d) $a_n = 2^n c + \sum_{k=1}^{n-1} 2^{n-k-1}3^k k^2$, $c \in \mathbb{C}$.

Problem 3.2.8. Solve the initial value problem $x_{n+1} - \frac{n+1}{n+2} \cdot x_n = 4$, $x_0 = \alpha \in \mathbb{C}$.

Solution 3.2.8. Let $\{a_n\}_{n\in\mathbb{N}_0}$ be one solution to the initial value problem. Then, $a_0 = \alpha$ and $a_1 = \frac{\alpha}{2} + 4$. For $n \geq 2$, by Problem 3.2.2(b),

$$a_n = \alpha \prod_{k=0}^{n-1} \frac{k+1}{k+2} + 4 \sum_{k=1}^{n-1} \prod_{m=k}^{n-1} \frac{m+1}{m+2} + 4$$

$$= \frac{\alpha}{n+1} + \frac{4}{n+1} \sum_{k=1}^{n-1}(k+1) + 4$$

$$= \frac{\alpha}{n+1} + \frac{4}{n+1} \cdot \left(\frac{n(n-1)}{2} + (n-1)\right) + 4$$

$$= \frac{\alpha}{n+1} + \frac{2(n-1)(n+2)}{n+1} + 4 = \frac{\alpha + 2n(n+3)}{n+1}.$$

Problem 3.2.9. Solve the non-homogeneous first-order difference equation $x_{n+1} - 2x_n = \cos 2n$.

Solution 3.2.9. Recall that $\cos 2n = \frac{e^{2ni}+e^{-2ni}}{2}$, where i is the imaginary unit. By Problem 3.2.1(c), the sequence $\left\{\frac{e^{2ni}}{2(e^{2i}-2)}\right\}_{n\in\mathbb{N}_0}$ is a solution to the equation $x_{n+1} = 2x_n + \frac{e^{2ni}}{2}$ and the sequence $\left\{\frac{e^{-2ni}}{2(e^{-2i}-2)}\right\}_{n\in\mathbb{N}_0}$ is a solution to the equation $x_{n+1} = 2x_n + \frac{e^{-2ni}}{2}$.

By the principle of superposition, the general solution is

$$x_n = 2^n c + \frac{e^{2ni}}{2(e^{2i} - 2)} + \frac{e^{-2ni}}{2(e^{-2i} - 2)}$$

$$= 2^n c + \frac{e^{2(n-1)i} - 2e^{2ni} + e^{-2(n-1)i} - 2e^{-2ni}}{2(5 - 2(e^{2i} + e^{-2i}))}$$

$$= 2^n c + \frac{\cos 2(n - 1) - 2\cos 2n}{5 - 4\cos 2}, \quad c \in \mathbb{C}.$$

Problem 3.2.10. Solve the following initial value problems:

(a) $x_{n+1} = (n + 1)x_n + 2^n(n + 1)!, \; x_0 = 1.$
(b) $x_{n+1} - (n + 1)x_n = 2n(n + 1)!, \; x_0 = \alpha.$

Answer 3.2.10. (a) $a_n = 2^n n!.$ (b) $a_n = (\alpha + n(n - 1))n!.$ See Problem 3.2.2(c).

Problem 3.2.11. Solve the initial value problem $x_{n+1} - x_n(1 - \alpha - \beta) = \alpha, \; \alpha, \; \beta \in [0, 1], \; x_0 = 0.$

Solution 3.2.11. If $\alpha^2 + \beta^2 \neq 0$, then $1 - \alpha - \beta \neq 1$ and, by Problem 3.2.1(b), the general solution is $x_n = c(1 - \alpha - \beta)^n + \frac{\alpha}{\alpha+\beta}, c \in \mathbb{C}.$ The solution to the initial value problem is $a_n = \frac{\alpha(1-(1-\alpha-\beta)^n)}{\alpha+\beta}.$ If $\alpha = \beta = 0$, then the initial value problem becomes $x_{n+1} = x_n,$ $x_0 = 0.$ Its solution is $a_n = 0, \; n \in \mathbb{N}_0.$

Problem 3.2.12. Solve the initial value problem $2x_{n+1} + (1 - 2\alpha) \cdot x_n = 1, \; \alpha \in [0, 1], \; x_0 = 0.$

Solution 3.2.12. Since, for any $\alpha \in [0, 1], \; 1 - 2\alpha \neq 2$, the general solution is $x_n = \frac{c(2\alpha-1)^n}{2^n} + \frac{\frac{1}{2}}{1-\frac{2\alpha-1}{2}} = \frac{c(2\alpha-1)^n}{2^n} + \frac{1}{3-2\alpha}, c \in \mathbb{C}.$ From the initial condition, it follows that $c = \frac{1}{2\alpha-3}.$ The solution to the initial value problem is $a_n = \frac{2^n - (2\alpha-1)^n}{2^n(3-2\alpha)}.$

Problem 3.2.13. The sequence $\{a_n\}_{n\in\mathbb{N}_0}$ is the solution to the initial value problem $x_{n+1} = 10 + 3x_n, \; x_0 = 0.$ Determine the smallest value of n for which a_n exceeds one million.

Solution 3.2.13. From $a_n = 5(3^n - 1) > 10^6$, it follows that $n > \frac{\ln 200{,}001}{\ln 3} \approx 11.11$. Hence, $a_{12} > 10^6$. Indeed, $a_{11} = 5(3^{11} - 1) = 885{,}730$ and $a_{12} = 5(3^{12} - 1) = 2{,}657{,}200$.

Problem 3.2.14. Determine α and β so that the solution to the initial value problem $x_{n+1} = \alpha x_n + \beta$, $x_0 = 1000$, is the sequence $1000, 100, -350, \ldots$.

Answer 3.2.14. $\alpha = \frac{1}{2}$ and $\beta = -400$.

3.3   Second-Order Linear Difference Equations

Problem 3.3.1. If $x_{n+2} = 2x_{n+1} + x_n$ and $x_1 = 1$, $x_2 = 5$, determine the values of x_3, x_4, and x_5.

Answer 3.3.1. $x_3 = 2x_2 + x_1 = 2 \cdot 5 + 1 = 11$, $x_4 = 27$, and $x_5 = 65$.

Problem 3.3.2. Determine $\alpha, \beta \in \mathbb{R}$ so that the sequence $1, 2, 8, 20, 68, \ldots$ is a solution to the second-order linear difference equation $x_{n+2} = \alpha x_{n+1} + \beta x_n$.

Solution 3.3.2. From $8 = 2\alpha + \beta$ and $20 = 8\alpha + 2\beta$, it follows that $\alpha = 1$ and $\beta = 6$. Observe that $1 \cdot 20 + 6 \cdot 8 = 68$.

Problem 3.3.3. Prove that the initial value problem

$$x_{n+k} = f(n, x_{n+k-1}, \ldots, x_n), \quad (x_0, x_1, \ldots, x_{k-1}) = (\alpha_0, \alpha_1, \ldots, \alpha_{k-1}),$$

cannot have two solutions.

Solution 3.3.3. Suppose that the sequences $\{a_n\}_{n \in \mathbb{N}_0}$ and $\{b_n\}_{n \in \mathbb{N}_0}$ are solutions to the initial value problem

$$x_{n+k} = f(n, x_{n+k-1}, \ldots, x_n), \quad (x_0, x_1, \ldots, x_{k-1}) = (\alpha_0, \alpha_1, \ldots, \alpha_{k-1}).$$

Let $M = \{m \in \mathbb{N}_0 : a_i = b_i \text{ for all } i \in \{m, \ldots, m + k - 1\}\}$. Then, since $\{a_n\}_{n \in \mathbb{N}_0}$ and $\{b_n\}_{n \in \mathbb{N}_0}$ are solutions to the initial value problem, $0 \in M$. Suppose that $m \in M$. Then, for all $i \in \{m, \ldots, m + k - 1\}$, $a_i = b_i$. From

$$a_{m+k} = f(m, a_{m+k-1}, \ldots, a_m) = f(m, b_{m+k-1}, \ldots, b_m) = b_{m+k},$$

it follows that for all $i \in \{m+1, \ldots, m+k\} = \{m+1, \ldots, (m+1) + k - 1\}$, $a_i = b_i$, i.e., $m + 1 \in M$. By the principle of mathematical induction, $M = \mathbb{N}_0$. Since for every $n \in \mathbb{N}_0$, $n \in \{n, \ldots, n + k - 1\}$, it follows that $a_n = b_n$, which completes the proof of the claim.

Problem 3.3.4. Suppose that $\{\alpha_n\}_{n \in \mathbb{N}_0}$ is a sequence of complex numbers and that and $\{\beta_n\}_{n \in \mathbb{N}_0}$ is a nonzero sequence of complex numbers. Suppose that $\{x_n^{(1)}\}_{n \in \mathbb{N}_0}$ and $\{x_n^{(2)}\}_{n \in \mathbb{N}_0}$ are two linearly independent solutions to the homogeneous equation $x_{n+2} + \alpha_n x_{n+1} + \beta_n x_n = 0$. Use Problem 3.3.3 to prove that any solution $\{a_n\}_{n \in \mathbb{N}_0}$ of the homogeneous equation is a linear combination of $\{x_n^{(1)}\}_{n \in \mathbb{N}_0}$ and $\{x_n^{(2)}\}_{n \in \mathbb{N}_0}$, i.e., there are c_1, $c_2 \in \mathbb{C}$ such that $a_n = c_1 x_n^{(1)} + c_2 x_n^{(2)}$, $n \in \mathbb{N}$.

Solution 3.3.4. Adjust Solution 2.3.12 to the case of a homogeneous second-order difference equation.

Problem 3.3.5. Let a, $b \in \mathbb{R}$, $b \neq 0$. Solve $x_{n+2} + a x_{n+1} + b x_n = 0$, a homogeneous second-order linear difference equation with constant coefficients.

Solution 3.3.5. Let r be a root, real or complex, of the quadratic equation[2] $t^2 + at + b = 0$. Observe that, since $b \neq 0$, $r \neq 0$. Let $a_n = r^n$, $n \in \mathbb{N}_0$. Then, for all $n \in \mathbb{N}_0$,

$$a_{n+2} + a a_{n+1} + b a_n = r^n(r^2 + ar + b) = 0,$$

which implies that $\{r^n\}_{n \in \mathbb{N}_0}$ is a solution to the difference equation. We distinguish the following three cases:

(a) If $a^2 - 4b > 0$, then the quadratic equation has two distinct real roots r_1 and r_2. Both $\{x_n^{(1)}\}_{n \in \mathbb{N}_0} = \{r_1^n\}_{n \in \mathbb{N}_0}$ and $\{x_n^{(2)}\}_{n \in \mathbb{N}_0} = \{r_2^n\}_{n \in \mathbb{N}_0}$ are solutions to the difference equation. From, for all $n \in \mathbb{N}_0$,

$$C_n = \begin{vmatrix} r_1^n & r_2^n \\ r_1^{n+1} & r_2^{n+1} \end{vmatrix} = (r_2 - r_1)(r_1 r_2)^n \neq 0,$$

it follows that $\{x_n^{(1)}\}_{n \in \mathbb{N}_0}$ and $\{x_n^{(2)}\}_{n \in \mathbb{N}_0}$ are linearly independent. By Problem 3.3.4, the general solution to the equation is $x_n = c_1 r_1^n + c_2 r_2^n$, $c_1, c_2 \in \mathbb{C}$.

---

[2]This is the *characteristic equation* of the difference equation $x_{n+2} + a x_{n+1} + b x_n = 0$.

(b) If $a^2 - 4b = 0$, then the quadratic equation has a single real root $r = -\frac{a}{2}$. The sequence $\{r^n\}_{n\in\mathbb{N}_0}$ is a solution to the difference equation. Another, linearly independent from $\{r^n\}_{n\in\mathbb{N}_0}$, solution is $\{nr^n\}_{n\in\mathbb{N}_0}$. The general solution to the equation is $x_n = (c_1 + c_2 n)r^n$, $c_1, c_2 \in \mathbb{C}$.

(c) If $a^2 - 4b < 0$, then the quadratic equation has two complex roots: $r_1 = e^{\alpha+i\beta} = -\frac{a}{2} + i\sqrt{b - \frac{a^2}{4}}$ and $r_2 = e^{\alpha-i\beta}$. It follows that two linearly independent solutions to the difference equation are, for $n \in \mathbb{N}_0$, $x_n^{(1)} = r_1^n = e^{\alpha n}(\cos\beta n + i\sin\beta n)$ and $x_n^{(2)} = r_2^n = e^{\alpha n}(\cos\beta n - i\sin\beta n)$. Two linearly independent real-valued solutions are obtained in the following way:

$$X_n^{(1)} = \frac{x_n^{(1)} + x_n^{(2)}}{2} = e^{\alpha n}\cos\beta n \text{ and } X_n^{(2)} = \frac{x_n^{(1)} - x_n^{(2)}}{2i} = e^{\alpha n}\sin\beta n.$$

The general solution to the equation is $x_n = (c_1\cos\beta n + c_2\sin\beta n)e^{\alpha n}$, $c_1, c_2 \in \mathbb{C}$.

Note: Observe that $e^\alpha = |r_1| = |r_2|$.

Problem 3.3.6. Determine an initial value problem, $x_{n+2} + ax_{n+1} + bx_n = 0$, $x_0 = \alpha_0$, $x_1 = \alpha_1$, with the solution:

(a) $\{a_n\}_{n\in\mathbb{N}_0} = \{2^n - 5^{n+2}\}_{n\in\mathbb{N}_0}$.

(b) $\{a_n\}_{n\in\mathbb{N}_0} = \{3\cos\frac{n\pi}{2} - \sin\frac{n\pi}{2}\}_{n\in\mathbb{N}_0}$.

Solution 3.3.6.

(a) Observe that $a_0 = -24$ and $a_1 = -121$.

Solution 1: Since $r_1 = 2$ and $r_2 = 5$ are the roots of the quadratic equation $(r-2)(r-5) = r^2 - 7r + 10 = 0$, the given sequence is the solution to the initial value problem $x_{n+2} - 7x_{n+1} + 10x_n = 0$, $x_0 = -24$, $x_1 = -121$.

Solution 2: From, for all $n \in \mathbb{N}_0$, $2^{n+2} - 5^{n+4} + a(2^{n+1} - 5^{n+3}) + b(2^n - 5^{n+2}) = 0$, it follows that $2^n(4+2a+b) = 5^{n+2}(25+5a+b)$. Hence, a and b satisfy the system of equations $4 + 2a + b = 0$ and $25 + 5a + b = 0$. Thus, $a = -7$ and $b = 10$.

Solution 3: From $a_n = 2^n - 5^{n+2}$ and $a_{n+1} = 2^{n+1} - 5^{n+3} = 2\cdot2^n - 5\cdot5^{n+2}$, it follows that $2^n = \frac{5a_n - a_{n+1}}{3}$ and $5^{n+2} = \frac{2a_n - a_{n+1}}{3}$. This, together with $a_{n+2} = 2^{n+2} - 5^{n+4} = 4\cdot2^n - 25\cdot5^{n+2}$, implies that $a_{n+2} = 4\cdot\frac{5a_n - a_{n+1}}{3} - 25\cdot\frac{2a_n - a_{n+1}}{3} = \frac{-30a_n + 21a_{n+1}}{3} = 7a_{n+1} - 10a_n$.

(b) Observe that $a_0 = 3$ and $a_1 = -1$.

Solution 1: Since $\cos \frac{n\pi}{2} = \frac{e^{\frac{n\pi i}{2}} + e^{-\frac{n\pi i}{2}}}{2}$ and $\sin \frac{n\pi}{2} = \frac{e^{\frac{n\pi i}{2}} - e^{-\frac{n\pi i}{2}}}{2i}$, we consider the quadratic equation with the roots $r_1 = e^{\frac{\pi i}{2}} = i$ and $r_2 = e^{-\frac{\pi i}{2}} = -i$, i.e., the equation $(r - i)(r + i) = r^2 + 1 = 0$. The sequence is the solution to the initial value problem $x_{n+2} + x_n = 0$, $x_0 = 3$, $x_1 = -1$.

Solution 2: Observe that for $n \in \mathbb{N}_0$, $a_{n+2} = 3\cos \frac{(n+2)\pi}{2} - \sin \frac{(n+2)\pi}{2} = 3\cos \left(\frac{n\pi}{2} + \pi \right) - \sin \left(\frac{n\pi}{2} + \pi \right) = -3\cos \frac{n\pi}{2} + \sin \frac{n\pi}{2} = -a_n$.

Problem 3.3.7. Determine a second-order linear difference equation with the general solution:

(a) $x_n = \frac{c_1}{n} + c_2$, $c_1, c_2 \in \mathbb{C}$.
(b) $x_n = c_1 \cos n\theta + c_2 \sin n\theta$, $c_1, c_2 \in \mathbb{C}$.
(c) $x_n = c_1 2^n + c_2 3^n + 0.5$, $c_1, c_2 \in \mathbb{C}$.

Solution 3.3.7.

(a) From $x_n = \frac{c_1}{n} + c_2$ and $x_{n+1} = \frac{c_1}{n+1} + c_2$, it follows that $c_1 = n(n+1)(x_n - x_{n+1})$ and $c_2 = (n+1)x_{n+1} - nx_n$. This, together with $x_{n+2} = \frac{c_1}{n+2} + c_2$, implies that $x_{n+2} = \frac{2(n+1)}{n+2}x_{n+1} - \frac{n}{n+2}x_n$. The sequence is the general solution to the homogeneous second-order linear difference equation $(n+2)x_{n+2} - 2(n+1)x_{n+1} + nx_n = 0$.

(b) *Solution 1:* Since $\cos n\theta = \frac{e^{n\theta i} + e^{-n\theta i}}{2}$ and $\sin n\theta = \frac{e^{n\theta i} - e^{-n\theta i}}{2i}$, we consider the quadratic equation with the roots $r_1 = e^{\theta i}$ and $r_2 = e^{-\theta i}$, i.e., the equation $(r - e^{\theta i})(r - e^{-\theta i}) = r^2 - 2r \cos \theta + 1 = 0$. The sequence is the general solution to the homogeneous second-order linear difference equation $x_{n+2} - 2x_{n+1} \cos \theta + x_n = 0$.

Solution 2: From $x_n = c_1 \cos n\theta + c_2 \sin n\theta$ and $x_{n+1} = c_1 \cos(n+1)\theta + c_2 \sin(n+1)\theta$, it follows that

$$c_1 = \frac{\begin{vmatrix} x_n & \sin n\theta \\ x_{n+1} & \sin(n+1)\theta \end{vmatrix}}{\begin{vmatrix} \cos n\theta & \sin n\theta \\ \cos(n+1)\theta & \sin(n+1)\theta \end{vmatrix}} = \frac{x_n \sin(n+1)\theta - x_{n+1} \sin n\theta}{\sin \theta}$$

and $c_2 = \frac{x_{n+1} \cos n\theta - x_n \cos(n+1)\theta}{\sin \theta}$. This, together with $x_{n+2} = c_1 \cos(n+2)\theta + c_2 \sin(n+2)\theta$, implies that $x_{n+2} = 2x_{n+1} \cos \theta - x_n$.

(c) *Solution 1:* Observe that $x_n^{(h)} = c_1 2^n + c_2 3^n$ is the general solution to the homogeneous second-order linear equation equation $x_{n+2} - 5x_{n+1} + 6x_n = 0$. One solution to the still unknown non-homogeneous linear equation is a constant sequence $\{0.5\}_{n\in\mathbb{N}_0}$. From $0.5 - 5 \cdot 0.5 + 6 \cdot 0.5 = 1$, it follows that the sequence is the general solution to the non-homogeneous second-order difference equation $x_{n+2} - 5x_{n+1} + 6x_n = 1$.

Solution 2: From $x_n = c_1 2^n + c_2 3^n + 0.5$ and $x_{n+1} = c_1 2^{n+1} + c_2 3^{n+1} + 0.5$, it follows that $c_1 = 2^{-n}(3x_n - x_{n+1} - 1)$ and $c_2 = 3^{-n}(-2x_n + x_{n+1} + 0.5)$. This, together with $x_{n+2} = c_1 2^{n+2} + c_2 3^{n+2}$, implies that $x_{n+2} = -6x_n + 5x_{n+1} + 1$.

Problem 3.3.8. Prove that if one of its solutions is known, then the problem of solving a homogeneous second-order difference equation reduces to the problem of solving two first-order difference equations.

Solution 3.3.8. Suppose that $\{x_n^{(1)}\}_{n\in\mathbb{N}_0}$, $x_n^{(1)} \neq 0$ for all $n \in \mathbb{N}_0$, is a solution to the equation $x_{n+2} + \alpha_n x_{n+1} + \beta_n x_n = 0$. The substitution $x_n = x_n^{(1)} y_n$ yields a homogeneous second-order equation in terms of the unknown sequence $\{y_n\}_{n\in\mathbb{N}_0}$:

$$0 = x_{n+2}^{(1)} y_{n+2} + \alpha_n x_{n+1}^{(1)} y_{n+1} + \beta_n x_n^{(1)} y_n$$

$$= y_{n+1}(x_{n+2}^{(1)} + \alpha_n x_{n+1}^{(1)} + \beta_n x_n^{(1)}) - x_{n+2}^{(1)} y_{n+1}$$

$$- \beta_n x_n^{(1)} y_{n+1} + x_{n+2}^{(1)} y_{n+2} + \beta_n x_n^{(1)} y_n$$

$$= x_{n+2}^{(1)}(y_{n+2} - y_{n+1}) - \beta_n x_n^{(1)}(y_{n+1} - y_n).$$

The substitution $z_n = y_{n+1} - y_n$ yields a homogeneous first-order difference equation $x_{n+2}^{(1)} z_{n+1} - \beta_n x_n^{(1)} z_n = 0$. After finding one solution $\{z_n^{(1)}\}_{n\in\mathbb{N}_0}$ to this equation (see Problem 3.2.2(a)), we find one solution $\{y_n^{(1)}\}_{n\in\mathbb{N}_0}$ to the first-order non-homogeneous linear equation $y_{n+1} - y_n = z_n^{(1)}$ (see Problem 3.2.1(d)). The sequence $\{x_n^{(1)} y_n^{(1)}\}_{n\in\mathbb{N}_0}$ is another solution, linearly independent from $\{x_n^{(1)}\}_{n\in\mathbb{N}_0}$, to the second-order equation.

Problem 3.3.9. Solve $x_{n+2} - (n+2)(n+1)x_n = 0$.

Solution 3.3.9. If n is an even number, then $x_{n+2} = (n+2)(n+1)x_n = \cdots = (n+2)(n+1)\cdots 2 \cdot 1 \cdot x_0$, and if n is an odd number, then $x_{n+2} = (n+2)(n+1)x_n = \cdots = (n+2)(n+1)\cdots 3\cdot 2\cdot x_1$. For $x_0 = x_1 = 1$, $x_n^{(1)} = n!$ is a solution to the given equation. The substitution $x_n = x_n^{(1)}y_n = n!y_n$ yields the equation $(n+2)!y_{n+2} - (n+2)(n+1)\cdot n!y_n = 0$, or, what is the same, $y_{n+2} - y_n = 0$. The substitution $z_n = y_{n+1} - y_n$ yields the difference equation $z_{n+1} + z_n = 0$. A particular solution, $z_n^{(1)} = (-1)^n$, yields $y_{n+1} - y_n = (-1)^n$. One solution to this equation is $y_n = \frac{1+(-1)^n}{2}$. This implies that $x_n^{(2)} = \frac{1+(-1)^n}{2}n!$ is another solution to the homogeneous second-order difference equation. The general solution to the given equation is, for $c_1, c_2 \in \mathbb{C}$,

$$x_n = n!\left(c_1 + c_2\frac{1+(-1)^n}{2}\right) = n!(c_1' + c_2'(-1)^n),\ c_1',\ c_2' \in \mathbb{C}.$$

Note: Since the roots of the quadratic equation $r^2 - 1 = 0$ are $r_1 = -1$ and $r_2 = 1$, it follows that the general solution to the homogeneous second-order difference equation with constant coefficients $y_{n+2} - y_n = 0$ is $y_n = c_1 + c_2(-1)^n$, $c_1, c_2 \in \mathbb{C}$.

Problem 3.3.10. Solve $x_{n+2} - x_{n+1} - (n+1)^2 x_n = 0$.

Solution 3.3.10. If $x_0 = x_1 = 1$, then $x_2 = x_1 + (0+1)^2 x_0 = 2$, $x_3 = x_2 + 2^2 x_1 = 2 + 2^2 = 2(1+2) = 2\cdot 3 = 3!$, and $x_4 = x_3 + 3^2 x_2 = 3! + 3^2 \cdot 2 = 3!(1+3) = 4!$. This suggests that a particular solution to the equation is $x_n^{(1)} = n!$, $n \in \mathbb{N}_0$. Indeed, $x_{n+1}^{(1)} + (n+1)^2 x_n^{(1)} = (n+1)! + (n+1)^2 n! = (n+1)!(n+2) = x_{n+2}^{(1)}$. The substitution $x_n = x_n^{(1)}y_n$ yields the equation $(n+2)!y_{n+2} - (n+1)!y_{n+1} - (n+1)^2 n!y_n = 0$, which is the same as $(n+2)y_{n+2} - y_{n+1} - (n+1)y_n = 0$. We rewrite this equation as $(n+2)(y_{n+2} - y_{n+1}) + (n+1)(y_{n+1} - y_n) = 0$ to obtain, by the substitution $y_{n+1} - y_n = z_n$, a homogeneous first-order linear difference equation $(n+2)z_{n+1} + (n+1)z_n = 0$. It follows (see Problem 3.2.2(a)) that $z_n = \frac{c_1(-1)^{n-1}}{n+1}$, $c_1 \in \mathbb{C}$. Next, we solve $y_{n+1} - y_n = \frac{c_1(-1)^n}{n+1}$. By Problem 3.2.1(d), $y_n = c_2 + c_1 \sum_{k=0}^{n-1}\frac{(-1)^k}{k+1}$, $n \geq 1$, $y_0 = c_2$, $c_1, c_2 \in \mathbb{C}$. The general solution to the given equation is $x_n = n!\left(c_1 + c_2\sum_{k=0}^{n-1}\frac{(-1)^k}{k+1}\right)$, $n \geq 1$, $x_0 = c_1$, $c_1, c_2 \in \mathbb{C}$.

Problem 3.3.11. Verify that $x_n^{(1)} = \frac{2^n}{n!}$ is a solution to the equation $x_{n+2} - \frac{n+3}{n+2}x_{n+1} + \frac{2}{n+2}x_n = 0$, and then solve the equation.

Answer 3.3.11. $x_n = \frac{2^n}{n!}\left(c_1 + c_2 \sum_{k=0}^{n-1} \frac{k!}{2^k}\right)$, $n \geq 1$, $x_0 = c_1$, $c_1, c_2 \in \mathbb{C}$.

Problem 3.3.12. Verify that $x_n^{(1)} = n+1$ is a solution to the equation $x_{n+2} - x_{n+1} - \frac{1}{n+1}x_n = 0$, and then solve the equation.

Answer 3.3.12. $x_n = (n+1)\left(c_1 + c_2 \sum_{k=0}^{n-1} \frac{(-1)^k}{(k+2)!}\right)$, $n \geq 1$, $x_0 = c_1$, $c_1, c_2 \in \mathbb{C}$.

Problem 3.3.13. Solve the following homogeneous second-order difference equations with constant coefficients:

(a) $x_{n+2} - 4x_{n+1} - x_n = 0.$ (b) $x_{n+2} - 6x_{n+1} + 14x_n = 0.$

Answer 3.3.13. (a) $x_n = c_1(2 - \sqrt{5})^n + c_2(2 + \sqrt{5})^n$, $c_1, c_2 \in \mathbb{C}$.
(b) $x_n = 14^{\frac{n}{2}}(c_1 \cos n\theta + c_2 \sin n\theta)$, where $\tan \theta = \frac{\sqrt{5}}{3}$, $c_1, c_2 \in \mathbb{C}$.

Problem 3.3.14. The sequence $\{a_n\}_{n\in\mathbb{N}_0}$ has the property that $a_0 = a_1 = 1$, and each term after the first two terms is the sum of the two immediately preceding terms multiplied by two. Determine a_n.

Solution 3.3.14. The question is to solve the initial value problem $x_{n+2} = 2(x_{n+1} + x_n)$, $x_0 = x_1 = 1$. Since the solutions to the characteristic equation are $r_1 = 1+\sqrt{3}$ and $r_2 = 1-\sqrt{3}$, the general solution is $x_n = c_1(1 + \sqrt{3})^n + c_2(1 - \sqrt{3})^n$, $c_1, c_2 \in \mathbb{C}$. From $a_0 = c_1 + c_2 = 1$ and $a_1 = c_1(1 + \sqrt{3}) + c_2(1 - \sqrt{3}) = 1$, it follows that $c_1 = c_2 = \frac{1}{2}$. Therefore, $a_n = \frac{1}{2}((1 + \sqrt{3})^n + (1 - \sqrt{3})^n)$.

Note: The sequence $\{a_n\}_{n\in\mathbb{N}_0}$, i.e., the sequence $1, 1, 4, 10, 28, 76, 208, \ldots$, is listed in the Online Encyclopedia of Integer Sequences as A026150.

Problem 3.3.15. Solve the following initial value problems:

(a) $x_{n+2} + 4x_{n+1} + 4x_n = 0$, $x_0 = -2$, $x_1 = 12.$
(b) $x_{n+2} - 3x_n = 0$, $x_0 = 0$, $x_1 = 3.$
(c) $x_{n+2} + 3x_n = 0$, $x_0 = 1$, $x_1 = 3.$

Answer 3.3.15. (a) $a_n = (-2)^{n+1}(1 + 2n)$. (b) $a_n = 3^{\frac{n+1}{2}}\frac{1-(-1)^n}{2}$.
(c) $a_n = 3^{\frac{n}{2}}\left(\cos \frac{n\pi}{2} + \sqrt{3}\cos \frac{n\pi}{2}\right).$

Problem 3.3.16. Determine the nth term of the sequence:

(a) $-3, 21, 3, 129, 147, \ldots$

(b) $2, 5, 11, 23, \ldots$

(c) $2, 5, 12, 27, 58, \ldots$

(d) $1, 2, 6, 16, 44, \ldots$

Solution 3.3.16.

(a) $a_n = 3^{n+1} + 3(-2)^{n+1}$. The sequence, which is the solution to the initial value problem $x_{n+2} - x_{n+1} - 6x_n = 0$, $x_0 = -3$, and $x_1 = 21$, is obtained from the general solution $x_n = c_1 3^n + c_2(-2)^n$, $c_1, c_2 \in \mathbb{C}$, for $c_1 = 3$ and $c_2 = -6$.

(b) $a_n = 3 \cdot 2^n - 1$. The sequence is the solution to the initial value problem $x_{n+2} - 3x_{n+1} + 2x_n = 0$, $x_0 = 2$ and $x_1 = 5$.

(c) $a_n = 2^{n+2} - (n + 2)$. The sequence is a solution to the non-homogeneous second-order linear equation $x_{n+2} - 3x_{n+1} + 2x_n = 1$. The general solution is $x_n = c_1 + c_2 2^n - n$, $c_1, c_2 \in \mathbb{C}$. The sequence is obtained for $c_1 = -2$ and $c_2 = 4$.

(d) $a_n = \frac{1}{2\sqrt{3}}\left((1 + \sqrt{3})^{n+1} - (1 - \sqrt{3})^{n+1}\right)$. The sequence is the solution to the initial value problem $x_{n+2} - 2x_{n+1} - 2x_n = 0$, $x_0 = 1$ and $x_1 = 2$.

Note: (b) See A083329 in the Online Encyclopedia of Integer Sequences.

(c) The sequence $\{2^n - n\}_{n \in \mathbb{N}} = \{1, 2, 5, 12, 27, \ldots\}$ is the number of ways to partition the set $\{1, 2, \ldots, n\}$ into arithmetic progressions of the same step and lengths of least one. See A000325 in the Online Encyclopedia of Integer Sequences.

(d) The term a_n, $n \in \mathbb{N}$, is the number of words of length $n - 1$ over the alphabet $\{0, 1, 2, 3\}$ in which 0 and 1 avoid runs of odd lengths. See A002605 in the Online Encyclopedia of Integer Sequences.

Problem 3.3.17. Solve the following homogeneous second-order linear difference equations and initial value problems:

(a) $x_{n+2} - \frac{2n+6}{n+2} x_{n+1} + \frac{n+3}{n+1} x_n = 0.$ (b) $x_{n+2} - 2x_{n+1} + 2x_n = 0.$

(c) $x_{n+2} - 2x_{n+1} - 15x_n = 0$, $x_0 = 1$, $x_1 = 77$.

Answer 3.3.17. For $c_1, c_2 \in \mathbb{C}$: (a) $x_n = (c_1 + nc_2)(n + 1)$. (b) $x_n = 2^{\frac{n}{2}}\left(c_1 \cos \frac{n\pi}{4} + c_2 \sin \frac{n\pi}{4}\right)$. (c) $a_n = 2 \cdot 5^{n+1} + 3(-3)^{n+1}$.

Problem 3.3.18. Let the sequence $\{a_n\}_{n\in\mathbb{N}_0}$ be a nonzero solution to the difference equation $x_{n+2} = x_{n+1} + 2x_n$. Determine $\lim\limits_{n\to\infty}\dfrac{a_n}{a_{n+1}}$.

Solution 3.3.18. Since $\{a_n\}_{n\in\mathbb{N}_0}$ is a solution to the equation $x_{n+2} - x_{n+1} - 2x_n = 0$, there are $c_1, c_2 \in \mathbb{C}$ such that $a_n = c_1(-1)^n + c_2 2^n$ for every $n \in \mathbb{N}_0$. If $c_2 = 0$, then $a_n = c_1(-1)^n$, $c_1 \neq 0$, and $\lim\limits_{n\to\infty}\dfrac{a_n}{a_{n+1}} = -1$. If $c_2 \neq 0$, then $\lim\limits_{n\to\infty}\dfrac{a_n}{a_{n+1}} =$

$$\lim_{n\to\infty}\frac{c_1(-1)^n + c_2 2^n}{c_1(-1)^{n+1} + c_2 2^{n+1}} = \lim_{n\to\infty}\frac{\frac{c_1(-1)^n}{2^n} + c_2}{\frac{c_1(-1)^{n+1}}{2^n} + 2c_2} = \frac{1}{2}.$$

Problem 3.3.19. Let $a, b, \alpha \in \mathbb{R}$, with $\alpha \neq 0$. Verify the method of undetermined coefficients in the case of the non-homogeneous difference equation $x_{n+2} + ax_{n+1} + bx_n = n\alpha^n$, under the assumption that α is a simple root of the equation $r^2 + ar + b = 0$.

Solution 3.3.19. Observe that if α is a simple root of the equation $r^2 + ar + b = 0$, then $a \neq -2\alpha$. Otherwise, from $0 = \alpha^2 + a\alpha + b = \alpha^2 - 2\alpha^2 + b$, it follows that $b = \alpha^2$, i.e., it follows that $0 = r^2 + ar + b = (r - \alpha)^2$, which contradicts the assumption.

Next, observe that for any $c_0, c_1 \in \mathbb{C}$ and any $n \in \mathbb{N}_0$, $y_n = (c_0 n + c_1)\alpha^n$ contains the term $c_1\alpha^n$ and that the sequence $\{c_1\alpha^n\}_{n\in\mathbb{N}_0}$ is a solution to the corresponding homogeneous equation. This implies that $\{y_n\}_{n\in\mathbb{N}_0}$ is not a solution to the non-homogeneous equation. Indeed, for any $c_0, c_1 \in \mathbb{C}$, there is $n \in \mathbb{N}_0$ such that

$$y_{n+2} + ay_{n+1} + by_n = \alpha^n(c_0((n+2)\alpha^2 + a(n+1)\alpha + bn)$$
$$+ c_1(\alpha^2 + a\alpha + b)) = c_0(2\alpha + a)\alpha^{n+1} \neq n\alpha^n.$$

We demonstrate that it is possible to determine $c_0, c_1 \in \mathbb{C}$ so that the sequence $x_n^{(p)} = n(c_0 n + c_1)\alpha^n$ is a particular solution to the equation, i.e., we establish that there are $c_0, c_1 \in \mathbb{C}$ such that, for any $n \in \mathbb{N}_0$,

$$n\alpha^n = (c_0(n+2)^2 + c_1(n+2))\alpha^{n+2} + a(c_0(n+1)^2 + c_1(n+1))\alpha^{n+1}$$
$$+ b(c_0 n^2 + c_1 n)\alpha^n.$$

This is the case if the coefficients c_0 and c_1 are such that, for any $n \in \mathbb{N}_0$,

$$n = 2n\alpha c_0(2\alpha + a) + \alpha c_0(4\alpha + a) + \alpha c_1(2\alpha + a).$$

It follows that $c_0 = \frac{1}{2\alpha(2\alpha+a)}$ and $c_1 = -\frac{4\alpha+a}{2\alpha(2\alpha+a)^2}$.

Problem 3.3.20. The sequence $a_0 = 0, a_1 = 2, a_2 = 5, a_3 = 5$, $a_4 = 14,\ldots$ is a solution to a difference equation $x_{n+2} + \alpha x_{n+1} + \beta x_n = m$, for some real numbers α, β, and m. Determine a_n, $n \in \mathbb{N}$.

Solution 3.3.20. From $a_2 + \alpha a_1 + \beta a_0 = 5 + 2\alpha = m$, $a_3 + \alpha a_2 + \beta a_1 = 5 + 5\alpha + 2\beta = m$, and $a_4 + \alpha a_3 + \beta a_2 = 14 + 5\alpha + 5\beta = m$, it follows that $\alpha = 2$, $\beta = -3$, and $m = 9$. The sequence is a solution to the equation $x_{n+2} + 2x_{n+1} - 3x_n = 9$. The characteristic equation is $r^2 + 2r - 3 = (r+3)(r-1) = 0$, which implies that the general solution to the corresponding homogeneous equation is $x_n^{(h)} = c_1 + c_2(-3)^n$. Since $r = 1$ is a root of the characteristic equation, by the method of undetermined coefficients, we are looking for a particular solution to the non-homogeneous equation that is in the form of $x_n^{(p)} = an$, for some constant a. From

$$9 = x_{n+2}^{(p)} + 2x_{n+1}^{(p)} - 3x_n^{(p)} = a(n+2) + 2a(n+1) - 3an = 4a,$$

it follows that $x_n^{(p)} = \frac{9n}{4}$. The general solution is $x_n = c_1 + c_2(-3)^n + \frac{9n}{4}$, $c_1, c_2 \in \mathbb{C}$. The solution to the initial value problem $x_{n+2} + 2x_{n+1} - 3x_n = 9$, $x_0 = 0$ and $x_1 = 2$, is $a_n = \frac{(-3)^n - 1}{16} + \frac{9n}{4}$, $n \in \mathbb{N}_0$.

Problem 3.3.21. Solve the initial value problem $\alpha x_{n+2} - x_{n+1} + \beta x_n = 0$, $x_0 = 0$, $x_1 = 1$, if $\alpha + \beta = 1$.

Solution 3.3.21. If $\alpha = 0$, then $\beta = 1$ and the equation becomes $x_{n+1} = x_n$. This implies, since $x_0 \neq x_1$, that $\alpha \neq 0$. Solutions to the characteristic equation $\alpha r^2 - r + \beta = \alpha r^2 - r + 1 - \alpha = 0$ are

$$r_{1,2} = \frac{1 \pm \sqrt{1 - 4\alpha(1-\alpha)}}{2\alpha} = \frac{1 \pm \sqrt{1 - 4\alpha + 4\alpha^2}}{2\alpha} = \frac{1 \pm |1 - 2\alpha|}{2\alpha}.$$

It follows that if $\alpha \neq \frac{1}{2}$, then $r_1 = 1$ and $r_2 = \frac{1-\alpha}{\alpha} = \frac{\beta}{\alpha} \neq 1$, and if $\alpha = \frac{1}{2}$, then $r_1 = r_2 = 1$. In the case of $\alpha \neq \frac{1}{2}$, the general solution is $x_n = c_1 + c_2\left(\frac{\beta}{\alpha}\right)^n$, $c_1, c_2 \in \mathbb{C}$. From $0 = x_0 = c_1 + c_2$ and $1 = c_1 + c_2\frac{\beta}{\alpha}$, it follows that $c_1 = -c_2 = \frac{\alpha}{\alpha-\beta}$. The solution to the initial value problem is $a_n = \frac{\alpha}{\alpha-\beta}\left(1 - \left(\frac{\beta}{\alpha}\right)^n\right)$, $n \in \mathbb{N}_0$. If $\alpha = \beta = \frac{1}{2}$ the general solution is $x_n = c_1 + c_2 n$, $c_1, c_2 \in \mathbb{C}$. From the initial conditions, $c_1 = 0$

and $c_2 = 1$. The solution to the initial value problem is $a_n = n$, $n \in \mathbb{N}_0$.

Note: For any $\alpha \neq 0$, $a_n = \sum_{k=0}^{n-1} \left(\frac{1}{\alpha} - 1\right)^k$, $n \in \mathbb{N}$.

Problem 3.3.22. Verify that $x_n^{(1)} = 3^{n-1}n$ and $x_n^{(2)} = 3^{n-1}(1+n)$ are solutions to the non-homogeneous second-order linear difference equation $x_{n+2} - x_{n+1} - 6x_n = 5 \cdot 3^n$ but that $y_n = \alpha x_n^{(1)} + \beta x_n^{(2)}$, $\alpha, \beta \in \mathbb{C}$, with $\alpha + \beta \neq 1$, is not.

Solution 3.3.22. From

$$x_{n+2}^{(1)} - x_{n+1}^{(1)} - 6x_n^{(1)} = 3^{n+1}(n+2) - 3^n(n+1) - 6 \cdot 3^{n-1}n$$
$$= 3^{n-1}(9n + 18 - 3n - 3 - 6n) = 15 \cdot 3^{n-1} = 5 \cdot 3^n$$

and

$$x_{n+2}^{(2)} - x_{n+1}^{(2)} - 6x_n^{(2)} = 3^{n+1}(n+3) - 3^n(n+2) - 6 \cdot 3^{n-1}(1+n)$$
$$= 15 \cdot 3^{n-1} = 5 \cdot 3^n,$$

it follows that $x_n^{(1)} = 3^{n-1}n$ and $x_n^{(2)} = 3^{n-1}(1+n)$ are solutions to the equation. Let $y_n = \alpha x_n^{(1)} + \beta x_n^{(2)}$, $\alpha, \beta \in \mathbb{C}$ with $\alpha + \beta \neq 1$. Then,

$$y_{n+2} - y_{n+1} - 6y_n = \alpha(x_{n+2}^{(1)} - x_{n+1}^{(1)} - 6x_n^{(1)}) + \beta(x_{n+2}^{(2)} - x_{n+1}^{(2)} - 6x_n^{(2)})$$
$$= 5 \cdot 3^n(\alpha + \beta) \neq 5 \cdot 3^n,$$

which implies that y_n is not a solution to the equation.

Note 1: Observe that for any $\alpha, \beta \in \mathbb{C}$ with $\alpha + \beta = 1$, the sequence $y_n = \alpha x_n^{(1)} + \beta x_n^{(2)}$, *is* a solution to the difference equation.

Note 2: A linear combination of two solutions to a non-homogeneous linear difference equation may or may not be a solution to the equation.

Problem 3.3.23. Solve the initial value problem $\alpha x_{n+2} - x_{n+1} + \beta x_n = -1$, $x_0 = x_1 = 0$, under the constraint $\alpha + \beta = 1$.

Solution 3.3.23. If $\alpha = 0$, then $\beta = 1$ and the equation becomes $x_{n+1} = x_n - 1$. Since $x_0 = x_1$, it follows that $\alpha \neq 0$. By Problem 3.3.21, if $\alpha \neq \frac{1}{2}$, the general solution to the corresponding

homogeneous equation is $x_n^{(h)} = c_1 + c_2 \left(\frac{\beta}{\alpha}\right)^n$, $c_1, c_2 \in \mathbb{C}$. Since $r = 1$ is a simple root of the characteristic equation, we take $x_n^{(p)} = bn$, which yields $-1 = b\alpha(n+2) - b(n+1) + b\beta n = (\alpha - 1 + \beta)bn + 2\alpha b - b = (2\alpha - 1)b$. It follows that $x_n^{(p)} = \frac{n}{1-2\alpha}$. The general solution to the non-homogeneous second-order linear equation $\alpha x_{n+2} - x_{n+1} + \beta x_n = -1$, $\alpha \neq \frac{1}{2}$, is $x_n = c_1 + c_2 \left(\frac{\beta}{\alpha}\right)^n + \frac{n}{1-2\alpha}$, $c_1, c_2 \in \mathbb{C}$. From the initial conditions $x_0 = x_1 = 0$, we obtain the solution to the initial value problem:
$$a_n = \frac{\alpha}{(\alpha-\beta)^2}\left(1 - \left(\frac{\beta}{\alpha}\right)^n\right) + \frac{n}{\beta - \alpha}, \ n \in \mathbb{N}_0.$$

If $\alpha = \beta = \frac{1}{2}$, then the initial value problem becomes $x_{n+2} - 2x_{n+1} + x_n = -2$. By Problem 3.3.21, the general solution to the corresponding homogeneous equation is $x_n^{(h)} = c_1 + c_2 n$, $c_1, c_2 \in \mathbb{C}$. For a particular solution, by the method of undetermined coefficients, we try $x_n^{(p)} = an^2$, which yields $-2 = a(n+2)^2 - 2a(n+1)^2 + an^2 = 4a - 2a = 2a$. Hence, $x_n = c_1 + c_2 n - n^2$, $c_1, c_2 \in \mathbb{C}$, is the general solution to the non-homogeneous second-order linear equation $x_{n+2} - 2x_{n+1} + x_n = -2$. From the initial conditions $x_0 = x_1 = 0$, we obtain that the solution to the initial value problem is the sequence $a_n = n(1 - n)$, $n \in \mathbb{N}_0$.

Problem 3.3.24. Solve the difference equation $x_{n+2} - 7x_{n+1} + 10x_n = 3e^{2n} - 3^n$.

Solution 3.3.24. Solutions to the characteristic equation $r^2 - 7r + 10 = (r - 2)(r - 5) = 0$ are $r_1 = 2$ and $r_2 = 5$. The general solution to the corresponding homogeneous equation is $x_n^{(h)} = c_1 2^n + c_2 5^n$, $c_1, c_2 \in \mathbb{C}$. To obtain a particular solution, we use the principle of superposition. By the method of undetermined coefficients, we first determine a particular solution to the equation $x_{n+2} - 7x_{n+1} + 10x_n = 3e^{2n}$ in the form of $u_n = ce^{2n}$, for some $c \in \mathbb{C}$. From $3e^{2n} = ce^{2(n+2)} - 7ce^{2(n+1)} + 10ce^{2n} = ce^{2n}(e^4 - 7e^2 + 10)$, it follows that $u_n = \frac{3e^{2n}}{(e^2-2)(e^2-5)}$, $n \in \mathbb{N}_0$, is the required particular solution. Next, we determine a particular solution to the equation $x_{n+2} - 7x_{n+1} + 10x_n = -3^n$ in the form of $v_n = 3^n d$, for some $d \in \mathbb{C}$. From $-3^n = 3^{n+2}d - 7 \cdot 3^{n+1}d + 10 \cdot 3^n d = d(9 - 21 + 10)3^n$, it follows that $v_n = \frac{3^n}{2}$, $n \in \mathbb{N}_0$. A particular solution to the given equation is $x_n^{(p)} = u_n + v_n = \frac{3e^{2n}}{e^4 - 7e^2 + 10} + \frac{3^n}{2}$. Its general solution is $x_n = x_n^{(h)} + x_n^{(p)} = c_1 2^n + c_2 5^n + \frac{3e^{2n}}{(e^2-2)(e^2-5)} + \frac{3^n}{2}$, $c_1, c_2 \in \mathbb{C}$.

Problem 3.3.25. Solve the difference equation $x_{n+2} - 4x_n = 2^n$.

Solution 3.3.25. Solutions to the characteristic equation $r^2 - 4 = (r-2)(r+2) = 0$ are $r_1 = -2$ and $r_2 = 2$. The general solution to the corresponding homogeneous equation is $x_n^{(h)} = c_1(-2)^n + c_2 2^n$, $c_1, c_2 \in \mathbb{C}$. Since each term of the sequence $\{2^n\}_{n \in \mathbb{N}_0}$ has a root of the characteristic equation as its base, we determine a particular solution in the form of $x_n^{(p)} = cn2^n$ for some $c \in \mathbb{C}$. From $2^n = c(n+2)2^{n+2} - 4cn2^n = 8c2^n$, it follows that $x_n^{(p)} = n2^{n-3}$. The general solution to the non-homogeneous second-order difference equation is $x_n = 2^n \left(c_1(-1)^n + c_2 + \frac{n}{8} \right)$, $c_1, c_2 \in \mathbb{C}$.

    Note: The substitution $x_n = 2^n y_n$ yields the equation $y_{n+2} - y_n = \frac{1}{4}$.

Problem 3.3.26. Solve the difference equation $x_{n+2} - 3x_{n+1} + 2x_n = 2\sin 3n$.

Solution 3.3.26. Solutions to the characteristic equation $r^2 - 3r + 2 = 0$ are $r_1 = 1$ and $r_2 = 2$. The general solution to the corresponding homogeneous equation is $x_n^{(h)} = c_1 + c_2 2^n$, $c_1, c_2 \in \mathbb{C}$. We determine a particular solution to the non-homogeneous equation in the form of $x_n^{(p)} = a\sin 3n + b\cos 3n$, $a, b \in \mathbb{C}$.
From, for all $n \in \mathbb{N}_0$,

$$2\sin 3n = a\sin 3(n+2) + b\cos 3(n+2) - 3(a\sin 3(n+1)$$

$$+ b\cos 3(n+1)) + 2(a\sin 3n + b\cos 3n)$$

$$= \sin 3n(a\cos 6 - b\sin 6 - 3a\cos 3 + 3b\sin 3 + 2a)$$

$$+ \cos 3n(a\sin 6 + b\cos 6 - 3a\sin 3 - 3b\cos 3 + 2b),$$

it follows that a and b satisfy the system of equations

$$2 = a(\cos 6 - 3\cos 3 + 2) - b(\sin 6 - 3\sin 3)$$

$$0 = a(\sin 6 - 3\sin 3) + b(\cos 6 - 3\cos 3 + 2).$$

The elimination of the unknown b yields the equation

$$2(\cos 3 - 1)(2\cos 3 - 1) = a((\cos 6 - 3\cos 3 + 2)^2 + (\sin 6 - 3\sin 3)^2)$$

$$= 2a(\cos 3 - 1)(4\cos 3 - 5).$$

Hence, $a = \dfrac{2\cos 3-1}{4\cos 3-5}$ and $b = -\dfrac{2\cos 3-1}{4\cos 3-5}\dfrac{\sin 6-3\sin 3}{(\cos 3-1)(2\cos 3-1)} = -\dfrac{\sin 6-3\sin 3}{(\cos 3-1)(4\cos 3-5)}$. A particular solution to the equation is

$$x_n^{(p)} = \frac{2\cos 3 - 1}{4\cos 3 - 5}\sin 3n - \frac{(2\cos 3 - 3)\sin 3}{(\cos 3 - 1)(4\cos 3 - 5)}\cos 3n.$$

The general solution is $x_n = c_1 + c_2 2^n + x_n^{(p)}$, $c_1, c_2 \in \mathbb{C}$.

Problem 3.3.27. Solve the difference equation $x_{n+2}+x_{n+1}-2x_n = \sin 2n$.

Solution 3.3.27. Solutions to the characteristic equation $r^2+r-2 = (r-1)(r+2) = 0$ are $r_1 = -2$ and $r_2 = 1$. The general solution to the corresponding homogeneous equation is $x_n^{(h)} = c_1(-2)^n + c_2$, $c_1, c_2 \in \mathbb{C}$. We determine a particular solution in the form of $x_n^{(p)} = a\cos 2n+b\sin 2n = \frac{a}{2}(e^{2ni}+e^{-2ni})-\frac{bi}{2}(e^{2ni}-e^{-2ni})$, for some $a, b \in \mathbb{C}$. From

$$\sin 2n = a(\cos(2n+4) + \cos(2n+2) - 2\cos 2n) + b(\sin(2n+4)$$

$$+ \sin(2n+2) - 2\sin 2n) = \frac{a - bi}{2}\left(e^{(2n+4)i} + e^{(2n+2)i} - 2e^{2ni}\right)$$

$$+ \frac{a + bi}{2}\left(e^{-(2n+4)i} + e^{-(2n+2)i} - 2e^{-2ni}\right)$$

$$= \frac{a - bi}{2}e^{2ni}\left(e^{4i} + e^{2i} - 2\right) + \frac{a + bi}{2}e^{-2ni}\left(e^{-4i} + e^{-2i} - 2\right)$$

$$= \frac{a - bi}{2}e^{2ni}(e^{2i} - 1)(e^{2i} + 2) + \frac{a + bi}{2}e^{-2ni}(e^{-2i} - 1)(e^{-2i} + 2),$$

it follows that $a - bi = \dfrac{1}{i(e^{2i}-1)(e^{2i}+2)}$ and $a + bi = -\dfrac{1}{i(e^{-2i}-1)(e^{-2i}+2)}$. Hence,

$$2a = \frac{1}{i(e^{2i} - 1)(e^{2i} + 2)} - \frac{1}{i(e^{-2i} - 1)(e^{-2i} + 2)}$$

$$= \frac{e^{-4i} + e^{-2i} - 2 - (e^{4i} + e^{2i} - 2)}{i(e^{2i} - 1)(e^{-2i} - 1)(e^{2i} + 2)(e^{-2i} + 2)} = -\frac{\sin 4 + \sin 2}{(1 - \cos 2)(5 + 4\cos 2)}$$

and

$$2b = \frac{1}{(e^{2i}-1)(e^{2i}+2)} + \frac{1}{(e^{-2i}-1)(e^{-2i}+2)}$$

$$= \frac{e^{-4i}+e^{-2i}-2+(e^{4i}+e^{2i}-2)}{(e^{2i}-1)(e^{-2i}-1)(e^{2i}+2)(e^{-2i}+2)} = \frac{\cos 4 + \cos 2 - 2}{(1-\cos 2)(5+4\cos 2)}$$

$$= \frac{(\cos 2 - 1)(3 + 2\cos 2)}{(1-\cos 2)(5+4\cos 2)} = -\frac{3+2\cos 2}{5+4\cos 2}.$$

It follows that

$$x_n^{(p)} = -\frac{(2\cos 2 + 1)\sin 2}{2(1-\cos 2)(5+4\cos 2)}\cos 2n - \frac{3+2\cos 2}{2(5+4\cos 2)}\sin 2n.$$

The general solution to the non-homogeneous second-order difference equation is $x_n = c_1(-2)^n + c_2 + x_n^{(p)}$, c_1, $c_2 \in \mathbb{C}$.

Note: Another way to determine the coefficients a and b is to use the fact that they satisfy the linear system obtained from the assumption that $x_n^{(p)}$ is a particular solution in the cases of $n = 0$ and $n = 1$:

$$0 = a(\cos 4 + \cos 2 - 2) + b(\sin 4 + \sin 2)$$

$$\sin 2 = a(\cos 6 + \cos 4 - 2\cos 2) + b(\sin 6 + \sin 4 - 2\sin 2).$$

Problem 3.3.28. For $\alpha \in (0,\pi)$, solve the difference equation $x_{n+2} - 2x_{n+1}\cos\alpha + x_n = \cos\alpha n$.

Solution 3.3.28. Solutions to the characteristic equation $r^2 - 2r\cos\alpha + 1 = 0$ are $r_1 = \cos\alpha + i\sin\alpha = e^{\alpha i}$ and $r_2 = \cos\alpha - i\sin\alpha = e^{-\alpha i}$. The general solution to the corresponding homogeneous equation is $x_n^{(h)} = c_1\cos\alpha n + c_2\sin\alpha n$, $c_1, c_2 \in \mathbb{C}$. Since $a_n = \cos\alpha n$, $n \in \mathbb{N}_0$, is a particular solution to the homogeneous equation, we find a particular solution to the non-homogeneous equation in the form of $x_n^{(p)} = n(a\cos\alpha n + b\sin\alpha n)$, for some $a, b \in \mathbb{C}$.

From $\cos\alpha n = x_{n+2}^{(p)} - 2x_{n+1}^{(p)}\cos\alpha + x_n^{(p)}$, it follows that a and b must be such that, for all $n \in \mathbb{N}_0$,

$$\cos\alpha n = a\cos\alpha(n+2) + b\sin\alpha(n+2) - a\cos\alpha n - b\sin\alpha n$$

$$= (a\cos 2\alpha + b\sin 2\alpha - a)\cos\alpha n$$

$$\quad - (a\sin 2\alpha - b\cos 2\alpha + b)\sin\alpha n$$

$$= (-2a\sin^2\alpha + b\sin 2\alpha)\cos\alpha n - (a\sin 2\alpha + 2b\sin^2\alpha)\sin\alpha n.$$

The coefficients a and b satisfy the system of equations $-2a\sin^2\alpha + b\sin 2\alpha = 1$ and $a\cos\alpha + b\sin\alpha = 0$. Hence,

$$a = \frac{\begin{vmatrix} 1 & \sin 2\alpha \\ 0 & \sin\alpha \end{vmatrix}}{\begin{vmatrix} -2\sin^2\alpha & \sin 2\alpha \\ \cos\alpha & \sin\alpha \end{vmatrix}} = \frac{\sin\alpha}{-2\sin^3\alpha - \sin 2\alpha\cos\alpha} = -\frac{1}{2}$$

and $b = \frac{1}{2}\cot\alpha$. The sequence $x_n^{(p)} = -\frac{n}{2}(\cos\alpha n - \cot\alpha\sin\alpha n)$, $n \in \mathbb{N}_0$, is a particular solution to the equation. The general solution is $x_n = \frac{1}{2}(c_1 - n)\cos\alpha n + \frac{1}{2}(c_2 + n\cot\alpha)\sin\alpha n$, $c_1, c_2 \in \mathbb{C}$.

Problem 3.3.29. Prove that $\{\{n + 2\}_{n\in\mathbb{N}_0}, \{2^{n+2}\}_{n\in\mathbb{N}_0}\}$ is a fundamental set of solutions to the equation $x_{n+2} - \frac{3n+4}{n+1}x_{n+1} + \frac{2(n+2)}{n+1}x_n = 0$.

Solution 3.3.29. Let $x_n^{(1)} = n + 2$ and $x_n^{(2)} = 2^{n+2}$. From

$$(n + 1)x_{n+2}^{(1)} - (3n + 4)x_{n+1}^{(1)} + 2(n + 2)x_n^{(1)} = (n + 1)(n + 4)$$
$$- (3n + 4)(n + 3) + 2(n + 2)^2 = 0$$

and

$$(n + 1)x_{n+2}^{(2)} - (3n + 4)x_{n+1}^{(2)} + 2(n + 2)x_n^{(2)} = (n + 1)2^{n+4} - (3n + 4)2^{n+3}$$
$$+ 2(n + 2)2^{n+2} = 2^{n+2}(4(n + 1) - 2(3n + 4) + 2(n + 2)) = 0,$$

it follows that both $x_n^{(1)} = n + 2$ and $x_n^{(2)} = 2^{n+2}$ are solutions to the given homogeneous second-order difference equation. Since, for $n \in \mathbb{N}_0$, their Casoratian is

$$C_n = \begin{vmatrix} n + 2 & 2^{n+2} \\ n + 3 & 2^{n+3} \end{vmatrix} = 2^{n+2}(n + 1) \neq 0,$$

it follows that these two solutions are linearly independent and, therefore, form a fundamental set of solutions to the equation.

Problem 3.3.30. Consider a non-homogeneous second-order difference equation $x_{n+2} + \alpha_n x_{n+1} + \beta_n x_n = \gamma_n$, and suppose that $x_n^{(h)} = c_1 x_n^{(1)} + c_2 x_n^{(2)}$, $c_1, c_2 \in \mathbb{C}$, is the general solution to the corresponding homogeneous equation. Prove that there is a particular solution to the non-homogeneous equation in the form of $x_n^{(p)} = c_n^{(1)} x_n^{(1)} + c_n^{(2)} x_n^{(2)}$, $n \in \mathbb{N}_0$, with $c_n^{(1)}$ and $c_n^{(2)}$ determined by the conjunction of the difference equations

$$c_{n+1}^{(1)} - c_n^{(1)} = -\frac{\gamma_n x_{n+1}^{(2)}}{x_{n+1}^{(1)} x_{n+2}^{(2)} - x_{n+2}^{(1)} x_{n+1}^{(2)}} \quad \text{and}$$

$$c_{n+1}^{(2)} - c_n^{(2)} = \frac{\gamma_n x_{n+1}^{(1)}}{x_{n+1}^{(1)} x_{n+2}^{(2)} - x_{n+2}^{(1)} x_{n+1}^{(2)}}.$$

Solution 3.3.30. Observe that $x_{n+1}^{(p)} = c_{n+1}^{(1)} x_{n+1}^{(1)} + c_{n+1}^{(2)} x_{n+1}^{(2)}$ may be rewritten as

$$x_{n+1}^{(p)} = c_n^{(1)} x_{n+1}^{(1)} + c_n^{(2)} x_{n+1}^{(2)} + \left(c_{n+1}^{(1)} - c_n^{(1)} \right) x_{n+1}^{(1)} + \left(c_{n+1}^{(2)} - c_n^{(2)} \right) x_{n+1}^{(2)}.$$

We restrict our search for the sequences $\left\{ c_n^{(1)} \right\}_{n \in \mathbb{N}_0}$ and $\left\{ c_n^{(2)} \right\}_{n \in \mathbb{N}_0}$ so that, for all $n \in \mathbb{N}_0$,

$$\left(c_{n+1}^{(1)} - c_n^{(1)} \right) x_{n+1}^{(1)} + \left(c_{n+1}^{(2)} - c_n^{(2)} \right) x_{n+1}^{(2)} = 0,$$

i.e., we are looking for a solution, $x_n^{(p)} = c_n^{(1)} x_n^{(1)} + c_n^{(2)} x_n^{(2)}$, to the non-homogeneous equation so that $x_{n+1}^{(p)} = c_n^{(1)} x_{n+1}^{(1)} + c_n^{(2)} x_{n+1}^{(2)}$, for all $n \in \mathbb{N}_0$. It follows that $x_n^{(p)}$ must be such that

$$\gamma_n = x_{n+2}^{(p)} + \alpha_n x_{n+1}^{(p)} + \beta x_n^{(p)}$$

$$= c_{n+1}^{(1)} x_{n+2}^{(1)} + c_{n+1}^{(2)} x_{n+2}^{(2)} + \alpha_n \left(c_n^{(1)} x_{n+1}^{(1)} + c_n^{(2)} x_{n+1}^{(2)} \right) + \beta_n \left(c_n^{(1)} x_n^{(1)} + c_n^{(2)} x_n^{(2)} \right)$$

$$= c_{n+1}^{(1)} x_{n+2}^{(1)} + c_{n+1}^{(2)} x_{n+2}^{(2)} + c_n^{(1)} \left(\alpha_n x_{n+1}^{(1)} + \beta_n x_n^{(1)} \right) + c_n^{(2)} \left(\alpha_n x_{n+1}^{(2)}) + \beta_n x_n^{(2)} \right)$$

$$= c_{n+1}^{(1)} x_{n+2}^{(1)} + c_{n+1}^{(2)} x_{n+2}^{(2)} - c_n^{(1)} x_{n+2}^{(1)} - c_n^{(2)} x_{n+2}^{(2)}$$

$$= \left(c_{n+1}^{(1)} - c_n^{(1)} \right) x_{n+2}^{(1)} + \left(c_{n+1}^{(2)} - c_n^{(2)} \right) x_{n+2}^{(2)}.$$

The solution $x_n^{(p)} = c_n^{(1)} x_n^{(1)} + c_n^{(2)} x_n^{(2)}$ is determined by the linear system of difference equations

$$\left(c_{n+1}^{(1)} - c_n^{(1)} \right) x_{n+1}^{(1)} + \left(c_{n+1}^{(2)} - c_n^{(2)} \right) x_{n+1}^{(2)} = 0$$

$$\left(c_{n+1}^{(1)} - c_n^{(1)} \right) x_{n+2}^{(1)} + \left(c_{n+1}^{(2)} - c_n^{(2)} \right) x_{n+2}^{(2)} = \gamma_n,$$

which establishes the claim.

Note: Problem 3.3.30 justifies the method of variation of parameters.

Problem 3.3.31. Solve the equation $(n + 4)x_{n+2} + x_{n+1} - (n + 1)x_n = 1$ by using the fact that the sequence $x_n^{(1)} = \frac{1}{(n+1)(n+2)}$, $n \in \mathbb{N}_0$, is a solution to the corresponding homogeneous equation.

Solution 3.3.31. To determine another solution to the corresponding homogeneous equation,[3] substitute $x_n = \frac{y_n}{(n+1)(n+2)}$ to obtain $\frac{(n+4)y_{n+2}}{(n+3)(n+4)} + \frac{y_{n+1}}{(n+2)(n+3)} - \frac{(n+1)y_n}{(n+1)(n+2)} = 0$. It follows that $(n+2)(y_{n+2}-y_{n+1})+(n+3)(y_{n+1}-y_n) = 0$, which, via the substitution $y_{n+1} - y_n = z_n$, yields $z_{n+1} = -\frac{n+3}{n+2}z_n$. The general solution to this first-order difference equation[4] is $z_n = c_1(-1)^n(n + 2)$, $c_1 \in \mathbb{C}$. The general solution to the non-homogeneous first-order difference equation $y_{n+1} - y_n = c_1(-1)^n(n + 2)$ is $y_n = c_2 + c_1 \sum_{k=0}^{n-1}(-1)^k (k + 2) = c_2 + c_1 \frac{(-1)^{n+1}(2n+3)+3}{4}$, $c_1, c_2 \in \mathbb{C}$. The general solution to the homogeneous equation $(n + 4)x_{n+2} + x_{n+1} - (n + 1)x_n = 0$ is $x_n^{(h)} = \frac{c_1+c_2(-1)^n(2n+3)}{(n+1)(n+2)}$, $c_1, c_2 \in \mathbb{C}$.

To establish a particular solution to the non-homogeneous difference equation, by the method of variation of parameters, we determine two sequences, $\left\{c_n^{(1)}\right\}_{n\in\mathbb{N}_0}$ and $\left\{c_n^{(2)}\right\}_{n\in\mathbb{N}_0}$, such that $x_n^{(p)} = c_n^{(1)} \frac{1}{(n+1)(n+2)} + c_n^{(2)} \frac{(-1)^n(2n+3)}{(n+1)(n+2)}$, $n \in \mathbb{N}_0$, is a particular solution to the given equation. By Problem 3.3.30, the sequences $c_n^{(1)}$ and $c_n^{(2)}$ satisfy the difference equations

$$c_{n+1}^{(1)} - c_n^{(1)} = -\frac{\frac{1}{n+4}\frac{(-1)^{n+1}(2n+5)}{(n+2)(n+3)}}{C_{n+1}} \quad \text{and} \quad c_{n+1}^{(2)} - c_n^{(2)} = \frac{\frac{1}{n+4}\frac{1}{(n+2)(n+3)}}{C_{n+1}},$$

where

$$C_{n+1} = \begin{vmatrix} \frac{1}{(n+2)(n+3)} & \frac{(-1)^{n+1}(2n+5)}{(n+2)(n+3)} \\ \frac{1}{(n+3)(n+4)} & \frac{(-1)^n(2n+7)}{(n+3)(n+4)} \end{vmatrix} = \frac{4(-1)^n}{(n + 2)(n + 3)(n + 4)}.$$

---

[3] See Problem 3.3.8.
[4] See Problem 3.2.1(a).

From

$$c_{n+1}^{(1)} - c_n^{(1)} = -\frac{\frac{(-1)^{n+1}(2n+5)}{(n+2)(n+3)(n+4)}}{\frac{4(-1)^n}{(n+2)(n+3)(n+4)}} = \frac{2n+5}{4},$$

it follows that we can take $c_0^{(1)} = 0$ and, for $n \in \mathbb{N}$, $c_n^{(1)} = \frac{1}{4}\sum_{k=0}^{n-1}(2k+5) = \frac{n(n+4)}{4}$. Similarly, $c_n^{(2)} = \frac{1}{4}\sum_{k=0}^{n-1}(-1)^k = \frac{1-(-1)^n}{8}$. The method of variation of parameters yields the solution $x_n^{(p)} = \frac{n(n+4)}{4(n+1)(n+2)} + \frac{(-1)^n(1-(-1)^n)(2n+3)}{8(n+1)(n+2)}$. From this solution, we can obtain a simpler particular solution to the given non-homogeneous equation in the following way. From

$$\begin{aligned}
x_n^{(p)} &= \frac{n(n+4)}{4(n+1)(n+2)} + \frac{(-1)^n(1-(-1)^n)(2n+3)}{8(n+1)(n+2)} \\
&= \frac{n(n+4)}{4(n+1)(n+2)} + \frac{(-1)^n(2n+3)-2n-3}{8(n+1)(n+2)} \\
&= \frac{n(n+3)}{4(n+1)(n+2)} + \frac{(-1)^n(2n+3)-3}{8(n+1)(n+2)}
\end{aligned}$$

and the fact that $a_n = \frac{-3+(-1)^n(2n+3)}{8(n+1)(n+2)}$ is a solution to the corresponding homogeneous equation (obtained from the general solution for $c_1 = -\frac{3}{8}$ and $c_2 = \frac{1}{8}$), we conclude that $y_n^{(p)} = \frac{n(n+3)}{4(n+1)(n+2)}$ is a particular solution to the non-homogeneous equation. The general solution is $x_n = \frac{c_1+c_2(-1)^n(2n+3)}{(n+1)(n+2)} + \frac{n(n+3)}{4(n+1)(n+2)}$, $c_1, c_2 \in \mathbb{C}$.

Problem 3.3.32. Solve $x_{n+2} - 2x_{n+1}\cos\alpha + x_n = \beta_n$, where $\alpha \in (0, \pi)$ and $\{\beta_n\}_{n\in\mathbb{N}_0}$ is a sequence of complex numbers.

Solution 3.3.32. Solutions to the characteristic equation $r^2 - 2r\cos\alpha + 1 = 0$ are $r_1 = \cos\alpha + i\sin\alpha = e^{\alpha i}$ and $r_2 = \cos\alpha - i\sin\alpha = e^{-\alpha i}$. The general solution to the corresponding homogeneous equation is $x_n^{(h)} = c_1\cos\alpha n + c_2\sin\alpha n$, $c_1, c_2 \in \mathbb{C}$. To establish a particular solution to the non-homogeneous difference equation, by the method of variation of parameters, we determine two sequences, $\left\{c_n^{(1)}\right\}_{n\in\mathbb{N}_0}$ and $\left\{c_n^{(2)}\right\}_{n\in\mathbb{N}_0}$, so that $x_n^{(p)} = c_n^{(1)}\cos n\alpha + c_n^{(2)}\sin n\alpha$ is a particular solution to the equation. We use the fact that $c_n^{(1)}$ and $c_n^{(2)}$ satisfy the difference equations

$$c_{n+1}^{(1)} - c_n^{(1)} = -\frac{\beta_n \sin(n+1)\alpha}{C_{n+1}} \quad \text{and} \quad c_{n+1}^{(2)} - c_n^{(2)} = \frac{\beta_n \cos(n+1)\alpha}{C_{n+1}},$$

where

$$C_{n+1} = \begin{vmatrix} \cos(n+1)\alpha & \sin(n+1)\alpha \\ \cos(n+2)\alpha & \sin(n+2)\alpha \end{vmatrix} = \sin\alpha.$$

From $c_{n+1}^{(1)} - c_n^{(1)} = -\frac{\beta_n \sin(n+1)\alpha}{\sin\alpha}$, it follows that we can take $c_0^{(1)} = 0$ and, for $n \in \mathbb{N}$, $c_n^{(1)} = -\frac{1}{\sin\alpha} \sum_{k=0}^{n-1} \beta_k \sin(k+1)\alpha$. From $c_{n+1}^{(2)} - c_n^{(2)} = \frac{\beta_n \cos(n+1)\alpha}{\sin\alpha}$, it follows that we can take $c_0^{(2)} = 0$ and, for $n \in \mathbb{N}$, $c_n^{(2)} = \frac{1}{\sin\alpha} \sum_{k=0}^{n-1} \beta_k \cos(k+1)\alpha$.

The method of variation of parameters yields the solution $x_0^{(p)} = 0$ and, for $n \in \mathbb{N}$,

$$x_n^{(p)} = -\frac{\cos n\alpha}{\sin\alpha} \sum_{k=0}^{n-1} \beta_k \sin(k+1)\alpha + \frac{\sin n\alpha}{\sin\alpha} \sum_{k=0}^{n-1} \beta_k \cos(k+1)\alpha$$

$$= \frac{1}{\sin\alpha} \sum_{k=0}^{n-1} \beta_k \left(\sin n\alpha \cos(k+1)\alpha - \cos n\alpha \sin(k+1)\alpha \right)$$

$$= \frac{1}{\sin\alpha} \sum_{k=0}^{n-1} \beta_k \sin(n-k-1)\alpha.$$

The general solution is $x_0 = c_1$ and, for $n \in \mathbb{N}$, $x_n = c_1 \cos n\alpha + c_2 \sin n\alpha + \frac{1}{\sin\alpha} \sum_{k=0}^{n-1} \beta_k \sin(n-k-1)\alpha$, $c_1, c_2 \in \mathbb{C}$.

Note: Problem 3.3.28 is a special case of Problem 3.3.32.

Problem 3.3.33. Solve the following non-homogeneous second-order difference equations:

(a) $x_{n+2} - 3x_{n+1} + 2x_n = 5^n$. (c) $6x_{n+2} - 5x_{n+1} + x_n = n$.

(b) $x_{n+2} - 2x_{n+1} + x_n = 1$. (d) $x_{n+2} + x_{n+1} + x_n = n^2$.

(e) $x_{n+2} - 6x_{n+1} + 8x_n = 2 + 3n^2 - 5 \cdot 3^n$.

(f) $x_{n+2} + 4x_n = 2^{n+3} \cos \frac{n\pi}{2}$.   (g) $x_{n+2} - 4x_{n+1} + 3a_n = n4^n$.

(h) $x_{n+2} - 5x_{n+1} + 6x_n = 3^n$.

(i) $x_{n+2} - 2x_{n+1} + 4x_n = 2^n \left(\cos \frac{n\pi}{3} + \sqrt{3} \sin \frac{n\pi}{3} \right)$.

Answer 3.3.33. For c_1, $c_2 \in \mathbb{C}$:

(a) $x_n = c_1 + c_2 2^n + \frac{5^n}{12}$.   (b) $x_n = c_1 + c_2 n + \frac{n^2}{2}$.

(c) $x_n = c_1 2^{-n} + c_2 3^{-n} + \frac{n}{2} - \frac{7}{4}$.

(d) $x_n = c_1 \cos \frac{2n\pi}{3} + c_2 \sin \frac{2n\pi}{3} + \frac{n^2}{3} - \frac{2n}{3} + \frac{1}{9}$.

(e) $x_n = c_1 2^n + c_2 4^n + \frac{44}{9} + \frac{8n}{3} + n^2 + 5 \cdot 3^n$.

(f) $x_n = 2^n \left((c_1 - n) \cos \frac{n\pi}{2} + c_2 \sin \frac{n\pi}{2} \right)$.

(g) $x_n = c_1 + c_2 3^n + \frac{(3n-16)4^n}{9}$.   (h) $x_n = c_1 2^n + c_2 3^n + n3^{n-1}$.

(i) Solutions to the characteristic equation $r^2 - 2r + 4 = 0$ are $r_1 = 1+i\sqrt{3}$ and $r_2 = 1-i\sqrt{3}$. The general solution to the corresponding homogeneous equation is $x_n^{(h)} = 2^n \left(c_1 \cos \frac{n\pi}{3} + c_2 \sin \frac{n\pi}{3} \right)$, $c_1, c_2 \in \mathbb{C}$. We use the method of undetermined coefficients to determine a particular solution to the equation. Since the sequence $\left\{ 2^n \left(\cos \frac{n\pi}{3} + \sqrt{3} \sin \frac{n\pi}{3} \right) \right\}_{n \in \mathbb{N}_0}$, i.e., the sequence on the right-hand side of the equation, is a solution to the corresponding homogeneous equation, we look for constants a and b such that $x_n^{(p)} = n2^n \left(a \cos \frac{n\pi}{3} + b \sin \frac{n\pi}{3} \right)$ is a particular solution to the equation. The assumption that $x_{n+2}^{(p)} - 2x_{n+1}^{(p)} + 4x_n^{(p)} = 2^n \left(\cos \frac{n\pi}{3} + \sqrt{3} \sin \frac{n\pi}{3} \right)$ implies that a and b are solutions to the linear system of equations

$$-3a + \sqrt{3}\, b = \frac{1}{2} \quad \text{and} \quad -\sqrt{3}\, a - 3b = \frac{\sqrt{3}}{2}.$$

It follows that $a = -\frac{1}{4}$, $b = -\frac{\sqrt{3}}{12}$, and

$$x_n^{(p)} = -\frac{n2^n}{12} \left(3 \cos \frac{n\pi}{3} + \sqrt{3} \sin \frac{n\pi}{3} \right) = -\frac{n2^n}{2\sqrt{3}} \sin \frac{(n+1)\pi}{3}.$$

The general solution to the equation is

$$x_n = 2^n \left(c_1 \cos \frac{n\pi}{3} + c_2 \sin \frac{n\pi}{3} \right) - \frac{n2^n}{2\sqrt{3}} \sin \frac{(n+1)\pi}{3}, \quad c_1, c_2 \in \mathbb{C}.$$

3.4 Special Sequences

Problem 3.4.1. The *Fibonacci sequence* is the sequence $\{F_n\}_{n\in\mathbb{N}_0}$ with the property that $F_0 = 0$ and $F_1 = 1$ and such that each term after the first two terms is the sum of the two immediately preceding terms.[5] It is named in honour of Leonardo Pisano, c. 1170–c. 1245, an Italian mathematician, better known as Fibonacci. It is a well established fact that the Fibonacci sequence was described by ancient Indian mathematicians as early as 200 BC [19].

(a) Solve the equation $x_{n+2} - x_{n+1} - x_n = 0$.
(b) For $n \in \mathbb{N}_0$, determine the Fibonacci number F_n.

Solution 3.4.1.

(a) Since the solutions to the corresponding quadratic equation $r^2 - r - 1 = 0$ are $\varphi = \frac{1+\sqrt{5}}{2}$ and $\psi = \frac{1-\sqrt{5}}{2}$, the general solution to the equation is $x_n = c_1 \left(\frac{1+\sqrt{5}}{2}\right)^n + c_2 \left(\frac{1-\sqrt{5}}{2}\right)^n$, $c_1, c_2 \in \mathbb{C}$.

(b) The Fibonacci sequence is the solution to the initial value problem $x_{n+2} - x_{n+1} - x_n = 0$, $x_0 = 0$ and $x_1 = 1$. Since the general solution to the equation is $x_n = c_1 \left(\frac{1+\sqrt{5}}{2}\right)^n + c_2 \left(\frac{1-\sqrt{5}}{2}\right)^n$, $c_1, c_2 \in \mathbb{C}$, the initial conditions imply that $F_n = \frac{1}{\sqrt{5}} \left(\left(\frac{1+\sqrt{5}}{2}\right)^n - \left(\frac{1-\sqrt{5}}{2}\right)^n\right)$, $n \in \mathbb{N}_0$.

Note: Recall that the number $\varphi = \frac{1+\sqrt{5}}{2} \approx 1.618033$ is called the *golden ratio*. The number $\psi = -\frac{1}{\varphi} = \frac{1-\sqrt{5}}{2} \approx -0.618033$ is its conjugate. The expression $F_n = \frac{\varphi^n - \psi^n}{\sqrt{5}}$ is called Binet's formula.[6] Since the sequence $\{\varphi^n\}_{n\in\mathbb{N}_0}$ is a solution to the difference equation, the golden ratio has the property that, for any $n \in \mathbb{N}_0$, $\varphi^{n+2} = \varphi^{n+1} + \varphi^n$.

---

[5] The Fibonacci sequence is listed in the Online Encyclopedia of Integer Sequences as A000045.

[6] Jacques Philippe Marie Binet, 1786–1856, French mathematician, physicist, and astronomer.

Problem 3.4.2. Establish the following properties of the Fibonacci numbers:

(a) If $r_n = \frac{F_n}{F_{n+1}}$, $n \in \mathbb{N}_0$, then $r_{n+1} = \frac{1}{1+r_n}$.

(b) If $r_n = \frac{F_n}{F_{n+1}}$, $n \in \mathbb{N}_0$, then $\lim_{n\to\infty} r_n = \dfrac{\sqrt{5}-1}{2}$.

(c) Consecutive Fibonacci numbers, F_n and F_{n+1}, are relatively prime.

(d) For all $n \in \mathbb{N}$, $F_n^2 - F_{n-1}F_{n+1} = (-1)^{n+1}$.

(e) For all $m \in \mathbb{N}_0$ and $n \in \mathbb{N}$, $F_{n-1}F_m + F_n F_{m+1} = F_{n+m}$.

(f) For all $m, n \in \mathbb{N}$, $F_m | F_{nm}$.

(g) For all $n \in \mathbb{N}$, $F_1 + F_2 + F_3 + \cdots + F_n = F_{n+2} - 1$.

(h) For all $n \in \mathbb{N}$, $F_1^2 + F_2^2 + \cdots + F_n^2 = F_n F_{n+1}$.

(i) For each $n \in \mathbb{N}_0$, the Fibonacci number F_n is the nearest integer to the number $\frac{(1+\sqrt{5})^n}{2^n \sqrt{5}}$.

Solution 3.4.2.

(a) By definition, $\dfrac{1}{1+r_n} = \dfrac{1}{1+\frac{F_n}{F_{n+1}}} = \dfrac{F_{n+1}}{F_{n+1}+F_n} = \dfrac{F_{n+1}}{F_{n+2}} = r_{n+1}$.

(b) Observe that $\left|\dfrac{\psi}{\varphi}\right| < 1$ implies $\lim_{n\to\infty} \left(\dfrac{\psi}{\varphi}\right)^n = 0$. From

$$F_n = \frac{\varphi^n - \psi^n}{\sqrt{5}}, \quad n \in \mathbb{N}_0, \quad \text{it follows that} \quad \lim_{n\to\infty} \frac{F_n}{\varphi^n} =$$

$$\frac{1}{\sqrt{5}} \lim_{n\to\infty} \left(1 - \left(\frac{\psi}{\varphi}\right)^n\right) = \frac{1}{\sqrt{5}}. \quad \text{Therefore,}$$

$$\lim_{n\to\infty} r_n = \lim_{n\to\infty} \frac{F_n}{F_{n+1}} = \lim_{n\to\infty} \frac{\varphi^{n+1}}{F_{n+1}} \frac{F_n}{\varphi^n} \frac{1}{\varphi} = \sqrt{5}\frac{1}{\sqrt{5}}\frac{1}{\varphi} = \frac{1}{\varphi} = \frac{\sqrt{5}-1}{2}.$$

(c) Suppose that the claim is not true and that a non-empty set S of natural numbers has the property that, for $n \in S$, F_n and F_{n+1} are not relatively prime. If $m = \min S > 1$, let $k \in \mathbb{N}\setminus\{1\}$ be such that $F_m = ka$ and $F_{m+1} = kb$, for some positive integers a and b, $a \neq b$. But then, by the definition of the Fibonacci sequence, $F_{m-1} = F_{m+1} - F_m = k(b - a)$, which implies that $m - 1 \in S$, contradicting $m = \min S$. Hence, such an S does not exist and the claim is true.

(d) Observe that $\varphi\psi = -1$ and $\varphi^2 + \psi^2 = \frac{(6+2\sqrt{5})+(6-2\sqrt{5})}{4} = 3$. It follows that, for any $n \in \mathbb{N}$,

$$
\begin{aligned}
5(F_n^2 - F_{n-1}F_{n+1}) &= (\varphi^n - \psi^n)^2 - (\varphi^{n-1} - \psi^{n-1})(\varphi^{n+1} - \psi^{n+1}) \\
&= \varphi^{2n} - 2\varphi^n\psi^n + \psi^{2n} - \varphi^{2n} + \varphi^{n-1}\psi^{n+1} \\
&\quad + \varphi^{n+1}\psi^{n-1} - \psi^{2n} \\
&= -2(-1)^n + (-1)^{n-1}(\varphi^2 + \psi^2) = 5(-1)^{n+1},
\end{aligned}
$$

which implies the claim.

(e) Recall that $\varphi\psi = -1$ and $\frac{1}{\varphi} + \varphi = -\frac{1}{\psi} - \psi = \sqrt{5}$. For all $m \in \mathbb{N}_0$ and $n \in \mathbb{N}$,

$$
\begin{aligned}
F_{n-1}F_m + F_n F_{m+1} &= \frac{\varphi^{n-1} - \psi^{n-1}}{\sqrt{5}} \frac{\varphi^m - \psi^m}{\sqrt{5}} \\
&\quad + \frac{\varphi^n - \psi^n}{\sqrt{5}} \frac{\varphi^{m+1} - \psi^{m+1}}{\sqrt{5}} \\
&= \frac{1}{5} \left(\varphi^{n+m-1} - \psi^{n-1}\varphi^m - \psi^m\varphi^{n-1} + \psi^{n+m-1} \right. \\
&\quad \left. + \varphi^{n+m+1} - \psi^n\varphi^{m+1} - \psi^{m+1}\varphi^n + \psi^{n+m+1} \right) \\
&= \frac{1}{5} \left(\varphi^{n+m} \left(\frac{1}{\varphi} + \varphi \right) + \psi^{n+m} \left(\frac{1}{\psi} + \psi \right) + (-1)^m \right. \\
&\quad \left. \cdot \left(\psi^{n-m-1} + \varphi^{n-m-1} - \psi^{n-m-1} - \varphi^{n-m-1} \right) \right) \\
&= \frac{\varphi^{n+m} - \psi^{n+m}}{\sqrt{5}} = F_{n+m}.
\end{aligned}
$$

(f) Let $m \in \mathbb{N}$ and Let $M = \{n \in \mathbb{N} : F_m | F_{nm}\}$. Clearly, $1 \in M$. Let $n \in M$ and let $a \in \mathbb{N}$ be such that $F_{nm} = aF_m$. Then,

$$
F_{(n+1)m} = F_{nm+m} = F_{nm-1}F_m + F_{nm}F_{m+1} = F_m \left(F_{nm-1} + aF_{m+1} \right),
$$

which implies that $n+1 \in M$. By the principle of mathematical induction, $M = \mathbb{N}$.

(g) *Solution 1:* Observe that $\varphi - \psi = \sqrt{5}$, $\varphi - 1 = -\psi$, $\psi - 1 = -\varphi$, and $\varphi\psi = -1$. It follows that, for any $n \in \mathbb{N}$,

$$\sqrt{5}\sum_{k=1}^{n} F_k = \sum_{k=1}^{n}(\varphi^k - \psi^k) = \frac{\varphi^{n+1} - 1}{\varphi - 1} - \frac{\psi^{n+1} - 1}{\psi - 1}$$

$$= \frac{\varphi^{n+1} - 1}{-\psi} - \frac{\psi^{n+1} - 1}{-\varphi} = \frac{-\varphi^{n+2} + \varphi + \psi^{n+2} - \psi}{\varphi\psi}$$

$$= \varphi^{n+2} - \psi^{n+2} - \sqrt{5},$$

which implies that $F_1 + F_2 + F_3 + \cdots + F_n = F_{n+2} - 1$.

Solution 2: From, for all $k \in \mathbb{N}$, $F_k = F_{k+1} - F_{k-1}$, it follows that, for all $n \in \mathbb{N}$,

$$\sum_{k=1}^{n} F_k = \sum_{k=1}^{n}(F_{k+1} - F_{k-1}) = \sum_{k=2}^{n+1} F_k - \sum_{k=1}^{n-1} F_k$$

$$= F_n + F_{n+1} - F_1 = F_{n+2} - 1.$$

(h) From, for all $k \in \mathbb{N}$, $F_k^2 = F_k(F_{k+1} - F_{k-1})$, it follows that, for all $n \in \mathbb{N}\backslash\{1\}$,

$$\sum_{k=1}^{n} F_k^2 = F_1^2 + \sum_{k=2}^{n}(F_k F_{k+1} - F_k F_{k-1}) = F_1 F_2 + \sum_{k=2}^{n} F_k F_{k+1}$$

$$- \sum_{k=1}^{n-1} F_k F_{k+1} = F_1 F_2 + F_n F_{n+1} - F_1 F_2 = F_n F_{n+1}.$$

(i) Observe that, for any $n \in \mathbb{N}_0$, $\frac{(1+\sqrt{5})^n}{2^n\sqrt{5}} = \frac{\varphi^n}{\sqrt{5}}$ and recall that $\varphi > 1$ and $\sqrt{5} > 2$. The question is to verify that, for any $n \in \mathbb{N}_0$,

$$\frac{1}{2} \geq \left| F_n - \frac{\varphi^n}{\sqrt{5}} \right| = \frac{1}{\sqrt{5}} \left| \varphi^n - \frac{(-1)^n}{\varphi^n} - \varphi^n \right| = \frac{1}{\sqrt{5}\varphi^n},$$

which is the same as verifying that $\sqrt{5}\varphi^n \geq 2$, $n \in \mathbb{N}_0$.

Problem 3.4.3. The *Lucas sequence* $\{L_n\}_{n\in\mathbb{N}_0}$, named after François Édouard Anatole Lucas, 1842–1891, a French mathematician, is defined as the solution to the initial value problem

$x_{n+2} = x_{n+1} + x_n$, $x_0 = 2$ and $x_1 = 1$. For $n \in \mathbb{N}_0$, determine the Lucas number L_n.[7]

Solution 3.4.3. Since the solutions to the corresponding quadratic equation $r^2 - r - 1 = 0$ are $\varphi = \frac{1+\sqrt{5}}{2}$ and $\psi = \frac{1-\sqrt{5}}{2}$, the general solution to the equation is $x_n = c_1 \left(\frac{1+\sqrt{5}}{2}\right)^n + c_2 \left(\frac{1-\sqrt{5}}{2}\right)^n$, $c_1, c_2 \in \mathbb{C}$. This, together with the initial conditions, implies that $L_n = \left(\frac{1+\sqrt{5}}{2}\right)^n + \left(\frac{1-\sqrt{5}}{2}\right)^n = \varphi^n + \psi^n$.

Problem 3.4.4. Establish the following properties of the Lucas numbers:

(a) For any $n \in \mathbb{N}$, $L_n = F_{n-1} + F_{n+1}$.

(b) $\lim\limits_{n \to \infty} \dfrac{L_n}{F_n} = \sqrt{5}$.

(c) For any $n \in \mathbb{N}$, $L_n^2 - L_{n-1}L_{n+1} = 5(-1)^n$.

(d) For any $m, n \in \mathbb{N}_0$, $2F_{m+n} = F_m L_n + F_n L_m$.

Solution 3.4.4.

(a) For all $n \in \mathbb{N}$,

$$\begin{aligned}
\sqrt{5}(F_{n-1} + F_{n+1}) &= \varphi^{n-1}\left(1 + \varphi^2\right) - \psi^{n-1}\left(1 + \psi^2\right) \\
&= \varphi^{n-1}\left(-\varphi\psi + \varphi^2\right) - \psi^{n-1}\left(-\varphi\psi + \psi^2\right) \\
&= \varphi^n\left(-\psi + \varphi\right) - \psi^n\left(-\varphi + \psi\right) \\
&= \sqrt{5}\varphi^n + \sqrt{5}\psi^n = \sqrt{5}L_n.
\end{aligned}$$

(b) By Problems 3.4.1(b) and 3.4.3,

$$\lim_{n \to \infty} \frac{L_n}{F_n} = \lim_{n \to \infty} \frac{\varphi^n + \psi^n}{\frac{\varphi^n - \psi^n}{\sqrt{5}}} = \sqrt{5} \lim_{n \to \infty} \frac{1 + \left(\frac{\psi}{\varphi}\right)^n}{1 - \left(\frac{\psi}{\varphi}\right)^n} = \sqrt{5} \cdot 1 = \sqrt{5}.$$

---

[7]The Lucas sequence is listed in the Online Encyclopedia of Integer Sequences as A000032.

(c) Recall that $\varphi\psi = -1$ and $\varphi^2 + \psi^2 = 3$. It follows that

$$\begin{aligned}
L_n^2 - L_{n-1}L_{n+1} &= (\varphi^n + \psi^n)^2 - (\varphi^{n-1} + \psi^{n-1})(\varphi^{n+1} + \psi^{n+1}) \\
&= \varphi^{2n} + 2\varphi^n\psi^n + \psi^{2n} - \varphi^{2n} - \varphi^{n-1}\psi^{n+1} \\
&\quad - \varphi^{n+1}\psi^{n-1} - \psi^{2n} = 2(-1)^n - (-1)^{n-1}(\varphi^2 + \psi^2) \\
&= 5(-1)^n.
\end{aligned}$$

(d) *Solution 1:* From, for all $n \in \mathbb{N}_0$, $F_n = \frac{1}{\sqrt{5}}(\varphi^n - \psi^n)$ and $L_n = \varphi^n + \psi^n$, it follows that, for all $n \in \mathbb{N}_0$,

$$\begin{aligned}
2\sqrt{5}F_{m+n} &= 2\varphi^m\varphi^n + \varphi^m\psi^n + \varphi^n\psi^m - \varphi^m\psi^n - \varphi^n\psi^m - 2\psi^m\psi^n \\
&= \varphi^m(\varphi^n + \psi^n) + \varphi^n(\varphi^m + \psi^m) - \psi^n(\varphi^m + \psi^m) \\
&\quad - \psi^m(\varphi^n + \psi^n) = L_n(\varphi^m - \psi^m) + L_m(\varphi^n - \psi^n) \\
&= \sqrt{5}L_nF_m + \sqrt{5}L_mF_n,
\end{aligned}$$

which implies the claim.

Solution 2: Let $M = \{n \in \mathbb{N}_0 : 2F_{n+m} = F_nL_m + F_mL_n$ for all $m \in \mathbb{N}_0\}$. From, for all $m \in \mathbb{N}_0$,

$$F_0L_m + L_0F_m = 0 \cdot L_m + 2F_m = 2F_{0+m}$$

and, by Part (a),

$$F_1L_m + L_1F_m = L_m + F_m = (F_{m-1} + F_{m+1}) + F_m = 2F_{1+m},$$

it follows that $0 \in M$ and $1 \in M$. Let $n \in M$ be such that if $k \leq n$, then $k \in M$. In particular, $n \in M$, $n \geq 1$ implies that $n - 1 \in M$. Then, by Problem 3.4.2(e), for all $m \in \mathbb{N}_0$,

$$\begin{aligned}
2F_{(n+1)+m} &= 2(F_{(n-1)+m} + F_{n+m}) \\
&= F_{n-1}L_m + F_mL_{n-1} + F_nL_m + F_mL_n \\
&= (F_{n-1} + F_n)L_m + F_m(L_{n-1} + L_n) \\
&= F_{n+1}L_m + F_mL_{n+1}.
\end{aligned}$$

Hence, if $n \in M$, then $n + 1 \in M$ and, by the principle of mathematical induction, $M = \mathbb{N}_0$.

Problem 3.4.5. Let $k \in \mathbb{N}$.

(a) Solve the equation $x_{n+2} - kx_{n+1} - x_n = 0$.
(b) Determine the k-*Fibonacci sequence* $\{F_{k,n}\}_{n\in\mathbb{N}_0}$, the solution to the initial value problem $x_{n+2} - kx_{n+1} - x_n = 0$, $x_0 = 0$, $x_1 = 1$.
(c) Determine the k-*Lucas sequence* $\{L_{k,n}\}_{n\in\mathbb{N}_0}$, the solution to the initial value problem $x_{n+2} - kx_{n+1} - x_n = 0$, $x_0 = 2$, $x_1 = k$.
(d) Determine the *Pell sequence* $\{P_n\}_{n\in\mathbb{N}_0}$, the solution to the initial value problem $x_{n+2} - 2x_{n+1} - x_n = 0$, $x_0 = 0$, $x_1 = 1$.[8]

Solution 3.4.5.

(a) Since the solutions to the corresponding quadratic equation $r^2 - kr - 1 = 0$ are $\sigma_k = \frac{k+\sqrt{k^2+4}}{2}$ and $\tau_k = \frac{k-\sqrt{k^2+4}}{2} = -\frac{1}{\sigma_k}$, the general solution to the equation is $x_n = c_1 \left(\frac{k+\sqrt{k^2+4}}{2}\right)^n + c_2 \left(\frac{k-\sqrt{k^2+4}}{2}\right)^n = c_1\sigma_k^n + c_2\tau_k^n$, $c_1, c_2 \in \mathbb{C}$.

(b) To determine $c_1, c_2 \in \mathbb{C}$ such that, for all $n \in \mathbb{N}_0$, $F_{k,n} = c_1 \left(\frac{k+\sqrt{k^2+4}}{2}\right)^n + c_2 \left(\frac{k-\sqrt{k^2+4}}{2}\right)^n$, we solve the system

$$F_{k,0} = 0 = c_1 + c_2 \text{ and } F_{k,1} = 1 = c_1 \frac{k + \sqrt{k^2 + 4}}{2} + c_2 \frac{k - \sqrt{k^2 + 4}}{2}.$$

From $c_2 = -c_1$, it follows that

$$1 = c_1 \left(\frac{k + \sqrt{k^2 + 4}}{2} - \frac{k - \sqrt{k^2 + 4}}{2}\right) = c_1\sqrt{k^2 + 4}.$$

Therefore, for all $n \in \mathbb{N}_0$,

$$F_{k,n} = \frac{1}{\sqrt{k^2 + 4}} \left(\left(\frac{k + \sqrt{k^2 + 4}}{2}\right)^n - \left(\frac{k - \sqrt{k^2 + 4}}{2}\right)^n\right)$$

$$= \frac{\sigma_k^n - \tau_k^n}{\sqrt{k^2 + 4}}.$$

---

[8] John Pell, 1611–1685, English mathematician.

(c) To determine $c_1, c_2 \in \mathbb{C}$ such that, for all $n \in \mathbb{N}_0$, $L_{k,n} = c_1 \left(\frac{k+\sqrt{k^2+4}}{2} \right)^n + c_2 \left(\frac{k-\sqrt{k^2+4}}{2} \right)^n$, we solve the system

$$2 = c_1 + c_2 \text{ and } k = c_1 \, \frac{k + \sqrt{k^2 + 4}}{2} + c_2 \, \frac{k - \sqrt{k^2 + 4}}{2}.$$

It follows that $c_1 = c_2 = 1$ and, for all $n \in \mathbb{N}_0$,

$$L_{k,n} = \left(\frac{k + \sqrt{k^2 + 4}}{2} \right)^n + \left(\frac{k - \sqrt{k^2 + 4}}{2} \right)^n = \sigma_k^n + \tau_k^n.$$

(d) To determine $c_1, c_2 \in \mathbb{C}$ such that, for all $n \in \mathbb{N}_0$, $P_n = c_1 \left(1 + \sqrt{2}\right)^n + c_2 \left(1 - \sqrt{2}\right)^n$, we solve the system

$$0 = c_1 + c_2 \text{ and } 1 = c_1 \left(1 + \sqrt{2}\right) + c_2 (1 - \sqrt{2}).$$

It follows that $c_1 = -c_2 = \frac{1}{2\sqrt{2}}$ and, for all $n \in \mathbb{N}_0$,

$$P_n = \frac{1}{2\sqrt{2}} \left(\left(1 + \sqrt{2}\right)^n - \left(1 - \sqrt{2}\right)^n \right).$$

Note: Observe that, for all $n \in \mathbb{N}_0$, $F_{1,n} = F_n$, $F_{2,n} = P_n$, and $L_{1,n} = L_n$.

Problem 3.4.6. The gamma function is defined as $\Gamma(x) = \int_0^\infty t^{x-1} e^{-t} \, dt$, $x > 0$.

(a) Establish that $\Gamma(1) = 1$ and $\Gamma(x + 1) = x\Gamma(x)$.
(b) Solve the initial value problem $x_{n+1} = (n + 1)x_n$, $x_0 = 1$. Conclude that, for any $n \in \mathbb{N}$, $\Gamma(n + 1) = n!$.
(c) Prove that for any $n \in \mathbb{N}$ and $x > n - 1$, $\prod_{k=0}^{n-1}(x - k) = x(x - 1) \cdots (x - n + 1) = \frac{\Gamma(x+1)}{\Gamma(x-n+1)}$.

Solution 3.4.6.

(a) Observe that $\Gamma(1) = \int_0^\infty e^{-t} \, dt = 1 - \lim_{t \to \infty} e^{-t} = 1$. Recall that, for any $x > 0$, $\lim_{t \to \infty} t^x e^{-t} = 0$. The integration-by-parts formula yields

$$\Gamma(x + 1) = \int_0^\infty t^x e^{-t} \, dtx = \left| \begin{matrix} u = t^x & du = xt^{x-1}dt \\ dv = e^{-t}dt & v = -e^{-t} \end{matrix} \right|$$

$$= \lim_{T \to \infty} \left(-t^x e^{-t} \big|_0^T + x \int_0^\infty t^{x-1} e^{-t} \, dt \right) = x\Gamma(x).$$

(b) By Problem 3.2.2(a), the solution $\{n!\}_{n\in\mathbb{N}_0}$ of the initial value problem is obtained from the general solution $x_n = c\prod_{k=0}^{n-1}(k+1) = c\cdot n!$ and $x_0 = 1 = c\cdot 0! = c$. If, for $n\in\mathbb{N}_0$, $y_n = \Gamma(n+1)$, then $y_{n+1} = \Gamma(n+2) = (n+1)\Gamma(n+1) = (n+1)y_n$ and $y_0 = \Gamma(1) = 1$. Since the solution to the initial value problem is unique, for all $n\in\mathbb{N}_0$, $y_n = \Gamma(n+1) = n!$.

(c) By Part (a), for any $n\in\mathbb{N}$ and $x > n-1$,

$$\prod_{k=0}^{n-1}(x-k) = \prod_{k=0}^{n-1}\frac{\Gamma(x-k+1)}{\Gamma(x-k)} = \frac{\Gamma(x+1)}{\Gamma(x)}\frac{\Gamma(x)}{\Gamma(x-1)}\cdots\frac{\Gamma(x-n)}{\Gamma(x-n+1)}$$

$$= \frac{\Gamma(x+1)}{\Gamma(x-n+1)}.$$

Problem 3.4.7. Let $\alpha\in(0,\pi)$. Consider the sequence $\{I_n(\alpha)\}_{n\in\mathbb{N}_0}$, where $I_n(\alpha) = \int_0^\pi \frac{\cos n\theta - \cos n\alpha}{\cos\theta - \cos\alpha}\,d\theta$, $n\in\mathbb{N}_0$.

(a) Solve the equation $x_{n+2} - 2x_{n+1}\cos\alpha + x_n = 0$.
(b) Establish that the sequence $\{I_n(\alpha)\}_{n\in\mathbb{N}_0}$ is the solution to the initial value problem $x_{n+2} - 2x_{n+1}\cos\alpha + x_n = 0$, $x_0 = 0$, $x_1 = \pi$.
(c) Evaluate $I_n(\alpha)$ for $n\in\mathbb{N}_0$.

Solution 3.4.7. For $\alpha\in(0,\pi)$, while passing through the interval $[0,\pi]$, the variable θ will take the value α. For $n\in\mathbb{N}$, the function $\theta\mapsto f_n(\theta) = \frac{\cos n\theta - \cos n\alpha}{\cos\theta - \cos\alpha}$ is continuous on its domain $D = [0,\alpha)\cup(\alpha,\pi]$. Since, by L'Hôpital's rule, $\lim\limits_{\theta\to\alpha}\dfrac{\cos n\theta - \cos n\alpha}{\cos\theta - \cos\alpha} = \lim\limits_{\theta\to\alpha}\dfrac{n\sin n\theta}{\sin\theta} = \dfrac{n\sin n\alpha}{\sin\alpha}$, it follows that the function f_n is bounded; therefore, for any $\alpha\in(0,\pi)$ and any $n\in\mathbb{N}_0$, the integral $I_n(\alpha)$ exists. This confirms that all the terms of the sequence $\{I_n(\alpha)\}_{n\in\mathbb{N}_0}$ are well-defined.

(a) Solutions to the characteristic equation $r^2 - 2r\cos\alpha + 1 = 0$ are $r_1 = \cos\alpha + i\sin\alpha = e^{\alpha i}$ and $r_2 = \cos\alpha - i\sin\alpha = e^{-\alpha i}$. The general solution to the difference equation is $x_n = c_1\cos\alpha n + c_2\sin\alpha n$, $c_1, c_2\in\mathbb{C}$.
(b) The fact that, for any $\varphi\in\mathbb{R}$ and any $n\in\mathbb{N}_0$, $\cos(n+2)\varphi + \cos n\varphi = 2\cos(n+1)\varphi\cos\varphi$, implies that, for any $\theta\in[0,\pi]$, $\alpha\in(0,\pi)$, and $n\in\mathbb{N}_0$, $\cos(n+2)\theta - 2\cos\alpha\cos(n+1)\theta + \cos n\theta =$

$2(\cos\theta - \cos\alpha)\cos(n+1)\theta$ and $\cos(n+2)\alpha - 2\cos\alpha\cos(n+1)\alpha + \cos n\alpha = 0$. It follows that

$$I_{n+2}(\alpha) - 2\cos\alpha I_{n+1}(\alpha) + I_n(\alpha)$$
$$= \int_0^\pi \frac{2(\cos\theta - \cos\alpha)\cos(n+1)\theta \; d\theta}{\cos\theta - \cos\alpha}$$
$$= \frac{2\sin(n+1)\theta}{n+1}\bigg|_0^\pi = 0.$$

In addition, $I_0(\alpha) = \int_0^\pi \frac{(1-1)\,d\theta}{\cos\theta - \cos\alpha} = 0$ and $I_1(\alpha) = \int_0^\pi \frac{(\cos\theta - \cos\alpha)\,d\theta}{\cos\theta - \cos\alpha} = \pi$. The sequence $\{I_n(\alpha)\}_{n\in\mathbb{N}_0}$ is the unique solution to the initial value problem $x_{n+2} - 2x_{n+1}\cos\alpha + x_n = 0$, $x_0 = 0$, $x_1 = \pi$.

(c) To determine $c_1, c_2 \in \mathbb{C}$ such that, for all $n \in \mathbb{N}_0$, $I_n(\alpha) = c_1\cos n\alpha + c_2\sin n\alpha$, we solve the system $I_0(\alpha) = 0 = c_1$ and $I_1(\alpha) = \pi = c_2\sin\alpha$. Thus, $c_2 = \frac{\pi}{\sin\alpha}$ and, for all $n \in \mathbb{N}_0$, $I_n(\alpha) = \int_0^\pi \frac{\cos(n\theta) - \cos(n\alpha)}{\cos\theta - \cos\alpha}\,d\theta = \frac{\pi\sin n\alpha}{\sin\alpha}$.

Problem 3.4.8. For $n \in \mathbb{N}_0$, the Chebyshev polynomial of the first kind of degree n is the function $T_n(x) = \cos(n\cos^{-1}(x))$, $|x| < 1$.[9]

(a) Establish that, for any $x \in (-1, 1)$, the sequence $\{T_n(x)\}_{n\in\mathbb{N}_0}$ is the solution to the initial value problem $y_{n+2} - 2xy_{n+1} + y_n = 0$, $y_0 = 1$, and $y_1 = x$.

(b) Solve the initial value problem in Part (a) to determine another expression for the Chebyshev polynomial $T_n(x)$.

(c) Determine the first four terms of the sequence $\{T_n(x)\}_{n\in\mathbb{N}_0}$.

Solution 3.4.8.

(a) Since, for any $\alpha \in \mathbb{R}$ and any $n \in \mathbb{N}_0$, $\cos(n+2)\alpha + \cos n\alpha = 2\cos(n+1)\alpha\cos\alpha$, it follows that, for $x \in (-1, 1)$,

$$T_{n+2}(x) + T_n(x) = 2\cos((n+1)\cos^{-1}(x))\cos(\cos^{-1}(x))$$
$$= 2x\cos((n+1)\cos^{-1}(x)) = 2xT_{n+1}(x),$$

---

[9]Pafnuty Lvovich Chebyshev, 1821–1894, Russian mathematician.

or, what is the same, $T_{n+2}(x) - 2xT_{n+1}(x) + T_n(x) = 0$. Also, $T_0(x) = \cos 0 = 1$ and $T_1(x) = \cos(\cos^{-1}(x)) = x$.

(b) For $x \in (-1,1)$, solutions to the characteristic equation $r^2 - 2xr + 1 = 0$ are $r_1 = x + \sqrt{x^2 - 1}$ and $r_2 = x - \sqrt{x^2 - 1}$. The general solution to the difference equation is $y_n = c_1 \left(x + \sqrt{x^2 - 1} \right)^n + c_2 \left(x - \sqrt{x^2 - 1} \right)^n$, $c_1, c_2 \in \mathbb{C}$. From the initial conditions $T_0(x) = 1$ and $T_1(x) = x$, it follows that $T_n(x) = \frac{1}{2} \left(\left(x + \sqrt{x^2 - 1} \right)^n + \left(x - \sqrt{x^2 - 1} \right)^n \right)$.

Note: Observe that $x \in (-1,1)$ implies $x^2 - 1 < 0$.

(c) As already established, $T_0(x) = 1$ and $T_1(x) = x$. Also,

$$T_2(x) = \frac{1}{2} \left(\left(x + \sqrt{x^2 - 1} \right)^2 + \left(x - \sqrt{x^2 - 1} \right)^2 \right)$$

$$= x^2 + (x^2 - 1) = 2x^2 - 1 \text{ and}$$

$$T_3(x) = \frac{1}{2} \left(\left(x + \sqrt{x^2 - 1} \right)^3 + \left(x - \sqrt{x^2 - 1} \right)^3 \right)$$

$$= x^3 + 3x(x^2 - 1) = 4x^3 - 3x.$$

Problem 3.4.9. For $n \in \mathbb{N}_0$, the Chebyshev polynomial of the second kind of degree n is the function $U_n(x) = \frac{1}{\sqrt{1-x^2}} \sin[(n + 1) \cdot \cos^{-1}(x)]$, $|x| < 1$.

(a) Establish that, for any $x \in (-1,1)$, the sequence $\{U_n(x)\}_{n \in \mathbb{N}_0}$ is the solution to the initial value problem $y_{n+2} - 2xy_{n+1} + y_n = 0$, $y_0 = 1$, and $y_1 = 2x$.

(b) Solve the initial value problem in Part (a) to determine another expression for the Chebyshev polynomial $U_n(x)$.

(c) Determine the first four terms of the sequence $\{U_n(x)\}_{n \in \mathbb{N}_0}$.

Solution 3.4.9.

(a) Since, for any $\alpha \in \mathbb{R}$ and any $n \in \mathbb{N}_0$, $\sin(n+3)\alpha + \sin(n+1)\alpha = 2\sin(n + 2)\alpha \cos \alpha$, it follows that, for $x \in (-1,1)$,

$$U_{n+2}(x) + U_n(x) = \frac{1}{\sqrt{1-x^2}} \cdot 2\sin((n+2)\cos^{-1}(x))\cos(\cos^{-1}(x))$$

$$= \frac{2x}{\sqrt{1-x^2}}\sin((n+2)\cos^{-1}(x)) = 2xU_{n+1}(x),$$

or, what is the same, $U_{n+2}(x) - 2xU_{n+1}(x) + U_n(x) = 0$. Also, $U_0(x) = \frac{1}{\sqrt{1-x^2}}\sin(\cos^{-1}(x)) = \frac{1}{\sqrt{1-x^2}}\sqrt{1-x^2} = 1$ and $U_1(x) = \frac{1}{\sqrt{1-x^2}}\sin(2\cos^{-1}(x)) = \frac{1}{\sqrt{1-x^2}}\cdot 2x\sqrt{1-x^2} = 2x$.

(b) For $x \in (-1,1)$, the general solution to the difference equation is $y_n = c_1(x + \sqrt{x^2-1})^n + c_2(x - \sqrt{x^2-1})^n$, $c_1, c_2 \in \mathbb{C}$. From the initial conditions $U_0(x) = 1$ and $U_1(x) = 2x$, it follows that $U_0(x) = 1 = c_1 + c_2$ and $U_1(x) = 2x = (c_1 + c_2)x + (c_1 - c_2)\sqrt{x^2-1} = x + (c_1 - c_2)\sqrt{x^2-1}$. From $c_1 + c_2 = 1$ and $c_1 - c_2 = \frac{x}{\sqrt{x^2-1}}$, it follows that $c_1 = \frac{x+\sqrt{x^2-1}}{2\sqrt{x^2-1}}$ and $c_2 = -\frac{x-\sqrt{x^2-1}}{2\sqrt{x^2-1}}$. Therefore,

$$U_n(x) = \frac{1}{2\sqrt{x^2-1}}\left(\left(x + \sqrt{x^2-1}\right)^{n+1} - \left(x - \sqrt{x^2-1}\right)^{n+1}\right).$$

Note: Observe that $x \in (-1,1)$ implies $x^2 - 1 < 0$.

(c) As already established, $U_0(x) = 1$ and $U_1(x) = 2x$. Also,

$$U_2(x) = \frac{1}{2\sqrt{x^2-1}}\left(\left(x + \sqrt{x^2-1}\right)^3 - \left(x - \sqrt{x^2-1}\right)^3\right)$$

$$= \frac{1}{\sqrt{x^2-1}}\left(3x^2\sqrt{x^2-1} + (x^2-1)\sqrt{x^2-1}\right) = 4x^2 - 1 \text{ and}$$

$$U_3(x) = \frac{1}{2\sqrt{x^2-1}}\left(\left(x + \sqrt{x^2-1}\right)^4 - \left(x - \sqrt{x^2-1}\right)^4\right)$$

$$= \frac{1}{2\sqrt{x^2-1}}8x\sqrt{x^2-1}\,(2x^2-1) = 8x^3 - 4x.$$

Problem 3.4.10. Establish that for $n \in \mathbb{N}_0$ and $\theta \in (0,\pi)$, $T_n(\cos\theta) = \cos n\theta$ and that $U_n(\cos\theta) = \frac{\sin(n+1)\theta}{\sin\theta}$.

Solution 3.4.10. From $T_n(x) = \frac{1}{2}\left(\left(x+\sqrt{x^2-1}\right)^n + \left(x-\sqrt{x^2-1}\right)^n\right)$, by De Moivre's theorem, it follows that

$$T_n(\cos\theta) = \frac{1}{2}\left(\left(\cos\theta + \sqrt{\cos^2\theta - 1}\right)^n + \left(\cos\theta - \sqrt{\cos^2\theta - 1}\right)^n\right)$$

$$= \frac{1}{2}\left((\cos\theta + i\sin\theta)^n + (\cos\theta - i\sin\theta)^n\right)$$

$$= \frac{1}{2}\left(\cos n\theta + i\sin n\theta + \cos n\theta - i\sin n\theta\right) = \cos n\theta.$$

Similarly, from $U_n(x) = \frac{1}{2\sqrt{x^2-1}}\left((x+\sqrt{x^2-1})^{n+1} - (x-\sqrt{x^2-1})^{n+1}\right)$, it follows that

$$U_n(\cos\theta) = \frac{1}{2i\sin\theta}\left((\cos\theta + i\sin\theta)^{n+1} - (\cos\theta - i\sin\theta)^{n+1}\right)$$

$$= \frac{1}{2i\sin\theta} \cdot 2i\sin(n+1)\theta = \frac{\sin(n+1)\theta}{\sin\theta}.$$

3.5 Nonlinear and Higher-Order Linear Difference Equations

Problem 3.5.1. Prove that the sequences $x_n^{(1)} = 2^n$, $x_n^{(2)} = (-2)^n$, and $x_n^{(3)} = (-3)^n$ form a fundamental set of solutions to the homogeneous third-order difference equation $x_{n+3} + 3x_{n+2} - 4x_{n+1} - 12x_n = 0$. What is the general solution to the equation?

Solution 3.5.1. From

$$x_{n+3}^{(1)} + 3x_{n+2}^{(1)} - 4x_{n+1}^{(1)} - 12x_n^{(1)} = 2^n(8 + 3\cdot 4 - 4\cdot 2 - 12) = 0,$$

$$x_{n+3}^{(2)} + 3x_{n+2}^{(2)} - 4x_{n+1}^{(2)} - 12x_n^{(2)} = (-2)^n(-8 + 3\cdot 4 - 4\cdot(-2) - 12) = 0,$$

$$x_{n+3}^{(3)} + 3x_{n+2}^{(3)} - 4x_{n+1}^{(3)} - 12x_n^{(3)} = (-3)^n(-27 + 3\cdot 9 - 4\cdot(-3) - 12) = 0,$$

it follows that $x_n^{(1)} = 2^n$, $x_n^{(2)} = (-2)^n$, and $x_n^{(3)} = (-3)^n$ are solutions to the equation. The Casoratian $\{C_n\}_{n\in\mathbb{N}_0}$ of the set of sequences

$$\left\{\{2^n\}_{n\in\mathbb{N}_0}, \{(-2)^n\}_{n\in\mathbb{N}_0}, \{(-3)^n\}_{n\in\mathbb{N}_0}\right\}$$

is, by definition, determined by

$$C_n = \begin{vmatrix} 2^n & (-2)^n & (-3)^n \\ 2^{n+1} & (-2)^{n+1} & (-3)^{n+1} \\ 2^{n+2} & (-2)^{n+2} & (-3)^{n+2} \end{vmatrix} = 12^n \begin{vmatrix} 1 & 1 & 1 \\ 2 & -2 & -3 \\ 4 & 4 & 9 \end{vmatrix} = -20 \cdot 12^n.$$

Since, for all $n \in \mathbb{N}_0$, $C_n \neq 0$, the set of solutions is a set of linearly independent sequences and, therefore, a fundamental set of solutions to the equation. The general solution to the equation is $x_n = c_1 2^n + c_2(-2)^n + c_3(-3)^n$, $c_1, c_2, c_3 \in \mathbb{C}$.

Note: The characteristic equation of the linear difference equation is $t^2 + 3t^2 - 4t - 12 = (t-2)(t+2)(t+3) = 0$.

Problem 3.5.2. Determine an initial value problem with the solution:

(a) $\{(1 + 2n + 3n^2)7^n\}_{n \in \mathbb{N}_0}$.   (c) $\{(n+2)5^n \sin \frac{n\pi}{4}\}_{n \in \mathbb{N}_0}$.

(b) $\{1 + 3n - 5n^2 + 6n^3\}_{n \in \mathbb{N}_0}$.

Solution 3.5.2.

(a) The sequence $\{(1+2n+3n^2)7^n\}_{n \in \mathbb{N}_0}$ is a particular solution to a homogeneous third-order difference equation with the characteristic equation $(t-7)^3 = t^3 - 21t^2 + 147t - 343 = 0$. The sequence is the solution to the initial value problem $x_{n+3} - 21x_{n+2} + 147x_{n+1} - 343x_n = 0$, $x_0 = 1$, $x_1 = 42$, and $x_2 = 833$.

(b) $x_{n+4} - 4x_{n+3} + 6x_{n+2} - 4x_{n+1} + x_n = 0$, $x_0 = 1$, $x_1 = 5$, $x_2 = 35$, and $x_3 = 127$.

(c) The general term of the sequence $a_n = 5^n(n+2)\sin \frac{n\pi}{4}$ indicates that $r_1 = 5e^{\frac{\pi i}{4}}$ and $r_2 = 5e^{-\frac{\pi i}{4}}$ are roots of multiplicity two of the characteristic equation. A possible characteristic equation of the, still undetermined, difference equation is

$$0 = \left(t - 5e^{\frac{\pi i}{4}}\right)^2 \left(t - 5e^{-\frac{\pi i}{4}}\right)^2 = t^4 - 10\sqrt{2}t^3 + 100t^2 - 250\sqrt{2}t + 625.$$

The sequence is the solution to the initial value problem $x_{n+4} - 10\sqrt{2}x_{n+3} + 100x_{n+2} - 250\sqrt{2}x_{n+1} + 625x_n = 0$, $x_0 = 0$, $x_1 = \frac{15\sqrt{2}}{2}$, $x_2 = 100$, and $x_3 = \frac{625\sqrt{2}}{2}$.

Problem 3.5.3. Solve the following homogeneous linear difference equations:

(a) $x_{n+3} - 2x_{n+2} - 5x_{n+1} + 6x_n = 0.$
(b) $x_{n+4} - 2x_{n+2} + x_n = 0.$
(c) $x_{n+3} - 2x_{n+1} + 4x_n = 0.$
(d) $x_{n+3} + x_{n+2} - 8x_{n+1} - 12x_n = 0.$
(e) $x_{n+4} - x_n = 0.$
(f) $x_{n+4} + x_n = 0.$

Solution 3.5.3.

(a) From $r^3 - 2r^2 - 5r + 6 = (r-3)(r-1)(r+2)$, it follows that the roots of the characteristic equation are $r_1 = -2$, $r_2 = 1$, and $r_3 = 3$. The general solution to the homogeneous third-order difference equation is $x_n = c_1 + c_2(-2)^n + c_3 3^n$, $c_1, c_2, c_3 \in \mathbb{C}$.

Note: If the polynomial $p(r) = r^3 - 2r^2 - 5r + 6$ has an integer root, then that root must be a factor of its constant term 6, i.e., the integers ± 1, ± 2, ± 3, and ± 6 are candidates for a root of the polynomial p. From $p(1) = 0$, by using long division, it follows that $p(r) = (r-1)(r^2 - r - 6)$.

(b) Solutions to the characteristic equation $0 = r^4 - 2r^2 + 1 = (r^2 - 1)^2 = (r+1)^2(r-1)^2$ are $r_1 = -1$ and $r_2 = 1$. Since both roots are of multiplicity two, the general solution to the homogeneous fourth-order difference equation is $x_n = c_1 + c_2 n + (c_3 + c_4 n)(-1)^n$, $c_1, c_2, c_3, c_4 \in \mathbb{C}$.

(c) Consider the integer factors of the constant term of the polynomial $p(r) = r^3 - 2r + 4$ to obtain $p(-2) = 0$ and $p(r) = (r+2)(r^2 - 2r + 2)$. The general solution to the difference equation is $x_n = c_1(-2)^n + 2^{\frac{n}{2}}(c_2 \cos \frac{n\pi}{4} + c_3 \sin \frac{n\pi}{4})$, $c_1, c_2, c_3 \in \mathbb{C}$.

(d) $x_n = (c_1 + c_2 n)(-2)^n + c_3 3^n.$

(e) Since solutions to the characteristic equation $r^4 - 1 = 0$ are $r_1 = 1$, $r_2 = -1$, $r_3 = i = e^{\frac{\pi i}{2}}$, and $r_4 = -i = e^{-\frac{\pi i}{2}}$, the general solution to the difference equation is $x_n = c_1 + c_2(-1)^n + c_3 \cos \frac{n\pi}{2} + c_4 \sin \frac{n\pi}{2}$, $c_1, c_2, c_3, c_4 \in \mathbb{C}$.

(f) Since solutions to the characteristic equation $r^4 + 1 = 0$ are $r_1 = e^{\frac{\pi i}{4}}$, $r_2 = e^{-\frac{\pi i}{4}}$, $r_3 = e^{\frac{3\pi i}{4}}$, and $r_4 = e^{-\frac{3\pi i}{4}}$, the general solution to the difference equation is $x_n = c_1 \cos \frac{n\pi}{4} + c_2 \sin \frac{n\pi}{4} + c_3 \cos \frac{3n\pi}{4} + c_4 \sin \frac{3n\pi}{4}$, $c_1, c_2, c_3, c_4 \in \mathbb{C}$.

Problem 3.5.4. Solve the initial value problem $x_{n+3} - 7x_{n+2} + 16x_{n+1} - 12x_n = 0$, $x_0 = x_1 = x_2 = 1$.

Solution 3.5.4. Since solutions to the characteristic equation $r^3 - 7r^2 + 16r - 12 = (r-2)^2(r-3) = 0$ are $r_1 = 2$, $r_2 = 3$, the general solution to the equation is $x_n = (c_1 + c_2 n)2^n + c_3 3^n$, $c_1, c_2, c_3 \in \mathbb{C}$. From $x_0 = c_1 + c_3 = 1$, $x_1 = 2c_2 + 3c_3 = 1$, and $x_2 = 8c_2 + 9c_3 = 1$, it follows that the solution to the initial value problem is $a_n = 3^n - n2^n$, $n \in \mathbb{N}_0$.

Problem 3.5.5. Solve the following non-homogeneous linear difference equations:

(a) $x_{n+3} - 7x_{n+2} + 16x_{n+1} - 12x_n = n2^n$.
(b) $x_{n+3} - x_{n+2} - 4x_{n+1} + 4x_n = 1 + n + 2^n$.

Solution 3.5.5.

(a) By Problem 3.5.4, the general solution to the corresponding homogeneous equation is $x_n^{(h)} = (c_1 + c_2 n)2^n + c_3 3^n$, $c_1, c_2, c_3 \in \mathbb{C}$. Since both $\{2^n\}_{n \in \mathbb{N}_0}$ and $\{n2^n\}_{n \in \mathbb{N}_0}$ are solutions to the homogeneous equation, by the method of undetermined coefficients, a particular solution to the equation is of the form $x_n^{(p)} = n^2(a + bn)2^n$ for some $a, b \in \mathbb{C}$. The general solution to the non-homogeneous third-order difference equation is $x_n = (c_1 + c_2 n)2^n + c_3 3^n - n^2\left(1 + \frac{n}{3}\right)2^{n-3}$, $c_1, c_2, c_3 \in \mathbb{C}$.

(b) The general solution to the corresponding homogeneous equation is $x_n^{(h)} = c_1 + c_2(-2)^n + c_3 2^n$, $c_1, c_2, c_3 \in \mathbb{C}$. By the method of undetermined coefficients, $a_n = -\frac{n(3n+7)}{18}$ is a particular solution to the equation $x_{n+3} - x_{n+2} - 4x_{n+1} + 4x_n = 1 + n$ and $b_n = n2^{n-3}$ is a particular solution to the equation $x_{n+3} - x_{n+2} - 4x_{n+1} + 4x_n = 2^n$. By the principle of superposition, the general solution to the equation is $x_n = c_1 + c_2(-2)^n + c_3 2^n - \frac{n(3n+7)}{18} + n2^{n-3}$, $c_1, c_2, c_3 \in \mathbb{C}$.

Problem 3.5.6. Establish that solving the Riccati difference equation

$$x_{n+1} = \frac{\alpha_n x_n + \beta_n}{\gamma_n x_n + \delta_n},$$

where $\gamma_n \neq 0$ and $\begin{vmatrix} \alpha_n & \beta_n \\ \gamma_n & \delta_n \end{vmatrix} \neq 0$, for all $n \geq 0$, reduces to solving a certain second-order linear difference equation.

Solution 3.5.6. The substitution $\gamma_n x_n + \delta_n = \frac{y_{n+1}}{y_n}$, $y_0 = 1$, i.e., the substitution[10] $x_n = \frac{y_{n+1} - \delta_n y_n}{\gamma_n y_n}$, $y_0 = 1$, transforms the Riccati equation into the equation

$$\frac{y_{n+2} - \delta_{n+1} y_{n+1}}{\gamma_{n+1} y_{n+1}} = \frac{\alpha_n \frac{y_{n+1} - \delta_n y_n}{\gamma_n y_n} + \beta_n}{\gamma_n \frac{y_{n+1} - \delta_n y_n}{\gamma_n y_n} + \delta_n} = \frac{\frac{\alpha_n y_{n+1} - \alpha_n \delta_n y_n + \beta_n \gamma_n y_n}{\gamma_n y_n}}{\frac{\gamma_n y_{n+1} - \gamma_n \delta_n y_n + \gamma_n \delta_n y_n}{\gamma_n y_n}}$$

$$= \frac{\alpha_n y_{n+1} - \alpha_n \delta_n y_n + \beta_n \gamma_n y_n}{\gamma_n y_{n+1}}.$$

Consequently, solving the Riccati difference equation reduces to solving the initial value problem

$$\gamma_n y_{n+2} - (\alpha_n \gamma_{n+1} + \gamma_n \delta_{n+1}) y_{n+1} + \gamma_{n+1}(\alpha_n \delta_n - \beta_n \gamma_n) y_n = 0, \quad y_0 = 1.$$

Since $\gamma_n \neq 0$ and $\begin{vmatrix} \alpha_n & \beta_n \\ \gamma_n & \delta_n \end{vmatrix} \neq 0$ for all $n \geq 0$, this is a homogeneous second-order linear difference equation.

Problem 3.5.7. Solve the equation $x_{n+1} = -\frac{2x_n + 3}{3x_n + 8}$.

Solution 3.5.7. This is a Riccati equation. By Problem 3.5.6, the substitution $3x_n + 8 = \frac{y_{n+1}}{y_n}$, $y_0 = 1$, i.e., the substitution $x_n = \frac{y_{n+1} - 8y_n}{3y_n}$, yields

$$\frac{y_{n+2} - 8y_{n+1}}{3y_{n+1}} = -\frac{2 \cdot \frac{y_{n+1} - 8y_n}{3y_n} + 3}{\frac{y_{n+1} - 8y_n}{y_n} + 8} = -\frac{2y_{n+1} - 7y_n}{3y_{n+1}}.$$

The general solution to the homogeneous second-order difference equation $y_{n+2} - 6y_{n+1} - 7y_n = 0$ is $y_n = c_1(-1)^n + c_2 7^n$, $c_1, c_2 \in \mathbb{C}$.

---

[10]Observe that we only require, in the new unknown sequence $\{y_n\}_{n \in \mathbb{N}_0}$, $y_0 \neq 0$. Hence, any nonzero value for y_0 would work.

The initial condition $y_0 = 1$ yields the family of solutions $y_n = c(-1)^n + (1-c)7^n$, $c \in \mathbb{C}$. The general solution to the Riccati equation is

$$x_n = \frac{c(-1)^{n+1} + (1-c)7^{n+1} - 8c(-1)^n - 8(1-c)7^n}{3(c(-1)^n + (1-c)7^n)}$$

$$= \frac{9c(-1)^{n+1} - (c-1)7^n}{3(c(-1)^n + (1-c)7^n)}, \quad c \in \mathbb{C}.$$

Problem 3.5.8. Suppose that $f : \mathbb{R}^2 \to \mathbb{R}$ is a homogeneous function, i.e., suppose that there is $s \in \mathbb{R}$ such that for any $t \neq 0$, x, and y, $f(tx, ty) = t^s f(x, y)$. Let $n \mapsto g(n)$, $n \in \mathbb{N}_0$, be a function such that $g(n) \neq 0$ and $f(g(n), 1) = 0$, for all $n \in \mathbb{N}_0$. Prove that the sequence $\{a_n\}_{n \in \mathbb{N}_0}$, where $a_0 = g(0)$ and $a_n = c \prod_{k=0}^{n-1} g(k)$, $n \in \mathbb{N}$, $c \in \mathbb{C} \setminus \{0\}$, is a solution to the difference equation $f(x_{n+1}, x_n) = 0$.

Solution 3.5.8. By Problem 3.2.1(a), the sequence $\{a_n\}_{n \in \mathbb{N}_0}$ is the general solution to the difference equation $x_{n+1} = g(n)x_n$ or, what is the same under our assumptions, the equation $\frac{x_{n+1}}{x_n} = g(n)$. It follows that for any $n \in \mathbb{N}_0$, $0 = f(g(n), 1) = f\left(\frac{a_{n+1}}{a_n}, 1\right) = a_n^s f(a_{n+1}, a_n)$. Since $a_n \neq 0$, $f(a_{n+1}, a_n) = 0$ and $\{a_n\}_{n \in \mathbb{N}_0}$ is a solution to the difference equation $f(x_{n+1}, x_n) = 0$.

Problem 3.5.9. Solve $x_{n+1}^3 - nx_{n+1}^2 x_n - (3+n)x_{n+1}x_n^2 + 2(n+1)x_n^3 = 0$.

Solution 3.5.9. The function $(x, y) \mapsto f(x, y) = x^3 - nx^2 y - (3 + n)xy^2 + 2(n+1)y^3$ has the property that $f(tx, xy) = t^3 f(x, y)$, i.e., f is a homogeneous function with the degree of homogeneity equal to three. By Solution 3.5.8, it is enough to solve the equation

$$f\left(\frac{x_{n+1}}{x_n}, 1\right) = \left(\frac{x_{n+1}}{x_n}\right)^3 - n\left(\frac{x_{n+1}}{x_n}\right)^2 - (3+n)\frac{x_{n+1}}{x_n} + 2(n+1)$$

$$= \left(\frac{x_{n+1}}{x_n} - 1\right)\left(\frac{x_{n+1}}{x_n} + 2\right)\left(\frac{x_{n+1}}{x_n} - (n+1)\right) = 0.$$

This equation is equivalent to the disjunction of three homogeneous first-order linear difference equations

$$x_{n+1} = x_n \text{ or } x_{n+1} = -2x_n \text{ or } x_{n+1} = (n+1)x_n.$$

The set of all solutions to the difference equation is

$$\{\{a\}_{n\in\mathbb{N}_0} : a \in \mathbb{C}\} \cup \{\{b(-2)^n\}_{n\in\mathbb{N}_0} : b \in \mathbb{C}\} \cup \{\{c \cdot n!\}_{n\in\mathbb{N}_0} : c \in \mathbb{C}\}.$$

Problem 3.5.10. Solve the initial value problem $x_{n+2} = \frac{x_{n+1}x_n}{x_{n+1}+x_n}$, $x_0 = x_1 = 1$.

Solution 3.5.10. *Solution 1:* The difference equation is equivalent to the equation $\frac{x_{n+2}(x_{n+1}+x_n)}{x_{n+1}x_n} = 1$. It follows that the initial value problem is equivalent to the problem

$$\frac{x_{n+2}}{x_{n+1}} + \frac{x_{n+2}}{x_n} = 1, \quad x_0 = x_1 = 1.$$

The substitution $y_n = \frac{x_{n+1}}{x_n}$ yields the initial value problem $y_{n+1} = \frac{1}{1+\frac{1}{y_n}}$, $y_0 = 1$. It follows that $y_1 = \frac{1}{1+1} = \frac{1}{2} = \frac{F_2}{F_3}$, $y_2 = \frac{1}{1+\frac{1}{y_1}} = \frac{2}{3} = \frac{F_3}{F_4}$ and, by induction, $y_n = \frac{F_{n+1}}{F_{n+2}}$. Therefore,

$$\frac{x_{n+1}}{x_n} = \frac{F_{n+1}}{F_{n+2}} = \frac{\frac{1}{F_{n+2}}}{\frac{1}{F_{n+1}}},$$

and the sequence $\left\{\frac{1}{F_{n+1}}\right\}_{n\in\mathbb{N}_0}$ is the solution to the initial value problem.

Solution 2: From

$$\frac{1}{x_{n+2}} = \frac{1}{x_{n+1}} + \frac{1}{x_n},$$

via the substitution $z_n = \frac{1}{x_n}$, we obtain the initial value problem $z_{n+2} = z_{n+1} + z_n$, $z_0 = z_1 = 1$.

3.6 Generating Functions and z-Transforms

Problem 3.6.1. Use the geometric series $\frac{1}{1-x} = \sum_{n=0}^{\infty} x^n$, $x \in (-1,1)$, to determine the power series of the following functions:

(a) $f(x) = \frac{1}{1-\alpha x}$, $\alpha \neq 0$.   (c) $h(x) = \frac{\alpha x(1+\alpha x)}{(1-\alpha x)^3}$, $\alpha \neq 0$.
(b) $g(x) = \frac{\alpha x}{(1-\alpha x)^2}$, $\alpha \neq 0$.

Solution 3.6.1.

(a) $f(x) = \sum_{n=0}^{\infty} \alpha^n x^n$, $x \in \left(-\frac{1}{|\alpha|}, \frac{1}{|\alpha|}\right)$.

(b) From, for $x \in \left(-\frac{1}{|\alpha|}, \frac{1}{|\alpha|}\right)$, $\frac{1}{1-\alpha x} = \sum_{n=0}^{\infty} \alpha^n x^n$, it follows that

$$\frac{\alpha}{(1-\alpha x)^2} = \left(\frac{1}{1-\alpha x}\right)' = \sum_{n=1}^{\infty} n\alpha^n x^{n-1}.$$

Hence, $g(x) = \sum_{n=1}^{\infty} n\alpha^n x^n$, $x \in \left(-\frac{1}{|\alpha|}, \frac{1}{|\alpha|}\right)$.

(c) From, for $x \in \left(-\frac{1}{|\alpha|}, \frac{1}{|\alpha|}\right)$, $\frac{\alpha x}{(1-\alpha x)^2} = \sum_{n=0}^{\infty} n\alpha^n x^n$, it follows that

$$\frac{\alpha(1+\alpha x)}{(1-\alpha x)^3} = \left(\frac{\alpha x}{(1-\alpha x)^2}\right)' = \sum_{n=1}^{\infty} n^2\alpha^n x^{n-1}.$$

Hence, $h(x) = \sum_{n=1}^{\infty} n^2\alpha^n x^n$, $x \in \left(-\frac{1}{|\alpha|}, \frac{1}{|\alpha|}\right)$.

Note: Problem 3.6.1 establishes generating functions associated with sequences $\{\alpha^n\}_{n\in\mathbb{N}_0}$, $\{n\alpha^n\}_{n\in\mathbb{N}_0}$, and $\{n^2\alpha^n\}_{n\in\mathbb{N}_0}$.

Problem 3.6.2. Determine a difference equation and the initial condition(s) satisfied by the sequence associated with the generating function $t \mapsto \frac{\ln(1+t)}{1-t}$, $t \in (-1, 1)$, i.e., obtain an initial value problem $F(n, x_n, \ldots, x_{n+k}) = 0$, $(x_0, \ldots, x_{n+k-1}) = (\alpha_0, \ldots, \alpha_{n-k+1})$, given

$$\frac{\ln(1+t)}{1-t} = \sum_{n=0}^{\infty} x_n t^n, \quad t \in (-1, 1),$$

Solution 3.6.2. *Solution 1:* From, for any $t \in (-1, 1)$,

$$\sum_{n=1}^{\infty} \frac{(-1)^{n-1}t^n}{n} = \ln(1+t) = (1-t)\sum_{n=0}^{\infty} x_n t^n = \sum_{n=0}^{\infty} x_n t^n - \sum_{n=0}^{\infty} x_n t^{n+1}$$

$$= \sum_{n=0}^{\infty} x_n t^n - \sum_{n=1}^{\infty} x_{n-1} t^n,$$

it follows that

$$\sum_{n=1}^{\infty} \frac{(-1)^{n-1}t^n}{n} = x_0 + \sum_{n=1}^{\infty} (x_n - x_{n-1})t^n, \quad t \in (-1, 1).$$

The sequence associated with the generating function $t \mapsto \frac{\ln(1+t)}{1-t}$, $t \in (-1,1)$, is the solution to the initial value problem $x_{n+1} - x_n = \frac{(-1)^n}{n+1}$, $x_0 = 0$.

Solution 2: From, for $t \in (-1,1)$,

$$\ln(1+t) \cdot \frac{1}{1-t} = \left(t - \frac{t^2}{2} + \frac{t^3}{3} - \frac{t^4}{4} + \cdots \right)(1 + t + t^2 + t^3 + \cdots)$$

$$= t + \left(1 - \frac{1}{2} \right) t^2 + \left(1 - \frac{1}{2} + \frac{1}{3} \right) t^3$$

$$+ \left(1 - \frac{1}{2} + \frac{1}{3} - \frac{1}{4} \right) t^4$$

$$+ \left(1 - \frac{1}{2} + \frac{1}{3} - \frac{1}{4} + \frac{1}{5} \right) t^5 + \cdots ,$$

we conclude that, for $n > 1$, $x_{n+1} = x_n + \frac{(-1)^n}{n+1}$.

Problem 3.6.3. Let $a, \alpha \in \mathbb{R}$, $a \neq 0$. Determine the generating function $f(t) = \sum_{n=0}^{\infty} a_n t^n$, $t \in I$, associated with the solution to the initial value problem $x_{n+1} + a x_n = 0$, $x_0 = \alpha$.

Solution 3.6.3. *Solution 1:* From, for $t \in I$,

$$f(t) = \alpha + \sum_{n=1}^{\infty} a_n t^n = \alpha + t \sum_{n=1}^{\infty} a_n t^{n-1} = \alpha + t \sum_{n=0}^{\infty} a_{n+1} t^n$$

$$= \alpha - at \sum_{n=0}^{\infty} a_n t^n = \alpha - at f(t),$$

it follows that $(1 + at)f(t) = \alpha$, i.e., $f(t) = \frac{\alpha}{1+at}$, $t \in \left(-\frac{1}{|a|}, \frac{1}{|a|} \right)$.

Solution 2: From, for all $n \in \mathbb{N}_0$ and all $t \in \mathbb{R}$, $a_{n+1} t^{n+1} + a a_n t^{n+1} = 0$, it follows that, for all $t \in I$,

$$0 = \sum_{n=0}^{\infty} a_{n+1} t^{n+1} + at \sum_{n=0}^{\infty} a_n t^n = f(t) - \alpha + at f(t).$$

Problem 3.6.4. Let a, b, α, and β be real numbers such that $b \neq 0$. Determine the generating function $f(t) = \sum_{n=0}^{\infty} a_n t^n$, $t \in I$, associated with the solution to the initial value problem $x_{n+2} + a x_{n+1} + b x_n = 0$, $x_0 = \alpha$ and $x_1 = \beta$.

Solution 3.6.4. Observe that, for $t \in I$, $tf(t) = \sum_{n=0}^{\infty} a_n t^{n+1} = \sum_{n=1}^{\infty} a_{n-1} t^n$ and $t^2 f(t) = \sum_{n=2}^{\infty} a_{n-2} t^n$. Then, from $a_n + a a_{n-1} + b a_{n-2} = 0$, $n \geq 2$, and $a_0 = \alpha$ and $a_1 = \beta$, for $t \in I$,

$$f(t) = \sum_{n=0}^{\infty} a_n t^n = \alpha + \beta t - \sum_{n=2}^{\infty} (a a_{n-1} + b a_{n-2}) t^n$$

$$= \alpha + \beta t - a \sum_{n=2}^{\infty} a_{n-1} t^n - b \sum_{n=2}^{\infty} a_{n-2} t^n$$

$$= \alpha + \beta t - a(tf(t) - \alpha t) - b t^2 f(t).$$

From $(1 + at + bt^2) f(t) = \alpha + (\beta + a\alpha)t$, it follows that $f(t) = \frac{\alpha + (\beta + a\alpha)t}{1 + at + bt^2}$, $t \in I = I(a, b)$.

Problem 3.6.5. Use Problems 3.6.3 and 3.6.4 to determine the generating function $f(t) = \sum_{n=0}^{\infty} a_n t^n$, $t \in I$, associated with the solution to each of the following initial value problems:

(a) $1, 2, 4, 8, 16, \ldots$
(b) $x_{n+2} = 2x_{n+1} + 8x_n$, $x_0 = 2$, $x_1 = 1$.
(c) $x_{n+2} + x_{n+1} + 3x_n = 0$, $x_0 = 2$, $x_1 = 5$.
(d) $x_{n+2} = 4x_n$, $x_0 = 1$, $x_1 = 3$.

Answer 3.6.5.

(a) $f(t) = \frac{1}{1-2t}$, $t \in \left(-\frac{1}{2}, \frac{1}{2}\right)$. The geometric progression $1, 2, 4, 8, 16, \ldots$ is the solution to the initial value problem $x_{n+1} = 2x_n$, $x_0 = 1$.

(b) $f(t) = \frac{3t-2}{8t^2+2t-1}$, $t \in \left(-\frac{1}{4}, \frac{1}{4}\right)$. The roots of the equation $8t^2 + 2t - 1 = 0$ are $t = -\frac{1}{2}$ and $t = \frac{1}{4}$.

(c) $f(t) = \frac{2+7t}{3t^2+t+1}$, $t \in \mathbb{R}$. The equation $3t^2 + t + 1 = 0$ has no real roots.

(d) $f(t) = \frac{1+3t}{1-4t^2}$, $t \in \left(-\frac{1}{2}, \frac{1}{2}\right)$.

Problem 3.6.6. Determine the generating function associated with the Fibonacci sequence $\{F_n\}_{n \in \mathbb{N}_0}$.

Solution 3.6.6. Since the Fibonacci sequence is the solution to the initial value problem $x_{n+2} - x_{n+1} - x_n = 0$, $x_0 = 0$, $x_1 = 1$, by Problem 3.6.4, the generating function is $f(t) = \frac{t}{1-t-t^2}$, $t \in \left(-\frac{\sqrt{5}+1}{2}, \frac{\sqrt{5}-1}{2}\right)$.

Problem 3.6.7. Determine the generating function associated with the Lucas sequence $\{L_n\}_{n\in\mathbb{N}_0}$.

Solution 3.6.7. Since the Lucas sequence is the solution to the initial value problem $x_{n+2}-x_{n+1}-x_n = 0$, $x_0 = 2$, $x_1 = 1$, by Problem 3.6.4, the generating function is $f(t) = \frac{2-t}{1-t-t^2}$, $t \in \left(-\frac{\sqrt{5}+1}{2}, \frac{\sqrt{5}-1}{2}\right)$.

Problem 3.6.8. Determine the generating function associated with the Pell sequence $\{P_n\}_{n\in\mathbb{N}_0}$.

Solution 3.6.8. Since the Pell sequence is the solution to the initial value problem $x_n - 2x_{n-1} - x_{n-2} = 0$, $x_0 = 0$, $x_1 = 1$, by Problem 3.6.4, the generating function is $f(t) = \frac{t}{1-2t-t^2}$, $t \in \left(-1 - \sqrt{2}, -1 + \sqrt{2}\right)$.

Problem 3.6.9. Determine the generating function associated with the sequence of Chebyshev polynomials $\{T_n(x)\}_{n\in\mathbb{N}_0}$, $x \in (-1, 1)$.

Solution 3.6.9. For a fixed $x \in (-1, 1)$, the sequence $\{T_n(x)\}_{n\in\mathbb{N}_0}$ is the solution to the initial value problem $y_{n+2} - 2xy_{n+1} + y_n = 0$, $y_0 = 1$, $y_1 = x$. Let, for a fixed $x \in (-1, 1)$, $t \mapsto f_x(t) = \sum_{n=0}^{\infty} a_n(x)t^n$, $t \in I$, be the generating function associated with the sequence $\{T_n(x)\}_{n\in\mathbb{N}_0}$. Since, for $t \in I \subseteq \mathbb{R}$, $tf_x(t) = \sum_{n=1}^{\infty} a_{n-1}(x)t^n$, and $t^2 f_x(t) = \sum_{n=2}^{\infty} a_{n-2}(x)t^n$, it follows that for $t \in I$,

$$f_x(t) = \sum_{n=0}^{\infty} a_n(x)t^n = 1 + xt + \sum_{n=2}^{\infty}(2xa_{n-1}(x) - a_{n-2}(x))t^n$$

$$= 1 + xt + 2x\sum_{n=2}^{\infty} a_{n-1}(x)t^n - \sum_{n=2}^{\infty} a_{n-2}(x)t^n$$

$$= 1 + xt + 2x(tf_x(t) - t) - t^2 f_x(t).$$

Hence, $(1 - 2xt + t^2)f_x(t) = 1 - xt$ and $f_x(t) = \frac{1-xt}{1-2xt+t^2}$, $t \in I$. Since $x \in (-1, 1)$, the equation $t^2 - 2xt + 1 = 0$ has no real roots, which implies that $I = \mathbb{R}$.

Note: We can use Problem 3.6.4 to obtain the generating function $f_x(t)$.

Problem 3.6.10. Solve the following initial value problems by using generating functions:

(a) $x_{n+1} = 4x_n$, $x_0 = 3$.
(b) $x_{n+2} = 3x_{n+1} + 4x_n$, $x_0 = 0$, $x_1 = 20$.
(c) $5x_{n+2} = 4x_{n+1} - x_n$, $x_0 = 0$, $x_1 = 2$.
(d) $x_{n+2} + 2x_{n+1} + x_n = 0$, $x_0 = 1$, $x_1 = -2$.
(e) $x_{n+2} - 5x_{n+1} + 6x_n = 0$, $x_0 = 1$, $x_1 = 3$.

Solution 3.6.10.

(a) By Problem 3.6.3, the associated generating function is $f(t) = \frac{3}{1-4t}$, $t \in \left(-\frac{1}{4}, \frac{1}{4}\right)$. From, for $t \in \left(-\frac{1}{4}, \frac{1}{4}\right)$, $\frac{3}{1-4t} = 3 \sum_{n=0}^{\infty} 4^n t^n$, it follows that the solution to the initial value problem is the sequence $a_n = 3 \cdot 4^n$, $n \in \mathbb{N}_0$.

(b) By Problem 3.6.4, the associated generating function is $f(t) = \frac{20t}{1-3t-4t^2}$, $t \in I$. From $\frac{20t}{1-3t-4t^2} = -\frac{4}{1+t} + \frac{4}{1-4t}$, it follows that for $t \in \left(-\frac{1}{4}, \frac{1}{4}\right)$

$$f(t) = -4 \sum_{n=0}^{\infty} (-1)^n t^n + 4 \sum_{n=0}^{\infty} 4^n t^n = \sum_{n=0}^{\infty} (4(-1)^{n+1} + 4^{n+1}) t^n.$$

The solution to the initial value problem is the sequence $a_n = 4(-1)^{n+1} + 4^{n+1}$, $n \in \mathbb{N}_0$.

(c) The associated generating function is $f(t) = \frac{2t}{1-\frac{4}{5}t+\frac{1}{5}t^2} = \frac{10t}{t^2-4t+5}$, $t \in I$. From

$$\frac{10t}{t^2 - 4t + 5} = \frac{10t}{(t-2-i)(t-2+i)} = \frac{5(1-2i)}{t-2-i} + \frac{5(1+2i)}{t-2+i}$$

$$= \frac{5i}{1-\frac{t}{2+i}} - \frac{5i}{1-\frac{t}{2-i}} = \frac{5i}{1-\frac{(2-i)t}{5}} - \frac{5i}{1-\frac{(2+i)t}{5}},$$

it follows that, for $t \in \left(-\sqrt{5}, \sqrt{5}\right)$,

$$f(t) = 5i \sum_{n=0}^{\infty} \frac{(2-i)^n}{5^n} t^n - 5i \sum_{n=0}^{\infty} \frac{(2+i)^n}{5^n} t^n$$

$$= \sum_{n=0}^{\infty} \frac{i}{5^{n-1}} \left((2-i)^n - (2+i)^n\right) t^n.$$

The solution to the initial value problem is the sequence $a_n = \frac{i}{5^{n-1}}\left((2-i)^n - (2+i)^n\right)$, $n \in \mathbb{N}_0$.

Note: Since the complex numbers $(2-i)^n$ and $(2+i)^n$ are conjugates to each other, $a_n \in \mathbb{R}$. Indeed, if $\theta = \arctan\frac{1}{2}$, then $(2 \pm i)^n = 5^{\frac{n}{2}}(\cos n\theta \pm i\sin n\theta)$, which implies that $a_n = \frac{i}{5^{n-1}} \cdot 5^{\frac{n}{2}} \cdot (-2i)\sin n\theta = 2 \cdot 5^{1-\frac{n}{2}}\sin n\theta$, $n \in \mathbb{N}_0$.

(d) The associated generating function is $f(t) = \frac{1}{t^2+2t+1}$, $t \in I$. From

$$\frac{1}{t^2+2t+1} = \frac{1}{(t+1)^2} = -\left(\frac{1}{1+t}\right)' \text{ it follows that, for } t \in (-1,1),$$

$$f(t) = -\left(\sum_{n=0}^{\infty}(-1)^n t^n\right)' = \sum_{n=1}^{\infty}(-1)^{n+1}nt^{n-1} = \sum_{n=0}^{\infty}(-1)^n(n+1)t^n.$$

The solution to the initial value problem is the sequence $a_n = (-1)^n(n+1)$, $n \in \mathbb{N}_0$.

(e) The associated generating function is

$$f(t) = \frac{1-2t}{6t^2-5t+1} = \frac{1-2t}{(1-2t)(1-3t)} = \frac{1}{1-3t} = \sum_{n=0}^{\infty}3^n t^n, \ |t| \le \frac{1}{3}.$$

The solution to the initial value problem is the sequence $a_n = 3^n$, $n \in \mathbb{N}_0$.

Problem 3.6.11. Let $a, \alpha \in \mathbb{R}$, $a \ne 0$, and let $t \mapsto g(t)$, $t \in I$, be the generating function associated with the sequence $\{\gamma_n\}_{n\in\mathbb{N}_0}$. Determine $f(t) = \sum_{n=0}^{\infty}a_n t^n$, $t \in J \subseteq I$, the generating function associated with the solution to the initial value problem $x_{n+1}+ax_n = \gamma_n$, $x_0 = \alpha$.

Solution 3.6.11. From, for $t \in J$,

$$f(t) = \alpha + \sum_{n=1}^{\infty}a_n t^n = \alpha + t\sum_{n=0}^{\infty}a_{n+1}t^n = \alpha + t\sum_{n=0}^{\infty}(\gamma_n - aa_n)t^n$$

$$= \alpha + tg(t) - atf(t),$$

it follows that $(1+at)f(t) = tg(t) + \alpha$, i.e., $f(t) = \frac{tg(t)+\alpha}{1+at}$, $t \in J \subseteq \left(-\frac{1}{|a|}, \frac{1}{|a|}\right)$.

Problem 3.6.12. Let $a, b, \alpha, \beta \in \mathbb{R}$, $b \ne 0$, and let $t \mapsto g(t)$, $t \in I$, be the generating function associated with the sequence

$\{\gamma_n\}_{n\in\mathbb{N}_0}$. Determine the generating function $f(t) = \sum_{n=0}^{\infty} a_n t^n$, $t \in J \subseteq I$, associated with the solution to the initial value problem $x_{n+2} + ax_{n+1} + bx_n = \gamma_n$, $x_0 = \alpha$ and $x_1 = \beta$.

Solution 3.6.12. From $a_{n+2} = \gamma_n - aa_{n+1} - ba_n$, $n \geq 0$, and $a_0 = \alpha$ and $a_1 = \beta$, for $t \in I$, it follows that

$$f(t) = \sum_{n=0}^{\infty} a_n t^n = \alpha + \beta t + t^2 \sum_{n=0}^{\infty} a_{n+2} t^n$$

$$= \alpha + \beta t + t^2 \sum_{n=0}^{\infty} (\gamma_n - aa_{n+1} - ba_n) t^n$$

$$= \alpha + \beta t + t^2 g(t) - at(f(t) - \alpha) - bt^2 f(t).$$

Hence, $(1 + at + bt^2)f(t) = t^2 g(t) + \alpha + (a\alpha + \beta)t$, and $f(t) = \frac{t^2 g(t) + \alpha + (a\alpha+\beta)t}{1+at+bt^2}$, $t \in J$.

Problem 3.6.13. Solve the following initial value problems by using generating functions:

(a) $x_{n+1} - 2x_n = 3^{n+1}$, $x_0 = 1$.
(b) $x_{n+2} - 2x_{n+1} - 8x_n = 8$, $x_0 = 0$, $x_1 = 8$.
(c) $x_{n+2} - x_{n+1} - 2x_n = (n+2)^2$, $x_0 = 0$, $x_1 = 1$.

Solution 3.6.13.

(a) Let $f(t) = \sum_{n=0}^{\infty} a_n t^n$, $t \in I$, be the generating function associated with the solution to the initial value problem. Since $g(t) = \sum_{n=0}^{\infty} 3^{n+1} t^n = \frac{3}{1-3t}$, $t \in \left(-\frac{1}{3}, \frac{1}{3}\right)$, by Problem 3.6.11,

$$f(t) = \frac{\frac{3t}{1-3t} + 1}{1 - 2t} = \frac{1}{(1-3t)(1-2t)} = \frac{3}{1-3t} - \frac{2}{1-2t}$$

$$= 3 \sum_{n=0}^{\infty} 3^n t^n - 2 \sum_{n=0}^{\infty} 2^n t^n = \sum_{n=0}^{\infty} (3^{n+1} - 2^{n+1}) t^n.$$

The sequence $a_n = 3^{n+1} - 2^{n+1}$, $n \in \mathbb{N}_0$, is the solution to the initial value problem.

(b) Let $f(t) = \sum_{n=0}^{\infty} a_n t^n$, $t \in I$, be the generating function associated with the solution to the initial value problem. Since $g(t) = \sum_{n=0}^{\infty} 8t^n = \frac{8}{1-t}$, $t \in (-1, 1)$, by Problem 3.6.12, for $t \in \left(-\frac{1}{4}, \frac{1}{4}\right)$,

$$f(t) = \frac{\frac{8t^2}{1-t} + 8t}{1 - 2t - 8t^2} = \frac{8t}{(1-t)(1+2t)(1-4t)} = -\frac{8}{9}\frac{1}{1-t} - \frac{8}{9}\frac{1}{1+2t}$$

$$+ \frac{16}{9}\frac{1}{1-4t} = -\frac{8}{9}\sum_{n=0}^{\infty} t^n - \frac{8}{9}\sum_{n=0}^{\infty}(-1)^n 2^n t^n + \frac{16}{9}\sum_{n=0}^{\infty} 4^n t^n$$

$$= \sum_{n=0}^{\infty} \frac{1}{9}\left(-8 + (-1)^{n+1}2^{n+3} + 4^{n+2}\right) t^n.$$

The sequence $a_n = \frac{1}{9}\left(-8 + (-1)^{n+1}2^{n+3} + 4^{n+2}\right)$, $n \in \mathbb{N}_0$, is the solution to the initial value problem.

(c) Let $f(t) = \sum_{n=0}^{\infty} a_n t^n$, $t \in I$, be the generating function associated with the solution to the initial value problem. Since $g(t) = \sum_{n=0}^{\infty}(n+2)^2 t^n = \frac{4-3t+t^2}{(1-t)^3}$, $t \in (-1, 1)$, by Problem 3.6.12, for $t \in \left(-\frac{1}{2}, \frac{1}{2}\right)$,

$$f(t) = \frac{\frac{t^2(4-3t+t^2)}{(1-t)^3} + t}{1 - t - 2t^2} = \frac{t(1+t)}{(1-t)^3(1+t)(1-2t)}$$

$$= \frac{t}{(1-t)^3(1-2t)} = \frac{-2}{1-t} - \frac{1}{(1-t)^2} - \frac{1}{(1-t)^3} + \frac{4}{1-2t}$$

$$= -2\sum_{n=0}^{\infty} t^n - \sum_{n=1}^{\infty} nt^{n-1} - \frac{1}{2}\sum_{n=2}^{\infty} n(n-1)t^{n-2} + 4\sum_{n=0}^{\infty} 2^n t^n$$

$$= \sum_{n=0}^{\infty}\left(-2 - (n+1) - \frac{(n+2)(n+1)}{2} + 2^{n+2}\right) t^n.$$

The sequence $a_n = 2^{n+2} - 3 - n - \frac{(n+2)(n+1)}{2}$, $n \in \mathbb{N}_0$, is the solution to the initial value problem.

Problem 3.6.14. Let $a, b, \alpha \in \mathbb{R}$, $b, \alpha \neq 0$. Use generating functions to verify the method of undetermined coefficients in the case of the non-homogeneous linear difference equation $x_{n+2} + ax_{n+1} + bx_n = \alpha^n$.

Solution 3.6.14. Let $f(t) = \sum_{n=0}^{\infty} a_n t^n$, $t \in I$, be the generating function associated with a particular solution to the non-homogeneous equation. Then, by Problems 3.6.12 and 3.6.3, from

$a_{n+2} + aa_{n+1} + ba_n = \alpha^n$, it follows that

$$f(t) = \frac{\frac{t^2}{1-\alpha t} + a_0 + (a_1 + aa_0)t}{1 + at + bt^2}$$

$$= \frac{(1 - \alpha(a_1 + aa_0))t^2 + (a_1 + a_0(a - \alpha))t + a_0}{(1 - \alpha t)(1 + at + bt^2)}.$$

Case 1: Suppose that α is not a root of the characteristic equation, i.e., suppose that $\alpha^2 + a\alpha + b \neq 0$. Then, for $a_0 = \frac{1}{\alpha^2 + a\alpha + b}$ and $a_1 = \frac{\alpha}{\alpha^2 + a\alpha + b}$,

$$(1 - \alpha(a_1 + aa_0))t^2 + (a_1 + a_0(a - \alpha))t + a_0 = a_0(1 + at + bt^2).$$

Hence, $f(t) = \frac{1}{\alpha^2 + a\alpha + b} \frac{1}{1 - \alpha t}$ is the generating function associated with the particular solution $a_n = \frac{\alpha^n}{\alpha^2 + a\alpha + b}$, $n \in \mathbb{N}_0$.

Case 2: Suppose that α is a simple root of the characteristic equation, i.e., suppose that $t^2 + at + b = (t - \alpha)(t - \beta)$, $\alpha \neq \beta = \frac{b}{\alpha}$. Then, from $1 + at + bt^2 = \frac{1}{\alpha}(1 - \alpha t)(\alpha - bt)$, for $a_0 = 0$, it follows that

$$f(t) = \alpha \frac{(1 - a_1\alpha)t^2 + a_1 t}{(1 - \alpha t)^2(\alpha - bt)} = \frac{\alpha t}{(1 - \alpha t)^2} \frac{(1 - a_1\alpha)t + a_1}{\alpha - bt}.$$

Since for $a_1 = \frac{\alpha}{\alpha^2 - b}$ we have $\frac{a_1}{\alpha} = \frac{1 - a_1\alpha}{-b}$, it follows that $f(t) = \frac{1}{\alpha^2 - b} \frac{\alpha t}{(1 - \alpha t)^2}$ is the generating function associated with the particular solution $a_n = \frac{n\alpha^n}{\alpha^2 - b}$, $n \in \mathbb{N}_0$.

Note: The assumption that α is a simple root of the characteristic equation implies that $\alpha^2 \neq b$.

Case 3: Suppose that α is a root of multiplicity two of the characteristic equation, i.e., suppose that $t^2 + at + b = (t - \alpha)^2$. Then, from $1 + at + bt^2 = (1 - \alpha t)^2$, for $a_0 = 0$ and $a_1 = \frac{1}{2\alpha}$, it follows that $f(t) = \frac{1}{2\alpha^2} \frac{\alpha t(\alpha t + 1)}{(1 - \alpha t)^3}$ is the generating function associated with the particular solution $a_n = \frac{n^2}{2}\alpha^{n-2}$, $n \in \mathbb{N}_0$.

Problem 3.6.15. Determine the z-transform of each of the following sequences:

(a) $a_n = \alpha$, $n \in \mathbb{N}_0$, $\alpha \in \mathbb{C}\setminus\{0\}$.
(b) $a_n = \alpha\beta^n$, $n \in \mathbb{N}_0$, $\alpha \in \mathbb{C}\setminus\{0\}$, $\beta \in \mathbb{C}\setminus\{0,1\}$.
(c) $\{a_n\}_{n\in\mathbb{N}_0}$ is a sequence with the property that there is $m \in \mathbb{N}_0$ such that $a_m \neq 0$ but, for all $n > m$, $a_n = 0$.
(d) $a_n = n$, $n \in \mathbb{N}_0$.
(e) $a_n = n^2$, $n \in \mathbb{N}_0$.
(f) $a_n = n\alpha^n$, $n \in \mathbb{N}_0$, $\alpha \in \mathbb{C}\setminus\{0,1\}$.

Solution 3.6.15.

(a) $\mathcal{Z}[a_n](z) = \sum_{n=0}^{\infty} \frac{\alpha}{z^n} = \frac{\alpha z}{z-1}$, $|z| > 1$.

(b) $\mathcal{Z}[a_n](z) = \sum_{n=0}^{\infty} \frac{\alpha\beta^n}{z^n} = \frac{\alpha z}{z-\beta}$, $|z| > |\beta|$.

(c) $\mathcal{Z}[a_n](z) = \sum_{n=0}^{m} \frac{a_n}{z^n} = \frac{1}{z^m}\sum_{n=0}^{m} a_n z^{m-n}$, $z \neq 0$.

(d) $\mathcal{Z}[n](z) = \frac{z}{(z-1)^2}$, $|z| > 1$. From, for $|z| > 1$, $\frac{z}{z-1} = \sum_{n=0}^{\infty} \frac{1}{z^n}$, it follows that

$$-\frac{z}{(z-1)^2} = z\left(1 + \frac{1}{z-1}\right)' = z\left(\frac{z}{z-1}\right)' = z\left(\sum_{n=0}^{\infty} \frac{1}{z^n}\right)'$$

$$= z\sum_{n=1}^{\infty} \frac{-n}{z^{n+1}} = -\sum_{n=1}^{\infty} \frac{n}{z^n} = -\mathcal{Z}[n](z).$$

(e) $\mathcal{Z}[n^2](z) = \frac{z(z+1)}{(z-1)^3}$, $|z| > 1$. By Part (d), for $|z| > 1$, $\frac{z}{(z-1)^2} = \sum_{n=1}^{\infty} \frac{n}{z^n}$. It follows that

$$-\frac{z(z+1)}{(z-1)^3} = z\left(\frac{z}{(z-1)^2}\right)' = z\left(\sum_{n=0}^{\infty} \frac{n}{z^n}\right)' = z\sum_{n=0}^{\infty} \frac{-n^2}{z^{n+1}} = -\sum_{n=0}^{\infty} \frac{n^2}{z^n}$$

$$= -\mathcal{Z}[n^2](z).$$

(f) $\mathcal{Z}[n\alpha^n](z) = \frac{\alpha z}{(z-\alpha)^2}$, $|z| > |\alpha|$. By Part (b), for $|z| > |\alpha|$, $\frac{z}{z-\alpha} = \sum_{n=1}^{\infty} \frac{\alpha^n}{z^n}$. It follows that

$$-\frac{\alpha z}{(z-\alpha)^2} = z\left(\frac{z}{z-\alpha}\right)' = z\left(\sum_{n=0}^{\infty} \frac{\alpha^n}{z^n}\right)' = z\sum_{n=0}^{\infty} \frac{-n\alpha^n}{z^{n+1}}$$

$$= -\sum_{n=0}^{\infty} \frac{n\alpha^n}{z^n} = -\mathcal{Z}[n\alpha^n](z).$$

Problem 3.6.16. Let $\{a_n\}_{n\in\mathbb{N}_0}$ and $\{b_n\}_{n\in\mathbb{N}_0}$ be two sequences of complex numbers such that, for some natural number k, $b_n = a_{n+k}$. Prove that $\mathcal{Z}[b_n](z) = z^k \mathcal{Z}[a_n](z) - \sum_{n=0}^{k-1} a_n z^{k-n}$, $|z| > R$, where the number $R > 0$ is determined by the sequence $\{a_n\}_{n\in\mathbb{N}_0}$.

Note: This property of the z-transformation is commonly written as
$$\mathcal{Z}[a_{n+k}](z) = z^k \mathcal{Z}[a_n](z) - \sum_{n=0}^{k-1} a_n z^{k-n}.$$

Solution 3.6.16. From, for $|z| > R$,

$$\mathcal{Z}[a_n](z) = \sum_{n=0}^{k-1} \frac{a_n}{z^n} + \sum_{n=k}^{\infty} \frac{a_n}{z^n} = \sum_{n=0}^{k-1} \frac{a_n}{z^n} + \frac{1}{z^k}\sum_{n=0}^{\infty} \frac{b_n}{z^n}$$

$$= \sum_{n=0}^{k-1} \frac{a_n}{z^n} + \frac{1}{z^k}\mathcal{Z}[b_n](z),$$

it follows that $\mathcal{Z}[b_n](z) = z^k \mathcal{Z}[a_n](z) - \sum_{n=0}^{k-1} a_n z^{k-n}$.

Problem 3.6.17. Let $\alpha \neq 0$ and β be complex numbers, and let $\{\gamma_n\}_{n\in\mathbb{N}_0}$ be a sequence of complex numbers. Solve the initial value problem $x_{n+1} - \alpha x_n = \gamma_n$, $x_0 = \beta$, by using the z-transformation if:

(a) $\gamma_n = 0$, $n \in \mathbb{N}_0$. $\qquad\qquad$ (b) $\gamma_n = \delta \neq 0$, $n \in \mathbb{N}_0$.

(c) $\gamma_n = \varepsilon \zeta^n$, $n \in \mathbb{N}_0$, $\varepsilon \in \mathbb{C}\backslash\{0\}$, $\zeta \in \mathbb{C}\backslash\{0,1\}$.

Solution 3.6.17. Let $\{a_n\}_{n\in\mathbb{N}_0}$ be the solution to the initial value problem, and let $F(z) = \mathcal{Z}[a_n](z)$. Observe that $a_0 = \beta$, and recall that, by Problem 3.6.16, for the sequences $\{a_n\}_{n\in\mathbb{N}_0}$ and $\{a_{n+1}\}_{n\in\mathbb{N}_0}$, $\mathcal{Z}[a_{n+1}](z) = z\mathcal{Z}[a_n](z) - \beta z$.

(a) From $a_{n+1} - \alpha a_n = 0$, it follows that $zF(z) - \beta z - \alpha F(z) = 0$. Hence,

$$F(z) = \frac{\beta z}{z - \alpha} = \frac{\beta}{1 - \frac{\alpha}{z}} = \beta \sum_{n=0}^{\infty} \frac{\alpha^n}{z^n}, \quad |z| > |\alpha|.$$

The solution to the initial value problem is the sequence $a_n = \beta\alpha^n$, $n \in \mathbb{N}_0$.

(b) From $a_{n+1} - \alpha a_n = \delta$, by Problem 3.6.15(a), it follows that $zF(z) - \beta z - \alpha F(z) = \frac{\delta z}{z-1}$. Hence, $F(z) = \frac{\beta z}{z-\alpha} + \frac{\delta z}{(z-1)(z-\alpha)}$. If $\alpha = 1$, then $F(z) = \frac{\beta z}{z-1} + \frac{\delta z}{(z-1)^2} = \beta Z[1](z) + \delta Z[n](z)$. The solution to the initial value problem is the sequence $a_n = \beta + n\delta$, $n \in \mathbb{N}_0$. If $\alpha \neq 1$ then $F(z) = \frac{\beta z}{z-\alpha} + \frac{\delta z}{(z-1)(z-\alpha)}$. Since $\frac{\delta z}{(z-1)(z-\alpha)} = \frac{\delta}{1-\alpha}\left(\frac{z}{z-1} - \frac{z}{z-\alpha}\right)$, it follows that $F(z) = \frac{\delta}{1-\alpha}Z[1](z) + \left(\beta - \frac{\delta}{1-\alpha}\right)Z[\alpha^n](z)$. The solution to the initial value problem is $a_n = \frac{\delta}{1-\alpha} + \left(\beta - \frac{\delta}{1-\alpha}\right)\alpha^n$, $n \in \mathbb{N}_0$.

(c) From $a_{n+1} - \alpha a_n = \varepsilon \zeta^n$, it follows that $zF(z) - \beta z - \alpha F(z) = \frac{\varepsilon z}{z-\zeta}$. Hence, $F(z) = \frac{\beta z}{z-\alpha} + \frac{\delta z}{(z-\zeta)(z-\alpha)}$. If $\alpha = \zeta$, then $F(z) = \frac{\beta z}{z-\alpha} + \frac{\varepsilon z}{(z-\alpha)^2} = \beta Z[\alpha^n](z)) + \varepsilon Z[n\alpha^{n-1}](z)$. The solution to the initial value problem is $a_n = \beta\alpha^n + \varepsilon n\alpha^{n-1}$, $n \in \mathbb{N}_0$. If $\alpha \neq \zeta$, then $F(z) = \frac{\beta z}{z-\alpha} + \frac{\varepsilon z}{(z-\zeta)(z-\alpha)}$. Since $\frac{\varepsilon z}{(z-\zeta)(z-\alpha)} = \frac{\epsilon}{\zeta-\alpha}\left(\frac{z}{z-\zeta} - \frac{z}{z-\alpha}\right)$, it follows that $F(z) = \frac{\varepsilon}{\zeta-\alpha}Z[\zeta^n](z) + \left(\beta - \frac{\varepsilon}{\zeta-\alpha}\right)Z[\alpha^n](z)$. The solution to the initial value problem is $a_n = \frac{\varepsilon\zeta^n}{\zeta-\alpha} + \left(\beta - \frac{\varepsilon}{\zeta-\alpha}\right)\alpha^n$, $n \in \mathbb{N}_0$.

Note: Compare with Problem 3.2.1.

Problem 3.6.18. Let α and β be complex numbers, and let $\{\gamma_n\}_{n\in\mathbb{N}_0}$ be a sequence of complex numbers. Solve the initial value problem $x_{n+2} - 5x_{n+1} + 6x_n = \gamma_n$, $x_0 = \alpha$, $x_1 = \beta$, by using the z-transforms if:

(a) $\gamma_n = 0$, $n \in \mathbb{N}_0$.
(b) $\gamma_n = \delta$, $n \in \mathbb{N}_0$, $\delta \in \mathbb{C}\backslash\{0\}$.
(c) $\gamma_n = \varepsilon\zeta^n$, $n \in \mathbb{N}_0$, $\varepsilon \in \mathbb{C}\backslash\{0\}$, $\zeta \in \mathbb{C}\backslash\{0, 1\}$.

Solution 3.6.18. Let $\{a_n\}_{n\in\mathbb{N}_0}$ be the solution to the initial value problem, and let $F(z) = Z[a_n](z)$. Observe that $a_0 = \alpha$, $a_1 = \beta$, and recall that, by Problem 3.6.16, for the sequences $\{a_n\}_{n\in\mathbb{N}_0}$, $\{a_{n+1}\}_{n\in\mathbb{N}_0}$, and $\{a_{n+2}\}_{n\in\mathbb{N}_0}$, $Z[a_{n+1}](z) = zZ[a_n](z) - \alpha z$ and $Z[a_{n+2}](z) = z^2 Z[a_n](z) - \alpha z^2 - \beta z$.

(a) From $a_{n+2} - 5a_{n+1} + 6a_n = 0$, it follows that $z^2 F(z) - \alpha z^2 - \beta z - 5(zF(z) - \alpha z) + 6F(z) = 0$. Hence,

$$F(z) = \frac{\alpha z^2 + (\beta - 5\alpha)z}{z^2 - 5z + 6} = \frac{(3\alpha - \beta)z}{z - 2} + \frac{(\beta - 2\alpha)z}{z - 3}$$

$$= \sum_{n=0}^{\infty} \frac{(3\alpha - \beta)2^n + (\beta - 2\alpha)3^n}{z^n},$$

for $|z| > 3$. The solution to the initial value problem is $a_n = (3\alpha - \beta)2^n + (\beta - 2\alpha)3^n$, $n \in \mathbb{N}_0$.

(b) From $a_{n+2} - 5a_{n+1} + 6a_n = \delta$, it follows that $z^2 F(z) - \alpha z^2 - \beta z - 5(zF(z) - \alpha z) + 6F(z) = \frac{\delta z}{z-1}$. Hence, $F(z) = \frac{(3\alpha - \beta)z}{z-2} + \frac{(\beta - 2\alpha)z}{z-3} + \frac{\delta z}{(z-1)(z-2)(z-3)}$. Since $\frac{1}{(z-1)(z-2)(z-3)} = \frac{1}{2}\frac{1}{z-1} - \frac{1}{z-2} + \frac{1}{2}\frac{1}{z-3}$, it follows that

$$F(z) = \frac{(3\alpha - \beta - \delta)z}{z - 2} + \frac{(\beta - 2\alpha + \frac{\delta}{2})z}{z - 3} + \frac{\delta}{2}\frac{z}{z - 1}$$

$$= \sum_{n=0}^{\infty} \frac{1}{z^n}\left((3\alpha - \beta - \delta)2^n + \left(\beta - 2\alpha + \frac{\delta}{2}\right)3^n + \frac{\delta}{2}\right),$$

for $|z| > 3$. The solution to the initial value problem is $a_n = (3\alpha - \beta - \delta)2^n + \left(\beta - 2\alpha + \frac{\delta}{2}\right)3^n + \frac{\delta}{2}$, $n \in \mathbb{N}_0$.

(c) From $a_{n+2} - 5a_{n+1} + 6a_n = \varepsilon\zeta^n$, it follows that $z^2 F(z) - \alpha z^2 - \beta z - 5(zF(z) - \alpha z) + 6F(z) = \frac{\varepsilon z}{z-\zeta}$. Hence, $F(z) = \frac{(3\alpha - \beta)z}{z-2} + \frac{(\beta - 2\alpha)z}{z-3} + \frac{\varepsilon z}{(z-\zeta)(z-2)(z-3)}$, $|z| > \max\{3, |\zeta|\}$. If $\zeta = 2$, then

$$F(z) = \frac{(3\alpha - \beta)z}{z - 2} + \frac{(\beta - 2\alpha)z}{z - 3} + \frac{\varepsilon z}{(z - 2)^2(z - 3)}$$

$$= \frac{(3\alpha - \beta)z}{z - 2} + \frac{(\beta - 2\alpha)z}{z - 3} - \frac{\varepsilon z}{z - 2} - \frac{\varepsilon z}{(z - 2)^2} + \frac{\varepsilon z}{z - 3}$$

$$= \frac{(3\alpha - \beta - \varepsilon)z}{z - 2} + \frac{(\beta - 2\alpha + \varepsilon)z}{z - 3} - \frac{\varepsilon z}{(z - 2)^2}, \quad |z| > 3.$$

The solution to the initial value problem is $a_n = (3\alpha - \beta - \varepsilon)2^n + (\beta - 2\alpha + \varepsilon)3^n - \varepsilon n 2^{n-1}$, $n \in \mathbb{N}_0$.

If $\zeta = 3$, then $F(z) = \frac{(3\alpha-\beta+\varepsilon)z}{z-2} + \frac{(\beta-2\alpha-\varepsilon)z}{z-3} + \frac{\varepsilon z}{(z-3)^2}$, $|z| > 3$. The solution to the initial value problem is $a_n = (3\alpha - \beta + \varepsilon)2^n + (\beta - 2\alpha - \varepsilon)3^n + \varepsilon n 3^{n-1}$, $n \in \mathbb{N}_0$.

If $\zeta \neq 2$ and $\zeta \neq 3$, then $F(z) = \frac{(3\alpha-\beta+\frac{\varepsilon}{\zeta-2})z}{z-2} + \frac{(\beta-2\alpha-\frac{\varepsilon}{\zeta-3})z}{z-3} + \frac{\varepsilon}{(\zeta-2)(\zeta-3)}\frac{z}{z-\zeta}$, $|z| > \max\{3, |\zeta|\}$. The solution to the initial value problem is

$$a_n = \left(3\alpha - \beta + \frac{\varepsilon}{\zeta - 2}\right)2^n + \left(\beta - 2\alpha - \frac{\varepsilon}{\zeta - 3}\right)3^n + \frac{\varepsilon}{(\zeta - 2)(\zeta - 3)}\zeta^n,$$

$n \in \mathbb{N}_0$.

Chapter 4

Laplace Transformation

4.1 Introduction

Use the following definitions, techniques, properties, and algorithms to solve the problems contained in this chapter. For more details, see [1, 6, 11, 20, 27, 36–38].

$\mathcal{L}$**-originals.** A function $f : \mathbb{R} \to \mathbb{R}$ is an $\mathcal{L}$-*original* if it satisfies the following conditions: (a) if $t < 0$, then $f(t) = 0$; (b) for any $a > 0$, the function f may have only a finite number of discontinuities in the interval $[0, a]$. At each of those discontinuities, both one-sided limits of f exist; and (c) the function f is of exponential order, i.e., there exist positive constants α, M, and T such that $|f(t)| < Me^{\alpha t}$, for all $t > T$.

Laplace Transformation. If the function $t \mapsto f(t)$ is an $\mathcal{L}$-original, with $|f(t)| < Me^{\alpha t}$, for $t > T$, $\alpha, M, T > 0$, then the function $s \mapsto \mathcal{L}[f(t)](s) = \int_0^\infty f(t)e^{-st}\, dt$, $\mathrm{Re}(s) > \alpha$, is called the *Laplace transform* of the function f.[1]
The function $s \mapsto \mathcal{L}[f(t)](s)$ is analytic in the half-plane $\mathrm{Re}(s) > \alpha$ [11].
The function $f(t) \mapsto \mathcal{L}[f(t)]$ is called the *Laplace transformation*.

Uniqueness of the Inverse Laplace Transform. Let f and g be two $\mathcal{L}$-originals such that $\mathcal{L}[f(t)](s) = \mathcal{L}[g(t)](s)$, $s > \alpha$, for

---

[1] Pierre-Simon, Marquis de Laplace, 1749–1827, French mathematician, engineer, physicist, astronomer, and philosopher.

some $\alpha > 0$. Then, $f(t) = g(t)$ at any $t \geq 0$ at which both functions are continuous [1].

4.2 Properties of Laplace Transformation

Problem 4.2.1. Suppose that the functions f and g are $\mathcal{L}$-originals. Establish the following properties of the Laplace transformation:

(a) (Limit of a Laplace Transform.) $\lim\limits_{\mathrm{Re}(s)\to\infty} \mathcal{L}[f(t)](s) = 0$.

(b) (Linearity Property.) $\mathcal{L}[af(t) + bg(t)] = a\mathcal{L}[f(t)] + b\mathcal{L}[g(t)]$, $a, b \in \mathbb{R}$.

(c) (First Shifting Property.) $\mathcal{L}[e^{at}f(t)](s) = \mathcal{L}[f(t)](s - a)$, $a \in \mathbb{R}$.

(d) (Second Shifting Property.) If a positive number a and $\mathcal{L}$-originals f and g are such that $g(t) = f(t - a)H(t - a) = \begin{cases} f(t-a) & \text{if } t \in [a, \infty), \\ 0 & \text{if } t \in [0, a), \end{cases}$ then $\mathcal{L}[g(t)](s) = e^{-as}\mathcal{L}[f(t)](s)$.

(e) (Change of Scale Property.) $\mathcal{L}[f(at)](s) = \frac{1}{a}\mathcal{L}[f(t)]\left(\frac{s}{a}\right)$, $a > 0$.

Solution 4.2.1. Let $\alpha, M, T \in \mathbb{R}^{+}$ be such that $|f(t)| < Me^{\alpha t}$, for $t > T$.

(a) Let $\beta > \alpha$ be such that $|f(t)| < Me^{\beta t}$, for $t \in [0, T]$, Then, for $s \in \mathbb{C}$ such that $\mathrm{Re}(s) > \beta$,

$$|\mathcal{L}[f(t)](s)| \leq \int_0^\infty |f(t)||e^{-st}|\, dt \leq M \int_0^\infty e^{\beta t}e^{-\mathrm{Re}(s)t}\, dt$$

$$= M \int_0^\infty e^{-(\mathrm{Re}(s)-\beta)t}\, dt = \frac{M}{\mathrm{Re}(s) - \beta},$$

which implies that $\lim\limits_{\mathrm{Re}(s)\to\infty} \mathcal{L}[f(t)](s) = 0$.

(b) The linearity of the Laplace transformation follows from the linearity of the integral.

(c) By definition, for $a \in \mathbb{R}$,

$$\mathcal{L}[e^{at}f(t)](s) = \int_0^\infty e^{at}f(t)e^{-st}\, dt = \int_0^\infty f(t)e^{-(s-a)t}\, dt$$

$$= \mathcal{L}[f(t)](s - a), \ \mathrm{Re}(s) > \alpha + a.$$

(d) By definition, for $a > 0$,

$$\mathcal{L}[g(t)](s) = \int_0^\infty g(t)e^{-st}\,dt = \int_a^\infty f(t-a)e^{-st}\,dt$$

$$= \int_0^\infty f(t)e^{-s(t+a)}\,dt = e^{-as}\int_0^\infty f(t)e^{-st}\,dt$$

$$= e^{-as}\mathcal{L}[f(t)](s),\ \mathrm{Re}(s) > \alpha.$$

(e) By definition, for $a > 0$,

$$\mathcal{L}[f(at)](s) = \int_0^\infty f(at)e^{-st}\,dt = \frac{1}{a}\int_0^\infty f(t)e^{-\frac{st}{a}}\,dt$$

$$= \frac{1}{a}\mathcal{L}[f(t)]\left(\frac{s}{a}\right),\ \mathrm{Re}(s) > \max\{\alpha, a\alpha\}.$$

Problem 4.2.2. Let f be an $\mathcal{L}$-original. Establish the following properties of the Laplace transformation:

(a) (Laplace Transforms of Derivatives.) Suppose that f is a differentiable function in $(0,\infty)$ and that f' is an $\mathcal{L}$-original. Then, $\mathcal{L}[f'(t)](s) = s\mathcal{L}[f(t)](s) - f(0)$. Here, $f(0) = \lim_{t\to 0+} f(t)$.

Let $n > 1$. Let $f, f', \ldots, f^{(n-1)}$ be functions of exponential order differentiable in $(0,\infty)$, and let $f^{(n)}$ be an $\mathcal{L}$-original. Then, $\mathcal{L}[f^{(n)}(t)](s) = s^n\mathcal{L}[f(t)](s) - s^{n-1}f(0) - \sum_{m=1}^{n-1} s^{n-m-1}f^{(m)}(0)$. Here, $f^{(m)}(0) = \lim_{t\to 0+} f^{(m)}(t)$.

(b) (Derivative of a Laplace Transform.) $\mathcal{L}[t^n f(t)](s) = (-1)^n \frac{d^n}{ds^n}\mathcal{L}[f(t)](s)$.

Solution 4.2.2. Let $\alpha, M, T \in \mathbb{R}^+$ be such that $|f(t)| < Me^{\alpha t}$, for $t > T$.

(a) Let $s \in \mathbb{C}$ be such that $\tau = \mathrm{Re}(s) > \alpha$. Then, using integration by parts,

$$\mathcal{L}[f'(t)](s) = \int_0^\infty f'(t)e^{-st}\,dt$$

$$= \lim_{R\to\infty} f(t)e^{-st}\Big|_0^R + s\int_0^\infty f(t)e^{-st}\,dt.$$

Since, for $R > T$, $|f(R)|\,|e^{-sR}| \le Me^{\alpha R}e^{-\tau R} = Me^{-(\tau-\alpha)R}$, it follows that

$$\lim_{R\to\infty} f(t)e^{-st}\Big|_0^R = \lim_{R\to\infty} f(R)e^{-sR} - f(0) = -f(0).$$

Hence, $\mathcal{L}[f'(t)](s) = s\mathcal{L}[f(t)](s) - f(0)$. If $n = 2$, then

$$\mathcal{L}[f''(t)](s) = s\mathcal{L}[f'(t)](s) - f'(0) = s^2\mathcal{L}[f(t)](s) - sf(0) - f'(0),$$

which establishes the claim in the case of $n = 2$. Suppose that $n \ge 2$ is such that $\mathcal{L}[f^{(n)}(t)](s) = s^n\mathcal{L}[f(t)](s) - s^{n-1}f(0) - \sum_{m=1}^{n-1} s^{n-m-1}f^{(m)}(0)$. Since $f^{(n+1)}(t) = \left(f^{(n)}(t)\right)'$, it follows that

$$\mathcal{L}[f^{(n+1)}(t)](s) = s\mathcal{L}[f^{(n)}(t)](s) - f^{(n)}(0)$$

$$= s^{n+1}\mathcal{L}[f(t)](s) - s^n f(0) - \sum_{m=1}^{n} s^{n-m}f^{(m)}(0).$$

By the principle of mathematical induction, the claim is true for all natural numbers n.

(b) We use the Leibniz rule for infinite regions[2] to interchange the order of differentiation and integration. It follows that

$$\frac{d}{ds}\left(\mathcal{L}[f(t)](s)\right) = \frac{d}{ds}\left(\int_0^\infty f(t)e^{-st}\,dt\right)$$

$$= \int_0^\infty \frac{\partial}{\partial s}\left(f(t)e^{-st}\right) dt = -\int_0^\infty tf(t)e^{-st}\,dt$$

$$= -\mathcal{L}[tf(t)](s), \ \operatorname{Re}(s) > \alpha.$$

Use mathematical induction to obtain the general case.

Problem 4.2.3. Let f be an $\mathcal{L}$-original. Establish the following properties of the Laplace transformation:

(a) (Laplace Transforms of Integrals.) $\mathcal{L}\left[\int_0^t f(u)\,du\right](s) = \frac{1}{s}\mathcal{L}[f(t)](s)$.

---

[2]See, for example, [18, Theorem 53.5].

(b) (Integral of a Laplace Transform.) If the function

$$g(t) = \begin{cases} \frac{f(t)}{t} & \text{if } t > 0, \\ 0 & \text{if } t \le 0, \end{cases}$$

is an $\mathcal{L}$-original, then $\mathcal{L}[g(t)](s) = \mathcal{L}\left[\frac{f(t)}{t}\right](s) = \int_s^\infty \mathcal{L}[f(t)](p)\,dp.$

Solution 4.2.3.

(a) Let $t \mapsto g(t) = \int_0^t f(u)\,du$. Since, for any $t < 0$, $f(t) = 0$, it follows that, for any $t < 0$, $g(t) = 0$. In addition, $g(0) = 0$. Since f is an $\mathcal{L}$-original, by the fundamental theorem of calculus, the function g' is, possibly piecewise, continuous in any interval $[0, a]$, $a > 0$. Finally, let α, M, $T \in \mathbb{R}^+$ be such that $|f(t)| \le Me^{\alpha t}$, $t > T$, and let $\beta > \alpha$ be such that $\int_0^T |f(u)|\,du \le \frac{M}{\beta}e^{\beta T}$.[3] Then, for $t > T$,

$$|g(t)| \le \int_0^t |f(u)|\,du = \int_0^T |f(u)|\,du + \int_T^t |f(u)|\,du$$

$$\le \frac{M}{\beta}e^{\beta T} + M\int_T^t e^{\beta u}\,du = \frac{M}{\beta}e^{\beta t},$$

which establishes that the function g is an $\mathcal{L}$-original.

Since $f(t) = g'(t)$, wherever the function g is differentiable, by Problem 4.2.2,

$$\mathcal{L}[f(t)](s) = s\mathcal{L}[g(t)](s) - g(0) = s\mathcal{L}\left[\int_0^t f(u)\,du\right](s), \ \text{Re}(s) > \beta,$$

which establishes the required property.

(b) From $\mathcal{L}[g(t)](s) = \int_0^\infty \frac{f(t)}{t}e^{-st}\,dt$, it follows that[4]

$$\frac{d}{ds}\left(\mathcal{L}[g(t)](s)\right) = \frac{d}{ds}\left(\int_0^\infty \frac{f(t)}{t}e^{-st}\,dt\right)$$

---

[3]To see that such a value of β exists, consider the function $\beta \mapsto e^{\beta T} - \frac{c\beta}{M}$, where c is a nonnegative constant and $\beta > \alpha$.

[4]Since the function g is an $\mathcal{L}$-original, we can interchange the order of differentiation and integration.

$$= \int_0^\infty \frac{\partial}{\partial s}\left(\frac{f(t)}{t}e^{-st}\right) dt = -\int_0^\infty f(t)e^{-st}\, dt$$

$$= -\mathcal{L}[f(t)](s).$$

Since the function $p \mapsto \mathcal{L}[f(t)](p)$ is analytic in the half-plane $\mathrm{Re}(p) > \sigma$, for some $\sigma > 0$, by the fundamental theorem of calculus, for $s \in \mathbb{C}$, $\mathrm{Re}(s) > \sigma$ and $R > \sigma$,

$$\int_s^R \mathcal{L}[f(t)](p)\, dp = -\int_s^R \frac{d}{dp}\left(\mathcal{L}\left[\frac{f(t)}{t}\right](p)\right) dp$$

$$= \mathcal{L}\left[\frac{f(t)}{t}\right](s) - \mathcal{L}\left[\frac{f(t)}{t}\right](R).$$

Since, by Problem 4.2.1(a), $\displaystyle\lim_{R\to\infty} \mathcal{L}\left[\frac{f(t)}{t}\right](R) = 0$, it follows that $\mathcal{L}\left[\frac{f(t)}{t}\right](s) = \int_s^\infty \mathcal{L}[f(t)](p)\, dp$, $\mathrm{Re}(s) > \sigma$.

Problem 4.2.4. (Laplace Transform of a Periodic Function.) Let f be an $\mathcal{L}$-original such that its restriction in $[0,\infty)$ is a periodic function with a period $T > 0$. Then, $\mathcal{L}\left[f(t)\right](s) = \frac{\int_0^T f(t)e^{-st}\, dt}{1-e^{-sT}}$.

Solution 4.2.4. By definition, since, for $t > 0$, $f(t+T) = f(t)$,

$$\mathcal{L}[f(t)](s) = \int_0^T f(t)e^{-st}\, dt + \int_T^\infty f(t)e^{-st}\, dt$$

$$= \int_0^T f(t)e^{-st}\, dt + \int_0^\infty f(t+T)e^{-s(t+T)}\, dt$$

$$= \int_0^T f(t)e^{-st}\, dt + e^{-sT}\int_0^\infty f(t)e^{-st}\, dt$$

$$= \int_0^T f(t)e^{-st}\, dt + e^{-sT}\mathcal{L}[f(t)](s),$$

where the real part of $s \in \mathbb{C}$ is large enough. It follows that $\mathcal{L}\left[f(t)\right](s) = \frac{\int_0^T f(t)e^{-st}\, dt}{1-e^{-sT}}$.

4.3 Laplace Transformation

4.3.1 *Transforms*

Determine the Laplace transform[5] of each of the following functions:

Problem 4.3.1. The Heaviside step function: $H(t) = \begin{cases} 1 & \text{if } t \geq 0, \\ 0 & \text{if } t < 0. \end{cases}$

Solution 4.3.1. By definition, for $\text{Re}(s) > 0$,

$$\mathcal{L}[H(t)](s) = \int_0^\infty H(t)e^{-st}\, dt = \int_0^\infty e^{-st}\, dt = \lim_{T\to\infty} \int_0^T e^{-st}\, dt$$

$$= \lim_{T\to\infty} \left(-\frac{1}{s}e^{-st} \Big|_0^T \right) = -\frac{1}{s}\lim_{T\to\infty} \left(e^{-sT} - 1 \right) = \frac{1}{s}.$$

Problem 4.3.2. For $a > 0$, $H_a(t) = H(t - a) = \begin{cases} 1 & \text{if } t \geq a, \\ 0 & \text{if } t < a. \end{cases}$

Solution 4.3.2. By definition, for $\text{Re}(s) > 0$,

$$\mathcal{L}[H(t-a)](s) = \int_0^\infty H(t-a)e^{-st}\, dt = \int_a^\infty e^{-st}\, dt$$

$$= \lim_{T\to\infty} \int_a^T e^{-st}\, dt = \lim_{T\to\infty} \left(-\frac{1}{s}e^{-st} \Big|_a^T \right)$$

$$= -\frac{1}{s}\lim_{T\to\infty} \left(e^{-sT} - e^{-as} \right) = \frac{e^{-as}}{s}.$$

Note: The obtained result confirms the second shifting property of the Laplace transformation: $\mathcal{L}[H(t - a)](s) = e^{-as}\mathcal{L}[H(t)](s)$. See Problem 4.2.1.

---

[5]Frequently there are multiple ways to obtain the required result.

Problem 4.3.3. For $k \in \mathbb{N}$, $f_k(t) = t^k H(t)$.

Solution 4.3.3. By definition, for $\mathrm{Re}(s) > 0$,

$$\mathcal{L}[f_k(t)](s) = \int_0^\infty t^k e^{-st} \, dt$$

$$= \lim_{T \to \infty} \int_0^T t^k e^{-st} \, dt = \begin{vmatrix} u = t^k & du = kt^{k-1}dt \\ dv = e^{-st}dt & v = -\frac{e^{-st}}{s} \end{vmatrix}$$

$$= \lim_{T \to \infty} \left(-\frac{t^k}{s} e^{-st} \Big|_0^T + \frac{k}{s} \int_0^T t^{k-1} e^{-st} \, dt \right)$$

$$= \frac{k}{s} \mathcal{L}[f_{k-1}(t)](s).$$

By Problem 3.2.2(a), $\mathcal{L}[f_k(t)](s) = \frac{k!}{s^k} \frac{1}{s} = \frac{k!}{s^{k+1}}$.

Problem 4.3.4. For $\alpha \in (-1, \infty)$, $f_\alpha(t) = t^\alpha H(t)$.

Solution 4.3.4. Recall that for $\alpha > -1$, $\Gamma(\alpha + 1) = \int_0^\infty t^\alpha e^{-t} \, dt$. To obtain the Laplace transform of the function f_α we use the fact[6] that, for any $\xi \in \mathbb{C}$ such that $\mathrm{Re}(\xi) > 0$ and the ray $C_\xi = \{\xi t : t \in [0, \infty)\}$, $\int_{C_\xi} z^\alpha e^{-z} \, dz = \Gamma(\alpha + 1)$. For $\mathrm{Re}(s) > 0$,

$$\mathcal{L}[f_\alpha(t)](s) = \int_0^\infty t^\alpha e^{-st} \, dt$$

$$= \begin{vmatrix} t = \frac{u}{s} \Rightarrow dt = \frac{du}{s} \\ t \in [0, \infty) \Rightarrow u \in C_s = \{st : t \in [0, \infty)\} \end{vmatrix}$$

$$= \int_{C_s} \left(\frac{u}{s} \right)^\alpha e^{-u} \frac{du}{s} = \frac{1}{s^{\alpha+1}} \int_{C_s} u^\alpha e^{-u} \, du = \frac{\Gamma(\alpha + 1)}{s^{\alpha+1}}.$$

---

[6]Here is a sketch of the proof. Let $\xi = a + ib$ be such that $\mathrm{Re}(\xi) = a > 0$, and suppose that $\varphi = \arctan \frac{b}{a} > 0$. For $0 < r < R$, consider the following curves: $C'_{r,R} = \left\{ z = \xi t : t \in \left[\frac{r}{|\xi|}, \frac{R}{|\xi|} \right] \right\}$, $C''_{r,R} = \{z = t : t \in [r, R]\}$, $c_r = \{re^{i\theta} : \theta \in [0, \varphi]\}$, and $c_R = \{Re^{i\theta} : \theta \in [0, \varphi]\}$. For $C_{r,R} = C''_{r,R} \cup c_R \cup (-C'_{r,R}) \cup (-c_r)$, we have $0 = \oint_{C_{r,R}} z^\alpha e^{-z} \, dz = \int_{C''_{r,R}} z^\alpha e^{-z} \, dz + \int_{c_R} z^\alpha e^{-z} \, dz - \int_{C'_{r,R}} z^\alpha e^{-z} \, dz - \int_{c_r} z^\alpha e^{-z} \, dz$. This, together with $\lim_{\substack{r \to 0 \\ R \to \infty}} \int_{C''_{r,R}} z^\alpha e^{-z} \, dz = \Gamma(\alpha + 1)$ and $\lim_{r \to 0} \int_{c_r} z^\alpha e^{-z} \, dz = \lim_{R \to \infty} \int_{c_R} z^\alpha e^{-z} \, dz = 0$, implies the claim.

Problem 4.3.5. For $k \in \mathbb{N}$ and $a > 0$, $f_{k,a}(t) = (t-a)^k H(t-a)$.

Solution 4.3.5. By definition, for $\mathrm{Re}(s) > 0$,

$$\mathcal{L}[f_{k,a}(t)](s) = \int_a^\infty (t-a)^k e^{-st}\, dt = \begin{vmatrix} u = t - a \Rightarrow du = dt \\ t = a \Rightarrow u = 0 \end{vmatrix}$$

$$= \int_0^\infty u^k e^{-s(u+a)}\, du = e^{-as}\frac{k!}{s^{k+1}}.$$

Note: The obtained result confirms the second shifting property of the Laplace transformation: $\mathcal{L}[f(t-a)H(t-a)](s) = e^{-as}\mathcal{L}[f(t)H(t)](s)$.

Problem 4.3.6. $f(t) = e^{at}H(t)$, $a \in \mathbb{C}\setminus\{0\}$.

Solution 4.3.6. By definition, for $\mathrm{Re}(s) > \mathrm{Re}(a)$,

$$\mathcal{L}[f(t)](s) = \int_0^\infty e^{-(s-a)t}\, dt = \lim_{T\to\infty}\left(-\frac{1}{s-a}e^{-(s-a)t}\Big|_0^T\right) = \frac{1}{s-a}.$$

Note: If $s = \sigma + i\tau$ and $a = \alpha + i\beta \in \mathbb{C}\setminus\{0\}$ then, for $\mathrm{Re}(s) = \sigma > \alpha = \mathrm{Re}(a)$, $\lim_{T\to\infty}\left|e^{-(s-a)T}\right| = \lim_{T\to\infty} e^{-(\sigma-\alpha)T}\left|\cos(\tau-\beta)T - i\sin(\tau-\beta)T\right| = 0$.

Problem 4.3.7. $f(t) = H(t)\cos\omega t$ and $g(t) = H(t)\sin\omega t$, $\omega \in \mathbb{R}\setminus\{0\}$.

Solution 4.3.7. Since for $\mathrm{Re}(s) > 0$

$$\frac{s+i\omega}{s^2+\omega^2} = \frac{1}{s-i\omega} = \mathcal{L}[e^{i\omega t}H(t)](s) = \int_0^\infty e^{-(s-i\omega)t}\, dt$$

$$= \int_0^\infty e^{-st}(\cos\omega t + i\sin\omega t)\, dt$$

$$= \mathcal{L}[f(t)](s) + i\mathcal{L}[g(t)](s),$$

it follows that $\mathcal{L}[f(t)](s) = \frac{s}{s^2+\omega^2}$ and $\mathcal{L}[g(t)](s) = \frac{\omega}{s^2+\omega^2}$.

Problem 4.3.8. $f(t) = H(t)\cosh\omega t$ and $g(t) = H(t)\sinh\omega t$, $\omega \in \mathbb{R}\setminus\{0\}$.

Solution 4.3.8. Since $\mathcal{L}$ is a linear operator, for $\mathrm{Re}(s) > |\omega|$,

$$\mathcal{L}[f(t)](s) = \frac{1}{2}\mathcal{L}[H(t)(e^{\omega t} + e^{-\omega t})](s)$$

$$= \frac{1}{2}\left(\frac{1}{s-\omega} + \frac{1}{s+\omega}\right) = \frac{s}{s^2 - \omega^2}$$

and

$$\mathcal{L}[g(t)](s) = \frac{1}{2}\mathcal{L}[H(t)(e^{\omega t} - e^{-\omega t})](s)$$

$$= \frac{1}{2}\left(\frac{1}{s-\omega} - \frac{1}{s+\omega}\right) = \frac{\omega}{s^2 - \omega^2}.$$

Problem 4.3.9. Determine $\mathcal{L}[f(t)](s)$ if the function $t \mapsto f(t)$ is given by its graph as follows:

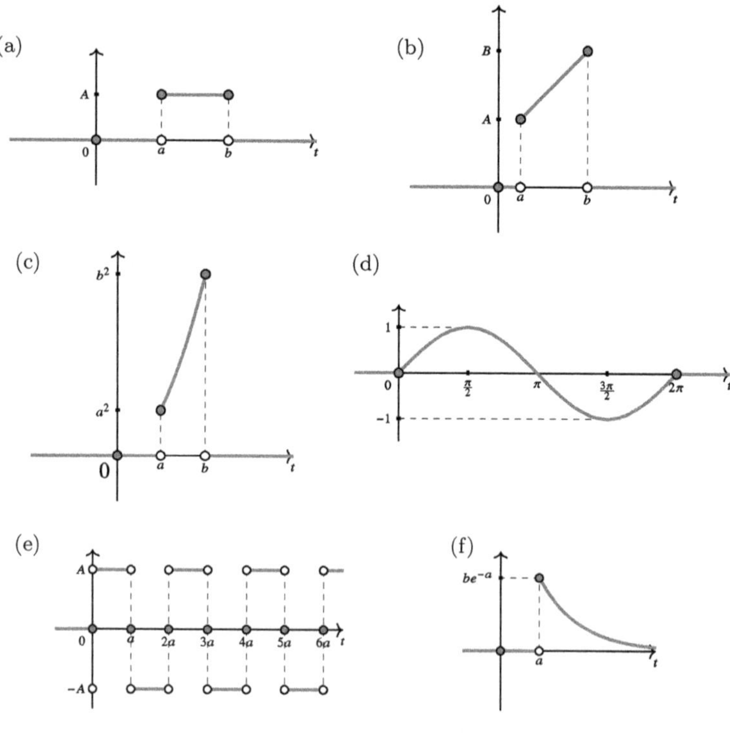

Solution 4.3.9.

(a) By definition, for $\mathrm{Re}(s) > 0$, $\mathcal{L}[f(t)](s) = \int_a^b Ae^{-st}\, dt = \frac{A}{s}(e^{-as} - e^{-bs})$.

(b) By definition, for $\mathrm{Re}(s) > 0$,

$$
\mathcal{L}[f(t)](s) = \int_a^b \left(A + \frac{B - A}{b - a}(t - a) \right) e^{-st}\, dt
$$

$$
= \frac{bA - aB}{b - a} \int_a^b e^{-st}\, dt + \frac{B - A}{b - a} \int_a^b te^{-st}\, dt
$$

$$
= \frac{bA - aB}{s(b - a)}(e^{-as} - e^{-bs})
$$

$$
+ \frac{B - A}{b - a}\left(\frac{ae^{-as}}{s} - \frac{be^{-bs}}{s} + \frac{e^{-as} - e^{-bs}}{s^2} \right)
$$

$$
= \frac{1}{s}(Ae^{-as} - Be^{-bs}) + \frac{B - A}{s^2(b - a)}(e^{-as} - e^{-bs}).
$$

(c) Since, for $0 < a < b$ and $\mathrm{Re}(s) > 0$, $f(t) = t^2(H(t - a) - H(t - b)) = ((t - a) + a)^2 H(t - a) - ((t - b) + b)^2 H(t - b)$, by Problem 4.2.1(d),

$$
\mathcal{L}[f(t)](s) = e^{-as}\mathcal{L}[(t + a)^2 H(t)](s) + e^{-bs}\mathcal{L}[(t + b)^2 H(t)](s)
$$

$$
= \frac{1}{s^3}\left(((as + 1)^2 + 1)e^{-as} - ((bs + 1)^2 + 1)e^{-bs} \right).
$$

(d) Recall that for $a \neq \pm i$, $\int e^{at} \sin t\, dt = \frac{e^{at}}{1+a^2}(a \sin t - \cos t) + C$. By definition, for $\mathrm{Re}(s) > 0$,

$$
\mathcal{L}[f(t)](s) = \int_0^{2\pi} e^{-st} \sin t\, dt = \frac{1 - e^{-2s\pi}}{1 + s^2}.
$$

Note: Another way to obtain this result is to observe that $f(t) = H(t) \sin t - H(t - 2\pi) \sin(t - 2\pi)$ and use Problem 4.3.1(d).

(e) If $\mathrm{Re}(z) > 0$, then $|e^{-z}| < 1$ and $\sum_{m=0}^{\infty}(-1)^m e^{-mz} = \frac{1}{1+e^{-z}}$. It follows that, for $a > 0$ and $\mathrm{Re}(s) > 0$,

$$\mathcal{L}[f(t)](s) = A\sum_{n=0}^{\infty}(-1)^n \int_{na}^{(n+1)a} e^{-st}\, dt$$

$$= \frac{A}{s}\sum_{n=0}^{\infty}(-1)^n(e^{-nas} - e^{-(n+1)as})$$

$$= \frac{A}{s}\left(\sum_{n=0}^{\infty}(-1)^n e^{-nas} + \sum_{n=0}^{\infty}(-1)^{n+1}e^{-(n+1)as}\right)$$

$$= \frac{A}{s}\left(\frac{2}{1+e^{-as}} - 1\right) = \frac{A}{s}\frac{1-e^{-as}}{1+e^{-as}}.$$

Note: Another way to determine $\mathcal{L}[f(t)](s)$ is to note that the restriction of the function f in $[0, \infty)$ is a periodic function with a period of $T = 2a$. By Problem 4.2.4, for $\mathrm{Re}(s) > 0$,

$$\mathcal{L}[f(t)](s) = \frac{A}{1-e^{-2as}}\left(\int_0^a e^{-st}\, dt - \int_a^{2a} e^{-st}\, dt\right)$$

$$= -\frac{A}{s(1-e^{-2as})}\left(e^{-st}\Big|_0^a - e^{-st}\Big|_a^{2a}\right) = \frac{A}{s}\frac{(1-e^{-as})^2}{1-e^{-2as}}$$

$$= \frac{A}{s}\frac{1-e^{-as}}{1+e^{-as}}.$$

(f) By definition, for $\mathrm{Re}(s) > -1$, $\mathcal{L}[f(t)](s) = \int_a^{\infty} be^{-t}e^{-st}\, dt = \frac{be^{-a(s+1)}}{s+1}$.

Note: Another way to obtain this result is to observe that $f(t) = be^{-t}H(t - a)$ and use Problem 4.3.1(d) and Problem 4.3.6.

Problem 4.3.10. For $\varepsilon > 0$, let $f_\varepsilon(t) = \begin{cases} \frac{1}{\varepsilon} & \text{if } t \in [0, \varepsilon], \\ 0 & \text{otherwise.} \end{cases}$ Observe that the area under the graph of f_ε is one unit.

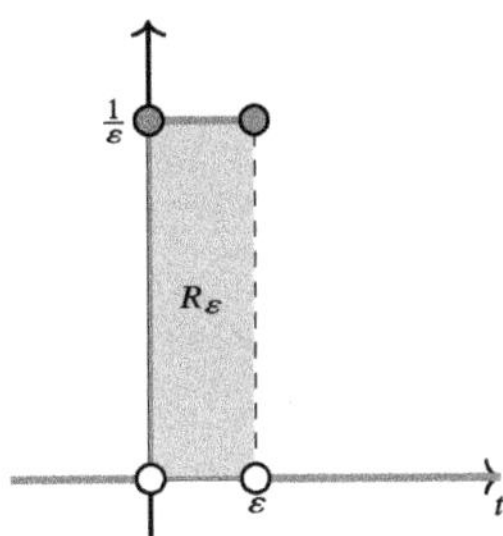

The notion of the *unit impulse function* $\delta_0(t)$ is obtained by considering what happens with rectangles R_ε, determined by functions f_ε, when $\varepsilon \to 0^+$, i.e., when the area of each rectangle R_ε remains equal to one while its height approaches infinity as its base approaches zero. Commonly, we think about $\delta_0(t)$ as a function that is zero everywhere but at the origin, where it is infinite. However, the area associated with $\delta_0(t)$ is one unit. Determine $\mathcal{L}[\delta_0(t)](s) = \lim\limits_{\varepsilon \to 0^+} \mathcal{L}[f_\varepsilon(t)](s)$, the Laplace transform of the unit impulse function.

Solution 4.3.10. Recall that for any $\varepsilon > 0$ and $\mathrm{Re}(s) > 0$, $\mathcal{L}[f_\varepsilon(t)](s) = \frac{1-e^{-s\varepsilon}}{s\varepsilon}$. By L'Hôpital's rule,

$$\mathcal{L}[\delta_0(t)](s) = \lim_{\varepsilon \to 0^+} \mathcal{L}[f_\varepsilon(t)](s) = \lim_{\varepsilon \to 0^+} \frac{1-e^{-s\varepsilon}}{s\varepsilon} = \lim_{\varepsilon \to 0^+} \frac{se^{-s\varepsilon}}{s} = 1.$$

Problem 4.3.11. Determine the Laplace transforms of the following functions:

(a) $f(t) = e^t(2\cos 4t - 3\sin 4t)H(t)$.

(b) $g(t) = e^{-2t}(3\sinh 2t - 2\cosh 2t)H(t)$.

(c) $h(t) = e^{3t}(t^4 + 3t^2 + 4)H(t)$.

Solution 4.3.11.

(a) Recall that $\mathcal{L}[(2\cos 4t - 3\sin 4t)H(t)](s) = \frac{2s}{s^2+16} - \frac{12}{s^2+16} = \frac{2(s-6)}{s^2+16}$, for $\mathrm{Re}(s) > 0$. By the first shifting property of Laplace transformation, referring to Problem 4.2.1, $\mathcal{L}[e^t(2\cos 4t - 3\sin 4t)H(t)](s) = \frac{2((s-1)-6)}{(s-1)^2+16} = \frac{2(s-7)}{(s-1)^2+16}$, for $\mathrm{Re}(s) > 1$.

(b) Recall that $\mathcal{L}[(3\sinh 2t - 2\cosh 2t)H(t)](s) = \frac{6}{s^2-4} - \frac{2s}{s^2-4} = \frac{2(3-s)}{s^2-4}$, for $\mathrm{Re}(s) > 2$. By the first shifting property of Laplace transformation, $\mathcal{L}[e^{-2t}(3\sinh 2t - 2\cosh 2t)H(t)](s) = -\frac{2(s-1)}{s(s+4)}$, for $\mathrm{Re}(s) > 2$.

(c) Recall that $\mathcal{L}[(t^4 + 3t^2 + 4)H(t)](s) = \frac{4!}{s^5} + \frac{3\cdot 2!}{s^3} + \frac{4}{s}$, for $\mathrm{Re}(s) > 0$. By the first shifting property of Laplace transformation, $\mathcal{L}[e^{3t}(t^4 + 3t^2 + 4)H(t)](s) = \frac{24}{(s-3)^5} + \frac{6}{(s-3)^3} + \frac{4}{s-3}$, for $\mathrm{Re}(s) > 3$.

Problem 4.3.12. $f(t) = \begin{cases} 0 & \text{if } t \in (-\infty, 0), \\ \sin t & \text{if } t \in [0, 2\pi), \\ \sin 2t & \text{if } t \in [2\pi, \infty). \end{cases}$

Solution 4.3.12. Observe that, for any $t \in \mathbb{R}$, $f(t) = H(t)\sin t + H(t - 2\pi)(\sin 2t - \sin t) = H(t)\sin t + H(t - 2\pi)(\sin 2(t - 2\pi) - \sin(t - 2\pi))$. By the linearity of Laplace transformation and the second shifting property, for $\mathrm{Re}(s) > 2$,

$$\mathcal{L}[f(t)](s) = \mathcal{L}[H(t)\sin t](s)$$
$$+ \mathcal{L}[H(t - 2\pi)(\sin 2(t - 2\pi) - \sin(t - 2\pi))](s)$$
$$= \frac{1}{s^2 + 1} + e^{-2\pi s}\left(\frac{2}{s^2 + 4} - \frac{1}{s^2 + 1}\right)$$
$$= \frac{1 - e^{-2\pi s}}{s^2 + 1} + \frac{2e^{-2\pi s}}{s^2 + 4}.$$

Problem 4.3.13. Use the transform of a derivative to determine the Laplace transform of $f(t) = t^n H(t)$, where n is a natural number.

Solution 4.3.13. For $n > 1$, the function f is continuous and differentiable in $\mathbb{R}$. In particular, $f(0) = 0$ and for any natural number $m < n$, from $f^{(m)}(t) = n(n-1)\cdots(n-m+1)t^{n-m}H(t)$, $f^{(m)}(0) = 0$. For $\mathrm{Re}(s) > 0$, if $n = 1$,

$$\frac{1}{s} = \mathcal{L}[f'(t)](s) = s\mathcal{L}[f(t)](s) - f(0) = s\mathcal{L}[f(t)](s),$$

and if $n > 1$,

$$\frac{n!}{s} = \mathcal{L}[f^{(n)}(t)](s) = s^n\mathcal{L}[f(t)](s) - s^{n-1}f(0) - \sum_{m=1}^{n-1} s^{n-m-1}f^{(m)}(0)$$
$$= s^n\mathcal{L}[f(t)](s).$$

Therefore, for any $n \in \mathbb{N}$, $\mathcal{L}[f(t)](s) = \frac{n!}{s^{n+1}}$.

Problem 4.3.14. Let $n \in \mathbb{N}$ and let $a > 0$. Determine the Laplace transforms of the following functions:

(a) $f(t) = (t-a)^n H(t)$.

(c) $h(t) = t^n H(t-a)$.

(b) $g(t) = (t-a)^n H(t-a)$.

Solution 4.3.14

(a) Since $f(t) = \sum_{k=0}^{n} \binom{n}{k}(-1)^{n-k} a^{n-k} t^k H(t)$, it follows that, for $\mathrm{Re}(s) > 0$, $L[f(t)](s) = (-1)^n a^n \sum_{k=0}^{n} \binom{n}{k} \frac{(-1)^k k!}{a^k s^{k+1}}$.

(b) By the second shifting property, for $\mathrm{Re}(s) > 0$, $\mathcal{L}[g(t)](s) = e^{-as} \frac{n!}{s^{n+1}}$.

(c) By the binomial formula and the second shifting property, for $\mathrm{Re}(s) > 0$,

$$\mathcal{L}[h(t)](s) = \mathcal{L}[((t-a)+a)^n H(t-a)](s)$$

$$= \mathcal{L}\left[\sum_{k=0}^{n} \binom{n}{k} a^{n-k}(t-a)^k H(t-a)\right](s)$$

$$= \sum_{k=0}^{n} \binom{n}{k} a^{n-k} \mathcal{L}\left[(t-a)^k H(t-a)\right](s)$$

$$= a^n e^{-as} \sum_{k=0}^{n} \binom{n}{k} \frac{k!}{a^k s^{k+1}}.$$

Problem 4.3.15. Use $\mathcal{L}[H(t)\cos t](s) = \frac{s}{s^2+1}$ to determine the Laplace transform of $f(t) = tH(t)\sin t$.

Solution 4.3.15. Observe that $f'(0) = f(0) = 0$. From, for any $t \in \mathbb{R}$, $f'(t) = H(t)(\sin t + t\cos t)$, it follows that, for any $t \neq 0$, $f''(t) = H(t)(2\cos t - t\sin t) = 2H(t)\cos t - f(t)$. Therefore, for $\mathrm{Re}(s) > 0$, by Problem 4.2.2(a),

$$\mathcal{L}[f(t)](s) = 2\mathcal{L}[H(t)\cos t](s) - \mathcal{L}[f''(t)](s) = \frac{2s}{s^2+1}$$

$$- s^2 \mathcal{L}[f(t)](s) + sf(0) + f'(0) = \frac{2s}{s^2+1} - s^2 \mathcal{L}[f(t)](s).$$

It follows that $\mathcal{L}[f(t)](s) = \frac{2s}{(s^2+1)^2}$, $\mathrm{Re}(s) > 0$.

Problem 4.3.16. Determine the Laplace transform of $f(t) = tH(t)\cos\omega t$, $\omega \in \mathbb{R}$.

Solution 4.3.16. For $\mathrm{Re}(s) > 0$, by Problem 4.2.2(b),

$$\mathcal{L}[f(t)](s) = -\frac{d}{ds}\mathcal{L}[H(t)\cos\omega t)](s) = -\frac{d}{ds}\left(\frac{s}{s^2 + \omega^2}\right) = \frac{s^2 - \omega^2}{(s^2 + \omega^2)^2}.$$

Problem 4.3.17. Determine the Laplace transform of $f(t) = t^n e^{at} H(t)$, $n \in \mathbb{N}$, $a \in \mathbb{C}\setminus\{0\}$.

Solution 4.3.17. For $\mathrm{Re}(s) > \mathrm{Re}(a)$, by Problem 4.2.2(b),

$$\mathcal{L}[f(t)](s) = (-1)^n \frac{d^n}{ds^n}\mathcal{L}[e^{at}H(t))](s)$$

$$= (-1)^n \frac{d^n}{ds^n}\left(\frac{1}{s - a}\right) = \frac{n!}{(s - a)^{n+1}}.$$

Problem 4.3.18. Determine the Laplace transform of $f(t) = \frac{\sinh t}{t}H(t)$.

Solution 4.3.18. Observe that $\displaystyle\lim_{t\to 0^+}\frac{\sinh t}{t} = 1$. For $s \in \mathbb{C}$, $\mathrm{Re}(s) > 1$, by Problem 4.2.3(b),

$$\mathcal{L}[f(t)](s) = \int_s^\infty \mathcal{L}[H(t)\sinh t](p)\,dp = \int_s^\infty \frac{dp}{p^2 - 1} = \frac{1}{2}\ln\left|\frac{s - 1}{s + 1}\right|.$$

Problem 4.3.19. Determine the Laplace transform of $f(t) = \frac{\sin^2 at}{t}H(t)$, $a > 0$, and then evaluate $\int_0^\infty \frac{\sin^2 t}{t^2}dt$.

Solution 4.3.19. From $\sin^2 at = \frac{1}{2}(1 - \cos 2at)$, it follows that, for $s \in \mathbb{C}$, $\mathrm{Re}(s) > 0$, by Problem 4.2.3(b),

$$\mathcal{L}[f(t)](s) = \frac{1}{2}\mathcal{L}\left[\frac{1 - \cos 2at}{t}H(t)\right](s)$$

$$= \frac{1}{2}\int_s^\infty \mathcal{L}[(1 - \cos 2at)H(t)](p)\,dp$$

$$= \frac{1}{2} \int_s^\infty \left(\frac{1}{p} - \frac{p}{p^2 + 4a^2} \right) dp = \frac{1}{4} \ln \frac{p^2}{p^2 + 4a^2} \bigg|_s^\infty$$

$$= \frac{1}{4} \ln \frac{s^2 + 4a^2}{s^2}.$$

In particular, for $a = 1$,

$$\int_0^\infty \frac{\sin^2 t}{t^2} dt = \mathcal{L}\left[\frac{\frac{\sin^2 t}{t}}{t} H(t) \right](0) = \int_0^\infty \mathcal{L}\left[\frac{\sin^2 t}{t} H(t) \right](p) \, dp$$

$$= \frac{1}{4} \int_0^\infty \ln \frac{p^2 + 4}{p^2} \, dp$$

$$= \begin{vmatrix} U = \ln \frac{p^2+4}{p^2} & dU = -\frac{8}{p(p^2+4)} \, dp \\ dV = dp & V = p \end{vmatrix}$$

$$= \frac{p}{4} \ln \frac{p^2 + 4}{p^2} \bigg|_0^\infty + 2 \int_0^\infty \frac{dp}{p^2 + 4} = \frac{\pi}{2}.$$

Note: Compare with Problem 1.3.5.

Problem 4.3.20. Determine the Laplace transform of $f(t) = \frac{\cos at - \cos bt}{t} H(t)$, $0 < b < a$, and then evaluate $\int_0^\infty \frac{\cos at - \cos bt}{t} dt$.

Solution 4.3.20. For $s \in \mathbb{C}$, $\mathrm{Re}(s) > 0$, by Problem 4.2.3(b),

$$\mathcal{L}[f(t)](s) = \int_s^\infty \mathcal{L}[(\cos at - \cos bt) H(t)](p) \, dp$$

$$= \int_s^\infty \left(\frac{p}{p^2 + a^2} - \frac{p}{p^2 + b^2} \right) dp = \frac{1}{2} \ln \frac{s^2 + b^2}{s^2 + a^2}.$$

It follows that

$$\int_0^\infty \frac{\cos at - \cos bt}{t} dt = \lim_{\substack{s \to 0 \\ \mathrm{Re}(s) > 0}} \mathcal{L}\left[\frac{\cos at - \cos bt}{t} H(t) \right](s)$$

$$= \lim_{\substack{s \to 0 \\ \mathrm{Re}(s) > 0}} \frac{1}{2} \ln \frac{s^2 + b^2}{s^2 + a^2} = \ln \frac{b}{a}.$$

Problem 4.3.21. Determine the Laplace transform of each of the following functions:

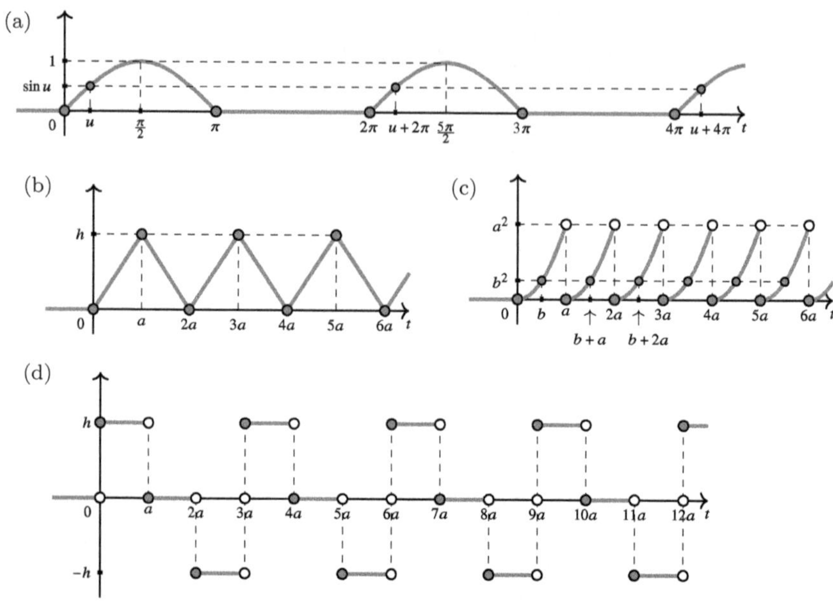

Solution 4.3.21.

(a) Since, for all $t \in [0, \infty)$, $f(t + 2\pi) = f(t)$, the restriction of f in $[0, \infty)$ is a periodic function with a period of $T = 2\pi$. Moreover, if $t \in [0, \pi]$, $f(t) = \sin t$, and if $t \in [\pi, 2\pi]$, $f(t) = 0$. From

$$I = \int_0^\pi e^{-st} \sin t \, dt = \begin{vmatrix} u = e^{-st} & du = -se^{-st}dt \\ dv = \sin t \, dt & v = -\cos t \end{vmatrix}$$

$$= -e^{-st} \cos t \Big|_0^\pi - s \int_0^\pi e^{-st} \cos t \, dt = \begin{vmatrix} u = e^{-st} \\ dv = \cos t \, dt \end{vmatrix}$$

$$= 1 + e^{-\pi s} - se^{-st} \sin t \Big|_0^\pi - s^2 \int_0^\pi e^{-st} \sin t \, dt$$

$$= 1 + e^{-\pi s} - s^2 I,$$

it follows that $I = \frac{1+e^{-\pi s}}{1+s^2}$. Therefore, by Problem 4.2.4, $\mathcal{L}[f(t)](s) = \frac{1+e^{-\pi s}}{(1+s^2)(1-e^{-2\pi s})} = \frac{1}{(1+s^2)(1-e^{-\pi s})}$, $\mathrm{Re}(s) > 0$.

(b) Since, for all $t \in [0, \infty)$, $f(t + 2a) = f(t)$, the restriction of f in $[0, \infty)$ is a periodic function with a period of $T = 2a$. Moreover, if $t \in [0, a]$, $f(t) = \frac{ht}{a}$, and if $t \in [a, 2a]$, $f(t) = -\frac{h(t-2a)}{a}$. From, for $s \neq 0$,

$$\int_a^{2a} (t - 2a)e^{-st}\, dt = \begin{vmatrix} t - a = u & dt = du \\ t = a \Rightarrow u = 0 & t = 2a \Rightarrow u = a \end{vmatrix}$$

$$= e^{-as} \int_0^a u e^{-su}\, du - a e^{-as} \int_0^a e^{-su}\, du$$

$$= e^{-as} \int_0^a u e^{-su}\, du + \frac{a e^{-as}}{s}(e^{-as} - 1),$$

it follows that

$$I = \int_0^{2a} f(t)e^{-st}\, dt = \frac{h}{a}\left(\int_0^a t e^{-st}\, dt - \int_a^{2a} (t - 2a)e^{-st} \right)$$

$$= \frac{h}{a}\left(1 - e^{-as}\right) \int_0^a t e^{-st}\, dt - \frac{h}{s} e^{-as}\left(e^{-as} - 1\right)$$

$$= \frac{h}{a}\left(1 - e^{-as}\right)\left(-\frac{1 + st}{s^2} e^{-st}\bigg|_0^a \right) - \frac{h}{s} e^{-as}\left(e^{-as} - 1\right)$$

$$= \frac{h}{as^2}\left(1 - e^{-as}\right)\left(-(1 + as)e^{-as} + 1 + ase^{-as}\right)$$

$$= \frac{h}{as^2}\left(1 - e^{-as}\right)^2.$$

Therefore, for $\text{Re}(s) > 0$,

$$\mathcal{L}[f(t)](s) = \frac{h\left(1 - e^{-as}\right)^2}{as^2(1 - e^{-2as})} = \frac{h\left(1 - e^{-as}\right)}{as^2(1 + e^{-as})} = \frac{h}{as^2}\tanh\frac{as}{2}.$$

(c) Since, for all $t \in [0, \infty)$, $f(t + a) = f(t)$, the restriction of f in $[0, \infty)$ is a periodic function with a period of $T = a$. Moreover, if $t \in [0, a)$, $f(t) = t^2$. From

$$I = \int_0^a t^2 e^{-st}\, dt = \begin{vmatrix} u = t^2 & du = 2t\,dt \\ dv = e^{-st}\, dt & v = -\frac{e^{-st}}{s} \end{vmatrix}$$

$$= -\frac{t^2}{s} e^{-st}\bigg|_0^a + \frac{2}{s}\int_0^a t e^{-st}dt = -\frac{a^2}{s} e^{-as} - \frac{2(1 + as)}{s^3} e^{-as} + \frac{2}{s^3}$$

$$= \frac{1}{s^3}\left(2 - e^{-as} - (as + 1)^2 e^{-as}\right),$$

it follows that, for $\text{Re}(s) > 0$, $\mathcal{L}[f(t)](s) = \frac{2 - e^{-as} - (as+1)^2 e^{-as}}{s^3(1 - e^{-as})}$.

(d) Since, for all $t \in [0, \infty)$, $f(t + 3a) = f(t)$, the restriction of f in $[0, \infty)$ is a periodic function with a period of $T = 3a$. Moreover, if $t \in [0, a)$, $f(t) = h$, if $t \in [a, 2a)$, $f(t) = 0$, and if $t \in [2a, 3a)$, $f(t) = -h$. From

$$\int_0^{3a} f(t)e^{-st}\, dt = h\int_0^a e^{-st}\, dt - h\int_{2a}^{3a} e^{-st}\, dt$$

$$= \frac{h}{s}\left(-e^{-as} + 1 + e^{-3as} - e^{-2as}\right)$$

$$= \frac{h}{s}(1 - e^{-as})(1 - e^{-2as}),$$

it follows that, for $\mathrm{Re}(s) > 0$,

$$\mathcal{L}[f(t)](s) = \frac{h(1 - e^{-as})(1 - e^{-2as})}{s(1 - e^{-3as})} = \frac{2h}{s}\frac{\sinh\frac{as}{2}\sinh as}{\sinh\frac{3as}{2}}.$$

Problem 4.3.22. $f(t) = H(t)\arccos(\cos t)$.

Solution 4.3.22. Since, for all $t \in [0, \infty)$, $f(t + 2\pi) = f(t)$, the restriction of f in $[0, \infty)$ is a periodic function with a period of $T = 2\pi$. Moreover, if $t \in [0, \pi]$, $f(t) = t$ and if $t \in [\pi, 2\pi]$, $f(t) = 2\pi - t$. By Problem 4.3.21(b), for $\mathrm{Re}(s) > 0$, $\mathcal{L}[f(t)](s) = \frac{1}{s^2}\tanh\frac{\pi s}{2}$.

Problem 4.3.23. $f(t) = \begin{cases} H(t)\arctan(\tan t) & \text{if } t \neq \frac{(2k-1)\pi}{2}, \ k \in \mathbb{Z}, \\[2mm] 0 & \text{if } t = \frac{(2k-1)\pi}{2}, \ k \in \mathbb{Z}. \end{cases}$

Solution 4.3.23. Since, for all $t \in [0, \infty)$, $f(t + \pi) = f(t)$, the restriction of f in $[0, \infty)$ is a periodic function with a period of $T = \pi$.

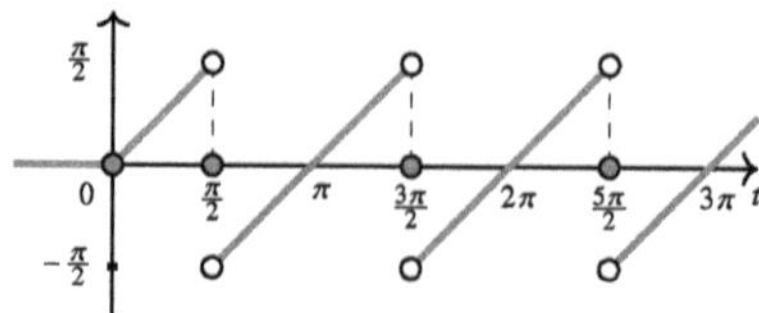

Moreover, if $t \in \left[0, \frac{\pi}{2}\right)$, $f(t) = t$, if $t \in \left(\frac{\pi}{2}, \pi\right]$, $f(t) = t - \pi$, $\lim_{t \to \frac{\pi}{2}-} f(t) = \frac{\pi}{2}$, and $\lim_{t \to \frac{\pi}{2}+} f(t) = -\frac{\pi}{2}$. From

$$\int_0^\pi f(t)e^{-st}\, dt = \int_0^{\frac{\pi}{2}} te^{-st} f(t)\, dt + \int_{\frac{\pi}{2}}^\pi (t - \pi)e^{-st} f(t)\, dt$$

$$= -\frac{1+st}{s^2} e^{-st}\Big|_0^\pi + \frac{\pi}{s} e^{-st}\Big|_{\frac{\pi}{2}}^\pi = \frac{1 - e^{-\pi s}}{s^2} - \frac{\pi}{s} e^{-\frac{\pi}{2} s},$$

it follows that, for $\operatorname{Re}(s) > 0$, $\mathcal{L}[f(t)](s) = \frac{1}{s^2} - \frac{\pi e^{-\frac{\pi}{2} s}}{s(1-e^{-\pi s})} = \frac{1}{s^2} - \frac{\pi}{2s \sinh \frac{\pi s}{2}}$.

Problem 4.3.24. The *error function* is defined as $\operatorname{erf}(t) = \frac{2}{\sqrt{\pi}} \int_0^t e^{-x^2}\, dx$, $t \in \mathbb{R}$. Determine the Laplace transform of the function

$$f(t) = \begin{cases} \operatorname{erf}(\sqrt{t}) & \text{if } t \geq 0, \\ 0 & \text{if } t < 0. \end{cases}$$

Solution 4.3.24. Let $g(t) = e^t f(t)$. Then, for $t > 0$, $g'(t) = e^t \operatorname{erf}(\sqrt{t}) + e^t \cdot \frac{2}{\sqrt{\pi}} e^{-t} \cdot \frac{1}{2\sqrt{t}} = g(t) + \frac{1}{\sqrt{\pi t}}$. Since $\lim_{t \to 0} g(t) = 0$, it follows that

$$s\mathcal{L}[g(t)](s) = \mathcal{L}[g(t)](s) + \frac{1}{\sqrt{\pi}} \mathcal{L}[H(t)t^{-\frac{1}{2}}](s) = \mathcal{L}[g(t)](s) + \frac{1}{\sqrt{\pi}} \frac{\Gamma\left(\frac{1}{2}\right)}{\sqrt{s}}.$$

The fact that $\Gamma\left(\frac{1}{2}\right) = \sqrt{\pi}$ implies $\mathcal{L}[g(t)](s) = \frac{1}{(s-1)\sqrt{s}}$, $\operatorname{Re}(s) > 1$. For $\operatorname{Re}(s) > 0$,

$$\mathcal{L}[f(t)](s) = \mathcal{L}[e^{-t} g(t)](s) = \frac{1}{s\sqrt{s+1}}.$$

Problem 4.3.25. Determine the Laplace transform of the function

$$f(t) = \begin{cases} -1 + \operatorname{erf}(\sqrt{t}) + \frac{e^{-t}}{\sqrt{\pi t}} & \text{if } t > 0, \\ 0 & \text{if } t \leq 0. \end{cases}$$

Solution 4.3.25. Since, for $\mathrm{Re}(s) > 0$,

$$\mathcal{L}[-H(t)](s) = -\frac{1}{s}, \quad \mathcal{L}[\mathrm{erf}(\sqrt{t})](s) = \frac{1}{s\sqrt{s+1}}$$

$$\text{and } \mathcal{L}\left[\frac{e^{-t}}{\sqrt{\pi t}}\right](s) = \frac{1}{\sqrt{s+1}},$$

it follows that

$$\mathcal{L}[f(t)](s) = -\frac{1}{s} + \frac{1}{s\sqrt{s+1}} + \frac{1}{\sqrt{s+1}} = -\frac{1}{s} + \frac{\sqrt{s+1}}{s} = \frac{1}{1+\sqrt{s+1}}.$$

Problem 4.3.26. Let $a > 0$. Determine the Laplace transform of the function $f(t) = H(t) \int_0^\infty \frac{\sin tx}{x^2+a^2}\, dx$.

Solution 4.3.26. Since, for any $t, x \in \mathbb{R}$, $\left|\frac{\sin tx}{x^2+a^2}\right| \le \frac{1}{x^2+a^2}$ and since $\int_0^\infty \frac{dx}{x^2+a^2} = \frac{\pi}{2a}$, the function f is defined in $\mathbb{R}$. Moreover, for any $t \in \mathbb{R}$, $\left|\int_0^\infty \frac{\sin tx}{x^2+a^2}\, dx\right| \le \frac{\pi}{2a}$. Let $s \in \mathbb{C}$ be such that $\sigma = \mathrm{Re}(s) > 0$. Since, for any $t > 0$,

$$\left| e^{-st} \int_0^\infty \frac{\sin tx}{x^2 + a^2}\, dx \right| = e^{-\sigma t} \left| \int_0^\infty \frac{\sin tx}{x^2 + a^2}\, dx \right| \le e^{-\sigma t}\frac{\pi}{2a}$$

and since $\int_0^\infty \frac{\pi}{2a} e^{-\sigma t}\, dt = \frac{\pi}{2a\sigma}$, the Laplace transform of the function f, $\mathcal{L}[f(t)](s) = \int_0^\infty e^{-st}\left(\int_0^\infty \frac{\sin tx}{x^2+a^2}\, dx\right) dt$, is defined in the half-plane $\mathrm{Re}(s) > 0$.

Let $\mathrm{Re}(s) = \sigma > 0$, $\varepsilon \in \left(0, \min\{\frac{\pi}{\sigma}, \frac{\pi}{a\sigma}\}\right)$, $m > a\tan\left(\frac{\pi}{2} - \frac{a\varepsilon\sigma}{2}\right)$, and $M > \frac{1}{\sigma}\ln\frac{\pi}{a\varepsilon\sigma}$. Then, for any $t > 0$,

$$\left| \int_m^\infty \frac{\sin tx}{x^2 + a^2}\, dx \right| \le \int_m^\infty \frac{dx}{x^2 + a^2} = \frac{1}{a}\left(\frac{\pi}{2} - \arctan\frac{m}{a}\right)$$

$$< \frac{1}{a}\left(\frac{\pi}{2} - \left(\frac{\pi}{2} - \frac{a\varepsilon\sigma}{2}\right)\right) = \frac{\varepsilon\sigma}{2},$$

implies that

$$\left| \int_0^M e^{-st}\left(\int_m^\infty \frac{\sin xt}{x^2 + a^2}\, dx\right) dt \right| \le \int_0^M e^{-\sigma t}\left|\int_m^\infty \frac{\sin xt}{x^2 + a^2}\, dx\right| dt$$

$$< \frac{\varepsilon\sigma}{2}\frac{1}{\sigma} = \frac{\varepsilon}{2}.$$

Also,

$$\left| \int_M^\infty e^{-st} \left(\int_0^\infty \frac{\sin tx}{x^2 + a^2} \, dx \right) dt \right| \leq \frac{\pi}{2a} \int_M^\infty e^{-\sigma t} \, dt = \frac{\pi}{2a\sigma} e^{-M\sigma}$$

$$< \frac{\pi}{2a\sigma} \frac{a\varepsilon\sigma}{\pi} = \frac{\varepsilon}{2}.$$

This, together with, for $\mathrm{Re}(s) > 0$,

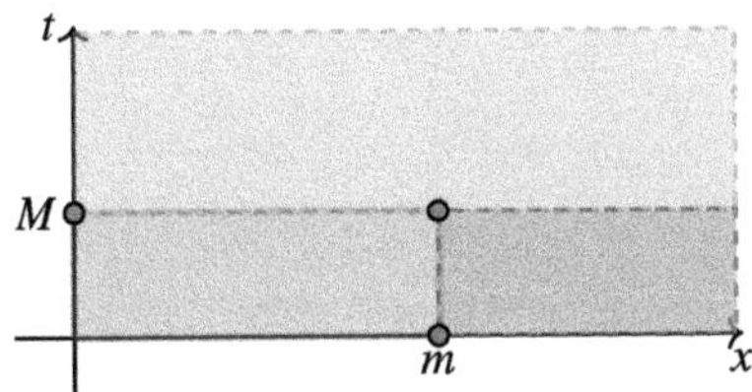

$$\int_0^\infty e^{-st} \left(\int_0^\infty \frac{\sin tx}{x^2+a^2} \, dx \right) dt = \int_0^M e^{-st} \left(\int_0^m \frac{\sin tx}{x^2+a^2} \, dx \right) dt + \int_0^M e^{-st} \cdot$$

$$\left(\int_m^\infty \frac{\sin tx}{x^2+a^2} \, dx \right) dt + \int_M^\infty e^{-st} \left(\int_0^\infty \frac{\sin tx}{x^2+a^2} \, dx \right) dt$$ implies that for any $\varepsilon > 0$, there are $A = A(\varepsilon) > 0$ and $B = B(\varepsilon) > 0$ such that, for any $M > A$ and $m > B$,

$$\left| \int_0^\infty e^{-st} \left(\int_0^\infty \frac{\sin tx}{x^2 + a^2} \, dx \right) dt - \int_0^m \left(\int_0^M \frac{e^{-st} \sin tx}{x^2 + a^2} \, dt \right) dx \right|$$

$$= \left| \int_0^\infty e^{-st} \left(\int_0^\infty \frac{\sin tx}{x^2 + a^2} \, dx \right) dt - \int_0^M e^{-st} \left(\int_0^m \frac{\sin tx}{x^2 + a^2} \, dx \right) dt \right|$$

$$= \left| \int_0^M e^{-st} \left(\int_m^\infty \frac{\sin tx}{x^2 + a^2} \, dx \right) dt + \int_M^\infty e^{-st} \left(\int_0^\infty \frac{\sin tx}{x^2 + a^2} \, dx \right) dt \right|$$

$$\leq \left| \int_0^M e^{-st} \left(\int_m^\infty \frac{\sin tx}{x^2 + a^2} \, dx \right) dt \right| + \left| \int_M^\infty e^{-st} \left(\int_0^\infty \frac{\sin tx}{x^2 + a^2} \, dx \right) dt \right|$$

$$< \frac{\varepsilon}{2} + \frac{\varepsilon}{2} = \varepsilon.$$

In other words, for $s \in \mathbb{C}$ such that $\mathrm{Re}(s) > 0$,

$$\int_0^\infty \left(\int_0^\infty \frac{e^{-st} \sin tx}{x^2 + a^2} \, dx \right) dt = \lim_{\substack{m \to \infty \\ M \to \infty}} \int_0^m \left(\int_0^M \frac{e^{-st} \sin tx}{x^2 + a^2} \, dt \right) dx$$

$$= \int_0^\infty \left(\int_0^\infty \frac{e^{-st} \sin tx}{x^2 + a^2} \, dt \right) dx.$$

Finally, for $\mathrm{Re}(s) > a$,

$$\mathcal{L}\left[\int_0^\infty \frac{\sin tx}{x^2 + a^2}\, dx\right](s) = \int_0^\infty \left(\int_0^\infty e^{-st} \frac{\sin xt}{x^2 + a^2}\, dt\right) dx$$

$$= \int_0^\infty \frac{\mathcal{L}[H(t)\sin tx](s)}{x^2 + a^2}\, dx = \int_0^\infty \frac{x}{(x^2 + a^2)(s^2 + x^2)}\, dx$$

$$= \frac{1}{2(s^2 - a^2)} \ln \frac{x^2 + a^2}{x^2 + s^2}\Big|_0^\infty = \frac{1}{a^2 - s^2} \ln \frac{a}{s}.$$

Note: The fact that in $\int_0^\infty e^{-st}\left(\int_0^\infty \frac{\sin tx}{x^2 + a^2}\, dx\right) dt$ one can reverse the order of integration may be established by using various adaptations of Fubini's theorem. Here is one from [10]:

> Let f be a continuous function on the set $[a, \infty) \times [b, \infty)$ and let both integrals $\int_a^\infty f(x, y)\, dx$ and $\int_b^\infty f(x, y)\, dy$ be uniformly convergent in every finite interval, i.e., for every $A > a$ and $B > b$ and every $\varepsilon > 0$, there are $M = M(A, \varepsilon)$ and $N = N(B, \varepsilon)$ such that for every $y \in [b, B]$ and every $\alpha > N$, we have $\left|\int_\alpha^\infty f(x, y)\, dx\right| < \varepsilon$, and for every $x \in [a, A]$ and every $\beta > M$, we have $\left|\int_\beta^\infty f(x, y)\, dy\right| < \varepsilon$. Then, if at least one of the iterated integrals $\int_b^\infty \left(\int_a^\infty |f(x, y)|\, dx\right) dy$ and $\int_a^\infty \left(\int_b^\infty |f(x, y)|\, dy\right) dx$ converges, the iterated integrals $\int_b^\infty \left(\int_a^\infty f(x, y)\, dx\right) dy$ and $\int_a^\infty \left(\int_b^\infty f(x, y)\, dy\right) dx$ are also convergent and their values are equal.

We use this criterion in Problem 4.3.27.

Problem 4.3.27. Let $a > 0$. Determine the Laplace transform of the function $f(t) = H(t) \int_0^\infty \frac{\cos tx}{x^2 + a^2}\, dx$, and then prove that $f(t) = \frac{\pi}{2a} e^{-at}$, $t \geq 0$.

Note: In Problem 1.4.5, we used properties of the Fourier transformation to establish that, for $t \geq 0$, $\frac{\pi}{2a} e^{-at} = \int_0^\infty \frac{\cos tx}{x^2 + a^2}\, dx$.

Solution 4.3.27. Let $s \in \mathbb{C}$ be such that $\sigma = \mathrm{Re}(s) > 0$, and let $g(x, t) = e^{-st} \frac{\cos tx}{x^2 + a^2}$, $(x, t) \in [0, \infty) \times [0, \infty)$. We start by establishing that $\int_0^\infty g(x, t)\, dx$ and $\int_0^\infty g(x, t)\, dt$ are uniformly convergent in every finite interval.

Let $\varepsilon \in \left(0, \min\left\{\frac{\pi}{2a}, \frac{1}{a^2\sigma}\right\}\right)$, and let $\alpha > N = N(\varepsilon) = -\frac{\ln(a^2\varepsilon\sigma)}{\sigma}$ and $\beta > M = M(\varepsilon) = a\tan\left(\frac{\pi}{2} - a\varepsilon\right)$. Then, for any $x > 0$,

$$\left|\int_\alpha^\infty \frac{e^{-st}\cos tx}{x^2 + a^2}\, dt\right| \le \frac{1}{a^2}\int_\alpha^\infty e^{-\sigma t}\, dt = \frac{1}{a^2\sigma}e^{-\alpha\sigma} < \frac{1}{a^2\sigma}\cdot a^2\varepsilon\sigma = \varepsilon$$

and, for any $t > 0$,

$$\left|\int_\beta^\infty \frac{e^{-st}\cos tx}{x^2 + a^2}\, dx\right| \le e^{-\sigma t}\int_\beta^\infty \frac{dx}{x^2 + a^2} < \frac{1}{a}\left(\frac{\pi}{2} - \arctan\frac{\beta}{a}\right) < \varepsilon.$$

Hence, both integrals are uniformly convergent which, together with

$$\int_0^\infty \left(\int_0^\infty \left|\frac{e^{-st}\cos tx}{x^2 + a^2}\right|\, dt\right) dx \le \frac{1}{\sigma}\int_0^\infty \frac{dx}{x^2 + a^2} \le \frac{\pi}{2a\sigma},$$

implies that, for $\mathrm{Re}(s) > 0$, we can reverse the order of integration in

$$\mathcal{L}[f(t)](s) = \int_0^\infty e^{-st}\left(\int_0^\infty \frac{\cos tx}{x^2 + a^2}\, dx\right) dt.$$

Therefore, for $\mathrm{Re}(s) > a$,

$$\begin{aligned}
\mathcal{L}[f(t)](s) &= \int_0^\infty \frac{1}{x^2 + a^2}\left(\int_0^\infty e^{-st}\cos tx\, dt\right) dx \\
&= \int_0^\infty \frac{\mathcal{L}[\cos tx](s)}{x^2 + a^2}\, dx = s\int_0^\infty \frac{dx}{(x^2 + a^2)(x^2 + s^2)} \\
&= \frac{s}{s^2 - a^2}\int_0^\infty \left(\frac{1}{x^2 + a^2} - \frac{1}{x^2 + s^2}\right) dx = \frac{\pi}{2a(s + a)}.
\end{aligned}$$

From, for $\mathrm{Re}(s) > a$, $\mathcal{L}[\frac{\pi}{2a}e^{-at}H(t)](s) = \frac{\pi}{2a}\frac{1}{s+a}$, because of the uniqueness of the inverse Laplace transform of a function, it follows that $\int_0^\infty \frac{\cos tx}{x^2+a^2}\, dx = \frac{\pi}{2a}e^{-at}$, $t \ge 0$.

4.3.2 *Inverses*

Problem 4.3.28. Determine the inverse Laplace transform $\mathcal{L}^{-1}[F(s)](t)$ for each of the following functions:

(a) $F(s) = \frac{2s}{s^2+4}$.

(c) $F(s) = \frac{4e^{-2s}}{s^2-16}$.

(b) $F(s) = \frac{6s}{s^2+4s+13}$.

(d) $F(s) = \frac{1}{s}\left(\frac{2}{s^2} + \frac{1}{s} - 2\right)$.

Answer 4.3.28.

(a) $\mathcal{L}^{-1}[F(s)](t) = 2H(t)\cos 2t$.

(b) From $F(s) = 6\left(\frac{s+2}{(s+2)^2+9} - \frac{2}{(s+2)^2+9}\right)$, it follows $\mathcal{L}^{-1}[F(s)](t) = e^{-2t}\left(6\cos 3t - 4\sin 3t\right)H(t)$.

(c) $\mathcal{L}^{-1}[F(s)](t) = H(t-2)\sinh 4(t-2)$.

(d) $\mathcal{L}^{-1}[F(s)](t) = (t^2 + t - 2)H(t)$.

Problem 4.3.29. Determine the inverse Laplace transform $\mathcal{L}^{-1}[F(s)](t)$ for each of the following functions:

(a) $F(s) = \frac{2s}{(s+3)^2}$.

(d) $F(s) = \frac{2s}{(s-1)^2(s+1)}$.

(b) $F(s) = \frac{1}{s(s+1)}$.

(e) $F(s) = \frac{e^{-s}}{s+1}$.

(c) $F(s) = \frac{1}{s^2(s-2)}$.

Solution 4.3.29.

(a) From $F(s) = \frac{2(s+3)-6}{(s+3)^2} = \frac{2}{s+3} - \frac{6}{(s+3)^2}$, it follows that $\mathcal{L}^{-1}[F(s)](t) = (2 - 6t)e^{-3t}H(t)$.

(b) From $F(s) = \frac{1}{s} - \frac{1}{s+1}$, it follows that $\mathcal{L}^{-1}[F(s)](t) = (1 - e^{-t})H(t)$.

Note: Another way to determine $\mathcal{L}^{-1}[F(s)](t)$ is as follows. Let $G(s) = sF(s) = \frac{1}{s+1}$. From $\mathcal{L}^{-1}[G(s)](t) = g(t) = e^{-t}H(t)$ and $F(s) = \frac{G(s)}{s}$, it follows that $\mathcal{L}^{-1}[F(s)](t) = H(t)\int_0^t g(u)\,du = H(t)\int_0^t e^{-u}\,du = (1 - e^{-t})H(t)$. See Problem 4.2.3(a).

(c) Since $F(s) = -\frac{1}{4}\frac{1}{s} - \frac{1}{2}\frac{1}{s^2} + \frac{1}{4}\frac{1}{s-2}$, it follows that $\mathcal{L}^{-1}[F(s)](t) = \left(-\frac{1}{4} - \frac{t}{2} + \frac{1}{4}e^{2t}\right)H(t)$.

(d) Since $F(s) = \frac{1}{2}\frac{1}{s-1} + \frac{1}{(s-1)^2} - \frac{1}{2}\frac{1}{s+1}$, it follows that $\mathcal{L}^{-1}[F(s)](t) = H(t)\left(\frac{1}{2}e^t + te^t - \frac{1}{2}e^{-t}\right)$.

(e) $\mathcal{L}^{-1}[F(s)](t) = e^{-(t-1)}H(t-1) = e^{1-t}H(t-1)$. See Problem 4.2.1(d).

Problem 4.3.30. Determine the inverse Laplace transform $\mathcal{L}^{-1}[F(s)](t)$ for each of the following functions:

(a) $F(s) = \frac{e^{-2s}}{s(s+1)^2}$.

(d) $F(s) = \frac{2}{s^2-2s-3}$.

(b) $F(s) = \frac{4}{s^2+2s+5}$.

(e) $F(s) = \frac{3s-1}{s^2-4s-5}$.

(c) $F(s) = \frac{4s+3}{s^2+4s+13}$.

Solution 4.3.30.

(a) Since $\frac{1}{s(s+1)^2} = \frac{1}{s} - \frac{1}{s+1} - \frac{1}{(s+1)^2}$, $F(s) = e^{-2s}\left(\frac{1}{s} - \frac{1}{s+1} - \frac{1}{(s+1)^2}\right)$
and $\mathcal{L}^{-1}[F(s)](t) = \left(1 - e^{-(t-2)} - (t-2)e^{-(t-2)}\right)H(t-2)$.

Note: Another way to determine $\mathcal{L}^{-1}[F(s)](t)$ is as follows. Let
$G(s) = sF(s) = \frac{e^{-2t}}{(s+1)^2}$. From $\mathcal{L}^{-1}[G(s)](t) = (t-2)e^{-(t-2)}$.
$H(t-2)$ and $F(s) = \frac{G(s)}{s}$, it follows that $\mathcal{L}^{-1}[F(s)](t) = H(t-$
$2)\int_0^t (u-2)e^{-(u-2)}\,du = \left(1 - e^{-(t-2)} - (t-2)e^{-(t-2)}\right)H(t-2)$.

(b) From $F(s) = \frac{4}{(s+1)^2+4}$, it follows that $\mathcal{L}^{-1}[F(s)](t) = 2e^{-t}H(t)\sin 2t$.

(c) From $F(s) = \frac{4(s+2)}{(s+2)^2+9} - \frac{5}{(s+2)^2+9}$, it follows that $\mathcal{L}^{-1}[F(s)](t) = e^{-2t}\left(4\cos 3t - \frac{5}{3}\sin 3t\right)H(t)$.

(d) From $F(s) = \frac{2}{(s-1)^2-4}$, it follows that $\mathcal{L}^{-1}[F(s)](t) = e^t H(t)\sinh 2t$.

(e) From $F(s) = \frac{3(s-2)}{(s-2)^2-9} + \frac{5}{(s-2)^2-9}$, it follows that $\mathcal{L}^{-1}[F(s)](t) = e^{2t}\left(3\cosh 3t + \frac{5}{3}\sinh 3t\right)H(t)$.

Problem 4.3.31. Determine the inverse Laplace transform $\mathcal{L}^{-1}[F(s)](t)$ for each of the following functions:

(a) $F(s) = \frac{4se^{-2\pi s}}{s^2+2s+5}$.

(c) $F(s) = \frac{4s}{(s^2+4)^2}$.

(b) $F(s) = \frac{2}{(s^2-1)(s^2+1)}$.

(d) $F(s) = \frac{2s+3}{(s^2+4s+13)^2}$.

Solution 4.3.31.

(a) From $F(s) = e^{-2\pi s}\left(\frac{4(s+1)}{(s+1)^2+4} - \frac{4}{(s+1)^2+4}\right)$, since

$$\mathcal{L}\left[e^{-t}(4\cos 2t - 2\sin 2t)H(t)\right](s) = \frac{4(s+1)}{(s+1)^2+4} - \frac{4}{(s+1)^2+4},$$

by Problem 4.2.1(d), it follows that

$$\mathcal{L}^{-1}[F(s)](t) = e^{-(t-2\pi)}\left(4\cos 2(t-2\pi) - 2\sin 2(t-2\pi)\right)H(t-2\pi)$$
$$= e^{-(t-2\pi)}\left(4\cos 2t - 2\sin 2t\right)H(t-2\pi).$$

(b) From $F(s) = \frac{1}{s^2-1} - \frac{1}{s^2+1}$, it follows that $\mathcal{L}^{-1}[F(s)](t) = (\sinh t - \sin t)H(t)$.

(c) From $F(s) = -\frac{d}{ds}\left(\frac{2}{s^2+4}\right)$, it follows that $\mathcal{L}^{-1}[F(s)](t) = tH(t)\sin 2t$. See Problem 4.2.2(b).

(d) From

$$F(s) = \frac{2(s+2)}{((s+2)^2+9)^2} - \frac{1}{((s+2)^2+9)^2}$$

$$= \frac{2(s+2)}{((s+2)^2+9)^2} - \frac{1}{18}\frac{18}{((s+2)^2+9)^2}$$

$$= \frac{2(s+2)}{((s+2)^2+9)^2} - \frac{1}{18}\frac{((s+2)^2+9) + ((s+2)^2+9 - 2(s+2)^2)}{((s+2)^2+9)^2}$$

$$= -\frac{d}{ds}\left(\frac{1}{(s+2)^2+9}\right) - \frac{1}{18}\frac{1}{(s+2)^2+9} - \frac{1}{18}\frac{d}{ds}\left(\frac{s+2}{(2+s)^2+9}\right),$$

it follows that $\mathcal{L}^{-1}[F(s)](t) = e^{-2t}\left(\frac{t}{3}\sin 3t - \frac{1}{54}\sin 3t + \frac{t}{18}\cos 3t\right) \cdot H(t)$.

Problem 4.3.32. Determine the inverse Laplace transform $\mathcal{L}^{-1}[F(s)](t)$ of the following functions:

(a) $F(s) = \ln\frac{s}{s-2}$.   (b) $F(s) = \ln\frac{s^2-4}{s^2+4}$.   (c) $F(s) = \ln\frac{s^2+6s+10}{s^2+2s+10}$.

Solution 4.3.32.

(a) From $\frac{dF}{ds} = \frac{1}{s} - \frac{1}{s-2}$, it follows that for $s \in \mathbb{C}$, $\mathrm{Re}(s) > 2$,

$$F(s) = \int_s^\infty \left(\frac{1}{p} - \frac{1}{p-2}\right)dp = -\int_s^\infty \mathcal{L}[(1 - e^{-2t})H(t)](p)\,du$$

$$= \mathcal{L}\left[\frac{e^{-2t} - 1}{t}H(t)\right](s).$$

Hence, $\mathcal{L}^{-1}[F(s)](t) = \frac{e^{-2t}-1}{t}H(t)$. See Problem 4.2.3(b).

(b) Use the fact that $\frac{dF}{ds} = \frac{2s}{s^2-4} - \frac{2s}{s^2+4}$ to conclude that $\mathcal{L}^{-1}[F(s)](t) = -2\frac{\cosh 2t - \cos 2t}{t}H(t)$.

(c) From $\frac{dF}{ds} = \frac{2(s+3)}{(s+3)^2+1} - \frac{2(s+1)}{(s+1)^2+9}$, it follows that for $s \in \mathbb{C}$, $\mathrm{Re}(s) > 0$,

$$F(s) = -\int_s^\infty \left(\frac{2(p+3)}{(p+3)^2+1} - \frac{2(p+1)}{(p+1)^2+9} \right) dp$$

$$= -2\int_s^\infty \mathcal{L}[(e^{-3t}\cos t - e^{-t}\cos 3t)H(t)](p)\, dp$$

$$= -2\mathcal{L}\left[\frac{e^{-3t}\cos t - e^{-t}\cos 3t}{t}H(t) \right](s).$$

Hence, $\mathcal{L}^{-1}[F(s)](t) = -2\frac{e^{-3t}\cos t - e^{-t}\cos 3t}{t}H(t)$.

Problem 4.3.33. Determine the inverse Laplace transform $\mathcal{L}^{-1}[F(s)](t)$ of the following functions:

(a) $F(s) = \frac{s^3+3s^2-2s+4}{s(s-1)(s-2)(s+1)(s+3)}$.

(b) $F(s) = \frac{s^2-1}{(s+3)(s-2)^3}$.

(c) $F(s) = \frac{s-1}{(s+1)(s^2+4)}$.

(d) $F(s) = \frac{1}{(s-1)(s^2+1)^2}$.

(e) $F(s) = \frac{s^2-1}{(s^2+4)(s^2+1)}$.

(f) $F(s) = \frac{s}{(s^2+1)^2(s^2+4)}$.

Solution 4.3.33

(a) From $\frac{s^3+3s^2-2s+4}{s(s-1)(s-2)(s+1)(s+3)} = \frac{A}{s} + \frac{B}{s-1} + \frac{C}{s-2} + \frac{D}{s+1} + \frac{E}{s+3}$, i.e., by evaluating[7] both sides of the identity $s^3 + 3s^2 - 2s + 4 = A(s-1)(s-2)(s+1)(s+3) + Bs(s-2)(s+1)(s+3) + Cs(s-1)(s+1)(s+3) + Ds(s-1)(s-2)(s+3) + Es(s-1)(s-2)(s+1)$ at $s = 0, 1, 2, -1$, and -3, we obtain $A = C = \frac{2}{3}$, $B = -\frac{3}{4}$, $D = -\frac{2}{3}$, and $E = \frac{1}{12}$.

Hence, $F(s) = \frac{2}{3}\frac{1}{s} - \frac{3}{4}\frac{1}{s-1} + \frac{2}{3}\frac{1}{s-2} - \frac{2}{3}\frac{1}{s+1} + \frac{1}{12}\frac{1}{s+3}$ and

$$\mathcal{L}^{-1}[F(s)](t) = \left(\frac{2}{3} - \frac{3}{4}e^t + \frac{2}{3}e^{2t} - \frac{2}{3}e^{-t} + \frac{1}{12}e^{-3t} \right) H(t).$$

---

[7]In strict terms, we are determining the limits.

(b) From $\frac{s^2-1}{(s+3)(s-2)^3} = \frac{A}{s+3} + \frac{B}{s-2} + \frac{C}{(s-2)^2} + \frac{D}{(s-2)^3}$, i.e., from $s^2-1 = A(s-2)^3 + B(s+3)(s-2)^2 + C(s+3)(s-2) + D(s+3) = (A+B)s^3 - (6A+B-C)s^2 + (12A-8B+C+D)s - (8A-12B+6C-3D)$, we obtain $A = -\frac{8}{125}$, $B = \frac{8}{125}$, $C = \frac{17}{25}$, and $D = \frac{3}{5}$.

Hence, $F(s) = -\frac{8}{125}\frac{1}{s+3} + \frac{8}{125}\frac{1}{s-2} + \frac{17}{25}\frac{1}{(s-2)^2} + \frac{3}{5}\frac{1}{(s-2)^3}$ and

$$\mathcal{L}^{-1}[F(s)](t) = \left(-\frac{8}{125}e^{-3t} + \frac{8}{125}e^{2t} + \frac{17}{25}te^{2t} + \frac{3}{10}t^2e^{2t} \right) H(t).$$

(c) From $\frac{s-1}{(s+1)(s^2+4)} = \frac{A}{s+1} + \frac{Bs+C}{s^2+4}$, i.e., $s - 1 = A(s^2+4) + (Bs+C)(s+1) = (A+B)s^2 + (B+C)s + 4A + C$, we obtain $A = -\frac{2}{5}$, $B = \frac{2}{5}$, and $C = \frac{3}{5}$.

Hence, $F(s) = -\frac{2}{5}\frac{1}{s+1} + \frac{1}{5}\frac{2s+3}{s^2+4}$ and

$$\mathcal{L}^{-1}[F(s)](t) = \left(-\frac{2}{5}e^{-t} + \frac{2}{5}\cos 2t + \frac{3}{10}\sin 2t \right) H(t).$$

(d) From $\frac{1}{(s-1)(s^2+1)^2} = \frac{A}{s-1} + \frac{Bs+C}{s^2+1} + \frac{Ds+E}{(s^2+1)^2}$, i.e., from $1 = A(s^2+1)^2 + (Bs+C)(s-1)(s^2+1) + (Ds+E)(s-1) = (A+B)s^4 + (-B+C)s^3 + (2A+B-C+D)s^2 + (-B+C-D+E)s + A - C - E$, we obtain $A = \frac{1}{4}$, $B = C = -\frac{1}{4}$, and $D = E = -\frac{1}{2}$. Hence, $F(s) = \frac{1}{4}\frac{1}{s-1} - \frac{1}{4}\frac{s+1}{s^2+1} - \frac{1}{2}\frac{s+1}{(s^2+1)^2}$. This, together with

$$\frac{s+1}{(s^2+1)^2} = \frac{s}{(s^2+1)^2} + \frac{1}{2}\frac{(s^2+1)+(s^2+1-2s^2)}{(s^2+1)^2}$$

$$= \frac{1}{2}\frac{1}{s^2+1} - \frac{1}{2}\frac{d}{ds}\left(\frac{1}{s^2+1} - \frac{s}{s^2+1} \right),$$

implies that $F(s) = \frac{1}{4}\frac{1}{s-1} - \frac{1}{4}\frac{s}{s^2+1} - \frac{1}{2}\frac{1}{s^2+1} - \frac{1}{4}\frac{d}{ds}\left(\frac{s}{s^2+1} - \frac{1}{s^2+1} \right)$
and $\mathcal{L}^{-1}[F(s)](t) = \frac{1}{4}(e^{-t} + (t-1)\cos t - (t+2)\sin t)H(t)$.

(e) From $\frac{s^2-1}{(s^2+4)(s^2+1)} = \frac{As+B}{s^2+4} + \frac{Cs+D}{s^2+1}$, i.e., from $s^2 - 1 = (As+B)(s^2+1) + (Cs+D)(s^2+4) = (A+B)s^3 + (B+D)s^2 + (A+4C)s + B + 4D$, we obtain $A = C = 0$, $B = \frac{5}{3}$, and $D = -\frac{2}{3}$. Hence, $F(s) = \frac{5}{3}\frac{1}{s^2+4} - \frac{2}{3}\frac{1}{s^2+1}$ and $\mathcal{L}^{-1}[F(s)](t) = \left(\frac{5}{6}\sin 2t - \frac{2}{3}\sin t \right) H(t)$.

(f) From $\frac{s}{(s^2+1)^2(s^2+4)} = \frac{As+B}{s^2+1} + \frac{Cs+D}{(s^2+1)^2} + \frac{Es+F}{s^2+4}$, i.e., from $s = (As+B)(s^2+1)(s^2+4) + (Cs+D)(s^2+4) + (Es+F)(s^2+1)^2 = (A+E)s^5+(B+F)s^4+(5A+C+2E)s^3+(5B+D+2F)s^2+(4A+4C+E)s+4B+4D+F$, we obtain $A = -\frac{1}{9}$, $B = D = F = 0$, $C = \frac{1}{3}$, and $E = \frac{1}{9}$. Hence,

$$F(s) = -\frac{1}{9}\frac{s}{s^2+1} + \frac{1}{3}\frac{s}{(s^2+1)^2} + \frac{1}{9}\frac{s}{s^2+4}$$

$$= -\frac{1}{9}\frac{s}{s^2+1} - \frac{1}{6}\frac{d}{ds}\left(\frac{1}{s^2+1}\right) + \frac{1}{9}\frac{s}{s^2+4}$$

and $\mathcal{L}^{-1}[F(s)](t) = -\frac{1}{18}\left(2\cos t - 3t\sin t - 2\cos 2t\right)H(t)$.

4.3.3 *Convolution*

Problem 4.3.34. In Problem 1.4.3, the convolution of two absolutely integrable functions $f : \mathbb{R} \to \mathbb{R}$ and $g : \mathbb{R} \to \mathbb{R}$ was defined as $(f * g)(t) = \int_{-\infty}^{\infty} f(x)g(t - x)\,dx$. Establish that, for $\mathcal{L}$-originals f and g, $(f * g)(t) = \int_0^t f(x)g(t - x)\,dx$.

Solution 4.3.34. Let f and g be $\mathcal{L}$-originals. Since, for $t \geq 0$,

$$f(x)g(t - x) = \begin{cases} 0 & \text{if } x < 0, \\ f(x)g(t - x) & \text{if } 0 \leq x \leq t, \\ 0 & \text{if } x > t, \end{cases}$$

and for $t < 0$ and $x \in \mathbb{R}$, $f(x)g(t - x) = 0$, it follows that, for all $t \in \mathbb{R}$,

$$(f * g)(t) = \int_{-\infty}^{\infty} f(x)g(t - x)\,dx = \int_0^t f(x)g(t - x)\,dx.$$

Problem 4.3.35. Determine the convolution $f * g$ if:

(a) $f(t) = H(t)\sin t$ and $g(t) = e^t H(t)$.
(b) $f(t) = e^{-at}H(t)$ and $g(t) = e^{-bt}H(t)$, $0 < b < a$.

Solution 4.3.35.

(a) By definition, for $t \geq 0$, $(f * g)(t) = \int_0^t e^{t-x} \sin x \; dx = e^t \int_0^t e^{-x} \sin x \; dx = e^t \cdot \frac{1}{2}(1 - e^{-t}(\sin t + \cos t)) = \frac{1}{2}(e^t - \sin t - \cos t)$.

(b) By definition, for $t \geq 0$, $(f * g)(t) = \int_0^t e^{-ax} e^{-b(t-x)} \; dx = e^{-bt} \int_0^t e^{(b-a)x} \; dx = \frac{e^{-bt} - e^{-at}}{a-b}$.

Problem 4.3.36. Establish that if f and g are $\mathcal{L}$-originals, then their convolution $(f * g)(t) = \int_0^t f(x)g(t-x) \; dx$ is an $\mathcal{L}$-original as well.

Solution 4.3.36. Since f and g are such that, for any $t < 0$, $f(t) = g(t) = 0$, it follows that, for $t < 0$, $(f * g)(t) = \int_0^t f(x)g(t-x) \; dx = 0$. In addition, since f and g are piecewise continuous in every interval $[0, a]$, $a \in \mathbb{R}$, by the fundamental theorem of calculus, the function $f * g$ is continuous in the interval $[0, a]$. If α_1, α_2, M_1, M_2, T_1, and T_2 are positive numbers such that $|f(t)| < M_1 e^{\alpha_1 t}$, if $t > T_1$ and $|g(t)| < M_2 e^{\alpha_2 t}$, if $t > T_2$, and if $\alpha = \max\{\alpha_1, \alpha_2\}$, $T = \max\{T_1, T_2\}$ and $M = \max\{M_1, M_2, \sup\{|f(t)| : t \in [0, T]\}, \sup\{|g(t)| : t \in [0, T]\}\}$, then, for $t \geq 0$, $|f(t)| < M e^{\alpha t}$ and $|g(t)| < M e^{\alpha t}$. It follows that, for $t \geq 0$,

$$|(f * g)(t)| \leq \int_0^t |f(x)||g(t-x)|dx \leq M^2 \int_0^t e^{\alpha x} e^{\alpha(t-x)} dx$$

$$= M^2 t e^{\alpha t} < M^2 e^{(\alpha+1)t},$$

which means that the convolution of two $\mathcal{L}$-originals is a function of exponential order and completes the proof of the claim.

Problem 4.3.37. Let C^1 denote the set of all $\mathcal{L}$-originals that are continuous in $[0, \infty)^8$ and with continuous first derivatives in $(0, \infty)$. Prove that $(C^1, *)$ is an Abelian semigroup, i.e., prove that for all $f, g, h \in C^1$, $f * g \in C^1$, $f * g = g * f$, and $(f * g) * h = f * (g * h)$.

Solution 4.3.37. Recall that, by Problem 4.3.36, the convolution of two $\mathcal{L}$-originals is an $\mathcal{L}$-original as well. For $f, g \in C^1$, by

---

[8]We assume that $g(0) = \lim\limits_{t \to 0^+} g(t)$.

the fundamental theorem of calculus,. the function $(f * g)(t) = \int_0^t f(x)g(t - x)\, dx$ is continuous in $[0, \infty)$. For $t > 0$, by Leibniz's rule of differentiation,

$$\frac{d}{dt}(f * g)(t) = f(t)g(t-t) + \int_0^t f(x)g'(t-x)\, dx = f(t)g(0) + (f * g')(t).$$

Since the function $t \mapsto f(t)g(0) + (f * g')(t)$ is continuous in $(0, \infty)$, it follows that if $f, g \in C^1$, then $f * g \in C^1$. From

$$\int_0^t f(x)g(t - x)\, dx = \begin{vmatrix} y = t - x \Rightarrow dy = -dx \\ x = 0 \Rightarrow y = t \\ x = t \Rightarrow y = 0 \end{vmatrix}$$

$$= -\int_t^0 f(t - y)g(y)\, dy = \int_0^t f(t - y)g(y)\, dy,$$

it follows that $f * g = g * f$. Lastly, for $t > 0$,

$$((f * g) * h)(t) = \int_0^t (f * g)(x)h(t - x)\, dx$$

$$= \int_0^t \left(\int_0^x f(y)g(x - y)\, dy \right) h(t - x)\, dx$$

$$= \iint_D f(y)g(x - y)h(t - x)\, dydx,$$

where $D = \{(x, y) : 0 \le y \le x \le t\}$. By changing the order of integration, we obtain

$$((f * g) * h)(t) = \int_0^t f(y) \left(\int_y^t g(x - y)h(t - x)\, dx \right) dy = |x = t - z|$$

$$= \int_0^t f(y) \left(\int_0^{t-y} g((t - y) - z)h(z)\, dz \right) dy$$

$$= \int_0^t f(y)(g * h)(t - y)\, dy = (f * (g * h))(t),$$

which completes the proof.

Problem 4.3.38. (Convolution Theorem.) If f and g are $\mathcal{L}$-originals, then $\mathcal{L}[(f * g)(t)] = \mathcal{L}[f(t)] \cdot \mathcal{L}[g(t)]$.

Solution 4.3.38. By Problem 4.3.36, the function $f * g$ is an $\mathcal{L}$-original.

Let $F(s) = \mathcal{L}[f(t)](s)$ and $G(s) = \mathcal{L}[g(t)](s)$, $s \in \mathbb{C}$, $\mathrm{Re}(s) > C$, where $C = C(f, g)$ is a positive constant. Then,

$$F(s)G(s) = \int_0^\infty e^{-sx} f(x) \, dx \int_0^\infty e^{-su} g(u) \, du$$

$$= \int_0^\infty \int_0^\infty e^{-s(x+u)} f(x)g(u) \, du dx = \begin{vmatrix} u = t - x \Rightarrow du = dt \\ u = 0 \Rightarrow t = x \\ u \to \infty \Rightarrow t \to \infty \end{vmatrix}$$

$$= \int_0^\infty \int_x^\infty e^{-st} f(x)g(t - x) \, dt dx.$$

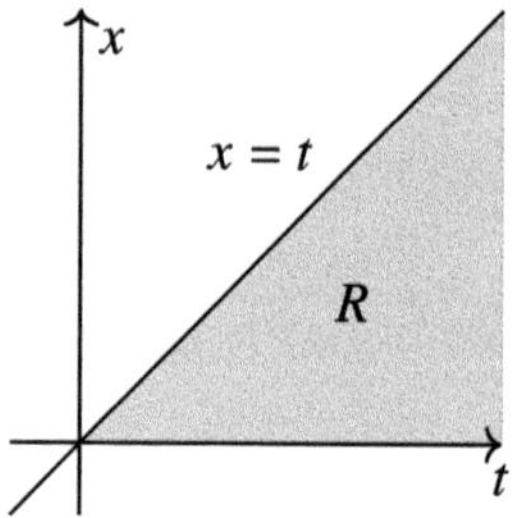

The region of integration is

$$R = \{(t, x) : x \geq 0 \text{ and } t \geq x\} = \{(t, x) : t \geq 0 \text{ and } 0 \leq x \leq t\}.$$

By Fubini's theorem, it is possible to interchange the order of integration. It follows that

$$F(s)G(s) = \int_0^\infty \int_x^\infty e^{-st} f(x)g(t - x) \, dt dx$$

$$= \int_0^\infty e^{-st} \left(\int_0^t f(x)g(t - x) \, dx \right) dt$$

$$= \int_0^\infty e^{-st} (f * g)(t) \, dt = \mathcal{L}[(f * g)(t)](s),$$

which completes the proof of the convolution theorem.

Problem 4.3.39. Let $a > 0$. Use the convolution theorem to determine the inverse Laplace transform $\mathcal{L}^{-1}[F(s)](t)$ for each of the following functions:

(a) $F(s) = \frac{1}{(s^2+a^2)^2}$. (b) $F(s) = \frac{1}{s(s^2-a^2)}$. (c) $F(s) = \frac{1}{s^2(s^2+a^2)^2}$.

Solution 4.3.39.

(a) From $F(s) = \frac{1}{a^2}\frac{a}{s^2+a^2}\frac{a}{s^2+a^2}$ and $\mathcal{L}[H(t)\sin at](s) = \frac{a}{s^2+a^2}$, by the convolution theorem, for $t \geq 0$,

$$\mathcal{L}^{-1}[F(s)](t) = \frac{1}{a^2}\int_0^t \sin ax \sin a(t-x)\, dx$$

$$= -\frac{1}{2a^2}\int_0^t (\cos at - \cos a(2x-t))\, dx$$

$$= -\frac{1}{2a^2}\left(t\cos at - \frac{1}{a}\sin at\right).$$

Note: From $F(s) = \frac{1}{2a^2}\frac{(s^2+a^2)+(s^2+a^2-2s^2)}{s^2+a^2} = \frac{1}{2a^3}\frac{a}{s^2+a^2} - \frac{1}{2a^2}\frac{d}{ds}\left(\frac{s}{s^2+a^2}\right)$, it follows that $\mathcal{L}^{-1}[F(s)](t) = \left(\frac{\sin at}{2a^3} - \frac{t\cos at}{2a^2}\right)H(t)$.

(b) Since $F(s) = \frac{1}{s}\frac{1}{s^2-a^2} = \frac{1}{a}\mathcal{L}[H(t)](s)\cdot\mathcal{L}[H(t)\sinh at](s)$, by the convolution theorem, for $t \geq 0$, $\mathcal{L}^{-1}[F(s)](t) = \frac{1}{a}\int_0^t \sinh ax\, dx = \frac{1}{a^2}(\cosh at - 1)$.

(c) From $\mathcal{L}^{-1}\left[\frac{1}{s^2}\right](t) = tH(t)$ and Part (a), by the convolution theorem, for $t \geq 0$

$$\mathcal{L}^{-1}[F(s)](t) = -\frac{1}{2a^2}\int_0^t (t-x)\left(x\cos ax - \frac{1}{a}\sin ax\right) dx$$

$$= \frac{1}{2a^5}(2at + at\cos at - 3\sin at).$$

Problem 4.3.40. Let $F(s) = \mathcal{L}[f(t)](s)$. Use the convolution theorem to establish the following facts:

(a) Let $a, b > 0$. For $t > 0$, $\mathcal{L}^{-1}\left[\frac{F(s)}{(s+a)^2+b^2}\right](t) = \frac{e^{-at}}{b}\int_0^t e^{ax}f(x)\sin b\cdot(t-x)\, dx$.

(b) Let $a > 0$. For $t > 0$, $\mathcal{L}^{-1}\left[\frac{F(s)}{(s+a)^2}\right](t) = \int_0^t xe^{-ax}f(t-x)\, dx$.

Solution 4.3.40.

(a) The claim follows from $\mathcal{L}^{-1}\left[\frac{1}{(s+a)^2+b^2}\right](t) = \frac{e^{-at}}{b}H(t)\sin bt$.

(b) The claim follows from $\mathcal{L}^{-1}\left[\frac{1}{(s+a)^2}\right](t) = tH(t)e^{-at}$.

4.4 Equations

4.4.1 *Linear differential equations*

Use Laplace transformation to solve the following initial value problems.[9]

In the remainder of this chapter, we write $Y(s) = \mathcal{L}[y(t)H(t)](s)$, $s \in D = \{z \in \mathbb{C} : \mathrm{Re}(z) > a\}$, where the number a depends on the given initial value problem and $y = y(t)$ is the solution to the initial value problem.

Problem 4.4.1. $y'' + 4y = 0$, $y(0) = 0$, $y'(0) = 10$.

Solution 4.4.1. From the initial conditions, it follows that $\mathcal{L}[y''(t)H(t)](s) = s^2 Y(s) - 10$. The function Y is determined by the identity $s^2 Y(s) - 10 + 4Y(s) = 0$, $s \in D$, i.e., $Y(s) = \frac{10}{s^2+4}$, $s \in D$. The solution to the initial value problem is $y(t) = 5\sin 2t$, $t \in \mathbb{R}$.

Note: As per convention (and convenience), we skip a step here. From $Y(s) = \frac{10}{s^2+4}$, it follows that $y(t) = 5H(t)\sin 2t$. This $\mathcal{L}$-original is not a differentiable function at $t = 0$, but it satisfies the initial value problem $y'' + 4y = 0$, $t > 0$, $y(0) = 0$, $\lim_{t \to 0+} y'(t) = 10$. The solution to the given initial value problem is the function $t \mapsto 5\sin 2t$, $t \in \mathbb{R}$. To simplify our work, by convention, we denote by $y = y(t)$ both the

---

[9]In strict terms, we use the Laplace transformation to determine an $\mathcal{L}$-original y that satisfies the given initial value problem. In light of this, the initial condition $y(0) = a$ actually means $\lim_{t \to 0+} y(t) = a$, and for $k \geq 1$, the initial condition $y^{(k)}(0) = a_k$ means $\lim_{t \to 0+} y^{(k)}(t) = a_k$. Since it is often possible to obtain, in a natural way, a solution $z = z(t)$, $t \in (-\varepsilon, \varepsilon)$ such that $z(t) = y(t)$ if $t \in [0, \varepsilon)$ and $z(t)$ are not identically equal to 0 in $(-\varepsilon, 0)$, it is common to use the phrase "use Laplace transform to solve the initial value problem" rather than "determine an $\mathcal{L}$-original that satisfies the initial value problem."

$\mathcal{L}$-original $\mathcal{L}^{-1}[Y(s)](t)$ and the solution to the given initial value problem.

Problem 4.4.2. $y'' + 5y' + 6y = 0$, $y(0) = 10$, $y'(0) = 0$.

Solution 4.4.2. From the initial conditions, it follows that $\mathcal{L}[y''(t)H(t)](s) = s^2Y(s) - 10s$ and $\mathcal{L}[y'(t)H(t)](s) = sY(s) - 10$. The function Y is determined by the identity $s^2Y(s) - 10s + 5sY(s) - 50 + 6Y(s) = 0$, $s \in D$. Therefore, $Y(s) = \frac{10(s+5)}{(s+2)(s+3)} = \frac{30}{s+2} - \frac{20}{s+3}$, $s \in D$, and the solution to the initial value problem is $y(t) = 30e^{-2t} - 20e^{-3t}$, $t \in \mathbb{R}$.

Problem 4.4.3. $y'' + 2ay' + a^2y = 0$, $y(0) = 0$, $y'(0) = 7$, $a > 0$.

Solution 4.4.3. From the initial conditions, it follows that $\mathcal{L}[y''(t)H(t)](s) = s^2Y(s) - 7$ and $\mathcal{L}[y'(t)H(t)](s) = sY(s)$. The function Y is determined by the identity $s^2Y(s) - 7 + 2asY(s) + a^2Y(s) = 0$, $s \in D$. Therefore, $Y(s) = \frac{7}{(s+a)^2}$, $s \in D$, and the solution to the initial value problem is $y(t) = 7te^{-at}$, $t \in \mathbb{R}$.

Problem 4.4.4. $y'' + 4y' + 8y = 0$, $y(0) = a$, $y'(0) = b$, $a, b \in \mathbb{R}$, $a^2 + b^2 \neq 0$.

Solution 4.4.4. From the initial conditions, it follows that $\mathcal{L}[y''(t)H(t)](s) = s^2Y(s) - as - b$ and $\mathcal{L}[y'(t)H(t)](s) = sY(s) - a$. The function Y is determined by the identity $s^2Y(s) - as - b + 4sY(s) - 4a + 8Y(s) = 0$, $s \in D$. Therefore, $Y(s) = \frac{as + 4a + b}{s^2 + 4s + 8} = \frac{a(s+2) + 2a + b}{(s+2)^2 + 4}$, $s \in D$, and the solution to the initial value problem is $y(t) = e^{-2t}\left(a\cos 2t + \frac{2a+b}{2}\sin 2t\right)$, $t \in \mathbb{R}$.

Problem 4.4.5. $y''' + y = 0$, $y(0) = 1$, $y'(0) = 2$, $y''(0) = 3$.

Solution 4.4.5. From the initial conditions, it follows that $\mathcal{L}[y'''(t)H(t)](s) = s^3Y(s) - s^2 - 2s - 3$. The function Y is determined by the identity $s^3Y(s) - s^2 - 2s - 3 + Y(s) = 0$, $s \in D$. Therefore, $Y(s) = \frac{s^2 + 2s + 3}{s^3 + 1} = \frac{2}{3}\frac{1}{s+1} + \frac{1}{3}\frac{s+7}{s^2 - s + 1} = \frac{2}{3}\frac{1}{s+1} + \frac{1}{3}\frac{s - \frac{1}{2}}{(s-\frac{1}{2})^2 + \frac{3}{4}} + \frac{5}{\sqrt{3}}\frac{\frac{\sqrt{3}}{2}}{(s-\frac{1}{2})^2 + \frac{3}{4}}$, $s \in D$. The solution to the initial value problem is $y(t) = \frac{2}{3}e^{-t} + e^{\frac{t}{2}}\left(\frac{1}{3}\cos\frac{\sqrt{3}t}{2} + \frac{5}{\sqrt{3}}\sin\frac{\sqrt{3}t}{2}\right)$, $t \in \mathbb{R}$.

Problem 4.4.6. $ty'' + (t-1)y' - y = 0$, $y(0) = a > 0$, $\lim_{t \to \infty} y(t) = 0$.

Solution 4.4.6. Suppose that $y'(0) = \alpha$, $\alpha \in \mathbb{R}$. Then, $\mathcal{L}[ty''(t)H(t)](s) = -\frac{d}{ds}\mathcal{L}[y''(t)H(t)](s) = -\frac{d}{ds}(s^2 Y(s) - as - \alpha) = -2sY(s) - s^2 Y'(s) + a$ and $\mathcal{L}[(t-1)y'(t)H(t)](s) = -\frac{d}{ds}\mathcal{L}[y'(t)H(t)](s) - \mathcal{L}[y'(t)H(t)](s) = -\frac{d}{ds}(sY(s) - a) - sY(s) + a = -(s+1)Y(s) - sY'(s) + a$. The function Y is determined by the identity $(s^2 + s)Y'(s) + (3s + 2)Y(s) = 2a$, $s \in D$, or, what is the same, by the identity $((s^3 + s^2)Y(s))' = 2as$, $s \in D$. For some constant C, $Y(s) = \frac{as^2 + C}{s^2(s+1)} = \frac{a+C}{s+1} - \frac{C}{s} + \frac{C}{s^2}$, $s \in D$, which implies that $y(t) = (a + C)e^{-t} + C(t-1)$. Since $\lim_{t \to \infty} y(t) = 0$, it follows that $C = 0$ and $y(t) = ae^{-t}$, $t \in \mathbb{R}$.

Note: Recall that, as a Laplace transform, the function Y is analytic in the region D.

Problem 4.4.7. $ty''' + 2y'' + y' + (t-1)y = 0$, $y(0) = y''(0) = 1$, $y'(0) = -1$.

Solution 4.4.7. From the initial conditions, $\mathcal{L}[ty'''(t)H(t)](s) = -\frac{d}{ds}(s^3 Y(s) - s^2 + s - 1) = -3s^2 Y(s) - s^3 Y'(s) + 2s - 1$, $\mathcal{L}[y''(t)H(t)](s) = s^2 Y(s) - s + 1$, $\mathcal{L}[y'(t)H(t)](s) = sY(s) - 1$, and $\mathcal{L}[(t-1)y(t)H(t)](s) = -Y'(s) - Y(s)$. The function Y is determined by the identity $(s^3 + 1)Y'(s) + (s^2 - s + 1)Y(s) = 0$, $s \in D$, or, what is the same, by the identity $0 = (s+1)Y'(s) + Y(s) = ((s+1)Y(s))'$, $s \in D$. For some constant C, $Y(s) = \frac{C}{s+1}$, $s \in D$, and $y(t) = Ce^{-t}$. Since $y(0) = 1$, it follows that $C = 1$ and the solution to the initial value problem is $y(t) = e^{-t}$, $t \in \mathbb{R}$.

Note: In the process of determining the function Y, we solve the separable differentiable equation $(s^3 + 1)z' + (s^2 - s + 1)z = 0$, where $z = z(s)$ is the unknown function.

Problem 4.4.8. $ty'' + (1-t)y' + y = 0$, $y(0) = a > 0$.

Solution 4.4.8. Suppose that $y'(0) = \alpha$, $\alpha \in \mathbb{R}$. From the initial conditions, $\mathcal{L}[ty''(t)H(t)](s) = -\frac{d}{ds}(s^2 Y(s) - as - \alpha) = -2sY(s) - s^2 Y'(s) + a$ and $\mathcal{L}[(1-t)y'(t)H(t)](s) = sY(s) - a + Y(s) + sY'(s)$. The function Y is determined by the identity $s(1-s)Y'(s) + (2-s)Y(s) = 0$, $s \in D$, or, what is the same, by the identity $\frac{Y'(s)}{Y(s)} = \frac{1}{s-1} - \frac{2}{s}$, $s \in D$. For some constant C, $Y(s) = \frac{C(s-1)}{s^2} = \frac{C}{s} - \frac{C}{s^2}$, $s \in D$, and

$y(t) = C(1 - t)$. Since $y(0) = a$ implies $C = a$, the solution to the initial value problem is $y(t) = a(1 - t)$, $t \in \mathbb{R}$.

Problem 4.4.9. $ty'' + (1 - t)y' + 2y = 0$.

Solution 4.4.9. Suppose that $y(0) = \alpha$ and $y'(0) = \beta$, $\alpha, \beta \in \mathbb{R}$. It follows that $\mathcal{L}[ty''(t)H(t)](s) = -\frac{d}{ds}(s^2 Y(s) - \alpha s - \beta) = -2sY(s) - s^2 Y'(s) + \alpha$ and $\mathcal{L}[(1 - t)y'(t)H(t)](s) = sY(s) - \alpha + Y(s) + sY'(s)$. The function Y is determined by the identity $s(1 - s)Y'(s) + (3 - s)Y(s) = 0$, $s \in D$, or, what is the same, by the identity $\frac{Y'(s)}{Y(s)} = \frac{2}{s-1} - \frac{3}{s}$, $s \in D$. For some constant C, $Y(s) = \frac{C(s-1)^2}{s^3} = \frac{C}{s} - \frac{2C}{s^2} + \frac{C}{s^3}$, $s \in D$, and $y(t) = C(1 - 2t + \frac{t^2}{2})$. Since $y(0) = \alpha$, it follows that $y(t) = \alpha(1 - 2t + \frac{t^2}{2})$, $t \in \mathbb{R}$, $\alpha \in \mathbb{R}$, is a family of solutions to the equation.

Note: The general solution to the equation is $y = c_1(2 - 4t + t^2) + c_2((2 - 4t + t^2)\mathrm{Ei}(t) - (t - 3)e^t)$, $c_1, c_2 \in \mathbb{R}$, where $\mathrm{Ei}(t)$ is the exponential integral function. By using Laplace transformation, we were able to determine only the family of solutions that were obtained from $\mathcal{L}$-originals.

Problem 4.4.10. $y'' + (t + 1)y' + ty = 0$, $y(0) = -y'(0) = a > 0$.

Solution 4.4.10. From the initial conditions, $\mathcal{L}[y''(t)H(t)](s) = s^2 Y(s) - as + a$, $\mathcal{L}[(t + 1)y'(t)H(t)](s) = sY(s) - a - Y(s) - sY'(s)$, and $\mathcal{L}[ty(t)H(t)](s) = -Y'(s)$. The function Y is a solution to the non-homogeneous first-order linear differential equation $-(s + 1)z' + (s^2 + s - 1)z = as$ or, what is the same, to the equation $z' - \left(s - \frac{1}{s+1}\right)z = -\frac{as}{s+1}$. By Problem 2.2.5, $Y(s) = \frac{C}{s+1}e^{\frac{s^2}{2}} + \frac{a}{s+1}$, $C \in \mathbb{R}$. Since, by Problem 4.2.1(a), $\lim_{s \to \infty} Y(s) = 0$, it follows that $C = 0$ and $Y(s) = \frac{a}{s+1}$, $s \in D$. The solution to the initial value problem is $y(t) = ae^{-t}$, $t \in \mathbb{R}$.

Problem 4.4.11. $y'' + 5y' + 6y = 12$, $y(0) = y'(0) = 0$.

Solution 4.4.11. From the initial conditions, $\mathcal{L}[y''(t)H(t)](s) = s^2 Y(s)$ and $\mathcal{L}[y'(t)H(t)](s) = sY(s)$. The function Y is determined by the identity $(s^2 + 5s + 6)Y(s) = \frac{12}{s}$, $s \in D$, which implies that $Y(s) = \frac{2}{s} - \frac{6}{s+2} + \frac{4}{s+3}$, $s \in D$. The solution to the initial value problem is $y(t) = 2 - 6e^{-2t} + 4e^{-3t}$, $t \in \mathbb{R}$.

Problem 4.4.12. $y'' + 2ay' + a^2 y = At$, $y(0) = y'(0) = 0$, $a, A > 0$.

Solution 4.4.12. From the initial conditions, $\mathcal{L}[y''(t)H(t)](s) = s^2 Y(s)$ and $\mathcal{L}[y'(t)H(t)](s) = sY(s)$. The function Y is determined by the identity $(s^2 + 2as + a^2)Y(s) = \frac{A}{s^2}$, $s \in D$, which implies that $Y(s) = -\frac{2A}{a^3}\frac{1}{s} + \frac{A}{a^2}\frac{1}{s^2} + \frac{2A}{a^3}\frac{1}{s+a} + \frac{A}{a^2}\frac{1}{(s+a)^2}$, $s \in D$. The solution to the initial value problem is $y(t) = -\frac{2A}{a^3} + \frac{A}{a^2}t + \frac{2A}{a^3}e^{-at} + \frac{A}{a^2}te^{-at}$, $t \in \mathbb{R}$.

Problem 4.4.13. $y'' + 2ay' + a^2 y = a^2 \sin at$, $y(0) = y'(0) = 0$, $a > 0$.

Solution 4.4.13. From the initial conditions, $\mathcal{L}[y''(t)H(t)](s) = s^2 Y(s)$ and $\mathcal{L}[y'(t)H(t)](s) = sY(s)$. The function Y is determined by the identity $(s^2 + 2as + a^2)Y(s) = \frac{a^3}{s^2+a^2}$, $s \in D$, which implies that $Y(s) = -\frac{1}{2}\frac{s}{s^2+a^2} + \frac{1}{2}\frac{1}{s+a} + \frac{a}{2}\frac{1}{(s+a)^2}$, $s \in D$. The solution to the initial value problem is $y(t) = -\frac{1}{2}\cos at + \frac{1}{2}e^{-at} + \frac{a}{2}te^{-at}$, $t \in \mathbb{R}$.

Problem 4.4.14. $y'' + a^2 y = b\cos \omega t$, $y(0) = y'(0) = c$, $a, b, c, \omega > 0$, $a \neq \omega$.

Solution 4.4.14. From the initial conditions, $\mathcal{L}[y''(t)H(t)](s) = s^2 Y(s) - cs - c$. The function Y is determined by the identity $(s^2 + a^2)Y(s) = c(s + 1) + \frac{bs}{s^2+\omega^2}$, $s \in D$, which implies that

$$Y(s) = c\left(\frac{s}{s^2 + a^2} + \frac{1}{s^2 + a^2}\right) + \frac{b}{\omega^2 - a^2}\left(\frac{s}{s^2 + a^2} - \frac{s}{s^2 + \omega^2}\right), \quad s \in D.$$

The solution to the initial value problem is $y(t) = c(\cos at + \frac{1}{a}\sin at) + \frac{b}{\omega^2-a^2}(\cos at - \cos \omega t)$, $t \in \mathbb{R}$.

Problem 4.4.15. $y'' + a^2 y = be^{\omega t} + c$, $y(0) = y'(0) = 0$, $a, b, c, \omega > 0$.

Solution 4.4.15. From the initial conditions, $\mathcal{L}[y''(t)H(t)](s) = s^2 Y(s)$. The function Y is determined by the identity $(s^2 + a^2)Y(s) = \frac{b}{s-\omega} + \frac{c}{s}$, $s \in D$, which implies that

$$Y(s) = \frac{b}{a^2 + \omega^2}\left(\frac{1}{s - \omega} - \frac{s + \omega}{s^2 + a^2}\right) + \frac{c}{a^2}\left(\frac{1}{s} - \frac{s}{s^2 + a^2}\right), \quad s \in D.$$

The solution to the initial value problem is

$$y(t) = \frac{b}{a^2 + \omega^2}\left(e^{\omega t} - \cos at - \frac{\omega}{a}\sin at\right) + \frac{c}{a^2}(1 - \cos at), \quad t \in \mathbb{R}.$$

Problem 4.4.16. $y'' + 2ay' + (a^2 + b^2)y = a\cos bt$, $y(0) = y'(0) = 0$, $a, b > 0$.

Solution 4.4.16. From the initial conditions, $\mathcal{L}[y''(t)H(t)](s) = s^2 Y(s)$ and $\mathcal{L}[y'(t)H(t)](s) = sY(s)$. The function Y is determined by the identity $((s+a)^2 + b^2)Y(s) = \frac{as}{s^2+b^2}$, $s \in D$, which implies that $Y(s) = \frac{a}{b}\frac{s}{s^2+b^2}\frac{b}{(s+a)^2+b^2}$. By the convolution theorem,

$$y(t) = \frac{a}{b}\int_0^t e^{-ax}\sin bx \cos b(t-x)\ dx$$

$$= \frac{a}{2b}\int_0^t e^{-ax}(\sin bt + \sin b(2x-t))\ dx$$

$$= \frac{1}{2b}(1 - e^{-at})\sin bt + \frac{a}{2b}e^{-\frac{at}{2}}\int_0^t e^{-a(x-\frac{t}{2})}\sin 2b\left(x - \frac{t}{2}\right)\ dx.$$

Since

$$\int_0^t e^{-a(x-\frac{t}{2})}\sin 2b\left(x - \frac{t}{2}\right)\ dx = \frac{1}{a^2 + 4b^2}\left(e^{-\frac{at}{2}}(-a\sin bt - 2b\cos bt)\right.$$

$$\left. - e^{\frac{at}{2}}(a\sin bt - 2b\cos bt)\right),$$

it follows that the solution to the initial value problem is

$$y(t) = \frac{1}{2b}(1 - e^{-at})\sin bt + \frac{a}{2b(a^2 + 4b^2)}\left(2b(1 - e^{-at})\cos bt\right.$$

$$\left. - a(1 + e^{-at})\sin bt\right),\ t \in \mathbb{R}.$$

Note: The fact $Y(s) = \frac{1}{a^2+4b^2}\left(\frac{as+2b^2}{s^2+b^2} - \frac{as+2(a^2+b^2)}{(s+a)^2+b^2}\right)$ yields this unique solution to the initial value problem but in a different form:

$$y(t) = \frac{1}{a^2 + 4b^2}\left(a\cos bt + 2b\sin bt + ae^{-at}\cos bt + \frac{2b^2 + a^2}{b}e^{-at}\sin bt\right).$$

Problem 4.4.17. $y^{(4)} = a$, $y(0) = y'(0) = y(b) = y'(b) = 0$, $a, b > 0$.

Solution 4.4.17. Let $y''(0) = \alpha$ and $y'''(0) = \beta$, $\alpha, \beta \in \mathbb{R}$. Then, $\mathcal{L}[y^{(4)}(t)H(t)](s) = s^4 Y(s) - \alpha s - \beta$, and Y is determined by the identity $s^4 Y(s) - \alpha s - \beta = \frac{a}{s}$, $s \in D$. Hence, $Y(s) = \frac{\alpha}{s^3} + \frac{\beta}{s^4} + \frac{a}{s^5}$, $s \in D$, and $y(t) = \frac{\alpha t^2}{2} + \frac{\beta t^3}{6} + \frac{a t^4}{24}$. From $y(b) = y'(b) = 0$, it follows that $\alpha = \frac{ab^2}{12}$ and $\beta = -\frac{ab}{2}$. The solution to the initial value problem is $y(t) = \frac{a t^2}{24}(b - t)^2$, $t \in \mathbb{R}$.

Problem 4.4.18. $y'' - 5y' + 4y = H(t) - H(t - 1)$, $y(0) = 0$, $\lim_{t \to 0^+} y'(t) = 1$.

Solution 4.4.18. From the initial conditions, $\mathcal{L}[y''(t)H(t)](s) = s^2 Y(s) - 1$ and $\mathcal{L}[y'(t)H(t)](s) = sY(s)$. The function Y is determined by the identity $(s^2 - 5s + 4)Y(s) - 1 = \frac{1}{s} - \frac{e^{-s}}{s}$, $s \in D$, which implies that $Y(s) = \frac{1}{(s-1)(s-4)} + \frac{1-e^{-s}}{s(s-1)(s-4)} = \frac{1}{4}\frac{1}{s} - \frac{2}{3}\frac{1}{s-1} + \frac{5}{12}\frac{1}{s-4} - e^{-s}\left(\frac{1}{4}\frac{1}{s} - \frac{1}{3}\frac{1}{s-1} + \frac{1}{12}\frac{1}{s-4}\right)$, $s \in D$. The solution to the initial value problem is $y(t) = \frac{1}{12}(3 - 8e^t + 5e^{4t})H(t) - \frac{1}{12}H(t - 1)\left(3 - 4e^{t-1} + e^{4(t-1)}\right)$, $t \in \mathbb{R}$.

Note: Observe that $y_h = c_1 e^t + c_2 e^{4t}$, $c_1, c_2 \in \mathbb{R}$, is the general solution to the corresponding homogeneous equation. The solution

$$y(t) = \begin{cases} 0 & \text{if } t < 0, \\[2mm] \frac{1}{4} - \frac{2}{3}e^t + \frac{5}{12}e^{4t} & \text{if } t \in [0, 1), \\[2mm] \frac{1}{3}(e^{-1} - 2)e^t + \frac{1}{12}(5 - e^{-4})e^{4t} & \text{if } t \geq 1, \end{cases}$$

is a continuous function in $\mathbb{R}$. From

$$y'(t) = \begin{cases} 0 & \text{if } t < 0, \\[2mm] -\frac{2}{3}e^t + \frac{5}{3}e^{4t} & \text{if } t \in (0, 1), \\[2mm] -\frac{2}{3}e + \frac{5}{3}e^4 & \text{if } t = 1, \\[2mm] \frac{1}{3}(e^{-1} - 2)e^t + \frac{1}{3}(5 - e^{-4})e^{4t} & \text{if } t > 1, \end{cases}$$

we conclude that the function $t \mapsto y(t)$ is differentiable everywhere except at $t = 0$, with $\lim_{t \to 0^-} y'(t) = 0$ and $\lim_{t \to 0^+} y'(t) = 1$.

Problem 4.4.19. $y'' + a^2 y = H(t) - H(t - b)$, $y(0) = 0$, $\lim_{t \to 0^+} y'(t) = c$, $a, b, c > 0$.

Solution 4.4.19. From the initial conditions, $\mathcal{L}[y''(t)H(t)](s) = s^2Y(s) - c$ and $\mathcal{L}[y'(t)H(t)](s) = sY(s)$. The function Y is determined by the identity $(s^2 + a^2)Y(s) = \frac{1}{s} - \frac{e^{-bs}}{s} + c$, $s \in D$, which implies that

$$Y(s) = \frac{1}{a^2}\left(\frac{1}{s} - \frac{s}{s^2 + a^2}\right) + \frac{c}{s^2 + a^2} - \frac{1}{a^2}\left(\frac{1}{s} - \frac{s}{s^2 + a^2}\right)e^{-bs}, \quad s \in D.$$

The solution to the initial value problem is

$$y(t) = \frac{1}{a^2}(1 - \cos at + ac\sin at)H(t) - \frac{1}{a^2}(1 - \cos a(t-b))H(t-b), \quad t \in \mathbb{R}.$$

Note: The solution

$$y(t) = \begin{cases} 0 & \text{if } t < 0, \\ \frac{1}{a^2}(1 - \cos at + ac\sin at) & \text{if } t \in [0, b), \\ \frac{1}{a^2}(\cos a(t-b) - \cos at + ac\sin at) & \text{if } t \geq b, \end{cases}$$

is a continuous function in $\mathbb{R}$. From

$$y'(t) = \begin{cases} 0 & \text{if } t < 0, \\ \frac{1}{a}(\sin at + ac\cos at) & \text{if } t \in (0, b), \\ \frac{1}{a}(\sin ab + ac\cos ab) & \text{if } t = b, \\ \frac{1}{a}(\sin at - \sin a(t-b) + ac\cos at) & \text{if } t > b, \end{cases}$$

we conclude that the function $t \mapsto y(t)$ is differentiable everywhere except at $t = 0$, with $\lim\limits_{t \to 0^-} y'(t) = 0$ and $\lim\limits_{t \to 0^+} y'(t) = c$.

Problem 4.4.20. $y'' + a^2 y = b(H(t) + H(t-c))$, $y(0) = d$, $\lim\limits_{t \to 0} y'(t) = 0$, $a, b, c, d > 0$.

Solution 4.4.20. From the initial conditions, $\mathcal{L}[y''(t)H(t)](s) = s^2Y(s) - ds$. The function Y is determined by the identity $(s^2 + a^2)Y(s) = ds + \frac{b}{s}(1 + e^{-cs})$, $s \in D$. Hence, $Y(s) = \frac{ds}{s^2 + a^2} + \frac{b}{a^2}\left(\frac{1}{s} - \frac{s}{s^2 + a^2}\right)(1 + e^{-cs})$, $s \in D$. The solution to the initial value problem is $y(t) = \frac{1}{a^2}((a^2 d - b)\cos at + b)H(t) + \frac{b}{a^2}(1 - \cos a(t - c))H(t - c)$.

Problem 4.4.21. $y'' + a^2 y = bt(H(t) - H(t-c))$, $y(0) = \alpha$, $y'(0) = 0$, $a, b, c, \alpha > 0$.

Solution 4.4.21. From the initial conditions, $\mathcal{L}[y''(t)H(t)](s) = s^2 Y(s) - \alpha s$. By Problem 4.2.2(b), the function Y is determined by the identity $(s^2 + a^2)Y(s) = \alpha s - b\frac{d}{ds}\left(\frac{1}{s} - \frac{e^{-cs}}{s}\right) = \alpha s + \frac{b}{s^2} - b\left(\frac{1}{s^2} + \frac{c}{s}\right)e^{-cs}$, $s \in D$. Hence, $Y(s) = \frac{\alpha s}{s^2 + a^2} + \frac{b}{a^2}\left(\frac{1}{s^2} - \frac{1}{s^2 + a^2}\right) - \frac{b}{a^2}\left(\frac{1}{s^2} - \frac{1}{s^2 + a^2} + \frac{c}{s} - \frac{cs}{s^2 + a^2}\right)e^{-cs}$, $s \in D$. The solution to the initial value problem is

$$y(t) = \frac{1}{a^3}\left(a^3 \alpha \cos at + abt - \sin at\right) H(t)$$

$$- \frac{b}{a^3}\left(a(t-c) - \sin a(t-c) + ac(1 - \cos a(t-c))\right) H(t-c), \ t \in \mathbb{R}.$$

Problem 4.4.22. $y'' - 2ay' + a^2 y = \sum_{k=1}^{n} kH(t - (k-1)b)$, $y(0) = 0$, $\lim_{t \to 0} y'(t) = 0$, $a, b > 0$, $n \in \mathbb{N}$.

Solution 4.4.22. From the initial conditions, $\mathcal{L}[y''(t)H(t)](s) = s^2 Y(s)$ and $\mathcal{L}[y'(t)H(t)](s) = sY(s)$.

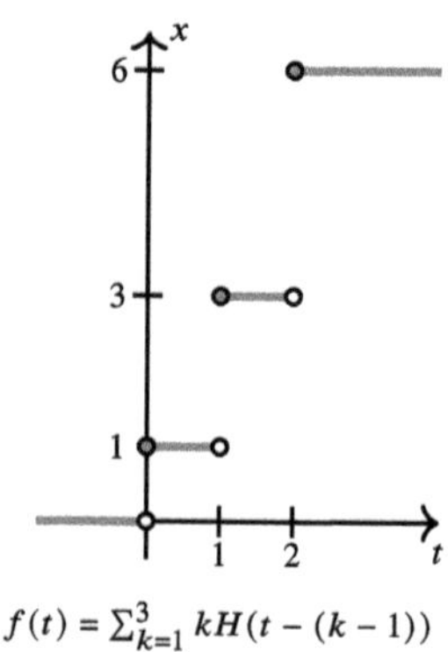

$$f(t) = \sum_{k=1}^{3} kH(t - (k-1))$$

The function Y is determined by the identity $(s - a)^2 Y(s) = \sum_{k=1}^{n} \frac{k}{s} e^{-(k-1)bs}$, $s \in D$, which implies that

$$Y(s) = \frac{1}{a^2} \sum_{k=1}^{n} k e^{-(k-1)bs} \left(\frac{1}{s} - \frac{1}{s-a} + \frac{a}{(s-a)^2}\right), \ s \in D.$$

The solution to the initial value problem is, for $t \in \mathbb{R}$,

$$y(t) = \frac{1}{a^2} \sum_{k=1}^{n} kH(t-(k-1)b) \left(1 + (a(t - (k-1)b) - 1)e^{-a(t-(k-1)b)}\right).$$

Problem 4.4.23. $y'' + a^2 y = H(t-1)\sin b(t-1)$ $y(0) = 1$, $y'(0) = 0$, $a, b > 0$.

Solution 4.4.23. From the initial conditions, $\mathcal{L}[y''(t)H(t)](s) = s^2 Y(s) - s$. The function Y is determined by the identity $(s^2 + a^2)Y(s) = s + \frac{be^{-s}}{s^2+b^2}$, $s \in D$. If $a \neq b$, then $Y(s) = \frac{s}{s^2+a^2} + \frac{be^{-s}}{b^2-a^2}\left(\frac{1}{s^2+a^2} - \frac{1}{s^2+b^2}\right)$, $s \in D$, and the solution to the initial value problem is

$$y(t) = H(t)\cos at + \frac{H(t-1)}{a(b^2 - a^2)}\left(b\sin a(t-1) - a\sin b(t-1)\right), \ t \in \mathbb{R}.$$

If $a = b$ then, for $s \in D$,

$$Y(s) = \frac{s}{s^2 + a^2} + \frac{e^{-s}}{2a}\frac{(s^2 + a^2) + (s^2 + a^2 - 2s^2)}{(s^2 + a^2)^2}$$

$$= \frac{s}{s^2 + a^2} + \frac{e^{-s}}{2a}\left(\frac{1}{s^2 + a^2} + \frac{d}{ds}\left(\frac{s}{s^2 + a^2}\right)\right).$$

The solution to the initial value problem is

$$y(t) = H(t)\cos at + \frac{H(t-1)}{2a^2}\left(\sin a(t-1) - a(t-1)\cos a(t-1)\right), \ t \in \mathbb{R}.$$

Problem 4.4.24. $y'' + 2ay' + (a^2 + k^2)y = f(t)H(t)$, where $f(t) = a\,\mathrm{sign}(\sin t)$, $y(0) = y'(0) = 0$, $a > 0$, and $k \in 2\mathbb{N} - 1$.

Solution 4.4.24. From the initial conditions, $\mathcal{L}[y''(t)H(t)](s) = s^2 Y(s)$ and $\mathcal{L}[y'(t)H(t)](s) = sY(s)$. The function f is periodic with a period $T = 2\pi$ and such that $f(t) = \begin{cases} a & \text{if } t \in (0, \pi), \\ -a & \text{if } t \in (\pi, 2\pi), \\ 0 & \text{if } t \in \{0, \pi\}. \end{cases}$ It follows that, referring to Problem 4.3.9(e), the function Y is determined by the identity

$$((s + a)^2 + k^2)Y(s) = \frac{a(1 - e^{-\pi s})}{s(1 + e^{-\pi s})}, \ s \in D,$$

which implies that $Y(s) = \frac{a(1-e^{-\pi s})}{s((s+a)^2+k^2)(1+e^{-\pi s})}$, $s \in D$. Since

$$\frac{1}{s((s+a)^2+k^2)} = \frac{1}{a^2+k^2}\left(\frac{1}{s} - \frac{s+a}{(s+a)^2+k^2} - \frac{a}{(s+a)^2+k^2}\right),$$

and, for $\mathrm{Re}(s) > 0$, $\frac{1-e^{-\pi s}}{1+e^{-\pi s}} = 1 + 2\sum_{n=1}^{\infty}(-1)^n e^{-n\pi s}$, it follows that, for $\mathrm{Re}(s) > 0$,

$$Y(s) = \frac{a}{a^2+k^2}\left(\frac{1}{s} - \frac{s+a}{(s+a)^2+k^2} - \frac{a}{(s+a)^2+k^2}\right)$$
$$\cdot\left(1 + 2\sum_{n=1}^{\infty}(-1)^n e^{-n\pi s}\right).$$

The solution to the initial value problem is, for $t \in \mathbb{R}$,

$$y(t) = \frac{a}{a^2+k^2}\left(1 - e^{-at}\left(\cos kt + \frac{a}{k}\sin kt\right)\right)H(t)$$
$$+ \frac{2a}{k(a^2+k^2)}\sum_{n=1}^{\infty}(-1)^n H(t-n\pi)(k - e^{-a(t-n\pi)}(k\cos k(t-n\pi)$$
$$+ a\sin k(t-n\pi))).$$

Problem 4.4.25. $y'' = f(t)$, $y(0) = c$, $y'(0) = 0$, where

$$f(t) = \begin{cases} 0 & \text{if } t < 0, \\ a & \text{if } t \in [0,b), \\ 0 & \text{if } t \in [b, 2b), \\ f(t-2b) & \text{if } t \geq 2b, \end{cases}$$

with $a, b, c > 0$.

Solution 4.4.25. For $t \geq 0$, the function f is periodic with a period $T = 2b$. From the initial conditions, $\mathcal{L}[y''(t)H(t)](s) = s^2 Y(s) - cs$. By Problem 4.2.4, the function Y is determined by the identity
$$s^2 Y(s) = cs + \frac{a}{1-e^{-2bs}}\int_0^b e^{-st}\,dt = cs + \frac{a(1-e^{-bs})}{s(1-e^{-2bs})} = cs + \frac{a}{s(1+e^{-bs})},$$

$s \in D$. Since, for $\mathrm{Re}(s) > 0$, $\frac{1}{1+e^{-bs}} = \sum_{n=0}^{\infty}(-1)^n e^{-bns}$, it follows that, for $\mathrm{Re}(s) > 0$,

$$Y(s) = \frac{c}{s} + \frac{a}{s^3} + a\sum_{n=1}^{\infty}(-1)^n \frac{e^{-bns}}{s^3}.$$

The solution to the initial value problem is

$$y(t) = \left(c + \frac{at^2}{2}\right)H(t) + \frac{a}{2}\sum_{n=1}^{\infty}(-1)^n(t-bn)^2 H(t-bn), \ t \in \mathbb{R}.$$

4.4.2 Systems of linear differential equations

Use Laplace transformation to solve the following homogeneous and non-homogeneous systems of differential equations and initial value problems. Take $X(s) = \mathcal{L}[x(t)H(t)](s)$, $Y(s) = \mathcal{L}[y(t)H(t)](s)$, and $Z(s) = \mathcal{L}[z(t)H(t)](s)$, $s \in D = \{z \in \mathbb{C} : \mathrm{Re}(z) > \alpha\}$, where the number α depends on the given system of equation or the initial value problem.

Problem 4.4.26. $\frac{dx}{dt} = 2x - y + z$, $\frac{dy}{dt} = x + z$, $\frac{dz}{dt} = -3x + y - 2z$, $x(0) = y(0) = 1$, $z(0) = 0$.

Solution 4.4.26. From the initial conditions, $\mathcal{L}[x'(t)H(t)](s) = sX(s) - 1$, $\mathcal{L}[y'(t)H(t)](s) = sY(s) - 1$, and $\mathcal{L}[z'(t)H(t)](s) = sZ(s)$. The functions $X(s)$, $Y(s)$, and $Z(s)$ satisfy the linear system of algebraic equations[10]:

$$(s-2)X(s) + Y(s) - Z(s) = 1,$$

$$-X(s) + sY(s) - Z(s) = 1,$$

$$3X(s) - Y(s) + (s+2)Z(s) = 0.$$

It follows that $X(s) = Y(s) = \frac{s+2}{s(s+1)} = \frac{2}{s} - \frac{1}{s+1}$ and $Z(s) = -\frac{2}{s(s+1)} = -\frac{2}{s} + \frac{2}{s+1}$, $\mathrm{Re}(s) > 0$. The solution to the initial value problem is $x(t) = y(t) = 2 - e^{-t}$, and $z(t) = -2 + 2e^{-t}$, $t \in \mathbb{R}$.

---

[10]It would be more precise to say that functions X, Y, and Z are determined by the conjunction of the three identities $sX(s) - 1 = 2X(s) - Y(s) + Z(s)$, $sY(s) - 1 = X(s) + Z(s)$, and $sZ(s) = X(s) + Z(s)$, $s \in D$. Writing and treating this conjunction as an algebraic system of equations simplifies our thinking and is an efficient way of determining the expressions for $X(s)$, $Y(s)$, and $Z(s)$.

Note: As before, we skip a step here. From $X(s) = Y(s) = \frac{2}{s} - \frac{1}{s+1}$ and $Z(s) = -\frac{2}{s} + \frac{2}{s+1}$, it follows that $\mathcal{L}^{-1}[X(s)](t) = \mathcal{L}^{-1}[Y(s)](t) = (2 - e^{-t})H(t)$ and $\mathcal{L}^{-1}[Z(s)](t) = (-2 + 2e^{-t})H(t)$. These $\mathcal{L}$-originals are not differentiable at $t = 0$, but they satisfy the given system of differential equations for $t \neq 0$. The unique solution to the given initial value problem is the triple of the functions $t \mapsto 2 - e^{-t}$, $t \in \mathbb{R}$, $t \mapsto 2 - e^{-t}$, $t \in \mathbb{R}$, and $t \mapsto -2 + 2e^{-t}$, $t \in \mathbb{R}$. To simplify our work, by convention, we denote by $x = x(t)$, $y = y(t)$, and $z = z(t)$, both the triple of the $\mathcal{L}$-originals $\mathcal{L}^{-1}[X(s)](t)$, $\mathcal{L}^{-1}[Y(s)](t)$, and $\mathcal{L}^{-1}[Z(s)](t)$, and the solution to the given initial value problem.

Problem 4.4.27. $\frac{dx}{dt} = 3x + z$, $\frac{dy}{dt} = 2x + y + z$, $\frac{dz}{dt} = -x + y + z$, $x(0) = 1$ $y(0) = z(0) = 0$.

Solution 4.4.27. From the initial conditions, $\mathcal{L}[x'(t)H(t)](s) = sX(s) - 1$, $\mathcal{L}[y'(t)H(t)](s) = sY(s)$, and $\mathcal{L}[z'(t)H(t)](s) = sZ(s)$. The functions $X(s)$, $Y(s)$, and $Z(s)$ satisfy the following linear system of algebraic equations:

$$(s - 3)X(s) - Z(s) = 1,$$

$$2X(s) - (s - 1)Y(s) + Z(s) = 0,$$

$$X(s) - Y(s) + (s - 1)Z(s) = 0.$$

It follows that, for $s \in D$,

$$X(s) = \frac{s^2 - 2s}{(s - 3)(s - 1)^2} = \frac{3}{4}\frac{1}{s - 3} + \frac{1}{4}\frac{1}{s - 1} + \frac{1}{2}\frac{1}{(s - 1)^2},$$

$$Y(s) = \frac{2s - 3}{(s - 3)(s - 1)^2} = \frac{3}{4}\frac{1}{s - 3} - \frac{3}{4}\frac{1}{s - 1} + \frac{1}{2}\frac{1}{(s - 1)^2},$$

$$Z(s) = -\frac{1}{(s - 1)^2}.$$

The solution to the initial value problem is $x(t) = \frac{3}{4}e^{3t} + \frac{1}{4}e^{t} + \frac{1}{2}te^{t}$, $y(t) = \frac{3}{4}e^{3t} - \frac{3}{4}e^{t} + \frac{1}{2}te^{t}$, and $z(t) = -te^{t}$, $t \in \mathbb{R}$.

Problem 4.4.28. $\frac{dx}{dt} = y - z$, $\frac{dy}{dt} = z - 2x$, $\frac{dz}{dt} = 2x - y$, $x(0) = y(0) = z(0) = 1$.

Solution 4.4.28. From the initial conditions, $\mathcal{L}[x'(t)H(t)](s) = sX(s) - 1$, $\mathcal{L}[y'(t)H(t)](s) = sY(s) - 1$, and $\mathcal{L}[z'(t)H(t)](s) = sZ(s) - 1$. The functions $X(s)$, $Y(s)$, and $Z(s)$ satisfy the following linear system of algebraic equations:

$$sX(s) - Y(s) + Z(s) = 1, \quad 2X(s) + sY(s) - Z(s) = 1,$$

$$-2X(s) + Y(s) + sZ(s) = 1.$$

It follows that, for $s \in D$,

$$X(s) = \frac{s^2 + 3}{s(s^2 + 5)} = \frac{3}{5}\frac{1}{s} + \frac{2}{5}\frac{s}{s^2 + 5},$$

$$Y(s) = \frac{s^2 - s + 6}{s(s^2 + 5)} = \frac{6}{5}\frac{1}{s} - \frac{1}{5}\frac{s}{s^2 + 5} - \frac{1}{s^2 + 5},$$

$$Z(s) = \frac{s^2 + s + 6}{s(s^2 + 5)} = \frac{6}{5}\frac{1}{s} - \frac{1}{5}\frac{s}{s^2 + 5} + \frac{1}{s^2 + 5}.$$

The solution to the initial value problem is $x(t) = \frac{3}{5} + \frac{2}{5}\cos t\sqrt{5}$, $y(t) = \frac{6}{5} - \frac{1}{5}\cos t\sqrt{5} - \frac{1}{\sqrt{5}}\sin t\sqrt{5}$, and $z(t) = \frac{6}{5} - \frac{1}{5}\cos t\sqrt{5} + \frac{1}{\sqrt{5}}\sin t\sqrt{5}$, $t \in \mathbb{R}$.

Problem 4.4.29. $\frac{d^2x}{dt^2} + 2y = 0$, $\frac{d^2y}{dt^2} + 2x = 0$, $x'(0) = 1$, $x(0) = y(0) = y'(0) = 0$.

Solution 4.4.29. From the initial conditions, $\mathcal{L}[x''(t)H(t)](s) = s^2 X(s) - 1$ and $\mathcal{L}[y''(t)H(t)](s) = s^2 Y(s)$. The functions $X(s)$ and $Y(s)$ satisfy the following linear system of algebraic equations

$$s^2 X(s) + 2Y(s) = 1, \quad 2X(s) + s^2 Y(s) = 0.$$

It follows that, $s \in D$,

$$X(s) = \frac{s^2}{s^4 - 4} = \frac{1}{2}\frac{1}{s^2 - 2} + \frac{1}{2}\frac{1}{s^2 + 2},$$

$$Y(s) = -\frac{2}{s^4 - 4} = \frac{1}{2}\frac{1}{s^2 + 2} - \frac{1}{2}\frac{1}{s^2 - 2}.$$

The solution to the initial value problem is $x(t) = \frac{1}{2\sqrt{2}}(\sinh t\sqrt{2} + \sin t\sqrt{2})$ and $y(t) = \frac{1}{2\sqrt{2}}(\sin t\sqrt{2} - \sinh t\sqrt{2})$, $t \in \mathbb{R}$.

Problem 4.4.30. $\frac{d^2x}{dt^2} = x - y - z$, $\frac{d^2y}{dt^2} = -x + y - z$, $\frac{d^2z}{dt^2} = -x - y + z$, $x(0) = 1$, $x'(0) = y(0) = y'(0) = z(0) = z'(0) = 0$.

Solution 4.4.30. From the initial conditions, $\mathcal{L}[x''(t)H(t)](s) = s^2 X(s) - s$, $\mathcal{L}[y''(t)H(t)](s) = s^2 Y(s)$, and $\mathcal{L}[z''(t)H(t)](s) = s^2 Z(s)$. The functions $X(s)$, $Y(s)$, and $Z(s)$ satisfy the following linear system of algebraic equations:

$$(s^2 - 1)X(s) + Y(s) + Z(s) = s,$$
$$X(s) + (s^2 - 1)Y(s) + Z(s) = 0,$$
$$X(s) + Y(s) + (s^2 - 1)Z(s) = 0.$$

It follows that, for $s \in D$,

$$X(s) = \frac{s^3}{(s^2 + 1)(s^2 - 2)} = \frac{1}{3}\frac{s}{s^2 + 1} + \frac{2}{3}\frac{s}{s^2 - 2},$$

$$Y(s) = Z(s) = -\frac{s}{(s^2 + 1)(s^2 - 2)} = \frac{1}{3}\frac{s}{s^2 + 1} - \frac{1}{3}\frac{s}{s^2 - 2}.$$

The solution to the initial value problem is $x(t) = \frac{1}{3}(\cos t + 2\cosh t\sqrt{2})$ and $y(t) = z(t) = \frac{1}{3}(\cos t - \cosh t\sqrt{2})$, $t \in \mathbb{R}$.

Problem 4.4.31. Let $a > 0$ and let the triple of the functions $(x(t), y(t), z(t))$, $t \in \mathbb{R}$, be a solution to the linear system of differential equations

$$\frac{dx}{dt} = -x, \quad \frac{dy}{dt} = -ax - y + az, \quad \frac{dz}{dt} = -3x - ay - z.$$

Prove the following claims:

(a) $\lim_{t\to\infty} x(t) = \lim_{t\to\infty} y(t) = \lim_{t\to\infty} z(t) = 0$.

(b) If the point $(x(0), y(0), z(0))$ lies on the line $3x + ay = 0$, $x = z$, then for any $\xi > 0$, the point $(x(\xi), y(\xi), z(\xi))$ also lies on the line.

Solution 4.4.31. From the first equation, it follows that $x(t) = \alpha e^{-t}$, for some $\alpha \in \mathbb{R}$. This implies that the function $y(t)$ and $z(t)$ satisfy the following system of differential equations:

$$\frac{dy}{dt} = -a\alpha e^{-t} - y + az, \quad \frac{dz}{dt} = -3\alpha e^{-t} - ay - z.$$

Let $\beta, \gamma \in \mathbb{R}$ be such that $y(0) = \beta$ and $z(0) = \gamma$, and let $Y(s) = \mathcal{L}[y(t)H(t)](s)$ and $Z(s) = \mathcal{L}[z(t)H(t)](s)$. Then, $\mathcal{L}[y'(t)H(t)](s) = sY(s) - \beta$ and $\mathcal{L}[z'(t)]H(t)(s) = sZ(s) - \gamma$. The functions $Y(s)$ and $Z(s)$ satisfy the following linear system of algebraic equations:

$$(s+1)Y(s) - aZ(s) = \beta - \frac{a\alpha}{s+1}, \quad aY(s) + (s+1)Z(s) = \gamma - \frac{3\alpha}{s+1}.$$

It follows that

$$Y(s) = \frac{(\gamma - \alpha)a}{(s+1)^2 + a^2} + \frac{3\alpha + a\beta}{a} \frac{s+1}{(s+1)^2 + a^2} - \frac{3\alpha}{a} \frac{1}{s+1},$$

$$Z(s) = \frac{\alpha}{s+1} - \frac{3\alpha + a\beta}{a} \frac{a}{(s+1)^2 + a^2} + \frac{(\gamma - \alpha)(s+1)}{(s+1)^2 + a^2}.$$

Therefore, $y(t) = \frac{e^{-t}}{a}(a(\gamma - \alpha)\sin at + (3\alpha + a\beta)\cos at - 3\alpha)$ and $z(t) = \frac{e^{-t}}{a}(a\alpha - (3\alpha + a\beta)\sin at + a(\gamma - \alpha)\cos at)$.

(a) By the squeeze theorem, for any $\alpha, \beta, \gamma \in \mathbb{R}$, $\lim\limits_{t \to \infty} x(t) = \lim\limits_{t \to \infty} y(t) = \lim\limits_{t \to \infty} z(t) = 0$.

(b) Observe that if $3\alpha + a\beta = 0$ and $\alpha = \gamma$, then $x(t) = \alpha e^{-t}$, $y(t) = -\frac{3\alpha e^{-t}}{a}$, and $z(t) = \alpha e^{-t}$. Hence, for any $\xi > 0$, $x(\xi) = \alpha e^{-\xi} = z(\xi)$ and $3x(\xi) + ay(\xi) = 3\alpha e^{-\xi} - 3\alpha e^{-\xi} = 0$.

Problem 4.4.32.

$\frac{dx}{dt} + y = H(t) - H(t-1)$, $\frac{dy}{dt} + x = H(t) - H(t-2)$, $x(0) = y(0) = 0$.

Solution 4.4.32. From the initial conditions, $\mathcal{L}[x'(t)H(t)](s) = sX(s)$ and $\mathcal{L}[y'(t)H(t)](s) = sY(s)$. Functions $X(s)$ and $Y(s)$ satisfy the following linear system of algebraic equations:

$$sX(s) + Y(s) = \frac{1}{s}(1 - e^{-s}), \quad X(s) + sY(s) = \frac{1}{s}(1 - e^{-2s}).$$

It follows that $X(s) = \frac{1}{s} - \frac{1}{s+1} - \frac{1}{s^2-1}e^{-s} - \left(\frac{1}{s} - \frac{s}{s^2-1}\right)e^{-2s}$ and $Y(s) = \frac{1}{s} - \frac{1}{s+1} - \left(\frac{1}{s} - \frac{s}{s^2-1}\right)e^{-s} - \frac{1}{s^2-1}e^{-2s}$. The solution to the initial value problem is $x(t) = H(t)(1 - e^{-t}) - H(t-1)\sinh(t-1) - H(t-2)(1 - \cosh(t-2))$ and $y(t) = H(t)(1 - e^{-t}) - H(t-1)(1 - \cosh(t-1)) - H(t-2)\sinh(t-2)$, $t \in \mathbb{R}$.

Problem 4.4.33.

$$\frac{dx}{dt} - \frac{dy}{dt} - 2x + 2y = 1 - 2t, \quad \frac{d^2x}{dt^2} + 2\frac{dy}{dt} + x = 0, \quad x(0) = y(0) = x'(0) = 0.$$

Solution 4.4.33. From the initial conditions, $\mathcal{L}[x''(t)H(t)](s) = s^2 X(s)$, $\mathcal{L}[x'(t)H(t)](s) = sX(s)$, and $\mathcal{L}[y'(t)H(t)](s) = sY(s)$. The functions $X(s)$ and $Y(s)$ satisfy the following linear system of algebraic equations:

$$X(s) - Y(s) = \frac{1}{s^2}, \quad (s^2 + 1)X(s) + 2sY(s) = 0.$$

It follows that $X(s) = \frac{2}{s} - \frac{2}{s+1} - \frac{2}{(s+1)^2}$ and $Y(s) = X(s) - \frac{1}{s^2} = \frac{2}{s} - \frac{1}{s^2} - \frac{2}{s+1} - \frac{2}{(s+1)^2}$. The solution to the initial value problem is $x(t) = 2 - 2(1 + t)e^{-t}$ and $y(t) = 2 - t - 2(1 + t)e^{-t}$, $t \in \mathbb{R}$.

Problem 4.4.34. $\frac{dx}{dt} = 3y - x$, $\frac{dy}{dt} = x + y + e^{2t}$, $x(0) = y(0) = 1$, $a > 0$.

Solution 4.4.34. From the initial conditions, $\mathcal{L}[x'(t)H(t)](s) = sX(s) - 1$ and $\mathcal{L}[y'(t)H(t)](s) = sY(s) - 1$. The functions $X(s)$ and $Y(s)$ satisfy the following linear system of algebraic equations:

$$(s+1)X(s) - 3Y(s) = 1, \quad -X(s) + (s-1)Y(s) = 1 + \frac{1}{s-2}.$$

It follows that $X(s) = \frac{13}{16}\frac{1}{s-2} + \frac{3}{16}\frac{1}{s+2} + \frac{3}{4}\frac{1}{(s-2)^2}$ and $Y(s) = \frac{17}{16}\frac{1}{s-2} - \frac{1}{16}\frac{1}{s+2} + \frac{3}{4}\frac{1}{(s-2)^2}$. The solution to the initial value problem is $x(t) = \frac{1}{16}(13e^{2t} + 3e^{-2t} + 12te^{2t})$ and $y(t) = \frac{1}{16}(17e^{2t} - e^{-2t} + 12te^{2t})$, $t \in \mathbb{R}$.

Problem 4.4.35. $\frac{d^2x}{dt^2} = y$, $\frac{d^2y}{dt^2} = x + H(t) - H(t-a)$, $a > 0$, $x(0) = x'(0) = y(0) = y'(0) = 0$.

Solution 4.4.35. From the initial conditions, $\mathcal{L}[x''(t)H(t)](s) = s^2 X(s)$ and $\mathcal{L}[y''(t)H(t)](s) = s^2 Y(s)$. The functions $X(s)$ and $Y(s)$

satisfy the following linear system of algebraic equations:

$$s^2 X(s) - Y(s) = 0, \quad -X(s) + s^2 Y(s) = \frac{1}{s}(1 - e^{-as}).$$

It follows that $X(s) = \frac{1-e^{-as}}{s(s^4-1)} = \left(-\frac{1}{s} + \frac{1}{2}\frac{s}{s^2-1} + \frac{1}{2}\frac{s}{s^2+1}\right)(1 - e^{-as})$ and $Y(s) = \frac{s(1-e^{-as})}{s^4-1} = \left(\frac{1}{2}\frac{s}{s^2-1} - \frac{1}{2}\frac{s}{s^2+1}\right)(1 - e^{-as})$. The solution to the initial value problem is $x(t) = \frac{H(t)}{2}(-2 + \cosh t + \cos t) - \frac{H(t-a)}{2}(-2 + \cosh(t-a) + \cos(t-a))$ and $y(t) = \frac{H(t)}{2}(\cosh t - \cos t) - \frac{H(t-a)}{2}(\cosh(t-a) - \cos(t-a))$, $t \in \mathbb{R}$.

Note: Observe that $x(t)$ and $y(t)$ are indeed differentiable at $t = 0$ with $x'(0) = y'(0) = 0$.

4.4.3 *Integral equations*

Use Laplace transformation and the convolution theorem, Problem 4.3.38, to solve the following integral equations and systems of integral equations for the unknown functions $x = x(t)$ and $y = y(t)$. Write $X(s) = \mathcal{L}[x(t)H(t)](s)$ and $Y(s) = \mathcal{L}[y(t)H(t)](s)$, $s \in D = \{z \in \mathbb{C} : \mathrm{Re}(z) > \alpha\}$, where the number α depends on the given equation.

Problem 4.4.36. $x(t) = t + 2\int_0^t \sin(t-u)x(u)\, du.$

Solution 4.4.36. Taking Laplace transforms of both sides of the integral equation yields the identity $X(s) = \frac{1}{s^2} + \frac{2}{s^2+1}X(s)$, $s \in D$. It follows that $X(s) = \frac{s^2+1}{s^2(s^2-1)} = \frac{2}{s^2-1} - \frac{1}{s^2}$ and $x(t) = 2\sinh t - t$, $t \in \mathbb{R}$.

Problem 4.4.37. $\int_0^t x(u)x(t-u)\, du = 2x(t) + \frac{t^{2\alpha-1}}{\Gamma(2\alpha)} - \frac{2t^{\alpha-1}}{\Gamma(\alpha)}$, $\alpha \geq 2$.

Solution 4.4.37. Taking Laplace transforms of both sides of the integral equation yields the identity $(X(s))^2 - 2X(s) = \frac{1}{s^{2\alpha}} - \frac{2}{s^\alpha}$, $s \in D$. By adding 1 to both sides of this identity, we obtain the identity $(X(s)-1)^2 = \left(\frac{1}{s^\alpha} - 1\right)^2$, $s \in D$. Hence, $X(s) = \frac{1}{s^\alpha}$ or $X(s) = 2 - \frac{1}{s^\alpha}$. We eliminate the later case because of $\lim\limits_{s\to\infty}\left(2 - \frac{1}{s^\alpha}\right) = 2 \neq 0$. It follows that $x(t) = \frac{t^{\alpha-1}}{\Gamma(\alpha)}$, $t \geq 0$.

Problem 4.4.38. $x(t) = 1 - \int_0^t y(u)e^{2(t-u)}\,du$, $y(t) = t + \int_0^t x(u)\,du$.

Solution 4.4.38. Taking Laplace transforms yields the linear system of algebraic equations $X(s) = \frac{1}{s} - \frac{1}{s-2}Y(s)$ and $Y(s) = \frac{1}{s^2} + \frac{1}{s}X(s)$. It follows that $X(s) = -\frac{1}{s} + \frac{2}{s-1} - \frac{2}{(s-1)^2}$ and $Y(s) = 2\left(-\frac{2}{s} + \frac{2}{s-1} - \frac{1}{(s-1)^2}\right)$. Hence, $x(t) = -1 + 2e^t - 2te^t$ and $y(t) = 2(-2 + 2e^t - te^t)$, $t \in \mathbb{R}$.

Problem 4.4.39. $x(t) = e^t + \int_0^t x(u)\,du + \int_0^t e^{t-u}y(u)\,du$, $y(t) = -t + \int_0^t (t-u)x(u)\,du + \int_0^t y(u)\,du$.

Solution 4.4.39. Taking Laplace transforms yields the linear system of algebraic equations $X(s) = \frac{1}{s-1} + \frac{1}{s}X(s) + \frac{1}{s-1}Y(s)$ and $Y(s) = -\frac{1}{s^2} + \frac{1}{s^2}X(s) + \frac{1}{s}Y(s)$. It follows that $X(s) = \frac{s^2-s-1}{(s-2)(s^2-s+1)} = \frac{1}{3}\frac{1}{s-2} + \frac{2}{3}\frac{s+1}{(s-\frac{1}{2})^2+\frac{3}{4}}$ and $Y(s) = -\frac{s^2-3s+1}{s(s-2)(s^2-s+1)} = \frac{1}{2}\frac{1}{s} + \frac{1}{6}\frac{1}{s-2} - \frac{2}{3}\frac{s+1}{(s-\frac{1}{2})^2+\frac{3}{4}}$. Hence, $x(t) = \frac{1}{3}\left(e^{2t} + 2e^{\frac{t}{2}}\left(\cos\frac{t\sqrt{3}}{2} + \sqrt{3}\sin\frac{t\sqrt{3}}{2}\right)\right)$ and $y(t) = \frac{1}{6}\left(3 + e^{2t} - 4e^{\frac{t}{2}}\left(\cos\frac{t\sqrt{3}}{2} + \sqrt{3}\sin\frac{t\sqrt{3}}{2}\right)\right)$, $t \in \mathbb{R}$.

Problem 4.4.40. $x'(t) = \sin at + \int_0^t x(u)\,du$, $x(0) = 1$, $a > 0$.

Solution 4.4.40. Taking Laplace transforms yields the identity $sX(s) - 1 = \frac{a}{s^2+a^2} + \frac{1}{s}X(s)$, $s \in D$. It follows that $X(s) = \frac{a^2+a+1}{a^2+1}\frac{s}{s^2-1} - \frac{a}{a^2+1}\frac{s}{s^2+a^2}$ and $x(t) = \frac{1}{a^2+1}((a^2+a+1)\cosh t - a\cos at)$, $t \in \mathbb{R}$.

Note: Observe that $x'(0) = 0$ and notice that the problem may be reduced to solving the initial value problem $x''(t) - x(t) = a\cos at$, $x(0) = 1$, $x'(0) = 0$.

Problem 4.4.41. $x'(t) = \cos at + \int_0^t (x'''(u) - x(u))e^{t-u}\,du$, $x(0) = x''(0) = 0$, $x'(0) = 1$, $a > 0$.

Solution 4.4.41. Taking Laplace transforms yields the algebraic equation $sX(s) = \frac{s}{s^2+a^2} + \frac{1}{s-1}(s^3X(s) - s - X(s))$. It follows that $X(s) = \frac{1}{2}\frac{1}{s-1} - \frac{1}{2}\frac{s-1}{s^2+1} - \frac{s}{(s^2+1)(s^2+a^2)}$. If $a = 1$, then $X(s) = \frac{1}{2}\frac{1}{s-1} - \frac{1}{2}\frac{s-1}{s^2+1} + \frac{1}{2}\frac{d}{ds}\left(\frac{1}{s^2+1}\right)$, and $x(t) = \frac{1}{2}(e^t - \cos t + \sin t - t\sin t)$. If

$a \neq 1$, then $X(s) = \frac{1}{2}\frac{1}{s-1} - \frac{1}{2}\frac{s-1}{s^2+1} + \frac{1}{1-a^2}\left(\frac{s}{s^2+1} - \frac{s}{s^2+a^2}\right)$ and
$x(t) = \frac{1}{2(1-a^2)}((1-a^2)(e^t + \sin t) + (1+a^2)\cos t - 2\cos at)$, $t \in \mathbb{R}$.

Note: Since the given equation is equivalent to the equation $(x'(t) - \cos at)e^{-t} = \int_0^t (x'''(u) - x(u))e^{-u}\, du$, the problem may be reduced to solving the initial value problem $x'''(t) - x''(t) + x'(t) - x(t) = a\sin at + \cos at$, $x(0) = x''(0) = 0$, $x'(0) = 1$.

Chapter 5

Vector Analysis

5.1 Introduction

Use the following definitions, techniques, properties, and algorithms to solve the problems contained in this chapter. For more details, see [13, 20, 23, 27, 33].

Hodograph of a Vector Function. For a given Cartesian coordinate system in $\mathbb{E}^3$, the *hodograph* of a vector function $\vec{a}(t) = \langle a_1(t), a_2(t), a_3(t) \rangle$, $t \in I \subseteq \mathbb{R}$, is the set of points $\Gamma = \{a(t) = (a_1(t), a_2(t), a_3(t)) \in \mathbb{E}^3 : t \in I\}$.[1] The hodograph of a vector function $\vec{a}(u, v) = \langle a_1(u, v), a_2(u, v), a_3(u, v) \rangle$, $(u, v) \in D \subseteq \mathbb{R}^2$ is the set of points $S = \{(a_1(u, v), a_2(u, v), a_3(u, v)) \in \mathbb{E}^3 : (u, v) \in D\}$.

Curve. A set $\Gamma \subseteq \mathbb{E}^3$ is a *curve* if there is an interval I and a continuous vector function $\vec{a} : I \to \vec{\mathbb{E}}^3$ such that Γ is its hodograph in a certain coordinate system.[2] We say that the vector function $\vec{a}$ is a *parametrisation* of the curve Γ. If $\vec{a}$ is one-to-one in I, except possibly at the end points of I, then Γ is a *simple curve*. If $I = [\alpha, \beta]$ and $\vec{a}(\alpha) = \vec{a}(\beta)$, then Γ is a *closed curve*. A *planar curve* is one that lies in a plane. A simple curve Γ is a *smooth curve* if it is the hodograph of

---

[1] The point $a(t) = (a_1(t), a_2(t), a_3(t))$ is the terminal point of the vector $\vec{a}(t)$.

[2] Additional conditions are required to guarantee that the hodograph Γ matches our intuitive perception of a curve as the trajectory of a moving point. This is commonly studied in a differential geometry course.

a continuously differentiable vector function $\vec{a} : I \to \mathbb{E}^3$ such that $|\vec{a}'(t)| \neq 0$, $t \in I$. A *piecewise smooth curve* is a curve that can be partition into finitely many smooth curves.

Surface. A set $S \subseteq \mathbb{E}^3$ is a *surface* if there is a connected set $D \subseteq \mathbb{R}^2$ and a continuous vector function $\vec{a} : D \to \vec{\mathbb{E}}^3$ such that S is its hodograph in a certain coordinate system. We say that the vector function $\vec{a}$ is a *parametrisation* of the surface S. If the vector function $\vec{a} : D \to \mathbb{E}^3$ is continuously differentiable with respect to both variables and such that $\left| \frac{\partial \vec{a}}{\partial u} \times \frac{\partial \vec{a}}{\partial v} \right| \neq$ $\vec{0}$, $(u, v) \in D$, then its hodograph is a *smooth surface*. A *piecewise smooth surface* is a surface that can be partition into finitely many smooth surfaces.

Arc Length. Let the smooth curve Γ be the hodograph of the continuously differentiable vector function $\vec{r}(t) = \langle x(t), y(t), z(t) \rangle$, $t \in [a, b]$. The length of Γ is

$$L = \int_a^b |\vec{r}'(t)|\, dt = \int_a^b \sqrt{\left(\frac{dx}{dt}\right)^2 + \left(\frac{dy}{dt}\right)^2 + \left(\frac{dz}{dt}\right)^2}\, dt.$$

If $A = r(a)$, then the point $S = r(u) \in \Gamma$, $u \in [a, b]$ is uniquely determined by $s = \int_a^u \sqrt{\left(\frac{dx}{dt}\right)^2 + \left(\frac{dy}{dt}\right)^2 + \left(\frac{dz}{dt}\right)^2}\, dt$, i.e., by the length of the arc of Γ with the initial point A and the terminal point S. Using s, the arc length, to obtain Γ as the hodograph of the vector function $\vec{R} = \vec{R}(s)$, $s \in [0, L]$, is commonly called the *natural parametrisation* of the curve.

Tangent Vector and Tangent Line. If a curve Γ is the hodograph of the vector function $t \mapsto \vec{a}(t)$, $t \in I \subseteq \mathbb{R}$, and if $t_0 \in I$ is such that $\vec{a}'(t_0)$ exists and is not the zero vector, then the *unit tangent vector* to Γ at the point $a(t_0) \in \Gamma$ is defined as $\mathbf{T}(t_0) = \frac{\vec{a}'(t_0)}{|\vec{a}'(t_0)|}$. The *tangent line* to Γ at the point $a(t_0) \in \Gamma$ is the Euclidean line that is the hodograph of the vector function $t \mapsto \vec{a}(t_0) + t\vec{a}'(t_0)$, $t \in \mathbb{R}$. If $\vec{a}(t) = \langle a_1(t), a_2(t), a_3(t) \rangle$, then the tangent line at the point $a(t_0) = (a_1(t_0), a_2(t_0), a_3(t_0)) \in \Gamma$ is determined by the equations $x(t) = a_1(t_0) + ta_1'(t_0)$, $y(t) = a_2(t_0) + ta_2'(t_0)$, and $z(t) = a_3(t_0) + ta_3'(t_0)$, $t \in \mathbb{R}$, or by the system of equations

$$\frac{x - a_1(t_0)}{a_1'(t_0)} = \frac{y - a_2(t_0)}{a_2'(t_0)} = \frac{z - a_3(t_0)}{a_3'(t_0)}.$$

Curvature. The curvature κ of the curve is the magnitude of the change of the unit tangent vector with respect to the arc length, $\kappa = |\frac{d\mathbf{T}}{ds}|$, where s is the arc length. The curvature at a point of the curve that is the hodograph of a vector function $\vec{r} = \vec{r}(t)$ is evaluated as $\kappa(t) = \frac{|\vec{r}'(t)\times\vec{r}''(t)|}{|\vec{r}'(t)|^3}$. The curvature of a curve that is the graph of a function $y = f(x)$, $x \in I$, at the point $(x, f(x))$ is $\kappa(x) = \frac{|f''(x)|}{[1+(f'(x))^2]^{3/2}}$.

Normal and Binormal Vectors. If a smooth curve Γ is a hodograph of the vector function $\vec{r} = \vec{r}(t)$, $t \in I$, then the *unit normal vector* to the curve Γ at the point $r(t)$ is defined as $\mathbf{N}(t) = \frac{\mathbf{T}(t)}{|\mathbf{T}'(t)|}$. The *binormal vector* to the curve Γ at the point $r(t)$ is defined as $\mathbf{B}(t) = \mathbf{T}(t) \times \mathbf{N}(t)$.

Normal and Osculating Planes. If a smooth curve Γ is the hodograph of the vector function $\vec{r} = \vec{r}(t)$, $t \in I$, then the *normal plane* to the curve Γ at the point $r(t_0)$, $t_0 \in I$, is defined as the Euclidean plane that is the hodograph of the vector function $\vec{R} = \vec{R}(u, v)$, $(u, v) \in \mathbb{R}^2$, such that $\vec{r}'(t_0) \cdot (\vec{R}(u, v) - \vec{r}(t_0)) = 0$, for all $(u, v) \in \mathbb{R}^2$. The *osculating plane* to the curve Γ at the point $r(t_0)$, $t_0 \in I$, is defined as the plane that is the hodograph of the vector function $\vec{R} = \vec{R}(u, v)$, $(u, v) \in \mathbb{R}^2$, such that $\mathbf{B}(t_0) \cdot (\vec{R}(u, v) - \vec{r}(t_0)) = 0$, for all $(u, v) \in \mathbb{R}^2$.

Gradient, Divergence, and Curl. Let $D \subseteq \mathbb{E}^3$ and let $Oxyz$ be a Cartesian coordinate system in $\mathbb{E}^3$. The *gradient* of a scalar field $f : D \to \mathbb{R}$ is the vector field $\nabla f = \text{grad } f = \left\langle \frac{\partial f}{\partial x}, \frac{\partial f}{\partial y}, \frac{\partial f}{\partial z} \right\rangle : D \to \vec{\mathbb{E}}^3$.

The *divergence* of a vector field $\vec{a} = \langle a_1, a_2, a_3 \rangle : D \to \vec{\mathbb{E}}^3$ is the scalar field

$$\nabla \cdot \vec{a} = \text{div } \vec{a} = \frac{\partial a_1}{\partial x} + \frac{\partial a_2}{\partial y} + \frac{\partial a_3}{\partial z} : D \to \mathbb{R}.$$

The curl of a vector field $\vec{a} = \langle a_1, a_2, a_3 \rangle : D \to \vec{\mathbb{E}}^3$, is the vector field

$$\nabla \times \vec{a} = \text{curl } \vec{a} = \begin{vmatrix} \vec{i} & \vec{j} & \vec{k} \\ \frac{\partial}{\partial x} & \frac{\partial}{\partial y} & \frac{\partial}{\partial z} \\ a_1 & a_2 & a_3 \end{vmatrix} : D \to \vec{\mathbb{E}}^3.$$

Note: For $A \in D$, the length and direction of vectors $(\nabla f)(A)$ and $(\nabla \times \vec{a})(A)$, and the number $(\nabla \cdot \vec{a})(A)$ are independent of the particular choice of Cartesian coordinate system in $\mathbb{E}^3$. See, for example, [20].

Directional Derivative. The directional derivative of a scalar field $f : D \to \mathbb{R}$, $D \subseteq \mathbb{E}^3$, along a unit vector $\vec{a}$ at a point $A \in D$ is the number $D_{\vec{a}}f(A) = \vec{a} \cdot \nabla f(A)$. If $Oxyz$ is a Cartesian coordinate system in $\mathbb{E}^3$, then, for $\vec{a} = \langle a_1, a_2, a_3 \rangle$ and $A = (x, y, z)$, $D_{\vec{a}}f(A) = \lim_{h \to 0} \frac{1}{h} \left(f(x + ha_1, y + ha_2, z + ha_3) - f(x, y, z) \right)$. The directional derivative measures the instantaneous rate of change of the function f at the point A in the direction of $\vec{a}$.

Conservative, Solenoidal, and Laplacian Fields. A vector field $\vec{F} : D \to \vec{\mathbb{E}}^3$, $D \subseteq \mathbb{E}^3$, is called a *conservative vector field* if it is the gradient of a scalar field, i.e., if there exists a function $f : D \to \mathbb{R}$ such that $\vec{F} = \mathrm{grad} f$. The field f is called a *scalar potential* for $\vec{F}$. If D is a simple connected region, then $\vec{F} : D \to \vec{\mathbb{E}}^3$ is a conservative vector field if and only if curl $\vec{F} = \vec{0}$.

A vector field $\vec{F} : D \to \vec{\mathbb{E}}^3$, $D \subseteq \mathbb{E}^3$, is called a *solenoidal vector field* or an *incompressible vector field* if div $\vec{F} = 0$.

A vector field $\vec{F} : D \to \vec{\mathbb{E}}^3$, $D \subseteq \mathbb{E}^3$, is called a *Laplacian vector field* if it is both conservative and solenoidal.

Line Integral of a Scalar Field. Let a smooth curve C be the hodograph of the vector function $\vec{r}(t) = \langle x(t), y(t), z(t) \rangle$, $t \in [a, b]$. Let $D \subseteq \mathbb{E}^3$ be a region that contains C. If $f : D \to \mathbb{R}$ is a continuous scalar field, then the *line integral* of f along C is

$$\int_C f(A) ds = \int_a^b f(r(t)) \, |\vec{r}\,'(t)| dt$$

$$= \int_a^b f(x(t), y(t), z(t)) \sqrt{\left(\frac{dx}{dt}\right)^2 + \left(\frac{dy}{dt}\right)^2 + \left(\frac{dz}{dt}\right)^2} \, dt.$$

The line integral of a scalar field is sometimes called the *line integral of the first kind*. The mass of the curve C with a linear density of $\rho : C \to \mathbb{R}^+$ is defined as $m = \int_C \rho(A) \, ds$.

Line Integral of a Vector Field. Let a smooth curve C be the hodograph of the vector function $\vec{r}(t) = \langle x(t), y(t), z(t) \rangle$, $t \in$

$[a, b]$, oriented[3] in the sense of increase of the variable t.[4] Let $D \subseteq \mathbb{E}^3$ be a region that contains C. If $\vec{F} : D \to \vec{\mathbb{E}}^3$ is a continuous vector field, then the *line integral* of $\vec{F}$ along C is

$$\int_C \vec{F} \cdot d\vec{r} = \int_a^b \vec{F}(r(t)) \cdot \vec{r}\,'(t)\ dt = \int_C \vec{F} \cdot \mathbf{T}\ ds.$$

If $\vec{F}(x, y, z) = \langle P(x, y, z), Q(x, y, z), R(x, y, z) \rangle$, $(x, y, z) \in D$, then

$$\int_C \vec{F} \cdot d\vec{r} = \int_C P\ dx + Q\ dy + R\ dz$$

$$= \int_a^b (P(x(t), y(t), z(t))x'(t) + Q(x(t), y(t), z(t))y'(t)$$

$$+ R(x(t), y(t), z(t))z'(t))\ dt.$$

The line integral of a vector field is sometimes called the *line integral of the second kind*. The work of the vector field $\vec{F}$ along the oriented curve C is defined as $W = \int_C \vec{F} \cdot d\vec{r}$.

Fundamental Theorem for Line Integrals. Let a smooth curve C be the hodograph of the vector function $\vec{r} = \vec{r}(t)$, $t \in [a, b]$, oriented in the sense of increase of the variable t. Let $D \subseteq \mathbb{E}^3$ be a region that contains C. If $f : D \to \mathbb{R}$ is a continuous scalar field such that its gradient $\nabla f : D \to \vec{\mathbb{E}}^3$ is a continuous vector field, then $\int_C \nabla f \cdot d\vec{r} = f(r(b)) - f(r(a))$ [33].

Surface Integral of a Scalar Field. Let a smooth surface S be the hodograph of the vector function $\vec{r}(u, v) = \langle x(u, v)y, y(u, v), z(u, v) \rangle$, $(u, v) \in D \subseteq \mathbb{R}^2$. Let $V \subseteq \mathbb{E}^3$ be a region that contains S. If $f : V \to \mathbb{R}$ is a continuous scalar field, then the *surface integral* of f over S is $\iint_S f(A)\ dS = \iint_D f(r(u, v))\ |\vec{r}_u \times \vec{r}_v|\ dA$. The surface integral of a scalar field is sometimes called the *surface integral of the first kind*. The mass of the surface S with a linear density of $\rho : S \to \mathbb{R}^+$ is defined as $m = \iint_S \rho(A)\ dS$.

---

[3] For all the details of the mathematical notion of an oriented curve, see [3].

[4] In this orientation of the curve C, if $a \leq t < t' \leq b$, the point $r(t)$ comes *before* the point $r(t')$. We say that $r(a)$ is the initial point and that $r(b)$ is the terminal point of C. If we think about C as a trajectory of a moving particle, this orientation corresponds to the motion of the particle starting at the point $r(a)$.

Surface Integral of a Vector Field. Let a smooth surface S be oriented[5] with the field of unit normal vectors $\vec{n} = \vec{n}(u, v)$, $(u, v) \in D \subseteq \mathbb{R}^2$. Let $V \subseteq \mathbb{E}^3$ be a region that contains S. If $\vec{F} : V \to \vec{\mathbb{E}}^3$ is a continuous vector field, then the *surface integral* of $\vec{F}$ over S is $\iint_S \vec{F} \cdot d\vec{S} = \iint_S \vec{F} \cdot \vec{n} \, dS$. The surface integral of a vector field is sometimes called the *surface integral of the second kind*. The *flux* of the vector field $\vec{F}$ through the oriented surface S is defined as $\Phi = \iint_S \vec{F} \cdot d\vec{S}$.

Green's Theorem. Let C be a positively oriented,[6] piecewise smooth, simple closed curve in the plane, and let D be the region bounded by C. If P and Q have continuous partial derivatives in an open region that contains D, then $\int_C P \, dx + Q \, dy = \iint_D \left(\frac{\partial Q}{\partial x} - \frac{\partial P}{\partial y} \right) dA$ [33].[7]

Divergence Theorem. Let E be a bounded, simple solid region, and let S be the boundary surface of E with outward orientation.[8] If $\vec{F}$ is a vector field whose component functions have continuous partial derivatives in an open region that contains E, then $\iint_S \vec{F} \cdot d\vec{S} = \iiint_E \operatorname{div} \vec{F} \, dV$ [33].

Stokes' Theorem. Let S be a piecewise smooth, orientable surface whose boundary $\partial S = C$ is a piecewise smooth curve. Let $\vec{n}$ be a unit vector normal to S, and let C be traversed with respect to $\vec{n}$ according to the right-hand rule.[9] If a vector field $\vec{F}(x, y, z)$ is defined and has continuous first-order partial derivatives in a region containing S, then $\iint_S (\operatorname{curl} \vec{F}) \cdot \vec{n} \, dS = \oint_C \vec{F} \cdot d\vec{C}$ [33].[10]

---

[5]A surface S is orientable if there exists a *continuous vector function* $\vec{n} : S \to \mathbb{E}^3 \backslash \{\vec{0}\}$ which assigns to all points $P \in S$ (not on the boundary) a normal vector to the tangent plane of S at P. The choice of $\vec{n}$ determines the *side* of S.

[6]If we visualise the curve as the trace of a moving particle, from the viewer's point of view, the particle is moving counterclockwise.

[7]George Green (1793–1841), British mathematical physicist.

[8]The *outward orientation* means that the unit normal vector field we choose is pointing towards the complement of the region E. The opposite of outward orientation of S is called *inward orientation*.

[9]This means that, seen from the tip of the normal vector $\vec{n}$, the orientation of the boundary curve is positive, i.e., counterclockwise. We will say that the orientation of the boundary curve is the *orientation induced by $\vec{n}$*.

[10]George Gabriel Stokes (1819–1903), Irish mathematician and physicist.

5.2 Vector Functions, Curves, and Surfaces

Problem 5.2.1. Determine a twice differentiable positive-valued increasing function $y = f(t)$, $t \in \mathbb{R}$, $f(1) = e^2$, $f'(1) = 2e^2$, such that $\vec{a}(t) = \langle t, -f(t), 0 \rangle$, $\vec{b}(t) = \langle 1, f'(t), f(t) \rangle$, and $\vec{b}'(t)$ are linearly dependent vectors, for all $t \in \mathbb{R}$.

Solution 5.2.1. The function f needs to be such that $f(t) > 0$ and $f'(t) > 0$, for all $t \in \mathbb{R}$. Since $\vec{b}'(t) = \langle 0, f''(t), f'(t) \rangle$, $t \in \mathbb{R}$, vectors $\vec{a}(t)$, $\vec{b}(t)$, and $\vec{b}'(t)$ are linearly dependent if and only if $(\vec{a}(t) \times \vec{b}(t)) \cdot \vec{b}'(t) = 0$, for all $t \in \mathbb{R}$, i.e., if and only if

$$0 = \begin{vmatrix} t & -f(t) & 0 \\ 1 & f'(t) & f(t) \\ 0 & f''(t) & f'(t) \end{vmatrix} = t((f'(t))^2 - f(t)f''(t)) + f(t)f'(t).$$

The function f is a solution to the initial value problem $yy' = t(y''y - (y')^2)$, $y(1) = e^2$, $y'(1) = 2e^2$. Since, for $t \in \mathbb{R}$ and $y(t) > 0$ and $y'(t) > 0$,

$$0 = yy' - t(y''y - (y')^2) = y^2 \left(\frac{y'}{y} - t \left(\frac{y'}{y} \right)' \right) = (y')^2 \left(\frac{t}{\frac{y'}{y}} \right)',$$

and since $\frac{y'(1)}{y(1)} = 2$, it follows that the function f is the solution to the initial value problem $y' - 2ty = 0$, $y(1) = e^2$, i.e., $f(t) = e^{t^2+1}$, $t \in \mathbb{R}$.

Problem 5.2.2.

(a) Sketch the hodograph of the function $t \mapsto \vec{a}(t) = \frac{t}{2}\vec{\imath} + |\sin t|\vec{\jmath}$.

(b) Use the Fourier series of the function $f(t) = \vec{a}(2t) \cdot \vec{\jmath}$ to evaluate $\sum_{k=1}^{\infty} \frac{1}{4k^2-1}$.

Solution 5.2.2.

(a) The hodograph of the function $\vec{a}$ is the set $\Gamma = \{(\frac{t}{2}, |\sin t|) : t \in \mathbb{R}\}$, i.e., for $x = \frac{t}{2}$ and $y = |\sin t|$, the graph of the function $y = |\sin 2x|$.

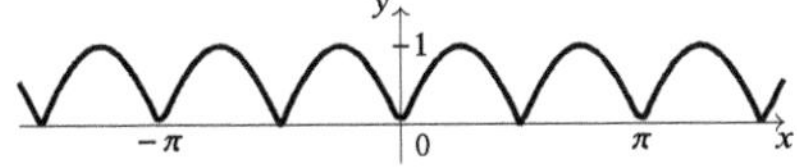

(b) The fundamental period of the function $f(t) = \vec{a}(2t) \cdot \vec{j} = |\sin 2t|$ is $T = \frac{\pi}{2}$. Since the function f is even, for any $k \in \mathbb{N}$, $b_k = 0$, and for any $k \in \mathbb{N}_0$, $a_k = \frac{8}{\pi} \int_0^{\frac{\pi}{4}} \sin 2t \cos 4kt \; dt = \frac{4}{\pi(1-4k^2)}$. By Dirichlet's theorem, for any $t \in \mathbb{R}$, $|\sin 2t| = \frac{2}{\pi} + \frac{4}{\pi} \sum_{k=1}^{\infty} \frac{\cos 4kt}{1-4k^2}$. For $t = 0$, $0 = \frac{2}{\pi} + \frac{4}{\pi} \sum_{k=1}^{\infty} \frac{1}{1-4k^2}$, i.e., $\sum_{k=1}^{\infty} \frac{1}{4k^2-1} = \frac{1}{2}$.

Problem 5.2.3. Let $a \in (0, \infty)$. On the sphere that is the hodograph of the vector function $\vec{r}(u, v) = a\langle \cos u \cos v, \; \sin u \cos v, \; \sin v \rangle$, $(u, v) \in [-\pi, \pi) \times [-\frac{\pi}{2}, \frac{\pi}{2}]$, lies the curve Γ determined by $u = v$. Determine the angles under which the curve Γ intersects the curves $u = c$, $c \in (-\pi, \pi)$, and $v = c$, $c \in (-\frac{\pi}{2}, \frac{\pi}{2})$, which also lie on the sphere.

Solution 5.2.3. The curve $u = v$ is the hodograph of the function $\vec{a}(t) = \vec{r}(t, t) = a\langle \cos^2 t, \; \sin t \cos t, \; \sin t \rangle$, $t \in [-\frac{\pi}{2}, \frac{\pi}{2}]$. Let $c \in (-\frac{\pi}{2}, \frac{\pi}{2})$. Then, $v = c$ determines the curve Γ_c that is the hodograph of the function $\vec{a}_c(t) = \vec{r}(t, c) = a\langle \cos t \cos c, \; \sin t \cos c, \; \sin c \rangle$, $t \in [-\pi, \pi)$.

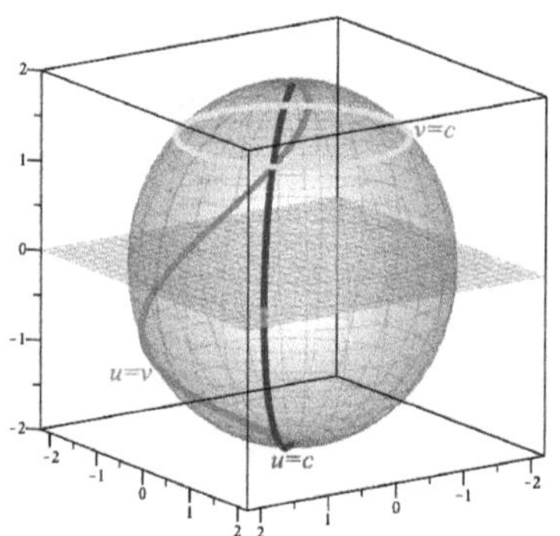

Let $t_c \in \left[-\frac{\pi}{2}, \frac{\pi}{2}\right]$ be such that $\vec{a}(t_c) = \vec{a}_c(t_c)$, i.e., the number t_c is such that $(\cos t_c)^2 = \cos t_c \cos c$, $\sin t_c \cos t_c = \sin t_c \cos c$, and $\sin t_c = \sin c$. Since $\cos c \neq 0$, it follows that $t_c \in (-\frac{\pi}{2}, \frac{\pi}{2})$. Therefore, $\cos t_c = \cos c$ and $\sin t_c = \sin c$, which implies that $t_c = c$. From $\vec{a}'(c) = a\langle -\sin 2c, \; \cos 2c, \cos c \rangle$, it follows that $|\vec{a}'(c)| = a\sqrt{1 + \cos^2 c}$. From $\vec{a}'_c(c) = a\left\langle -\frac{1}{2}\sin 2c, \; \cos^2 c, \; 0 \right\rangle$, it follows that $|\vec{a}'_c(c)| = a\cos c$. Let α_c be the angle between the curves Γ and Γ_c at the intersection point of the two curves. Then,

$$\cos \alpha_c = \frac{\vec{a}'(c) \cdot \vec{a}'_c(c)}{|\vec{a}'(c)| \cdot |\vec{a}'_c(c)|} = \frac{\cos c}{\sqrt{1 + \cos^2 c}}.$$

For $c \in (-\pi, \pi)$, let Γ^c be the curve determined by $u = c$, i.e., let Γ^c be the hodograph of the function $\vec{b}_c(t) = \vec{r}(c, t) = a\langle \cos c \cos t, \ \sin c \cos t, \ \sin t \rangle$, $t \in \left(-\frac{\pi}{2}, \frac{\pi}{2}\right)$. If β_c is the angle between the curves Γ and Γ^c, then $\cos \beta_c = \frac{1}{\sqrt{1+\cos^2 c}}$.

Note: The curve Γ is part of Viviani's curve, a curve on the sphere determined by $|u| = |v|$, $v \in \left[-\frac{\pi}{2}, \frac{\pi}{2}\right]$.[11] The circle Γ_c is the intersection of the sphere $x^2 + y^2 + z^2 = a^2$ and the plane $z = a \cos c$. The open semi-circle Γ^c lies in the intersection of the sphere $x^2 + y^2 + z^2 = a^2$ and the plane $x \sin c - y \cos c = 0$.

Problem 5.2.4. Determine an equation of the tangent line to the hodograph of the vector function $\vec{r}(t) = \langle x(t), y(t), z(t) \rangle$ at the point $r(0) = (5, 2, 3)$, if the functions x, y, and z are such that $x' = -x + y$, $y' = -y + 4z$, and $z' = x - 4z$.

Solution 5.2.4. Since $x'(0) = -5 + 2 = -3$, $y'(0) = 10$, and $z'(0) = -7$, the equation of the tangent line is $\frac{x-5}{-3} = \frac{y-2}{10} = \frac{z-3}{-7}$.

Problem 5.2.5. The curve Γ is the intersection of the plane $x + 2y - 1 = 0$ and the part of the paraboloid $4 - z = x^2 + y^2$ that lies in the first octant.

(a) Determine a parametrisation of Γ and establish if the curve is smooth or not.

(b) Determine the point at which Γ has a tangent line parallel to the plane $2x + 6y + z = 1$.

Solution 5.2.5.

(a) If the point $(x, y, z) \in \Gamma$, then $4 - z \geq 0$, $x \geq 0$, $y \geq 0$, and $z \geq 0$. By taking $y = t$ as the parameter, we obtain that Γ is the hodograph of the vector function $\vec{a}(t) = \langle 1 - 2t, \ t, \ 3 + 4t - 5t^2 \rangle$, $t \in I$, where the interval $I = \{t \in \mathbb{R} : 1 - 2t \geq 0 \text{ and } t \geq 0 \text{ and } 3 + 4t - 5t^2 \in [0, 4]\} = \left[0, \frac{1}{2}\right]$. Since, for any $t \in \left(0, \frac{1}{2}\right)$, $\vec{a}'(t) = \langle -2, \ 1, \ 2(2 - 5t) \rangle$ and $|\vec{a}'(t)|^2 = 5 + 4(2 - 5t)^2$, it follows that $\vec{a}$ is a continuously differentiable one-to-one vector function such that $|\vec{a}'(t)| \neq 0$ for all $t \in \left(0, \frac{1}{2}\right)$. Therefore, Γ is a smooth curve.

---

[11] Vincenzo Viviani (1622–1703), Italian mathematician and scientist.

(b) The tangent line to Γ at the point $a(t) = (1 - 2t, t, 3 + 4t - 5t^2)$ is parallel to the plane $2x + 6y + z = 1$ if $\vec{a}'(t) \cdot \langle 2,\, 6,\, 1 \rangle = 0$. But, for any $t \in \left(0, \frac{1}{2} \right)$, $\vec{a}'(t) \cdot \langle 2,\, 6,\, 1 \rangle = 2(3 - 5t) \in (1, 6)$. The curve Γ has no tangent line that is parallel to the plane $2x + 6y + z = 1$.

Problem 5.2.6. The closed curve Γ lies in the intersection of the cylinders $x^2 + y^2 = 5$ and $x^2 + z^2 = 5$ and passes through the point $A = \left(0, \sqrt{5}, -\sqrt{5} \right)$. Determine an equation of the osculating plane to the curve Γ at the point A.

Solution 5.2.6. The curve Γ is the hodograph of the function $\vec{a}(t) = \sqrt{5} \langle \cos t,\ \sin t,\ -\sin t \rangle$, $t \in [0, 2\pi]$.

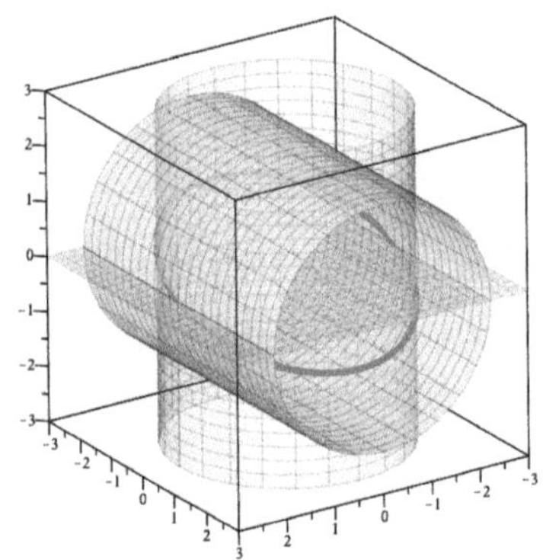

For $t \in (0, 2\pi)$, $\vec{a}'(t) = \sqrt{5} \langle -\sin t,\ \cos t,\ -\cos t \rangle$ and $\vec{a}''(t) = -\vec{a}(t)$ imply $a\left(\frac{\pi}{2} \right) = \left(0, \sqrt{5}, -\sqrt{5} \right)$, $\vec{a}'\left(\frac{\pi}{2} \right) = -\sqrt{5} \langle 1, 0, 0 \rangle$, and $\vec{a}''\left(\frac{\pi}{2} \right) = -\sqrt{5} \langle 0, 1, -1 \rangle$.

Hence, $\vec{a}'\left(\frac{\pi}{2} \right) \times \vec{a}''\left(\frac{\pi}{2} \right) = 5\langle 0, 1, 1 \rangle$, and the equation of the osculating plane is $5(y - \sqrt{5}) + 5(z + \sqrt{5}) = 0$, i.e., $y + z = 0$.

Note: Since the ellipse Γ is a planar curve, its osculating plane is the plane in which the curve Γ lies.

Problem 5.2.7. The closed curve Γ lies in the intersection of the sphere $x^2 + y^2 + z^2 = 6$ and the one-sheeted hyperboloid $x^2 - y^2 + z^2 = 4$ and passes through the point $A = (1, 1, 2)$. Determine an equation of the osculating plane to the curve Γ at the point A.

Answer 5.2.7. $y = 1$.

Problem 5.2.8. Let $I \subseteq \mathbb{R}$ be an interval, and let $f : I \to \mathbb{R}$ be a differentiable function. Let S be the surface determined by

$z = xf\left(\frac{y}{x}\right)$. Prove that each tangent plane to the surface S contains the origin.

Solution 5.2.8. The surface S is the hodograph of the function $\vec{a}(u, v) = \langle u, v, uf\left(\frac{v}{u}\right)\rangle$, $(u, v) \in D$, where $D = \{(u, v) \in \mathbb{R}^2 : u \neq 0$ and $\frac{v}{u} \in I\}$. Since, for each $(u, v) \in D$,

$$\frac{\partial \vec{a}(u, v)}{\partial u} = \left\langle 1, 0, f\left(\frac{v}{u}\right) - \frac{v}{u}f'\left(\frac{v}{u}\right)\right\rangle \text{ and}$$

$$\frac{\partial \vec{a}(u, v)}{\partial v} = \left\langle 0, 1, f'\left(\frac{v}{u}\right)\right\rangle,$$

it follows that

$$\frac{\partial \vec{a}(u, v)}{\partial u} \times \frac{\partial \vec{a}(u, v)}{\partial v} = \left\langle \frac{v}{u}f'\left(\frac{v}{u}\right) - f\left(\frac{v}{u}\right), \ -f'\left(\frac{v}{u}\right), 1\right\rangle.$$

The equation of the tangent plane to the surface S at the point $a(u, v)$, $(u, v) \in D$, is

$$\left(\frac{v}{u}f'\left(\frac{v}{u}\right) - f\left(\frac{v}{u}\right)\right)(x - u) - f'\left(\frac{v}{u}\right)(y - v) + z - uf\left(\frac{v}{u}\right) = 0,$$

i.e., $(\frac{v}{u}f'(\frac{v}{u}) - f(\frac{v}{u}))x - f'(\frac{v}{u})y + z = 0$, which proves the claim.

Problem 5.2.9. Let S be the surface obtained by rotating the hodograph[12] of the vector function $\vec{a}(t) = \langle t - \sin t, 1 - \cos t, 0\rangle$, $t \in [0, 2\pi]$, about the x-axis. Determine all points on the surface at which the normal line to S is parallel to the plane $y - z\sqrt{3} = 0$.

Solution 5.2.9. Let $A = (x, y, z) \in S \backslash \{(0, 0, 0), (2\pi, 0, 0)\}$. Let $t \in (0, 2\pi)$ be such that $x = t - \sin t$, and let $B = (x, 0, 0)$ and $C = (x, 1 - \cos t, 0) = (t - \sin t, 2\sin^2 \frac{t}{2}, 0) \in S$. Since A and C are points with the same first coordinate that belong to the surface, they both lie on a circle with its centre at B and radius $2\sin^2 \frac{t}{2}$. Thus, $|\overline{AB}| = |\overline{BC}|$ and $y^2 + z^2 = 4\sin^4 \frac{t}{2}$. The surface S is a hodograph of the vector function $\vec{r}(t, u) = \langle t - \sin t, 2\sin^2 \frac{t}{2} \cos u, 2\sin^2 \frac{t}{2} \sin u\rangle$, $(t, u) \in D = [0, 2\pi] \times [0, 2\pi]$. The direction of the normal line to S

---

[12] The hodograph is an arch of the cycloid.

at the point $r(t, u)$, $(t, u) \in (0, 2\pi) \times [0, 2\pi]$, is determined by the vector

$$\vec{N}(t, u) = \frac{\partial \vec{r}(t, u)}{\partial t} \times \frac{\partial \vec{r}(t, u)}{\partial u}$$

$$= 2\sin^2 \frac{t}{2} \left\langle \sin t, \ -2\sin^2 \frac{t}{2} \cos u, \ -2\sin^2 \frac{t}{2} \sin u \right\rangle.$$

The normal line to the surface S is parallel to the plane $y - z\sqrt{3} = 0$ at all points $r(t, u)$ for which $\vec{N}(t, u) \cdot \langle 0, 1, \ -\sqrt{3} \rangle = 0$, i.e., at all points $r(t, u)$, $(t, u) \in (0, 2\pi) \times [0, 2\pi]$, such that $4\sin^4 \frac{t}{2}(-\cos u + \sqrt{3}\sin u) = 0$. This equation is equivalent to the equation $\cos u - \sqrt{3}\ \sin u = 2\sin(\frac{\pi}{6} - u) = 0$. The set of all points on the surface at which the normal line to S is parallel to the plane $y - z\sqrt{3} = 0$ is

$$M = \left\{ r\left(t, \frac{\pi}{6} + k\pi\right) : t \in (0, 2\pi) \text{ and } k \in \{0, 1\} \right\}$$

$$= \left\{ \left(t - \sin t, \ \pm\sqrt{3}\sin^2 \frac{t}{2}, \ \pm\sin^2 \frac{t}{2} \right) : t \in (0, 2\pi) \right\}.$$

The set M is the intersection of the surface and the plane $y - z\sqrt{3} = 0$, without points $(0, 0, 0)$ and $(2\pi, 0, 0)$.

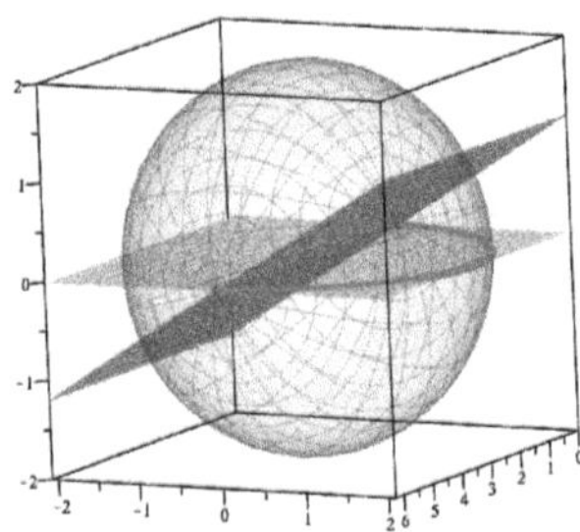

Note: Once it is established that the surface S has a tangent line at any point, except $(0, 0, 0)$ and $(2\pi, 0, 0)$, the fact that the set of points with the required property is at the intersection of the plane and the surface follows from the fact that the surface is obtained by rotating a planar curve about the x-axis.

Problem 5.2.10. Let S be the hodograph of the vector function $\vec{r}(u, v) = \langle u^2 + v^2, uv, u^2 + v^2 \rangle$, $(u, v) \in \mathbb{R}^2$, and let $\alpha = \alpha(u, v)$ be the acute angle between the normal line to S at the point $r(u, v)$ and the line $\frac{x - (u^2 + v^2)}{2} = \frac{y - uv}{1} = \frac{z - (u^2 + v^2)}{1}$. Determine the Fourier series of a periodic function f, with the fundamental period $T = 1$, such that, for $t \in (1, 2)$, $f(t) = t \cos \alpha(t, 2t)$.

Solution 5.2.10. The normal line to S at the point $r(u, v)$ is determined by the vector $\vec{N}(u, v) = \frac{\partial \vec{r}(u,v)}{\partial u} \times \frac{\partial \vec{r}(u,v)}{\partial v} = 2(v^2 - u^2)\langle 1, 0, -1 \rangle$. Since $|\vec{N}(u, v)| = 2\sqrt{2}\,|v^2 - u^2|$, it follows that there is a normal line to S at the point $r(u, v)$ if and only if $u^2 \neq v^2$. Consequently, the domain of the function α is the set $D = \{(u, v) \in \mathbb{R}^2 : u^2 \neq v^2\}$. For $(u, v) \in D$, since $\alpha(u, v)$ is an acute angle,

$$\cos \alpha(u, v) = \frac{\vec{N}(u, v) \cdot \langle 2, 1, 1 \rangle}{|\vec{N}(u, v)||\langle 2, 1, 1 \rangle|} = \frac{\sqrt{3}(v^2 - u^2)}{6|v^2 - u^2|} = \frac{\sqrt{3}}{6}.$$

From $f(t) = \frac{t\sqrt{3}}{6}$, for $t \in (1, 2)$, it follows that $f(t) \sim \frac{\sqrt{3}}{4} - \frac{\sqrt{3}}{6\pi} \sum_{k=1}^{\infty} \frac{\sin 2k\pi t}{k}$.

Note: Observe that S is part of the half-plane $x - z = 0$, $x \geq 0$, squeezed between two planes, $x - 4y + z = 0$ and $x + 4y + z = 0$. The condition $u^2 = v^2$ determines the boundary of S, the union of two rays in the intersection of the half-plane and the two planes.

5.3 Gradient, Divergence, and Curl

Problem 5.3.1. Let D be a region in $\mathbb{R}^3$. Let $C^0(D)$ be the set of all continuous functions $f : D \to \mathbb{R}$, and let $C^1(D)$ be the set of all functions $f : D \to \mathbb{R}$ with continuous first-order partial derivatives. Prove that the gradient, i.e., the function $\nabla : C^1(D) \to C^0(D)$ defined, for $f \in C^1(D)$ and all $(x, y, z) \in D$, by $(\nabla f)(x, y, z) = \langle \frac{\partial f}{\partial x}(x, y, z), \frac{\partial f}{\partial y}(x, y, z), \frac{\partial f}{\partial z}(x, y, z) \rangle$, is a differential operator. In other words, prove that for any $a, b \in \mathbb{R}$ and any $f, g \in C^1(D)$, $\nabla(a\,f + b\,g) = a\nabla f + b\nabla g$ and $\nabla(fg) = f\nabla g + g\nabla f$.

Solution 5.3.1. For $(x, y, z) \in D$,

$$(\nabla(a\,f + b\,g))(x, y, z) = \left\langle \frac{\partial(af + bg)}{\partial x}(x, y, z), \frac{\partial(af + bg)}{\partial y}(x, y, z), \right.$$

$$\left. \frac{\partial(af + bg)}{\partial z}(x, y, z) \right\rangle = a\left\langle \frac{\partial f}{\partial x}(x, y, z), \frac{\partial f}{\partial y}(x, y, z), \frac{\partial f}{\partial z}(x, y, z) \right\rangle$$

$$+ b\left\langle \frac{\partial g}{\partial x}(x, y, z), \frac{\partial g}{\partial y}(x, y, z), \frac{\partial g}{\partial z}(x, y, z) \right\rangle$$

$$= a(\nabla f)(x, y, z) + b(\nabla g)(x, y, z),$$

which establishes that ∇ is a linear operator. For $(x, y, z) \in D$,

$$(\nabla(fg))(x,y,z) = \left\langle \frac{\partial(fg)}{\partial x}(x,y,z), \; \frac{\partial(fg)}{\partial y}(x,y,z), \; \frac{\partial(fg)}{\partial z}(x,y,z) \right\rangle$$

$$= g(x,y,z) \left\langle \frac{\partial f}{\partial x}(x,y,z), \; \frac{\partial f}{\partial y}(x,y,z), \; \frac{\partial f}{\partial z}(x,y,z) \right\rangle$$

$$+ f(x,y,z) \left\langle \frac{\partial g}{\partial x}(x,y,z), \; \frac{\partial g}{\partial y}(x,y,z), \; \frac{\partial g}{\partial z}(x,y,z) \right\rangle$$

$$= g(x,y,z)(\nabla f)(x,y,z) + f(x,y,z)(\nabla g)(x,y,z),$$

which completes the proof that ∇ is a differential operator.[13]

In Problems 5.3.2–5.3.4, $\vec{a} = \langle a_1, a_2, a_3 \rangle$ is a nonzero vector and $\vec{r}(x,y,z) = \langle x, y, z \rangle$ is the radius vector of the point (x, y, z) in the given rectangular coordinate system.

Problem 5.3.2. Establish that, for $\vec{r} \neq \vec{0}$:

(a) $\operatorname{grad} |\vec{r}| = \frac{1}{|\vec{r}|}\vec{r}$.

(b) $\operatorname{grad} \frac{1}{|\vec{r}|} = -\frac{1}{|\vec{r}|^3}\vec{r}$.

(c) $\operatorname{grad} (\vec{a} \cdot \vec{r}) = \vec{a}$.

Solution 5.3.2.

(a) From $|\vec{r}| = \sqrt{x^2 + y^2 + z^2}$, it follows that

$$\operatorname{grad} |\vec{r}| = \left\langle \frac{x}{\sqrt{x^2+y^2+z^2}}, \; \frac{y}{\sqrt{x^2+y^2+z^2}}, \; \frac{z}{\sqrt{x^2+y^2+z^2}} \right\rangle$$

$$= \frac{\langle x, y, z \rangle}{\sqrt{x^2+y^2+z^2}} = \frac{1}{|\vec{r}|}\vec{r}.$$

(b) Since the gradient is a differential operator, the chain rule applies:

$$\operatorname{grad} \frac{1}{|\vec{r}|} = -\frac{1}{|\vec{r}|^2}\operatorname{grad} |\vec{r}| = -\frac{1}{|\vec{r}|^3}\vec{r}.$$

---

[13]We will interchangeably use ∇f and $\operatorname{grad} f$ to denote the gradient of the function f. For simplicity, we will denote the gradient of the scalar field f at the point (x, y, z) by $\nabla f(x, y, z)$ or by $\operatorname{grad} f(x, y, z)$ instead of the possibly less ambiguous $(\nabla f)(x, y, z)$ and $(\operatorname{grad} f)(x, y, z)$.

(c) From $\vec{a} \cdot \vec{r} = a_1 x + a_2 y + a_3 z$, it follows that

$$\operatorname{grad}\,(\vec{a} \cdot \vec{r}) = \operatorname{grad}\,(a_1 x + a_2 y + a_3 z) = \langle a_1,\, a_2,\, a_3 \rangle = \vec{a}.$$

Problem 5.3.3. Prove that $\operatorname{grad} |\vec{a} \times \vec{r}|^2 = 2(\vec{a} \times (\vec{r} \times \vec{a}))$.

Solution 5.3.3. *Solution 1*: From $|\vec{a} \times \vec{r}|^2 = (a_2 z - a_3 y)^2 + (a_1 z - a_3 x)^2 + (a_1 y - a_2 x)^2$, it follows that

$$\operatorname{grad} |\vec{a} \times \vec{r}|^2 = 2\langle (a_2^2 + a_3^2)x - a_1 a_2 y - a_1 a_3 z,\ (a_1^2 + a_3^2)y$$
$$- a_1 a_2 x - a_2 a_3 z,\ (a_1^2 + a_2^2)z - a_1 a_3 x - a_2 a_3 y \rangle.$$

The claim follows from

$$(\vec{a} \times \vec{r}) \times \vec{a} = (\vec{a} \cdot \vec{a})\vec{r} - (\vec{a} \cdot \vec{r})\vec{a} = (a_1^2 + a_2^2 + a_3^2)\langle x,\, y,\, z \rangle$$
$$- \langle a_1^2 x + a_1 a_2 y + a_1 a_3 z,\ a_2^2 y + a_1 a_2 x + a_2 a_3 z,$$
$$a_3^2 z + a_1 a_3 x + a_2 a_3 y \rangle.$$

Solution 2: For $\vec{r} \neq \vec{0}$, by Problem 5.3.2,

$$\operatorname{grad} |\vec{a} \times \vec{r}|^2 = \operatorname{grad} |\vec{a}|^2 |\vec{r}|^2 \sin^2 \angle(\vec{a}, \vec{r}) = \operatorname{grad} |\vec{a}|^2 |\vec{r}|^2 \left(1 - \frac{(\vec{a} \cdot \vec{r})^2}{|\vec{a}|^2 |\vec{r}|^2}\right)$$
$$= \operatorname{grad}\,\left(|\vec{a}|^2 |\vec{r}|^2 - (\vec{a} \cdot \vec{r})^2\right)$$
$$= 2|\vec{a}|^2 |\vec{r}|\operatorname{grad} |\vec{r}| - 2(\vec{a} \cdot \vec{r})\operatorname{grad}\,(\vec{a} \cdot \vec{r})$$
$$= 2|\vec{a}|^2 \vec{r} - 2(\vec{a} \cdot \vec{r})\vec{a} = 2((\vec{a} \cdot \vec{a})\vec{r} - (\vec{a} \cdot \vec{r})\vec{a})$$
$$= 2(\vec{a} \times (\vec{r} \times \vec{a})).$$

Problem 5.3.4. Establish that, for $\vec{r} \neq \vec{0}$:

(a) if $\vec{a} \times \vec{r} \neq \vec{0}$, then $\operatorname{grad} \sqrt{(\vec{a} \times \vec{r})^2} = \dfrac{\vec{a} \times (\vec{r} \times \vec{a})}{\sqrt{(\vec{a} \times \vec{r})^2}}$.

(b) $\operatorname{grad} \dfrac{\vec{a} \cdot \vec{r}}{|\vec{r}|^3} = \dfrac{\vec{a}}{|\vec{r}|^3} - \dfrac{3(\vec{a} \cdot \vec{r})\,\vec{r}}{|\vec{r}|^5}$.

Answer 5.3.4. Use Problem 5.3.3 and the fact that the gradient is a differential operator.

Problem 5.3.5. Determine the rate of change of the function $f(x, y) = e^{x^2 y} - 3xy$ at the point $(2, 3)$ in the direction of the vector $\langle -1, 2 \rangle$.

Solution 5.3.5. The vector $\vec{u} = \langle -\frac{1}{\sqrt{5}}, \frac{2}{\sqrt{5}} \rangle$ is the unit vector in the direction of the vector $\langle -1, 2 \rangle$.

Solution 1: By the limit definition of the directional derivative,

$$D_{\vec{u}}f(2,3) = \lim_{h \to 0} \frac{f\left(2 - \frac{h}{\sqrt{5}}, 3 + \frac{2h}{\sqrt{5}}\right) - f(2,3)}{h}$$

$$= \lim_{h \to 0} \frac{1}{h}\left(e^{\left(2 - \frac{h}{\sqrt{5}}\right)^2 \left(3 + \frac{2h}{\sqrt{5}}\right)} - 3\left(2 - \frac{h}{\sqrt{5}}\right)\left(3 + \frac{2h}{\sqrt{5}}\right)\right.$$

$$\left. - e^{12} + 18\right),$$

which, after some effort, yields $D_{\vec{u}}f(2,3) = \frac{-4e^{12} - 3}{\sqrt{5}}$.

Solution 2: From $\nabla f(x,y) = \langle 2xye^{x^2 y} - 3y, x^2 e^{x^2 y} - 3x \rangle$, it follows that $\nabla f(2,3) = \langle 12e^{12} - 9, 4e^{12} - 6 \rangle$. Therefore, $D_{\vec{u}}f(2,3) = \nabla f(2,3) \cdot \vec{u} = -\frac{12}{\sqrt{5}}e^{12} + \frac{9}{\sqrt{5}} + \frac{8}{\sqrt{5}}e^{12} - \frac{12}{\sqrt{5}} = \frac{-4e^{12} - 3}{\sqrt{5}}$.

Problem 5.3.6. Let $O \in \mathbb{E}^3$, and let a scalar field $\mu : \mathbb{E}^3 \backslash \{O\} \to \mathbb{R}$ be defined, for any $P \in \mathbb{E}^3 \backslash \{O\}$, as the reciprocal of the distance between P and O, i.e., $\mu(P) = \frac{1}{|\overrightarrow{OP}|}$. Evaluate the directional derivative of μ in the direction of grad $\mu(P)$.

Solution 5.3.6. Let $Oxyz$ be a rectangular coordinate system with the origin at the point O. Then, for $P \in \mathbb{E}^3 \backslash \{O\}$, $\vec{r} = \overrightarrow{OP}$ is the radius vector of the point P. By Problem 5.3.2, grad $\mu(P) =$ grad $\frac{1}{|\vec{r}|} = -\frac{1}{|\vec{r}|^3}\vec{r}$. Let the vector field $\vec{g} : \mathbb{E}^3 \backslash \{O\} \to \vec{\mathbb{E}}^3$ be defined by

$$\vec{g}(P) = \frac{\text{grad } \mu(P)}{|\text{grad } \mu(P)|} = -\frac{1}{|\vec{r}|}\vec{r}.$$

The directional derivative of μ in the direction of grad $\mu(P)$ is

$$D_{\vec{g}(P)}\mu(P) = \text{grad } \mu(P) \cdot \vec{g}(P) = \frac{1}{|\vec{r}|^2} = \frac{1}{|\overrightarrow{OP}|^2}.$$

Problem 5.3.7. Let S be a sphere in $\mathbb{E}^3$ with its centre at the point O and radius r. Let the scalar field $F : \mathbb{E}^3 \to \mathbb{R}$ be defined by $F(A) = d(A, S)$, where $d(A, S)$ is the distance between the point A

and the sphere S. Let A_1 be a point inside the sphere, and let A_2 be a point outside the sphere S. Evaluate the directional derivative of F in the direction of a fixed unit vector $\vec{u}$ at the points A_1 and A_2.

Solution 5.3.7. The distance between the point $A \in \mathbb{E}^3$ and the sphere is the absolute value of the difference between the length of the vector $\overrightarrow{OA}$ and the radius of the sphere, i.e., $F(A) = \left| |\overrightarrow{OA}| - r \right|$.

If the point A is inside the sphere, then $F(A) = r - |\overrightarrow{OA}|$ and grad $F(A) = -\dfrac{\overrightarrow{OA}}{|\overrightarrow{OA}|}$. If the point A is outside the sphere, then $F(A) = |\overrightarrow{OA}| - r$ and grad $F(A) = \dfrac{\overrightarrow{OA}}{|\overrightarrow{OA}|}$. Since the point A_1 is inside the sphere, $D_{\vec{u}}F(A_1) = \text{grad } F(A_1) \cdot \vec{u} = -\cos \angle \left(\vec{u}, \overrightarrow{OA_1} \right)$. Since the point A_2 is outside the sphere, $D_{\vec{u}}F(A_2) = \cos \angle \left(\vec{u}, \overrightarrow{OA_2} \right)$.

Problem 5.3.8. Let $A = (2, 1, 1)$, and let a function $F : \mathbb{R}^3 \to \mathbb{R}$ with continuous partial derivatives be such that $F(A) = 5$, $\frac{\partial F}{\partial x}(A) = 1$, $\frac{\partial F}{\partial y}(A) = 2$, and $\frac{\partial F}{\partial z}(A) = 3$. Let the function $G : \mathbb{R}^3 \to \mathbb{R}$ be determined by $G(x, y, z) = x + z$.

(a) Use the linearisation of F at the point A to approximate the number $F(1.9, 1, 1.2)$.
(b) What is the rate of change of the function F at the point A in the direction of the unit vector pointing towards the point $B = (3, 2, 2)$?
(c) Determine a unit vector so that the rates of change of F and G at the point A are both zero in the direction of that vector.

Solution 5.3.8.

(a) The linearisation of the function F in a neighbourhood of the point A is $L(x, y, z) = 5 + (x - 2) + 2(y - 1) + 3(z - 1)$. Hence, $F(1.9, 1, 1.2) \approx L(1.9, 1, 1.2) = 5 - 0.1 + 3 \cdot 0.2 = 5.5$.
(b) The unit vector in the direction of $\overrightarrow{AB} = \langle 1, 1, 1 \rangle$ is $\vec{\ell} = \frac{\langle 1,1,1 \rangle}{\sqrt{3}}$. The directional derivative of the function F in the direction of $\vec{\ell}$ is

$$D_{\vec{\ell}}F(A) = \nabla F(A) \cdot \vec{\ell} = \langle 1, 2, 3 \rangle \cdot \frac{\langle 1, 1, 1 \rangle}{\sqrt{3}} = 2\sqrt{3}.$$

(c) The question is to determine all unit vectors $\vec{u}$ such that $D_{\vec{u}}F(A) = \vec{u} \cdot \nabla F(A) = 0$ and $D_{\vec{u}}G(A) = \vec{u} \cdot \nabla G(A) = 0$. In other words, $\vec{u}$ must be a unit vector such that $\vec{u}$ is parallel to $\nabla F(A) \times \nabla G(A) = \langle 1, 2, 3 \rangle \times \langle 1, 0, 1 \rangle = \langle 2, 2, -2 \rangle$. Therefore, $\vec{u} = \pm \frac{\langle 1, 1, -1 \rangle}{\sqrt{3}}$.

Problem 5.3.9. Let $T(x, y, z) = x^3 y + y^3 z + xz^3$, $(x, y, z) \in \mathbb{R}^3$, and let $P = (2, -1, 0)$.

(a) Evaluate the directional derivative of the function $T = T(x, y, z)$ at the point P in the direction from P to the point $Q = (1, 1, 2)$.

(b) Suppose that $T = T(x, y, z)$ is the temperature in degrees Celsius (°C) at the point (x, y, z) in a neighbourhood of the origin. A housefly is flying through space at a constant speed of 5 km/h in the direction of increasing temperature. If the housefly's direction of flight at any given point is perpendicular to the level surface of $f(x, y, z) = 2x^2 + 3y^2 + z^2$ passing through this point, evaluate the rate of change of temperature experienced by the housefly at the point P.

Solution 5.3.9.

(a) From $\nabla T(x, y, z) = \langle 3x^2 y + z^3, \ x^3 + 3y^2 z, \ y^3 + 3xz^2 \rangle$, $\nabla T(P) = \langle -12, 8, -1 \rangle$, and $\vec{u} = \frac{\overrightarrow{PQ}}{|\overrightarrow{PQ}|} = \frac{\langle -1, 2, 2 \rangle}{3}$, it follows that $D_{\vec{u}}T(P) = \langle -12, 8, -1 \rangle \cdot \frac{\langle -1, 2, 2 \rangle}{3} = \frac{26}{3}$.

(b) Let $\vec{r}(t) = \langle x(t), y(t), z(t) \rangle$ be the position vector of the housefly at time t. The housefly's velocity at time t is $\vec{v} = \frac{d\vec{r}}{dt} = \langle x'(t), y'(t), z'(t) \rangle$. It is given that at any time t, $\vec{v}(t) \parallel \nabla f(x(t), y(t), z(t)) = \langle 4x(t), 6y(t), 2z(t) \rangle$. At the instant t_0, when the housefly is at the point P, the velocity vector $\vec{v}(t_0)$ is parallel to the vector $\langle 8, -6, 0 \rangle$, which, together with $|\vec{v}(t_0)| = 5$, implies that $\vec{v}(t_0) = \pm \langle 4, -3, 0 \rangle$. Let $\vec{v}_0 = \frac{\vec{v}(t_0)}{|\vec{v}(t_0)|}$. The rate of change of temperature experienced by the housefly at the point $P = (2, -1, 0)$ is

$$D_{\vec{v}_0}T(P) = \nabla T(P) \cdot \frac{\vec{v}(t_0)}{5} = \mp 14.4.$$

Since the housefly is flying in the direction of increasing temperature, the rate must be positive, i.e., equal to 14.4°C/h.

Problem 5.3.10. The scalar field f is defined by $f(x, y, z) = 3z + e^{x^2 - y^2}$. Let C be the set of terminal points of all unit vectors $\vec{v}$ with the property that, at the point $(0, 0, 1)$ and in the direction $\vec{v}$, the function f increases at one-third of its maximum possible directional rate of change at $(0, 0, 1)$. Determine the set C.

Solution 5.3.10. Since $\nabla f(x, y, z) = \langle 2xe^{x^2 - y^2}, -2ye^{x^2 - y^2}, 3 \rangle$ and $\nabla f(0, 0, 1) = \langle 0, 0, 3 \rangle$, the maximum possible directional rate of change of f at $(0, 0, 1)$ is $|\nabla f(0, 0, 1)| = 3$. The question is to determine the set

$$C = \left\{ (v_1, v_2, v_3) : v_1^2 + v_2^2 + v_3^2 = 1 \text{ and} \right.$$

$$\left. D_{\langle v_1, v_2, v_3 \rangle} f(0, 0, 1) = \frac{1}{3} \cdot 3 = 1 \right\}.$$

Since $D_{\langle v_1, v_2, v_3 \rangle} f(0, 0, 1) = \langle v_1, v_2, v_3 \rangle \cdot \langle 0, 0, 3 \rangle = 3v_3$, it follows that $v_3 = \frac{1}{3}$. This means that $v_1^2 + v_2^2 = \frac{8}{9}$. The set C is the circle in the plane $z = \frac{1}{3}$ with centre at $(0, 0, \frac{1}{3})$ and radius $\frac{2\sqrt{2}}{3}$.

Problem 5.3.11. Suppose that a scalar field f and a vector field $\vec{a} = \langle a_1, a_2, a_3 \rangle$ share the same domain. Establish that $\operatorname{div}(f\,\vec{a}) = \vec{a} \cdot \operatorname{grad} f + f \operatorname{div} \vec{a}$.

Solution 5.3.11. Let $D \subseteq \mathbb{R}^3$ be the domain of f and $\vec{a}$. Then, in D,

$$\operatorname{div}(f\,\vec{a}) = \operatorname{div}(\langle f\,a_1, f\,a_2, f\,a_3 \rangle) = a_1 \frac{\partial f}{\partial x} + f \frac{\partial a_1}{\partial x} + a_2 \frac{\partial f}{\partial y} + f\,\frac{\partial a_2}{\partial y}$$

$$+ a_3 \frac{\partial f}{\partial z} + f\,\frac{\partial a_3}{\partial z} = \left(a_1 \frac{\partial f}{\partial x} + a_2 \frac{\partial f}{\partial y} + a_3 \frac{\partial f}{\partial z} \right)$$

$$+ f \left(\frac{\partial a_1}{\partial x} + \frac{\partial a_2}{\partial y} + \frac{\partial a_3}{\partial z} \right) = \vec{a} \cdot \operatorname{grad} f + f \operatorname{div} \vec{a}.$$

Problem 5.3.12. Determine all twice differentiable functions $f : \mathbb{R}^+ \to \mathbb{R}$ with the property that, for any $P \in \mathbb{R}^3 \setminus \{(0, 0, 0)\}$ and $\vec{r} = \vec{r}(P)$, the radius vector of the point P, $2|\vec{r}| \operatorname{div}(\operatorname{grad} f(|\vec{r}|)) = \operatorname{div}(\operatorname{grad} |\vec{r}|)$.

Solution 5.3.12. For $\vec{r} \neq \langle 0,0,0 \rangle$, $\mathrm{grad} f(|\vec{r}|) = \frac{f'(|\vec{r}|)}{|\vec{r}|} \vec{r}$ and $\mathrm{div}(\mathrm{grad}\ f(|\vec{r}|)) = \vec{r} \cdot \mathrm{grad}\ \frac{f'(|\vec{r}|)}{|\vec{r}|} + \frac{f'(|\vec{r}|)}{|\vec{r}|} \mathrm{div}\ \vec{r} = f''(|\vec{r}|) + \frac{2}{|\vec{r}|} f'(|\vec{r}|)$. Since, for $\vec{r} \neq \vec{0}$, $\mathrm{div}(\mathrm{grad}\ |\vec{r}|) = \frac{2}{|\vec{r}|}$, the statement $2|\vec{r}|\mathrm{div}(\mathrm{grad}\ f(|\vec{r}|)) = \mathrm{div}(\mathrm{grad}\ |\vec{r}|)$ is equivalent to the statement

$$f''(|\vec{r}|) + \frac{2}{|\vec{r}|} f'(|\vec{r}|) = \frac{2}{|\vec{r}|}.$$

The function f is a solution to a non-homogeneous second-order differential equation $y'' + \frac{2}{t} y' = \frac{2}{t}$, $t > 0$. It follows that, for $a, b \in \mathbb{R}$, $f(t) = a + t + \frac{b}{t}$, $t \in \mathbb{R}^+$.

Problem 5.3.13. Let $\vec{c}$ be a nonzero vector. Evaluate $\mathrm{div}(\mathrm{grad}\ a)$ if $a(P) = \vec{c} \cdot \vec{r} + \frac{1}{2}\ln(\vec{c} \times \vec{r})^2$, where $\vec{r}$ is the radius vector of the point $P \in D$ and $D \subseteq \mathbb{R}^3$ is a set that needs to be determined.

Solution 5.3.13. The domain of the scalar field a is the set $D = \{(x,y,z) \in \mathbb{R}^3 : \vec{c} \times \langle x,y,z \rangle \neq \vec{0}\}$, i.e., the set of all points in $\mathbb{R}^3$ that does not belong to the line that passes through the origin and is parallel to $\vec{c}$. It follows that, in D,

$$\mathrm{grad}\ a = \mathrm{grad}\ (\vec{c} \cdot \vec{r}) + \frac{1}{2}\mathrm{grad}\ (\ln(\vec{c} \times \vec{r})^2)$$

$$= \vec{c} + \frac{1}{2|\vec{c} \times \vec{r}|^2}\ \mathrm{grad}\ (\vec{c} \times \vec{r})^2$$

$$= \vec{c} + \frac{1}{|\vec{c} \times \vec{r}|^2}\ (\vec{c} \times (\vec{r} \times \vec{c})).$$

By Problems 5.3.3 and 5.3.11,

$$\mathrm{div}(\mathrm{grad}\ a) = \mathrm{div}(\vec{c}) + \mathrm{div}\left(\frac{1}{|\vec{c} \times \vec{r}|^2}\ (\vec{c} \times (\vec{r} \times \vec{c}))\right)$$

$$= -\frac{2(\vec{c} \times (\vec{r} \times \vec{c}))^2}{|\vec{c} \times \vec{r}|^4} + \frac{2|\vec{c}|^2}{|\vec{c} \times \vec{r}|^2}$$

$$= -\frac{2|\vec{c}|^2|\vec{r} \times \vec{c}|^2}{|\vec{c} \times \vec{r}|^4} + \frac{2|\vec{c}|^2}{|\vec{c} \times \vec{r}|^2} = 0.$$

Problem 5.3.14. Let the surface S be the hodograph of the vector function $\vec{R}(u,v) = \langle 1 + \cos u \cos v, \sin u \cos v, \sin v \rangle$, $(u,v) \in D = [0, 2\pi] \times (0, \frac{\pi}{2}]$.

(a) Determine the geometrical meaning of the parameters u and v.

(b) Let $\vec{n}$ be the vector field which, to a point $P \in S$, associates the unit normal vector of S at P that makes an obtuse angle with the unit vector $\vec{k} = \langle 0,\, 0,\, 1 \rangle$. Determine the domain of the field $\vec{n}$. Determine div $\vec{n}$.

Solution 5.3.14.

(a) The parameter v is the angle which the vector $\vec{R}(u, v) - \vec{\imath}$ makes with its projection into the xy-plane. The parameter u is the angle that this projection makes with the vector $\vec{\imath}$. If $(x, y, z) \in \mathbb{R}^3$ and $(u, v) \in D$ are such that $x = 1 + \cos u \cos v$, $y = \sin u \cos v$, and $z = \sin v$, then $z = \sqrt{1 - (x - 1)^2 - y^2}$. The surface S is the hemisphere with its centre at $O_1 = (1, 0, 0)$, radius 1, and located above the xy-plane.

(b) A unit vector normal to the hemisphere S at the point $P(x, y, z) \in S$ is $\vec{N}(P) = \overrightarrow{O_1 P} = \langle x - 1, y, z \rangle = \vec{r} - \vec{\imath}$. Here, $\vec{r} = \vec{r}(P) = \langle x, y, z \rangle$ is the radius vector of the point P. Since $\vec{N}(P) \cdot \vec{k} = z > 0$ and since $\vec{N}(1, 0, 1) = \vec{k}$ is parallel to the z-axis, it follows that $\vec{n}(P) = -\vec{N}(P) = \vec{\imath} - \vec{r}$, $P \in S' = S \setminus \{(1, 0, 1)\}$. For any $P \in S'$, div $\vec{n}(P) = $ div $(\vec{\imath} - \vec{r}(P)) = -3$.

Problem 5.3.15. Let $D \subseteq \mathbb{R}$, and let $f : D \to \mathbb{R}$ be a scalar field. Let $\vec{c} = \langle c_1, c_2, c_3 \rangle$ be a nonzero vector, and let $\vec{r} = \langle x, y, z \rangle$ be the radius vector of the point $(x, y, z) \in \mathbb{R}^3$. Prove that for any $(x, y, z) \in D$:

(a) $\mathrm{curl}(f(x, y, z)\, \vec{c}) = (\mathrm{grad}\ f(x, y, z)) \times \vec{c}$.

(b) $\mathrm{curl}(f(x, y, z)\, \vec{r}) = (\mathrm{grad}\ f(x, y, z)) \times \vec{r}$.

(c) $\mathrm{curl}(\vec{c} \times \vec{r}) = 2\vec{c}$.

Solution 5.3.15.

(a) By the definition of the curl of a vector field,

$$
\mathrm{curl}(f\, \vec{c}) = \begin{vmatrix} \vec{\imath} & \vec{\jmath} & \vec{k} \\ \dfrac{\partial}{\partial x} & \dfrac{\partial}{\partial y} & \dfrac{\partial}{\partial z} \\ c_1 f & c_2 f & c_3 f \end{vmatrix}
$$

$$
= \left\langle c_3 \frac{\partial f}{\partial y} - c_2 \frac{\partial f}{\partial z},\, c_1 \frac{\partial f}{\partial z} - c_3 \frac{\partial f}{\partial x},\, c_2 \frac{\partial f}{\partial x} - c_1 \frac{\partial f}{\partial y} \right\rangle
$$

$$
= (\mathrm{grad}\ f) \times \vec{c}.
$$

Note: Observe that for a scalar field f and a vector field $\vec{a} = \langle a_1(x), a_2(y), a_3(z) \rangle$, $\mathrm{curl}(f(x, y, z)\, \vec{a}) = (\mathrm{grad}\ f(x, y, z)) \times \vec{a}$.

Problem 5.3.16. Consider vector fields $\vec{a}(x, y, z) = \langle xy, 0, -xz \rangle$ and $\vec{b}(x, y, z) = \langle xz, 0, yz \rangle$, $(x, y, z) \in \mathbb{R}^3$. Determine all points at which the gradients of the scalar fields $\mathrm{div}\ \vec{a} \cdot \mathrm{div}\ \vec{b}$ and $\mathrm{curl}\ \vec{a} \cdot \mathrm{curl}\ \vec{b}$ are opposite to each other.[14]

Solution 5.3.16. From $\mathrm{div}\ \vec{a}(x, y, z) = y - x$ and $\mathrm{div}\ \vec{b}(x, y, z) = y + z$, for all $(x, y, z) \in \mathbb{R}^3$, it follows that

$$\mathrm{grad}\ (\mathrm{div}\ \vec{a}(x, y, z) \cdot \mathrm{div}\ \vec{b}(x, y, z)) = \langle -(z + y), z + 2y - x, y - x \rangle.$$

For all $(x, y, z) \in \mathbb{R}^3$, from $\mathrm{curl}\ \vec{a}(x, y, z) = \begin{vmatrix} \vec{i} & \vec{j} & \vec{k} \\ \frac{\partial}{\partial x} & \frac{\partial}{\partial y} & \frac{\partial}{\partial z} \\ xy & 0 & -xz \end{vmatrix} = \langle 0, z, -x \rangle$

and $\mathrm{curl}\ \vec{b}(x, y, z) = \langle z, x, 0 \rangle$, it follows that

$$\mathrm{grad}\ (\mathrm{curl}\ \vec{a}(x, y, z) \cdot \mathrm{curl}\ \vec{b}(x, y, z)) = \langle z, 0, x \rangle.$$

The condition $\mathrm{grad}\ (\mathrm{div}\ \vec{a} \cdot \mathrm{div}\ \vec{b}) = -\mathrm{grad}\ (\mathrm{curl}\ \vec{a} \cdot \mathrm{curl}\ \vec{b})$ is satisfied at any point $(x, y, z) \in \mathbb{R}^3$ such that

$$-(z + y) = -z \quad \text{and} \quad z + 2y - x = 0 \quad \text{and} \quad y - x = -x.$$

The gradients of the scalar fields $\mathrm{div}\ \vec{a} \cdot \mathrm{div}\ \vec{b}$ and $\mathrm{curl}\ \vec{a} \cdot \mathrm{curl}\ \vec{b}$ are opposite to each other at any point of the set $\{(x, 0, x) : x \in \mathbb{R}\}$, i.e., at any point that lies on the line $\frac{x}{1} = \frac{y}{0} = \frac{z}{1}$.

Problem 5.3.17. Let $\vec{c}$ be a nonzero vector, let $D \subseteq \mathbb{R}^3$, and let $\vec{a} : D \to \mathbb{E}^3$ be a continuously differentiable vector field, i.e., a field whose scalar components have continuous partial derivatives. Consider the vector field $\vec{A} = \mathrm{grad}\ (\vec{c} \cdot \vec{a}) + \mathrm{curl}\ (\vec{c} \times \vec{a})$. For any $(x, y, z) \in D$, evaluate the scalar projection of the vector $\vec{A}(x, y, z)$ in the direction of $\vec{c}$.

---

[14]The symbol "$\cdot$" is used to denote two different operations: the function $\mathrm{div}\ \vec{a} \cdot \mathrm{div}\ \vec{b}$ is obtained as the product of two real scalar fields, while $\mathrm{curl}\ \vec{a} \cdot \mathrm{curl}\ \vec{b}$ is a real-valued function obtained as the scalar product of two vector fields.

Solution 5.3.17. Let $\vec{c} = \langle c_1, c_2, c_3 \rangle \neq \vec{0}$ and $\vec{a} = \langle a_1, a_2, a_3 \rangle$. Then,

$$
\vec{A} \cdot \vec{c} = \vec{c} \cdot \mathrm{grad}\ (\vec{c} \cdot \vec{a}) + \vec{c} \cdot \mathrm{curl}\ (\vec{c} \cdot \vec{a})
$$

$$
= \vec{c} \cdot \mathrm{grad}\ (c_1 a_1 + c_2 a_2 + c_3 a_3)
$$

$$
+ \vec{c} \cdot \mathrm{curl}\ \langle c_2 a_3 - c_3 a_2, c_3 a_1 - c_1 a_3, c_1 a_2 - c_2 a_1 \rangle
$$

$$
= c_1 \left(c_1 \frac{\partial a_1}{\partial x} + c_2 \frac{\partial a_2}{\partial x} + c_3 \frac{\partial a_3}{\partial x} \right) + c_2 \left(c_1 \frac{\partial a_1}{\partial y} + c_2 \frac{\partial a_2}{\partial y} + c_3 \frac{\partial a_3}{\partial y} \right)
$$

$$
+ c_3 \left(c_1 \frac{\partial a_1}{\partial z} + c_2 \frac{\partial a_2}{\partial z} + c_3 \frac{\partial a_3}{\partial z} \right) + c_1 \left(c_1 \frac{\partial a_2}{\partial y} - c_2 \frac{\partial a_1}{\partial y} - c_3 \frac{\partial a_1}{\partial z} \right.
$$

$$
\left. + c_1 \frac{\partial a_3}{\partial z} \right) - c_2 \left(c_1 \frac{\partial a_2}{\partial x} - c_2 \frac{\partial a_1}{\partial x} - c_2 \frac{\partial a_3}{\partial z} + c_3 \frac{\partial a_2}{\partial z} \right)
$$

$$
+ c_3 \left(c_3 \frac{\partial a_1}{\partial x} - c_1 \frac{\partial a_3}{\partial x} - c_2 \frac{\partial a_3}{\partial y} + c_3 \frac{\partial a_2}{\partial y} \right)
$$

$$
= (c_1^2 + c_2^2 + c_3^2) \cdot \left(\frac{\partial a_1}{\partial x} + \frac{\partial a_2}{\partial y} + \frac{\partial a_3}{\partial z} \right) = |\vec{c}|^2 \mathrm{div}\ \vec{a}.
$$

It follows that, for any $(x, y, z) \in D$, $\mathrm{comp}_{\vec{c}} \vec{A}(x, y, z) = \frac{\vec{A}(x,y,z) \cdot \vec{c}}{|\vec{c}|} = |\vec{c}| \mathrm{div}\ \vec{a}$.

Problem 5.3.18.

(a) Determine a scalar field F with the property that, for any $(x, y, z) \in \mathbb{R}^3$, $\mathrm{curl}(\langle F(x, y, z), 0, 0 \rangle) = \langle 0, 1, -1 \rangle$ and $\mathrm{div}\ (\mathrm{grad}\ F(x, y, z)) = F(0, 1, 1) = F(1, 1, 0) = 2$.

(b) Use the Fourier series of the function $f(t) = F(t, t, t)$, $t \in [-2, 2)$, to establish that $\sum_{k=1}^{\infty} (-1)^{k+1} \frac{2\pi k^2 - 2k - 1}{k^2(2k-1)} = \frac{5\pi^2}{12}$.

Solution 5.3.18.

(a) Suppose that the scalar field F has the required property. Since, for any $(x, y, z) \in \mathbb{R}^3$, $\mathrm{curl}(\langle F(x, y, z), 0, 0 \rangle) = (\mathrm{grad}\ F(x, y, z)) \times \vec{\imath} = \langle 0, \frac{\partial F(x,y,z)}{\partial z}, -\frac{\partial F(x,y,z)}{\partial y} \rangle$, it follows that $\frac{\partial F(x,y,z)}{\partial y} = \frac{\partial F(x,y,z)}{\partial z} = 1$. This implies that there is a twice differentiable function $g : \mathbb{R} \to \mathbb{R}$ such that, for any $(x, y, z) \in \mathbb{R}^3$, $F(x, y, z) = g(x) + y + z$. From $\mathrm{div}(\mathrm{grad}\ F(x, y, z)) = 2$, it follows that, for all $x \in \mathbb{R}$, $g''(x) = 2$, i.e., $g(x) = x^2 + ax + b$, for some $a, b \in \mathbb{R}$. Since $F(0,1,1) = F(1,1,0) = 2$, it follows that $a = b = 0$ and $F(x,y,z) = x^2 + y + z$.

(b) Since $f(t) = F(t,t,t) = t^2 + 2t$, for all $k \in \mathbb{N}_0$,

$$a_k = \frac{1}{2} \int_{-2}^{2} (t^2 + 2t) \cos \frac{k\pi t}{2} \, dt = \int_{0}^{2} t^2 \cos \frac{k\pi t}{2} \, dt$$

$$= \begin{cases} \frac{8}{3} & \text{if } k = 0, \\ \frac{16(-1)^k}{k^2\pi^2} & \text{if } k \neq 0. \end{cases}$$

For all $k \in \mathbb{N}$,

$$b_k = \frac{1}{2} \int_{-2}^{2} (t^2 + 2t) \sin \frac{k\pi t}{2} \, dt = 2 \int_{0}^{2} t \sin \frac{k\pi t}{2} \, dt = \frac{8(-1)^{k+1}}{k\pi}.$$

By Dirichlet's theorem, for all $t \in (-2, 2)$,

$$t^2 + 2t = \frac{4}{3} + \frac{8}{\pi^2} \sum_{k=1}^{\infty} \frac{(-1)^k}{k^2} \left(2\cos \frac{k\pi t}{2} - k\pi \sin \frac{k\pi t}{2} \right).$$

For $t = 1$,

$$3 = \frac{4}{3} + \frac{8}{\pi^2} \sum_{k=1}^{\infty} \frac{(-1)^k}{k^2} \left(2\cos \frac{k\pi}{2} - k\pi \sin \frac{k\pi}{2} \right)$$

$$= \frac{4}{3} + \frac{8}{\pi^2} \sum_{k=1}^{\infty} \left(\frac{(-1)^k}{2k^2} + \frac{(-1)^{k+1}\pi}{2k-1} \right)$$

$$= \frac{4}{3} + \frac{4}{\pi^2} \sum_{k=1}^{\infty} (-1)^{k+1} \frac{2\pi k^2 - 2k + 1}{k^2(2k-1)}.$$

Therefore, $\sum_{k=1}^{\infty} (-1)^{k+1} \frac{2\pi k^2 - 2k + 1}{k^2(2k-1)} = \frac{5\pi^2}{12}$.

Problem 5.3.19. Determine the divergence and the curl of the vector field $\vec{a} = \vec{r} \times (\vec{c} \times \vec{r})$, where $\vec{r} = \langle x, y, z \rangle$ is the radius vector of the point $(x, y, z) \in \mathbb{R}^3$ and $\vec{c}$ is a nonzero vector.

Solution 5.3.19. By Problems 5.3.2 and 5.3.11,

$$\text{div } \vec{a} = \text{div } ((\vec{r} \cdot \vec{r})\, \vec{c} - (\vec{r} \cdot \vec{c})\, \vec{r}) = \text{div } |\vec{r}|^2 \vec{c} - \text{div } (\vec{r} \cdot \vec{c})\, \vec{r}$$

$$= \vec{c} \cdot \text{grad } |\vec{r}|^2 - \vec{r} \cdot \text{grad } (\vec{r} \cdot \vec{c}) - (\vec{r} \cdot \vec{c})\, \text{div } \vec{r} = -2(\vec{r} \cdot \vec{c}).$$

By Problems 5.3.2, 5.3.11, and 5.3.15,

$$\text{curl } \vec{a} = \text{curl } (|\vec{r}|^2 \vec{c} - (\vec{r} \cdot \vec{c})\, \vec{r})$$

$$= \text{grad } |\vec{r}|^2 \times \vec{c} + \vec{r} \times \text{grad } (\vec{r} \cdot \vec{c}) = 3(\vec{r} \times \vec{c}).$$

Problem 5.3.20. Let $D \subseteq \mathbb{R}^3$. Let $\vec{a} : D \to \vec{\mathbb{E}}^3$ be a continuously differentiable vector field such that, for each $(x, y, z) \in D$, the vector $\vec{a}(x, y, z)$ is parallel to a certain constant nonzero vector $\vec{c}$. Prove that, for each $(x, y, z) \in D$, the vector curl $\vec{a}(x, y, z)$ is perpendicular to the vector $\vec{c}$.

Solution 5.3.20. Since $\vec{c} \parallel \vec{a}(x, y, z)$ for all $(x, y, z) \in D$, there is a scalar function $f : D \to \mathbb{R}$ such that $\vec{a}(x, y, z) = f(x, y, z)\, \vec{c}$, for all $(x, y, z) \in D$. By Problem 5.3.15, curl $\vec{a}(x, y, z) = (\text{grad } f(x, y, z)) \times \vec{c}$, which establishes the claim.

Problem 5.3.21. Let $\vec{c}$ be a nonzero vector, and let $\vec{a} = \vec{c} \sin^2 |\vec{r}|$ be the vector field, where $\vec{r}$ is the radius vector. Determine the fundamental period of the function $f(t) = ((\text{grad } |\vec{a}| + \text{curl } \vec{a}) \cdot (\vec{c}\, \text{div } \vec{a}))(0, 0, t)$, $t \in \mathbb{R} \backslash \{0\}$, and $f(0) = 0$.

Solution 5.3.21. For $\vec{r} \neq \vec{0}$, grad $|\vec{a}| = (\frac{|\vec{c}|}{|\vec{r}|} \sin 2|\vec{r}|)\vec{r}$, curl $\vec{a} = (\frac{|\vec{c}|}{|\vec{r}|} \sin 2|\vec{r}|)(\vec{r} \times \vec{c})$, and div $\vec{a} = \frac{|\vec{c}|}{|\vec{r}|}(\vec{r} \cdot \vec{c}) \sin 2|\vec{r}|$. It follows that, for $\vec{r} \neq \vec{0}$,

$$(\text{grad } |\vec{a}| + \text{curl } \vec{a}) \cdot (\vec{c}\, \text{div } \vec{a}) = \frac{|\vec{c}|^2}{|\vec{r}|^2}(\vec{r} \cdot \vec{c})^2 \sin^2 2|\vec{r}|.$$

Hence, for $t \neq 0$ and $\vec{c} = \langle c_1, c_2, c_3 \rangle$, $f(t) = \frac{|\vec{c}|^2\, c_3^2}{2}(1 - \cos 4t)$. The fundamental period of the function f is $T = \frac{2\pi}{4} = \frac{\pi}{2}$.

Problem 5.3.22. Determine all continuously differentiable functions $a = a(x)$ and $b = b(x)$, $x \in \mathbb{R}$, such that the vector field $\vec{R}(x, y, z) = \langle (2a(x) + b(x))z - a(x)y, b(x) - 2\cosh x, a(x) + 2\sinh x \rangle$ is conservative.

Solution 5.3.22. Since curl $\vec{R} = \langle 0, -(a'(x) + 2\cosh x - 2a(x) - b(x)), b'(x) - 2\sinh x + a(x)\rangle$, it follows that curl $\vec{R} = \vec{0}$, i.e., the vector field $\vec{R}$ is conservative, if and only if the functions a and b are solutions of the non-homogeneous system of linear differential equations

$$a'(x) = 2a(x) + b(x) - 2\cosh x \quad \text{and} \quad b'(x) = -a(x) + 2\sinh x.$$

Since by differentiating $a'(x) = 2a(x) + b(x) - 2\cosh x$ we obtain $b'(x) = a''(x) - 2a'(x) + 2\sinh x$, solving this system reduces to solving the homogeneous second-order linear differential equation with constant coefficients $a''(x) - 2a'(x) + a(x) = 0$. It follows that $a(x) = (c_1 + c_2 x)e^x$ and $b(x) = (c_2 - c_1 - c_2 x)e^x + 2\cosh x$, $x \in \mathbb{R}$, for $c_1, c_2 \in \mathbb{R}$.

Problem 5.3.23. Let $f : \mathbb{R}^+ \to \mathbb{R}$ be a continuously differentiable function, and let $\vec{a} = f(|\vec{r}|)\,\vec{r}$, where $\vec{r} \neq \vec{0}$ is the radius vector. Determine if the vector field $\vec{a}$ is conservative.

Solution 5.3.23. Since, by Problems 5.3.2 and 5.3.15, for $\vec{r} \neq \vec{0}$, curl $\vec{a} = $ curl $(f(|\vec{r}|)\,\vec{r}) = \frac{f'(|\vec{r}|)}{|\vec{r}|}\,(\vec{r} \times \vec{r}) = \vec{0}$, it follows that the vector field $\vec{a}$ is conservative in any simply connected region that does not contain the origin.

Problem 5.3.24. Determine a scalar potential for the gravitational field caused by mass m situated at the origin.

Solution 5.3.24. The gravitational field is determined by $\vec{a} = -\frac{gm}{|\vec{r}|^3}\,\vec{r}$, where $\vec{r} \neq \vec{0}$ is the radius vector and g is the gravitational constant. By Problem 5.3.23, the vector field $\vec{a}$ is conservative in any simply connected region that does not contain the origin. In such a region, there is a scalar potential for the field $\vec{a}$, i.e., a scalar field U such that grad $U = \vec{a}$. The fact that $-\frac{gm}{|\vec{r}|^3}\,\vec{r} = gm\,\mathrm{grad}\frac{1}{|\vec{r}|}$ implies that $U = \frac{gm}{|\vec{r}|} + c$, for some $c \in \mathbb{R}$. Since $\lim\limits_{|\vec{r}|\to\infty} U = 0$, it follows that $c = 0$ and $U = \frac{gm}{|\vec{r}|}$.

Problem 5.3.25. Let $D \subseteq \mathbb{R}^3$ be a simply connected region, and let $\vec{a}(x, y, z) = \langle P(x, y, z), Q(x, y, z), R(x, y, z)\rangle$, $(x, y, z) \in D$, be a conservative field with continuous partial derivatives. Let $\alpha, \beta, \gamma, \delta, \varepsilon, \zeta \in \mathbb{R}$ be such that $D' = (\alpha, \beta) \times (\gamma, \delta) \times (\varepsilon, \zeta) \subseteq D$. Let (a, b, c) be a point in D'. Confirm that the function $U : D' \to \mathbb{R}$

defined by

$$U(x,y,z) = \int_a^x P(t,\,y,\,z)\,dt + \int_b^y Q(a,\,t,\,z)\,dt + \int_c^z R(a,\,b,\,t)\,dt$$

is a scalar potential for the field $\vec{a}$ in D'.

Solution 5.3.25. Since for any $(x,\,y,\,z) \in D'$, any $t \in (a,x)$ (or $t \in (x,a)$ if $x < a$) belongs to (α,β), any $t \in (b,y)$ (or $t \in (y,b)$ if $y < b$) belongs to (γ,δ), and any $t \in (c,z)$ (or $t \in (z,c)$ if $z < c$) belongs to (ε,ζ), the function U is well defined. Also, since $\vec{a}(x,\,y,\,z) = \langle P(x,\,y,\,z), Q(x,\,y,\,z), R(x,\,y,\,z)\rangle$ is a conservative field with continuous partial derivatives, in the region D,

$$\frac{\partial P}{\partial y} = \frac{\partial Q}{\partial x}, \; \frac{\partial P}{\partial z} = \frac{\partial R}{\partial x}, \quad \text{and} \quad \frac{\partial Q}{\partial z} = \frac{\partial R}{\partial y}.$$

By the fundamental theorem of calculus, $\frac{\partial U(x,y,z)}{\partial x} = P(x,y,z)$. Next,

$$\begin{aligned}
\frac{\partial U(x,y,z)}{\partial y} &= \int_a^x \frac{\partial P(t,y,z)}{\partial y}\,dt + Q(a,y,z)\\
&= \int_a^x \frac{\partial Q(t,y,z)}{\partial x}\,dt + Q(a,y,z)\\
&= Q(x,y,z) - Q(a,y,z) + Q(a,y,z) = Q(x,y,z).
\end{aligned}$$

Similarly,

$$\begin{aligned}
\frac{\partial U(x,y,z)}{\partial z} &= \int_a^x \frac{\partial P(t,y,z)}{\partial z}\,dt + \int_b^x \frac{\partial Q(a,t,z)}{\partial z}\,dt + R(a,b,z)\\
&= \int_a^x \frac{\partial R(t,y,z)}{\partial x}\,dt + \int_b^y \frac{\partial R(a,t,z)}{\partial y}\,dt + R(a,b,z)\\
&= R(x,y,z) - R(a,y,z) + R(a,y,z) - R(a,b,z) + R(a,b,z)\\
&= R(x,y,z).
\end{aligned}$$

It follows that $\vec{a} = \operatorname{grad} U$, which confirms that the function U is a scalar potential for the conservative field $\vec{a}$ in D'.

Note: Observe that $U(a,b,c) = 0$. For any $d \in \mathbb{R}$, the scalar potential for $\vec{a}$ that takes the value d at the point (a,b,c) is determined by $U_d(x,y,z) = U(x,y,z) + d, \; (x,y,z) \in D'$.

Problem 5.3.26. Determine a continuously differentiable function, i.e., a function whose partial derivatives are continuous, $f : \mathbb{R}^2 \to \mathbb{R}$ so that the vector field $\vec{a} = \langle yz, xz, f(x,y)\rangle$ is conservative and that it has a scalar potential U such that $U(0,0,0) = U(1,1,1) = 0$.

Solution 5.3.26. Since curl $\vec{a}(x,y,z) = \left\langle \frac{\partial f(x,y)}{\partial y} - x, y - \frac{\partial f(x,y)}{\partial x}, 0 \right\rangle$, it follows that curl $\vec{a}(x,y,z) = \vec{0}$, i.e., the field is conservative, if and only if $\frac{\partial f(x,y)}{\partial y} = x$ and $\frac{\partial f(x,y)}{\partial x} = y$, $(x,y) \in \mathbb{R}^2$. Consequently, the field $\vec{a}$ is conservative if and only if $f(x,y) = xy + c$, for some $c \in \mathbb{R}$. The scalar potential for the field $\vec{a} = \langle yz, xz, xy + c \rangle$, $c \in \mathbb{R}$, is

$$U(x,y,z) = \int_0^x yz \, dt + \int_0^z c \, dt + d = xyz + cz + d,$$

for some $d \in \mathbb{R}$. From $U(0,0,0) = U(1,1,1) = 0$, it follows that $c = -1$ and $d = 0$. Hence, $f(x,y) = xy - 1$.

Problem 5.3.27. Consider the vector field $\vec{a} = \langle e^{-x}, 1, z^2 \rangle$.

(a) Establish that $\vec{a}$ is a conservative vector field. Determine the scalar potential U for the field $\vec{a}$ such that $U(0,0,0) = -1$.
(b) Determine the Fourier transform of the function $f : [0, \infty) \to \mathbb{R}$ defined by $U(x, f(x), 0) = 0$, for all $x \in [0, \infty)$.

Answer 5.3.27.

(a) The scalar potential is $U(x,y,z) = -e^{-x} + y + \frac{z^3}{3}$.
(b) $f(x) = e^{-x}$ and $\mathcal{F}_{\sin}[f(x)](u) = \sqrt{\frac{2}{\pi}} \frac{u}{1+u^2}$.

Problem 5.3.28. Let $\vec{a}$ and $\vec{b}$ be nonzero vectors, and let $\vec{g} = \vec{a} \times (\vec{r} \times \vec{b})$, where $\vec{r}$ is the radius vector. Determine curl $\vec{g}$ and div $\vec{g}$. Establish the condition(s) under which $\vec{g}$ is a conservative field and condition(s) under which $\vec{g}$ is a solenoidal field. Is it possible for the field $\vec{g}$ to be a Laplacian field?

Solution 5.3.28. From curl $\vec{g} = \mathrm{curl}\left((\vec{a} \cdot \vec{b})\, \vec{r} - (\vec{a} \cdot \vec{r})\, \vec{b}\right) = -\vec{a} \times \vec{b}$, it follows that the field is conservative if and only $\vec{a}$ and $\vec{b}$ are collinear vectors. From div $\vec{g} = 2(\vec{a} \cdot \vec{b})$ it follows that the field is solenoidal if and only if the vectors $\vec{a}$ and $\vec{b}$ are perpendicular to each other. Since a Laplacian field is both conservative and solenoidal and since $\vec{a}$ and $\vec{b}$ are nonzero vectors, the vector field $\vec{g}$ cannot be a Laplacian field.

Problem 5.3.29. Determine a continuously differentiable function $x \mapsto f(x)$, $x \in I$, such that $f(-2) = 1$ and such that $\vec{a} = f(x)\langle 1, \frac{y}{x+1}, f(x)\, z \rangle$ is a solenoidal field.

Solution 5.3.29. A continuously differentiable vector field $\vec{a} = f(x)\langle 1, \frac{y}{x+1}, f(x)\,z\rangle$ is solenoidal if and only if div $\vec{a} = 0$. i.e., if and only if, for any $x \neq -1$ in the domain of the function f, $f'(x) + \frac{f(x)}{1+x} + (f(x))^2 = 0$. Hence, f is the solution of the initial value problem $u' + \frac{u}{x+1} = -u^2$, $u(-2) = 1$, that contains a Bernoulli differential equation. The substitution $v = u^{1-2} = \frac{1}{u}$ transforms this initial value problem into the initial value problem $v' - \frac{v}{x+1} = 1$, $v(-2) = 1$, with the solution

$$v = e^{\int_{-2}^{x} \frac{dt}{1+t}} \left(1 + \int_{-2}^{x} e^{-\int_{-2}^{t} \frac{ds}{s+1}}\,dt\right) = |1+x|\left(1 + \int_{-2}^{x} \frac{dt}{|1+t|}\right)$$

$$= -(x+1)(1 - \ln|x+1|).$$

It follows that $f(x) = -\frac{1}{(x+1)(1-\ln|x+1|)}$, $x \in I = (-e-1, -1)$.

Problem 5.3.30. Determine a continuously differentiable function f for which $f(1) = \frac{3}{2}$ and such that $\vec{a} = \langle f(x), \frac{2xyf(x)}{x^2+1}, \frac{-3z}{x^2+1}\rangle$ is a solenoidal field.

Answer 5.3.30. $f(x) = \frac{3x}{x^2+1}$.

Problem 5.3.31. Let f and g be twice differentiable functions in $\mathbb{R}$, and let a vector field be defined by $\vec{a}(x,y,z) = \langle f(x) - g(x) + x^2 - 1 - (x+y)z, 2y - g(x) - x^2 - xz, f(x) - xy + z\rangle$. Use Laplace transformation to determine f and g so that grad (div $\vec{a}$) $= \langle 2f(x), 0, -1\rangle$, (curl $\vec{a}$) $\cdot \langle 1, -1, 2\rangle = 0$, and $f(0) = f'(0) = g'(0) = 0$, $g(0) = 1$.

Solution 5.3.31. Since grad (div $\vec{a}$) $= \langle f''(x) - g''(x) + 2, 0, -1\rangle$ and curl $\vec{a} = \langle 0, -f'(x) - x, -g'(x) - 2x\rangle$, the pair (f,g) is the solution to the initial value problem $f''(x) - g''(x) = 2f(x) - 2$, $f'(x) - 2g'(x) = 3x$, $f(0) = f'(0) = g'(0) = 0$, $g(0) = 1$. Let $F(s) = \mathcal{L}[f(x)](s)$ and $G(s) = \mathcal{L}[g(x)](s)$.[15] From the initial conditions, $\mathcal{L}[f''(t)](s) = s^2 F(s)$, $\mathcal{L}[f'(t)](s) = sF(s)$, $\mathcal{L}[g''(t)](s) = s^2 G(s) - s$, and $\mathcal{L}[g'(t)](s) = sG(s) - 1$. Functions $F(s)$ and $G(s)$ satisfy the linear system of algebraic equations: $(s^2 - 2)F(s) - s^2 G(s) = -s - \frac{2}{s}$,

---

[15]Recall that F and G are actually the Laplace transforms of $\mathcal{L}$-originals that coincide, respectively, with f and g in the interval $(0, \infty)$.

$sF(s) - 2sG(s) = -2 + \frac{3}{s^2}$. It follows that

$$F(s) = -\frac{7}{s(s^2-4)} = \frac{7}{4}\left(\frac{1}{s} - \frac{s}{s^2-4}\right) \text{ and}$$

$$G(s) = \frac{15}{8}\frac{1}{s} - \frac{3}{4}\frac{2}{s^3} - \frac{7}{8}\frac{s}{s^2-4}.$$

Therefore, $f(x) = \frac{7}{4}(1 - \cosh 2x)$ and $g(x) = \frac{1}{8}(15 - 6x^2 - 7\cosh 2x)$, $x \in \mathbb{R}$.

5.4   Line Integrals

Problem 5.4.1. Evaluate $\int_C ye^{-x}\, ds$ if C is the hodograph of the function $\vec{r}(t) = \ln(1+t^2)\,\vec{i} + (2\arctan t - t + 3)\,\vec{j}$, $t \in [0,1]$.

Solution 5.4.1. Since, for all $t \in (0,1)$, $|\vec{r}\,'(t)| = \sqrt{(\frac{2t}{1+t^2})^2 + (\frac{2}{1+t^2} - 1)^2} = 1$, the curve C is smooth, as a hodograph of a continuously differentiable one-to-one vector function with a derivative that is never zero. Since the function $(x,y) \mapsto ye^{-x}$ is continuous in $\mathbb{R}^2$, the line integral of the first kind $\int_C ye^{-x}\, ds$ exists and

$$\int_C ye^{-x}\, ds = \int_0^1 (2\arctan t - t + 3)e^{-\ln(1+t^2)}|\vec{r}\,'(t)|dt$$

$$= \frac{\pi(\pi+12)}{16} - \frac{\ln 2}{2}.$$

Problem 5.4.2. Evaluate $\int_C (x+y)\, ds$ if C is the union of a semicircle and a line segment, as shown in the figure.

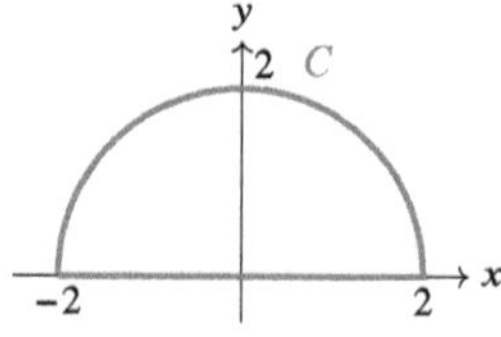

Solution 5.4.2. Evaluate the integral as the sum of two line integrals, one over the line segment $C_1 = \{(t,0) : t \in [-2,2]\}$ and the other over the semicircle $C_2 = \{(2\cos t, 2\sin t) : t \in [0,\pi]\}$. From

$\int_{C_1}(x+y)\,ds = \int_{-2}^{2} t\,dt = 0$ and $\int_{C_2}(x+y)\,ds = \int_0^\pi (2\cos t + 2\sin t)\sqrt{(-2\cos t)^2 + (2\sin t)^2}\,dt = 8$, it follows that $\int_C(x+y)\,ds = \int_{C_1}(x+y)\,ds + \int_{C_2}(x+y)\,ds = 8$.

Problem 5.4.3. Let $a \in \mathbb{R}^+$, and let C be the curve that is the intersection of the surfaces $z = \sqrt{a^2 - x^2 - y^2}$ and $\cosh(\arctan\frac{y}{x}) = \frac{a}{\sqrt{x^2+y^2}}$, $x > 0$. Evaluate the arc length of the curve C between the points $(a, 0, 0)$ and (x, y, z).

Solution 5.4.3. The curve C is the intersection of the hemisphere $z = \sqrt{a^2 - x^2 - y^2}$ and the cylinder with the generating curve $\Gamma = \{(x, y, 0) : \cosh(\arctan\frac{y}{x}) = \frac{a}{\sqrt{x^2+y^2}}, x > 0\}$. Since $r = \frac{a}{\cosh\theta}$, $\theta \in (-\frac{\pi}{2}, \frac{\pi}{2})$, is the polar equation of Γ, the curve C is the hodograph of the vector function

$$\vec{r}(\theta) = \left\langle \frac{a\cos\theta}{\cosh\theta}, \frac{a\sin\theta}{\cosh\theta}, \sqrt{a^2 - \left(\frac{a\cos\theta}{\cosh\theta}\right)^2 - \left(\frac{a\sin\theta}{\cosh\theta}\right)^2} \right\rangle$$

$$= \frac{a}{\cosh\theta}\langle \cos\theta, \sin\theta, |\sinh\theta|\rangle, \quad \theta \in \left(-\frac{\pi}{2}, \frac{\pi}{2}\right),$$

with $|\vec{r}'(\theta)| = \frac{a\sqrt{2}}{\cosh\theta}$. Let $T \in (-\frac{\pi}{2}, \frac{\pi}{2})$, and let C_T be the arc of the curve C with the initial point $r(0) = (a, 0, 0)$ and the terminal point $r(T) = (x, y, z)$. The arc length of C_T is

$$\int_{C_T} ds = \int_0^T |\vec{r}'(\theta)|\,d\theta = 2a\sqrt{2}\left(\arctan e^T - \frac{\pi}{4}\right).$$

Problem 5.4.4. A disc of diameter $2r$ centimetres is placed on the xy-plane so that its centre is at the origin. The disc is rotating counterclockwise about its centre at a constant angular speed of ω revolutions per second. Starting at the centre of the disc, a ladybug crawls along a fixed radius towards the edge of the disc at a constant speed of s cm/s. Assume that at time $t = 0$, the fixed radius lies on the positive ray of the x-axis.

(a) At what time T will the ladybug reach the edge of the disc?
(b) Let $(x(t), y(t))$ be the coordinates of the point in the xy-plane where the ladybug is located t seconds after it starts crawling. Determine the vector function $\vec{a} = \vec{a}(t) = \langle x(t), y(t)\rangle$, $t \in [0, T]$.

Note: The coordinate axis is fixed, so at any instant of time t, the length of $\vec{a} = \vec{a}(t)$ is changing at a rate of s cm/s and the angle between $\vec{a}(t)$ and $\vec{\imath}$ is changing at a rate of $2\pi\omega$ radians/s.

(c) Sketch the path $c = \{(x(t), y(t)) : t \in [0, T]\}$ traced by the ladybug.

(d) What is the arc length of c during the first $t \in (0, T)$ seconds of the ladybug's journey.

Solution 5.4.4.

(a) Since the ladybug is crawling along a line segment of length r cm at the constant rate of s cm/s, it will reach the edge of the disc after $T = \frac{r}{s}$ s.

(b) At the instant $t \in [0, T]$, the ladybug is at a distance of st from the origin, and the fixed radius, which the ladybug is crawling over, forms an angle of $2\pi\omega t$ with the positive direction of the x-axis, i.e., with its initial position. It follows that $\vec{a} = \vec{a}(t) = st\langle \cos 2\pi\omega t, \sin 2\pi\omega t \rangle$, $t \in [0, T]$.

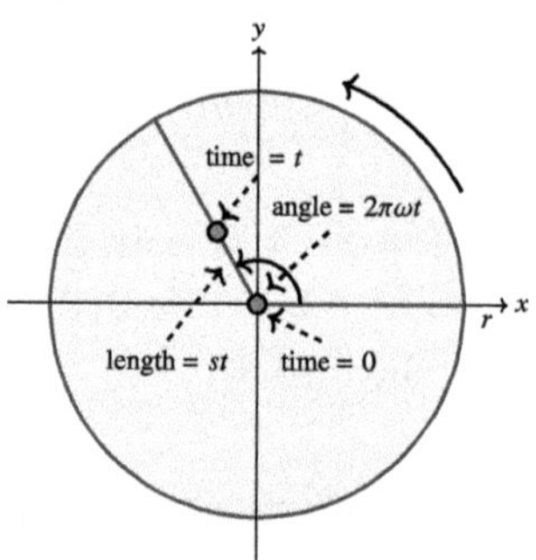

(c) See the adjacent image for a sketch of the spiral c traced by the ladybug on the xy-plane while the disc was rotating, i.e., for the hodograph of the vector function $\vec{a} = \vec{a}(t)$, $t \in [0, T]$. Note that $|\vec{a}(\frac{r}{s})| = r$, which agrees with the result obtained in Part (a).

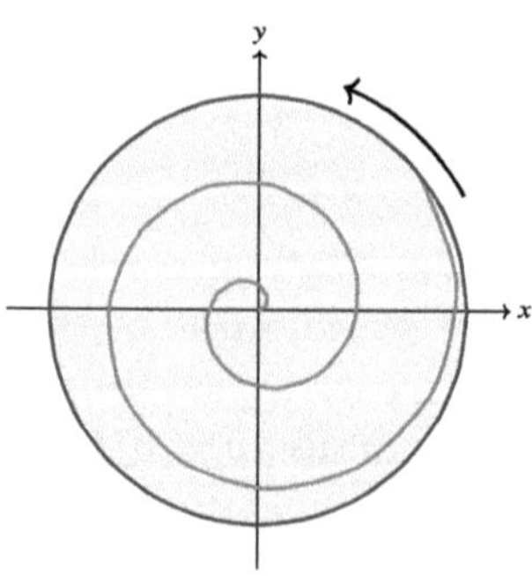

(d) Since $\frac{d\vec{a}(t)}{dt} = \langle s\cos 2\pi\omega t - 2\pi s\omega t\sin 2\pi\omega t, s\sin 2\pi\omega t + 2\pi s\omega t\cos 2\pi\omega\rangle$ and

$$\left|\frac{d\vec{a}(t)}{dt}\right|^2 = s^2(\cos 2\pi\omega t - 2\pi\omega t\sin 2\pi\omega t)^2$$
$$+ s^2(\sin 2\pi\omega t + 2\pi\omega t\cos 2\pi\omega t)^2 = s^2(1 + 4\pi^2\omega^2 t^2),$$

the arc length of the curve c during the first $t \in (0, T)$ seconds of the ladybug's journey is

$$\ell(t) = \int_0^t \left|\frac{d\vec{a}(u)}{du}\right| du = s\int_0^t \sqrt{1 + 4\pi^2\omega^2 u^2}\; du \text{ cm}.$$

Problem 5.4.5. The *total curvature* of a curve C is defined as $\int_C \kappa(s)\, ds$, where $s \mapsto \kappa(s)$ is the curvature in terms of the natural parametrisation of the curve C. Evaluate the total curvature of the circle $C = \{(x, y, 0) : x^2 + y^2 = R^2\}$, $R > 0$.

Solution 5.4.5. The natural parametrisation of the circle C is $\vec{c}(s) = \langle R\cos\frac{s}{R}, R\sin\frac{s}{R}, 0\rangle$, $s \in [0, 2R\pi]$. The curvature is $\kappa(s) = |\vec{c}''(s)| = \frac{1}{R}$. The total curvature is $\int_C \kappa\, ds = \frac{1}{R}\int_C ds = \frac{1}{R} \cdot 2R\pi = 2\pi$.

Problem 5.4.6. Evaluate the mass of the arc of the curve, determined by $y = \frac{2}{3}x^{\frac{3}{2}}$, between the point $O = (0, 0)$ and the point $A = (1, \frac{2}{3})$, if the linear density of the curve at the point M is the length of the arc with O as its initial point and M as its terminal point.

Solution 5.4.6. The curve is the hodograph of the vector function $\vec{r}(t) = \langle t, \frac{2}{3}t^{\frac{3}{2}}\rangle$, $t \in [0, \infty)$. Since, for $t \in \mathbb{R}^+$, $|\vec{r}(t)| = \sqrt{1 + t} > 1$, this is a smooth curve. Observe that $O = r(0)$ and $A = r(1)$. Let ℓ be the arc between O and A. Let $t \in [0, 1]$, let $M = r(t)$, and let ℓ_M be the arc with the endpoints O and M. The arc length of ℓ_M is

$$g(M) = g(r(t)) = \int_0^t |\vec{r}'(u)|\, du = \int_0^t \sqrt{1 + u}\; du = \frac{2}{3}((1 + t)^{\frac{3}{2}} - 1).$$

The mass of the arc ℓ is

$$m = \int_c g\, ds = \frac{2}{3}\int_0^1 \left((1 + t)^{\frac{3}{2}} - 1\right)|\vec{r}'(t)|\, dt = \frac{2}{3}\left(3 - \frac{4\sqrt{2}}{3}\right).$$

Problem 5.4.7. Evaluate the mass of the curve c determined[16] by $z = 1 - \sqrt{x^2 + y^2}$, $z \geq 0$, and $(x - 1)^2 + y^2 = 1$ if the linear density at any point of the curve is the distance between the point and the xz-plane.

Solution 5.4.7. The curve c is the intersection of the cone $z = 1 - \sqrt{x^2 + y^2}$, $z \geq 0$, and the cylinder, with the circle $(x-1)^2 + y^2 = 1$ as the generating curve.

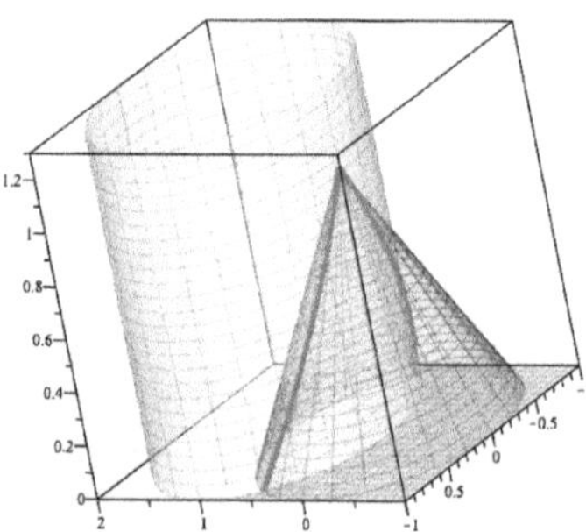

Since the polar equation of this circle is $r = 2\cos\theta$, $\theta \in \left[-\frac{\pi}{2}, \frac{\pi}{2}\right]$, a parametrisation of the curve c is $\vec{r}(\theta) = \langle 2\cos^2\theta, 2\sin\theta\cos\theta, 1 - 2\cos\theta\rangle$, $\theta \in I \subseteq \left[-\frac{\pi}{2}, \frac{\pi}{2}\right]$. Because of $0 \leq z = 1 - \sqrt{x^2 + y^2}$, it follows that, for any $(x, y, z) \in c$, $z \in [0, 1]$. This implies that $\theta \in I$ if and only if $1 - 2\cos\theta \in [0, 1]$. Hence, $\theta \in I$ if and only if $\cos\theta \in [0, \frac{1}{2}]$; therefore, $I = \left[-\frac{\pi}{2}, -\frac{\pi}{3}\right] \cup \left[\frac{\pi}{3}, \frac{\pi}{2}\right]$. For $\alpha \in [0, \frac{\pi}{6}]$,

$$\vec{r}\left(-\frac{\pi}{2} + \alpha\right) = \langle 2\sin^2\alpha, \ -\sin 2\alpha, \ 1 - 2\sin\alpha\rangle$$

and

$$\vec{r}\left(\frac{\pi}{2} - \alpha\right) = \langle 2\sin^2\alpha, \ \sin 2\alpha, \ 1 - 2\sin\alpha\rangle,$$

which demonstrates that points $r(-\frac{\pi}{2} + \alpha)$ and $r(\frac{\pi}{2} - \alpha)$ are symmetric with respect to the xz-plane. This means that the linear density at the point $r(-\frac{\pi}{2} + \alpha)$ is equal to the linear density at the point $r(\frac{\pi}{2} - \alpha)$. Since the curve c is symmetric with respect to the xz-plane, it is enough to evaluate the mass of the part of the curve that is determined by $\theta \in [\frac{\pi}{3}, \frac{\pi}{2}]$. Since, for $\theta \in [\frac{\pi}{3}, \frac{\pi}{2}]$, the linear

---

[16]This means that c is the intersection of the two surfaces.

density at the point $r(\theta)$ is $g(\theta) = \sin 2\theta$ and $|\vec{r}\,'(\theta)| = 2\sqrt{1 + \sin^2 \theta}$, it follows that the mass of the curve is

$$m = \int_c g \, ds = 2 \int_{\frac{\pi}{3}}^{\frac{\pi}{2}} g(\theta)|\vec{r}\,'(\theta)| \, d\theta = 4 \int_{\frac{\pi}{3}}^{\frac{\pi}{2}} \sin 2\theta \cdot \sqrt{1 + \sin^2 \theta} \, d\theta$$

$$= \frac{1}{3}(16\sqrt{2} - 7\sqrt{7}).$$

Problem 5.4.8. The function $y = f(x)$, $x \in I$, is the solution of the initial value problem $2x \, dx + (1 - x^2) \, dy = 0$, $y(0) = 0$. Let the planar curve Γ be the hodograph of the vector function $\vec{r}(t) = \langle t, f(t) \rangle$, $t \in I$. Evaluate the mass of the curve if the linear density at the point on Γ is the geometric mean of the distances between the point and the lines $x = 1$ and $x = -1$.

Solution 5.4.8. For $x \in (-1, 1)$, the initial value problem $2x \, dx + (1 - x^2) \, dy = 0$, $y(0) = 0$, is equivalent to the initial value problem $y' = \frac{2x}{x^2 - 1}$, $y(0) = 0$. It follows that, for $x \in (-1, 1)$, $f(x) = \int_0^x \frac{2u}{u^2 - 1} \, du = \ln(1 - x^2)$. Since, for all $t \in (-1, 1)$, $|\vec{r}\,'(t)| = \frac{1 + t^2}{1 - t^2} > 0$, the curve Γ is smooth. Let $t \in (-1, 1)$ and $A(t) = (t, \ln(1 - t^2)) \in \Gamma$. The distance between the point $A(t)$ and the line $x = 1$ is $d_1(A(t)) = 1 - t$. Similarly, $d_{-1}(A(t)) = 1 + t$. The linear density at the point $A(t)$ is $g(t) = \sqrt{(1 - t)(1 + t)} = \sqrt{1 - t^2}$. The mass of the curve is

$$m = \int_\Gamma g \, ds = \int_{-1}^1 \sqrt{1 - t^2} \, \frac{1 + t^2}{1 - t^2} \, dt = 2 \int_0^1 \frac{1 + t^2}{\sqrt{1 - t^2}} \, dt$$

$$= \begin{vmatrix} t = \sin u \Rightarrow dt = \cos u \, du \\ t = 0 \Rightarrow u = 0 \\ t = 1 \Rightarrow u = \frac{\pi}{2} \end{vmatrix} = 2 \int_0^{\frac{\pi}{2}} (1 + \sin^2 u) \, du = \frac{3\pi}{2}.$$

Note: Observe that $\int_{-1}^1 \sqrt{1 - t^2} \, \frac{1 + t^2}{1 - t^2} \, dt$ is an improper integral.

Problem 5.4.9. Let $\vec{F}(x, y) = \langle x + 2y, x^2 - y^2 \rangle$. Evaluate $\int_C \vec{F} \cdot d\vec{r}$ if:

(a) C is the line segment C_1 from $(0, 0)$ to $(1, 0)$, followed by the line segment C_2 from $(1, 0)$ to $(1, 1)$.

(b) C is the arc of the parabola $y = x^2$ from $(0, 0)$ to $(1, 1)$.

Are the results in Parts (a) and (b) the same? Explain why or why not.

Solution 5.4.9.

(a) From $C_1 = \{(x,0) : 0 \le x \le 1\}$ and $C_2 = \{(1,x) : 0 \le x \le 1\}$, it follows that $\int_C \vec{F} \cdot d\vec{r} = \int_{C_1} \vec{F} \cdot d\vec{r} + \int_{C_2} \vec{F} \cdot d\vec{r} = \int_0^1 x \, dx + \int_0^1 (1 - x^2) \, dx = \frac{7}{6}$.

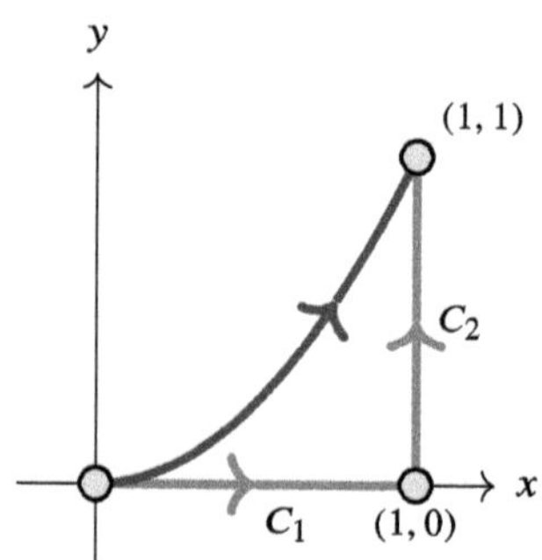

(b) In this case, $C = \{(x, x^2) : 0 \le x \le 1\}$ and $\int_C \vec{F} \cdot d\vec{r} = \int_0^1 (x + 2x^2 + 2x(x^2 - x^4)) \, dx = \frac{4}{3}$.

Since $\frac{\partial}{\partial y}(x + 2y) = 2 \ne \frac{\partial}{\partial x}(x^2 - y^2) = 2x$, the vector field $\vec{F}$ is not conservative. The line integral depends on the path between the points $(0,0)$ and $(1,1)$.

Problem 5.4.10. Let $f(x,y) = \sqrt{x^2 + y^2 + 1}$ and let $\vec{F} = \nabla f$. Evaluate the line integral $\int_C \vec{F} \cdot d\vec{r}$ along a piecewise smooth curve C starting at $P = (0,0)$ and ending at $Q = (2,1)$.

Solution 5.4.10. By definition, $\vec{F}(x,y) = \nabla f(x,y) = \frac{\langle x, y \rangle}{\sqrt{x^2 + y^2 + 1}}$. By the fundamental theorem for line integrals, $\int_C \vec{F} \cdot d\vec{r} = f(2,1) - f(0,0) = \sqrt{6} - 1$.

Problem 5.4.11. Determine the scalar components of a vector field $\vec{a} : \mathbb{R} \to \mathbb{E}^3$ such that $\vec{a}''(t) = \langle 6, 0, 0 \rangle$ for all t, $\vec{a}(0) = \langle 0, 3, 4 \rangle$, and $\vec{a}'(0) = \langle 0, 0, 1 \rangle$. Evaluate the line integral of the vector field $\vec{F}(x,y) = \langle \pi \cos \pi x, 3y^2 + z, 4z^3 + y \rangle$ along the hodograph of $\vec{a}(t)$, $t \in [0, 1]$.

Solution 5.4.11. From $\vec{a}'(t) = \langle 6t, 0, 0 \rangle + \vec{c}$ and $\vec{a}'(0) = \langle 0, 0, 1 \rangle$, it follows that $\vec{c} = \langle 0, 0, 1 \rangle$ and $\vec{a}'(t) = \langle 6t, 0, 1 \rangle$. Similarly, $\vec{a}(t) = \langle 3t^2, 3, 4 + t \rangle$, which implies $\vec{a}(1) = \langle 3, 3, 5 \rangle$.

Since curl $\vec{F} = \begin{vmatrix} \vec{i} & \vec{j} & \vec{k} \\ \frac{\partial}{\partial x} & \frac{\partial}{\partial y} & \frac{\partial}{\partial z} \\ \pi \cos(\pi x) & 3y^2 + z & 4z^3 + y \end{vmatrix} = \vec{0}$, the vector field $\vec{F}$ is conservative. By Problem 5.3.25, a scalar potential for the field $\vec{F}$ is the function $f(x, y, z) = \int_0^x \pi \cos \pi t \, dt + \int_0^y (3t^2 + z) \, dt + \int_0^z 4t^3 \, dt = \sin \pi x + y^3 + yz + z^4$. By the fundamental theorem for line integrals, $\int_C \vec{F} \cdot d\vec{r} = f(3, 3, 5) - f(0, 3, 4) = 372$.

Problem 5.4.12. Establish that $\vec{F}(x, y) = \langle 3 + 2xy, x^2 - 3y^2 \rangle$ is a conservative field. Evaluate $\int_C \vec{F} \cdot d\vec{r}$ if C is the hodograph of the vector function $\vec{r}(t) = \langle e^t \sin t, e^t \cos t \rangle$, $t \in [0, \pi]$, oriented in the sense of increase of the variable t.

Solution 5.4.12. Since $\frac{\partial}{\partial y}(3 + 2xy) = 2x = \frac{\partial}{\partial x}(x^2 - 3y^2)$, for all $(x, y) \in \mathbb{R}^2$, the vector field $\vec{F}$ is conservative. By Problem 5.3.25, a scalar potential for the field $\vec{F}$ is the function $f(x, y) = \int_0^x (3 + 2ty) \, dt - 3 \int_0^y t^2 \, dt = 3x + x^2 y - y^3$. Since $\vec{a}(0) = \langle 0, 1 \rangle$ and $\vec{a}(\pi) = \langle 0, -e^\pi \rangle$, by the fundamental theorem for line integrals, $\int_C \vec{F} \cdot d\vec{r} = f(0, -e^\pi) - f(0, 1) = e^{3\pi} + 1$.

Problem 5.4.13. Establish that $\vec{F}(x, y, z) = \langle y + y^2 z, x - z + 2xyz, -y + xy^2 \rangle$ is a conservative field. Evaluate $\int_C \vec{F} \cdot d\vec{r}$, where C is the line segment with the initial point $(2, 2, 1)$ and the terminal point $(1, -1, 2)$.

Answer 5.4.13. $\int_C \vec{F} \cdot d\vec{r} = -7$.

Problem 5.4.14. Let $\vec{F} = \langle \frac{1}{1+y^2}, -\frac{2xy}{(y^2+1)^2} + ze^{yz}, ye^{yz} + 2z \rangle$. Evaluate $\int_C \vec{F} \cdot d\vec{r}$, where C is the part of the helix determined by $\vec{r}(t) = \langle \cos t, \sin t, t \rangle$, $t \in [0, 2\pi]$.

Solution 5.4.14. Let $P(x, y, z) = \frac{1}{1+y^2}$, $Q(x, y, z) = -\frac{2xy}{(y^2+1)^2} + ze^{yz}$, and $R(x, y, z) = ye^{yz} + 2z$. Since, for all $(x, y, z) \in \mathbb{R}^3$, $\frac{\partial P(x,y,z)}{\partial y} = -\frac{2y}{(1+y^2)^2} = \frac{\partial Q(x,y,z)}{\partial x}$, $\frac{\partial P(x,y,z)}{\partial z} = 0 = \frac{\partial R(x,y,z)}{\partial x}$, and $\frac{\partial Q(x,y,z)}{\partial z} = (1 + yz)e^{yz} = \frac{\partial R(x,y,z)}{\partial y}$, the vector field $\vec{F}$ is conservative. A scalar potential for the field $\vec{F}$ is the function

$$f(x,y,z) = \int_0^x \frac{dt}{1+y^2} + \int_0^y ze^{tz}\, dt + \int_0^z 2t\, dt$$

$$= \frac{x}{1+y^2} + e^{yz} + z^2 - 1.$$

By the fundamental theorem for line integrals, $\int_C \vec{F} \cdot d\vec{r} = f(1,0,2\pi) - f(1,0,0) = 4\pi^2$.

Problem 5.4.15. Establish that the work done by the vector field $\vec{a}(x,y) = \langle 2xy, x^2 \rangle$ along a curve c depends only on the initial and terminal points of c, i.e., establish that the work along two curves with the same initial and terminal points is the same. Evaluate the work of the field along a curve with the initial point $A = (1,1)$ and the terminal point $B = (2,5)$.

Solution 5.4.15. Since, for any $x,y \in \mathbb{R}^2$, curl $\vec{a}(x,y) = \vec{0}$, the field $\vec{a}$ is conservative. By the fundamental theorem for line integrals, for a curve c with the initial point A and the terminal point B, $\int_c \vec{a} \cdot d\vec{r} = U(B) - U(A)$, where $U(x,y) = x^2 y + \alpha$, $\alpha \in \mathbb{R}$, is a scalar potential for $\vec{a}$. If c is a curve with the initial point $A = (1,1)$ and the terminal point $B = (2,5)$, then $\int_c \vec{a} \cdot d\vec{r} = U(2,5) - U(1,1) = 19$.

Problem 5.4.16. Evaluate $I = \int_{(1,1)}^{(1,-1)} (x-y)(dy - dx)$.

Answer 5.4.16. $I = -2$.

Problem 5.4.17. Let a, b, and c be positive numbers. Evaluate the work done by the field $\vec{a} = \langle y, x, 0 \rangle$ along the curve Γ, the shorter arc of the great ellipse of the ellipsoid $\frac{x^2}{a^2} + \frac{y^2}{b^2} + \frac{z^2}{c^2} = 1$, with the initial point $A = (0,0,c)$ and the terminal point $B = \left(\frac{a}{\sqrt{2}}, -\frac{b}{\sqrt{2}}, 0\right)$.

Solution 5.4.17. *Solution 1*: The great ellipse that contains the curve Γ is the intersection of the ellipsoid and the plane determined by the origin and points A and B, i.e., the plane $bx + ay = 0$. Since the point B belongs to the fourth quadrant of the xy-plane and since the point A lies on the positive ray of the z-axis, it follows that, for any $(x,y,z) \in \Gamma$, $x \geq 0$, $y \leq 0$, and $z \geq 0$. Thus, the curve Γ

is the hodograph of the vector function $\vec{r}(t) = \langle at, -bt, c\sqrt{1 - 2t^2}\,\rangle$, $t \in [0, \frac{\sqrt{2}}{2}]$. It follows that

$$\int_{\Gamma} \vec{a} \cdot d\vec{r} = \int_{0}^{\frac{\sqrt{2}}{2}} \langle -bt,\, at,\, 0 \rangle \cdot \left\langle a,\, -b,\, -\frac{2tc}{\sqrt{1 - 2t^2}} \right\rangle \, dt = -\frac{ab}{2}.$$

Solution 2: Since curl $\vec{a} = \vec{0}$, the field $\vec{a}$ is conservative with a scalar potential $U(x, y, z) = xy$. By the fundamental theorem for line integrals, $\int_{\Gamma} \vec{a} \cdot d\vec{r} = U(\frac{a}{\sqrt{2}}, -\frac{b}{\sqrt{2}}, 0) - U(0, 0, 0) = -\frac{ab}{2}$.

Problem 5.4.18. Evaluate the work done by the field $\vec{a} = \langle x - 1, y - 2, 2x - y \rangle$ along the curve c that is the intersection of the paraboloid $z = x^2 + y^2$, $z \in [0, 10]$, and the plane $2x + 4y - z = 0$. When viewed from a point that lies on the positive ray of the y-axis, the curve c is oriented clockwise.

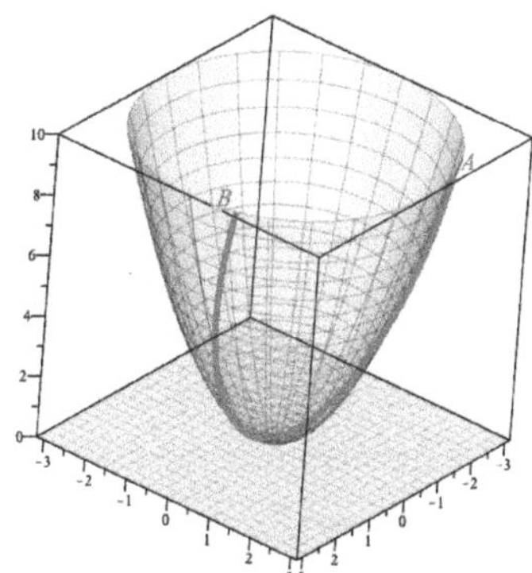

Solution 5.4.18. If $(x, y, 10) \in c$, then $x^2 + y^2 = 10$ and $2x + 4y = 10$. This system of equations has two solutions: $x = -1$, $y = 3$ and $x = 3$, $y = 1$. It follows that the end points of the curve are $A = (-1, 3, 10)$ and $B = (3, 1, 10)$. The curve is the hodograph of the vector function $\vec{r}(t) = \langle 1 + \sqrt{5}\ \cos t, 2 + \sqrt{5}\ \sin t, 2\sqrt{5}\ (\sqrt{5} + \cos t + 2\sin t) \rangle$, $t \in [\pi - T, 2\pi - T]$, where $T = \arccos \frac{2}{\sqrt{5}}$. When viewed from a point that lies on the positive ray of the y-axis, in the clockwise orientation of c, $A = r(\pi - T)$ is the initial point and $B = r(2\pi - T)$ is the terminal point of c. The work done by the field $\vec{a}$ is $\int_{c} \vec{a} \cdot d\vec{r} = \int_{\pi - T}^{2\pi - T} \langle \sqrt{5}\ \cos t, \sqrt{5}\ \sin t, \sqrt{5}\ (2\cos t - \sin t) \rangle \cdot \langle -\sqrt{5}\ \sin t, \sqrt{5}\ \cos t, 2\sqrt{5}\ (-\sin t + 2\cos t) \rangle \, dt = 25\pi$.

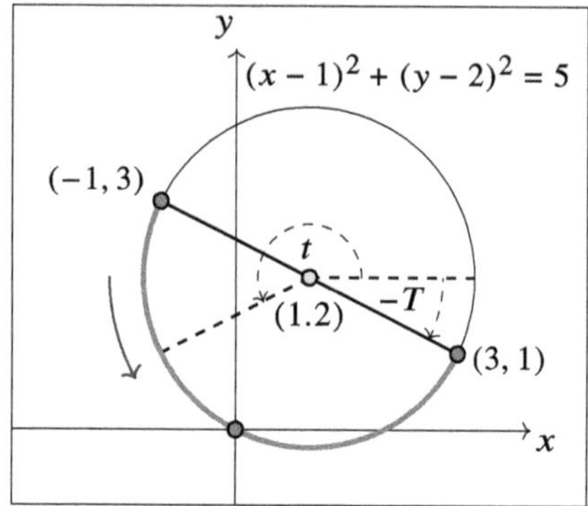

Note: Since curl $\vec{a} = -2\vec{j}$, the vector field $\vec{a}$ is not conservative. We cannot use the fundamental theorem for line integrals.

5.5   Surface Integrals

Problem 5.5.1. Evaluate $\iint_S \frac{dS}{x^2+y^2+z^2}$, where S is the part of the cylinder $x^2 + y^2 = 1$ that lies in the first octant below the plane $z = 1$.

Solution 5.5.1. The surface S is the hodograph of the vector function $\vec{r}(t, z) = \langle \cos t, \sin t, z \rangle$, $(t, z) \in D = [0, \frac{\pi}{2}] \times [0, 1]$. The fact that, for all $(t, z) \in (0, \frac{\pi}{2}) \times (0, 1)$, $\left| \frac{\partial \vec{r}(t,z)}{\partial t} \times \frac{\partial \vec{r}(t,z)}{\partial z} \right| = 1$ implies that S is a smooth surface. Since the function $(x, y, z) \mapsto \frac{1}{x^2+y^2+z^2}$ is continuous in $\mathbb{R}^3 \setminus \{(0, 0, 0)\}$, it follows that

$$\iint_S \frac{dS}{x^2 + y^2 + z^2} = \iint_D \frac{\left| \frac{\partial \vec{r}(t,z)}{\partial t} \times \frac{\partial \vec{r}(t,z)}{\partial z} \right|}{\cos^2 t + \sin^2 t + z^2} \, dt\, dz$$

$$= \int_0^{\frac{\pi}{2}} dt \int_0^1 \frac{dz}{1 + z^2} = \frac{\pi^2}{8}.$$

Problem 5.5.2. Evaluate the area of the portion of the unit sphere inside the cylinder $x^2 + y^2 = \frac{1}{2}$ and $z > 0$.

Solution 5.5.2. Let S be the portion of the unit sphere inside the cylinder. Then, $S = \{(\cos\theta \cos\phi, \sin\theta \cos\phi, \sin\phi) : \theta \in [0, 2\pi]$ and $\phi \in [\frac{\pi}{4}, \frac{\pi}{2}]\}$. The area is $A = \iint_S dS = \int_0^{2\pi} d\theta \int_{\frac{\pi}{4}}^{\frac{\pi}{2}} \cos\phi \, d\phi = \pi(2 - \sqrt{2})$.

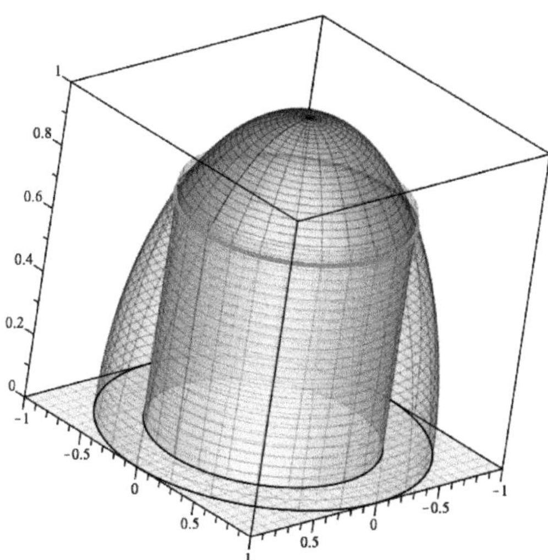

Problem 5.5.3. Evaluate the surface area of the spider web surface parametrised by $\vec{r}(u,v) = \langle \sqrt{2}v\cos u, v\sin u, v\sin u \rangle$, $(u,v) \in [0,2\pi] \times [0,1]$.

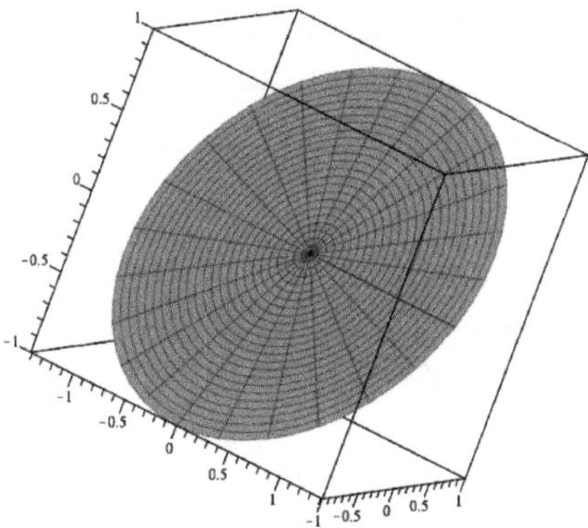

Solution 5.5.3. Since

$$\frac{\vec{r}(u,v)}{\partial u} \times \frac{\vec{r}(u,v)}{\partial v} = \langle -\sqrt{2}v\sin u, v\cos u, v\cos u \rangle$$

$$\times \langle \sqrt{2}\cos u, \sin u, \sin u \rangle$$

$$= \sqrt{2}v\langle 0, 1, -1 \rangle,$$

the surface area is $\int_0^{2\pi}\int_0^1 |\frac{\vec{r}(u,v)}{\partial u} \times \frac{\vec{r}(u,v)}{\partial v}| \, dudv = 2\int_0^{2\pi} du \int_0^1 v\, dv = 2\pi$.

Problem 5.5.4. Evaluate the surface area of the part of the surface $(x+y)^2 + z = 1$ that lies in the first octant.

Solution 5.5.4. The question is to evaluate the surface area of the surface S that is the hodograph of the function $\vec{a}(x,y) = \langle x, y, 1 - (x+y)^2 \rangle$, $(x,y) \in D = \{(s,t) : s \in [0,1] \text{ and } t \in [0,1] \text{ and } s+t \le 1\}$.

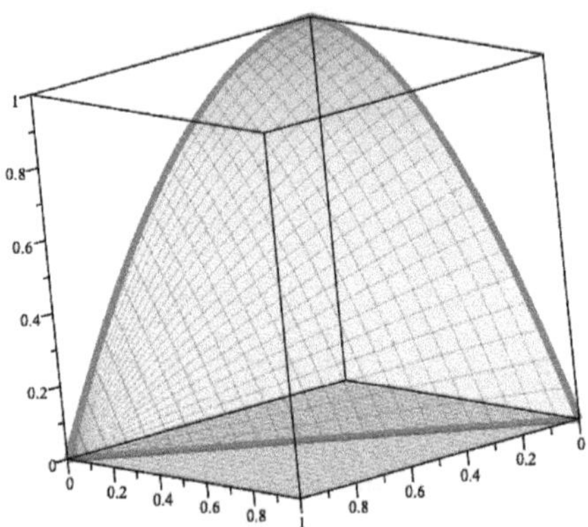

The substitutions $x = \frac{u+v}{2}$ and $y = \frac{u-v}{2}$ yield another parametrisation of the surface S: $\vec{b}(u,v) = \langle \frac{u+v}{2}, \frac{u-v}{2}, 1 - u^2 \rangle$, $(u,v) \in E = \{(s,t) : |t| \le s \le 1\}$. Since, for each (u,v) in the interior of E, $|\frac{\partial \vec{b}(u,v)}{\partial u} \times \frac{\partial \vec{b}(u,v)}{\partial v}| = \frac{1}{2}\sqrt{1 + 8u^2} > 0$, S is a smooth surface. The surface area of S is

$$P = \iint_S dS = \iint_E \left| \frac{\partial \vec{b}(u,v)}{\partial u} \times \frac{\partial \vec{b}(u,v)}{\partial v} \right| du\, dv$$

$$= \frac{1}{2} \int_0^1 \sqrt{1 + 8u^2}\, du \int_{-u}^{u} dv = \frac{3\sqrt{3} - 1}{24}.$$

Problem 5.5.5. Evaluate the surface area of the part of the hyperbolic paraboloid $z = xy$ that lies inside the cylinder $(x^2 + y^2)^2 = 2xy$.

Solution 5.5.5. The generating curve of the cylinder is the lemniscate determined by $r^2 = \sin 2\theta$, $\theta \in [0, \frac{\pi}{2}] \cup [\pi, \frac{3\pi}{2}]$. The surface S, the part of the hyperbolic paraboloid that lies inside the cylinder, is the hodograph of the vector function $\vec{R}(r,\theta) = \langle r\cos\theta, r\sin\theta, r^2\sin\theta\cos\theta \rangle$, $(r,\theta) \in D = \{(\rho,\varphi) : \varphi \in [0, \frac{\pi}{2}] \cup [\pi, \frac{3\pi}{2}], \rho \in [0, \sqrt{\sin 2\varphi}]\}$. Since, for any (r,θ) in the interior of D, $|\frac{\partial \vec{R}(r,\theta)}{\partial r} \times \frac{\partial \vec{R}(r,\theta)}{\partial \theta}| = r(1 + r^2)$, S is a piecewise smooth surface. Its surface area is

$$P = \iint_D dS = \int_E \left| \frac{\partial \vec{R}(r, \theta)}{\partial r} \times \frac{\partial \vec{R}(r, \theta)}{\partial \theta} \right| dr d\theta$$

$$= 2 \int_0^{\frac{\pi}{2}} d\theta \int_0^{\sqrt{\sin 2\theta}} r(1 + r^2) \, dr = \frac{\pi + 4}{8}.$$

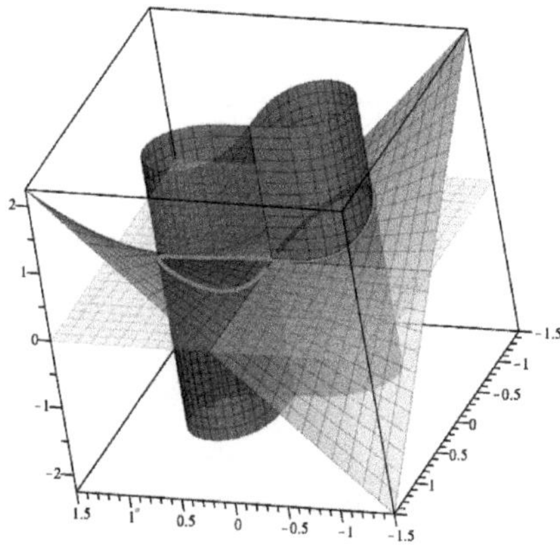

Problem 5.5.6. Evaluate the surface area of the torus S that is the hodograph of the vector function $\vec{r}(u, v) = \langle (2 + \cos u) \cos v, (2 + \cos u) \sin v, \sin u \rangle$, $(u, v) \in [0, 2\pi] \times [0, 2\pi]$.

Solution 5.5.6. From

$$\frac{\partial \vec{r}(u, v)}{\partial u} \times \frac{\partial \vec{r}(u, v)}{\partial v} = \begin{vmatrix} \vec{i} & \vec{j} & \vec{k} \\ -\sin u \cos v & -\sin u \sin v & \cos u \\ -(2 + \cos u) \sin v & (2 + \cos u) \cos v & 0 \end{vmatrix}$$

$$= \langle -(2 + \cos u) \cos u \cos v, -(2 + \cos u) \cos u \sin v, -(2 + \cos u) \sin u \rangle,$$

it follows that $|\frac{\partial \vec{r}(u,v)}{\partial u} \times \frac{\partial \vec{r}(u,v)}{\partial v}| = 2 + \cos u$. The surface area of the torus is $\iint_S dS = \iint_D |\frac{\partial \vec{r}(u,v)}{\partial u} \times \frac{\partial \vec{r}(u,v)}{\partial v}| dA = \int_0^{2\pi} (2 + \cos u) \, du \int_0^{2\pi} dv = 8\pi^2$.

Problem 5.5.7. Establish that the definition of the surface area as the value of the surface integral $\iint_S dS$ is consistent with the formula for the surface area of the surface of revolution obtained by revolving the graph of a function around the x-axis from single-variable calculus.

Solution 5.5.7. Let $y = f(x) \geq 0$, $x \in [a, b]$, be a continuously differentiable function, and let the surface S be obtained by revolving the graph of the function f about the x-axis. The intersection of the surface S and a plane $x = u$, for any $u \in [a, b]$, is a circle with its centre at the point $(u, 0, 0)$ and radius $f(u)$. Consequently, the surface S is the hodograph of the vector function $\vec{r}(u, v) = \langle u, f(u) \cos v, f(u) \sin v \rangle$, $(u, v) \in [a, b] \times [0, 2\pi]$. From

$$\frac{\partial \vec{r}(u, v)}{\partial u} \times \frac{\partial \vec{r}(u, v)}{\partial v} = \begin{vmatrix} \vec{i} & \vec{j} & \vec{k} \\ 1 & f'(u) \cos v & f'(u) \sin v \\ 0 & -f(u) \sin v & f(u) \cos v \end{vmatrix}$$

$$= \langle f(u) f'(u), -f(u) \cos v, -f(u) \sin v \rangle,$$

it follows that the surface area is

$$\iint_S dS = \iint_D \left| \frac{\partial \vec{r}(u, v)}{\partial u} \times \frac{\partial \vec{r}(u, v)}{\partial v} \right| dA$$

$$= \int_0^{2\pi} dv \int_a^b f(u) \sqrt{1 + (f'(u))^2} \, du$$

$$= 2\pi \int_a^b f(u) \sqrt{1 + (f'(u))^2} \, du.$$

Problem 5.5.8. Let the surface S be the hodograph of the vector function $\vec{R}(u, v) = \langle \cos u, \sin u, \frac{uv}{2} \rangle$, $u, v \in [0, \pi]$. Evaluate the mass of the surface S if the surface density is $f(x, y, z) = e^z$, $(x, y, z) \in S$.

Solution 5.5.8. The question is to evaluate $\iint_S f(x, y, z) \, dS$. From $\frac{\partial \vec{R}(u,v)}{\partial u} \times \frac{\partial \vec{R}(u,v)}{\partial v} = \langle \frac{u \cos u}{2}, \frac{u \sin u}{2}, 0 \rangle$, it follows that $dS = |\frac{\partial \vec{R}(u,v)}{\partial u} \times \frac{\partial \vec{R}(u,v)}{\partial v}| \, dudv = \frac{u}{2} \, dudv$. Therefore, $\iint_S f(x, y, z) \, dS = \frac{1}{2} \int_0^\pi u \, du \int_0^\pi e^{\frac{uv}{2}} \, dv = \int_0^\pi (e^{\frac{\pi u}{2}} - 1) du = \frac{2}{\pi}(e^{\frac{\pi^2}{2}} - 1) - \pi.$

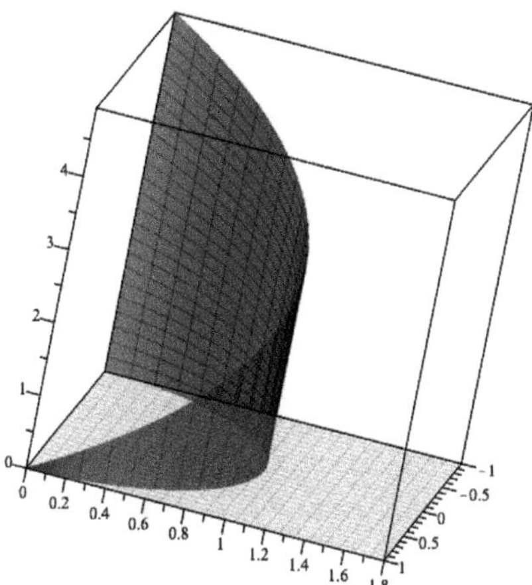

Problem 5.5.9. Evaluate the mass of the sphere S if its surface density at a point is the distance between the point and some fixed diameter of the sphere.

Solution 5.5.9. Let the rectangular coordinate system $Oxyz$ be such that the origin O is the centre of the sphere and that the fixed diameter of the sphere lies on the z-axis. Let $a > 0$ be the radius of the sphere. The equation of the sphere is $x^2 + y^2 + z^2 = a^2$. The distance between the point $(x, y, z) \in S$ and the z-axis is $\sqrt{x^2 + y^2}$. By using the spherical coordinates, we obtain that S is the hodograph of the vector function $\vec{r}(u, v) = \langle a \cos u \sin v, \sin u \sin v, a \cos v \rangle$, $(u, v) \in D = [0, 2\pi] \times [0, \pi]$. Since $\left| \frac{\partial \vec{r}(u,v)}{\partial u} \times \frac{\partial \vec{r}(u,v)}{\partial v} \right| = a^2 \sin v$ and since, for $(u, v) \in D$, the surface density at the point $r(u, v)$ is $g(u, v) = \sqrt{a^2 \cos^2 u \sin^2 v + a^2 \sin^2 u \sin^2 v} = a \sin v$. The mass of S is

$$m = \iint_S g \, dS = \iint_D (a \sin v)(a^2 \sin v) \, dudv = a^3 \pi^2.$$

Problem 5.5.10. Evaluate the mass of the surface of the tetrahedron S whose vertices are the origin, $(1, 0, 0)$, $(0, 1, 0)$, and $(0, 0, 2)$ if its surface density is determined by $f(x, y, z) = xz$, $(x, y, z) \in S$.

Solution 5.5.10. The question is to evaluate $\iint_S xz \, dS$. Let S_1, S_2, S_3, and S_4 denote the faces of the tetrahedron in the xz-plane,

yz-plane, xy-plane, and the plane $2x + 2y + z = 2$, respectively. Since $f(x, y, z) = 0$ for all $(x, y, z) \in S_2 \cup S_3$, it follows that

$$\iint_S xz\,dS = \iint_{S_1} xz\,dS + \iint_{S_4} xz\,dS.$$

Since $S_1 = \{(x, 0, z) : 0 \le x \le 1 \text{ and } 0 \le z \le 2 - 2x\}$, it follows that $\iint_{S_1} xz\,dS = \int_0^1 x\,dx \int_0^{2-2x} z\,dz = \frac{1}{2} \int_0^1 x(2 - 2x)^2\,dx = \frac{1}{6}$. Since $S_4 = \{(x, y, 2 - 2x - 2y) : 0 \le x \le 1 \text{ and } 0 \le y \le 1 - x\}$, it follows that $\iint_{S_4} xz\,dS = 3\int_0^1 x\,dx \int_0^{1-x}(2 - 2x - 2y)\,dy = \frac{1}{4}$. The mass of the surface of the tetrahedron is $\iint_S xz\,dS = \frac{5}{12}$.

Problem 5.5.11. Let a be a positive number. Evaluate the flux of the field $\vec{E} = \frac{1}{|\vec{r}|^{\frac{3}{2}}}\,\vec{r}$, where $\vec{r} \ne \vec{0}$ is the radius vector, through the side of the surface $z = \sqrt{a^2 - x^2 - y^2}$ for which the unit normal vector at the point $(0, 0, a)$ is $\vec{k}$.

Solution 5.5.11. As a hemisphere with its centre at the origin and radius a, the given surface S is smooth. The field of unit normal vectors determined by $\vec{k}$ is $\vec{n} = \frac{1}{a}\vec{r}$. The flux is

$$\Phi = \iint_S \vec{E} \cdot \vec{n}\,dS = \iint_S \left(\frac{1}{|\vec{r}|^{\frac{3}{2}}}\,\vec{r}\right) \cdot \left(\frac{1}{a}\vec{r}\right)dS$$

$$= \frac{1}{\sqrt{a}} \iint_S dS = 2a\sqrt{a}\,\pi.$$

Problem 5.5.12. Evaluate the flux of the vector field $\vec{a} = \langle x^2, 2xy, 0\rangle$, through the side of the hodograph S of the vector function $\vec{r}(u, v) = \langle \cos u \sin v, \sin u \sin v, \cos v\rangle$, $(u, v) \in D = [0, \frac{\pi}{4}]\times[0, \frac{\pi}{2}]$, for which the unit normal vector at the point $r(\frac{\pi}{6}, \frac{\pi}{3})$ is $\frac{1}{4}\langle 3, \sqrt{3}, 2\rangle$.

Solution 5.5.12. The surface S is part of the unit sphere, which implies that the field of unit normal vectors determined by $\frac{1}{4}\langle 3, \sqrt{3}, 2\rangle$ is $\vec{n} = \vec{n}(u, v) = \vec{r}(u, v)$. Since, for any $(u, v) \in (0, \frac{\pi}{4}) \times (0, \frac{\pi}{2})$, $\left|\frac{\partial \vec{r}}{\partial u} \times \frac{\partial \vec{r}}{\partial v}\right| = \sin v$, it follows that the flux is $\Phi = \iint_S \vec{a} \cdot \vec{n}\,dS = \iint_D \langle \cos^2 u \sin^2 v, 2\cos u \sin u \sin^2 v, 0\rangle \cdot \langle \cos u \sin v, \sin u \sin v, \cos v\rangle \sin v\,du\,dv = \frac{7\pi\sqrt{2}}{64}$.

Problem 5.5.13. Evaluate the flux of the vector field $\vec{F}(x, y, z) = \langle 8x, 8y, 4 \rangle$ outward through the part of the paraboloid $z = x^2 + y^2 + 5$ that is below the plane $z = 7$.

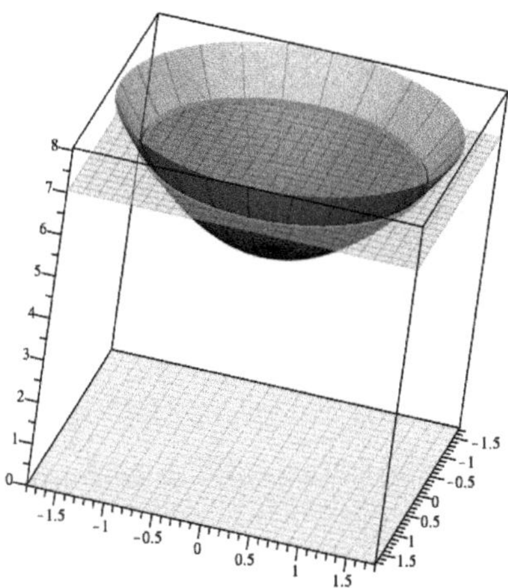

Solution 5.5.13. Let $S = \{(x, y, z) : z = x^2 + y^2 + 5$ and $5 \leq z \leq 7\}$. A vector normal to the surface S at the point (x, y, z) is $\vec{N} = \vec{N}(x, y, z) = \mathrm{grad}(x^2 + y^2 - z + 5) = \langle 2x, 2y, -1 \rangle$. The projection of the surface S into the xy-plane is the disc $D = \{(x, y) : x^2 + y^2 \leq 2\}$. Since the $\vec{k}$-component of $\vec{N}$ is negative, the direction of $\vec{N}$ corresponds to the required *outward* orientation. The flux is $\iint_S \vec{F} \cdot \vec{n}\, dS = \iint_D \langle 8x, 8y, 4 \rangle \cdot \langle 2x, 2y, -1 \rangle\, dxdy = \iint_D (16x^2 + 16y^2 - 4)\, dxdy = 4 \int_0^{2\pi} d\theta \int_0^{\sqrt{2}} r(4r^2 - 1)\, dr = 24\pi$.

Problem 5.5.14. Let S be the portion of the plane $x + y + z = 4$ that lies in the first octant. Evaluate the flux of the vector field $\vec{F}(x, y, z) = \langle x, y, z \rangle$ through the surface S in the upward direction.[17]

Solution 5.5.14. The projection of the surface into the xy-plane is the set $\{(x, y) : 0 \leq x \leq 4$ and $0 \leq y \leq 4 - x\}$. Since $z = 4 - x - y$, the required flux is $\iint_S \vec{F} \cdot \vec{n}\, dS = \iint_D \langle x, y, 4 - x - y \rangle \cdot \langle 1, 1, 1 \rangle\, dA = 4 \iint_D dA = 32$.

---

[17]This means that the $\vec{k}$ component of the normal vector is positive.

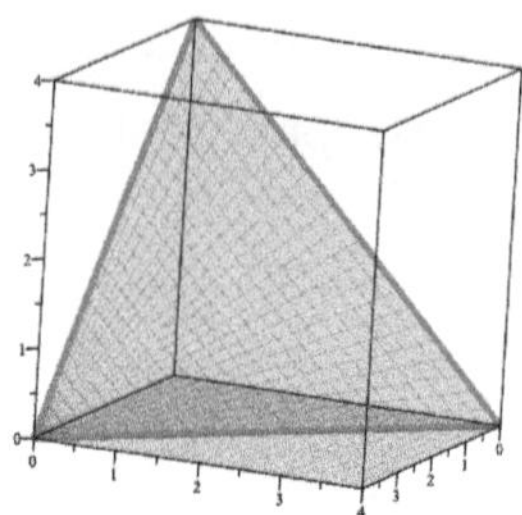

5.6   Green's Theorem

Problem 5.6.1. Evaluate $\int_C (7x - 5y)\, dx + (6x - 7y)\, dy$, where C is the counterclockwise-oriented circle $(x - 3)^2 + (y - 3)^2 = 9$.

Solution 5.6.1. For $P(x, y) = 7x - 5y$, $Q(x, y) = 6x - 7y$, $\frac{\partial P}{\partial y} = -5$, $\frac{\partial Q}{\partial x} = 6$, and $D = \{(x, y) : (x - 3)^2 + (y - 3)^2 \leq 9\}$, by Green's theorem,

$$\int_C (7x - 5y)\, dx + (6x - 7y)\, dy = \iint_D (6 - (-5))\, dx\, dy = 11 \cdot 3^2 \pi = 99\pi.$$

Problem 5.6.2. Evaluate $\oint_C \vec{F} \cdot d\vec{r}$, where $\vec{F}(x, y) = \langle \cos^2 x + y^2 x, x^2 y + x - e^{y^2} \rangle$ and C is the path starting at $(0, 4)$, going down the y-axis to $(0, -4)$, travelling counterclockwise along $x^2 + y^2 = 16$ to $(4, 0)$, then following $y + x = 4$ back to $(0, 4)$.

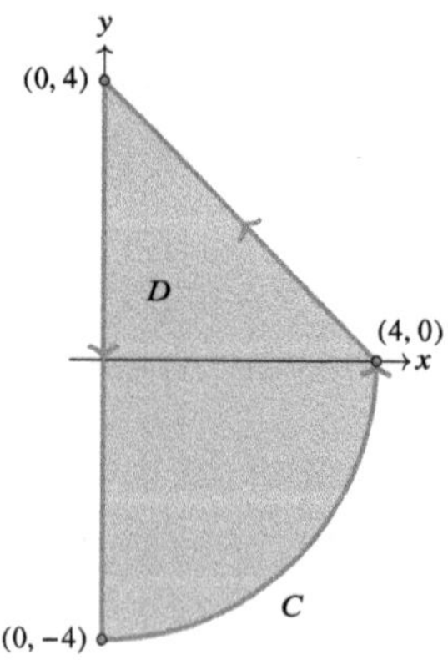

Solution 5.6.2. The closed piecewise smooth curve C is positively oriented. By Green's theorem, $\oint_C \vec{F} \cdot d\vec{r} = \iint_D (2xy + 1 - 2xy)\, dA = \int_D dA = \frac{4 \cdot 4}{2} + \frac{4^2 \pi}{4} = 4(2 + \pi).$

Problem 5.6.3. Evaluate the area of the Starfleet logo bounded by $\vec{r}(t) = \langle 3\cos t + 3\sin 2t, 4\sin t + 3\sin 2t \rangle$, with $0 \le t \le 2\pi$.

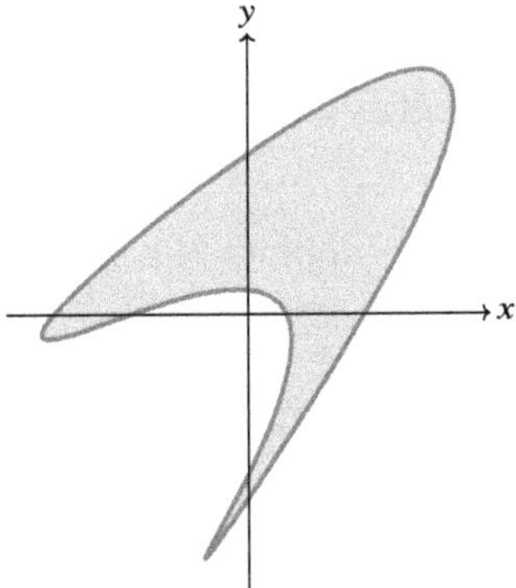

Solution 5.6.3. By Green's theorem, the area is $A = \iint_D dA = \oint_c x\,dy = \int_0^{2\pi} (3\cos t + 3\sin 2t)(4\cos t + 6\cos 2t)dt = 6\int_{-\pi}^{\pi}(\cos t + \sin 2t)(2\cos t + 3\cos 2t)\ dt = 12\int_0^{\pi}(2\cos^2 t + 3\cos t\cos 2t)\ dt = 6\int_0^{\pi}(2 + 2\cos 2t + 3\cos 3t + 3\cos t)\ dt = 12\pi$.

Problem 5.6.4. Evaluate the circulation of the vector field $\vec{F}(x, y) = \langle e^{x^2+1} - y, e^{y^3} + x^2 \rangle$ along the counterclockwise-oriented closed curve C consisting of three line segments with the end points $(0,0)$ and $(2,0)$; $(2,0)$ and $(0,2)$; and $(0,2)$ and $(0,0)$.

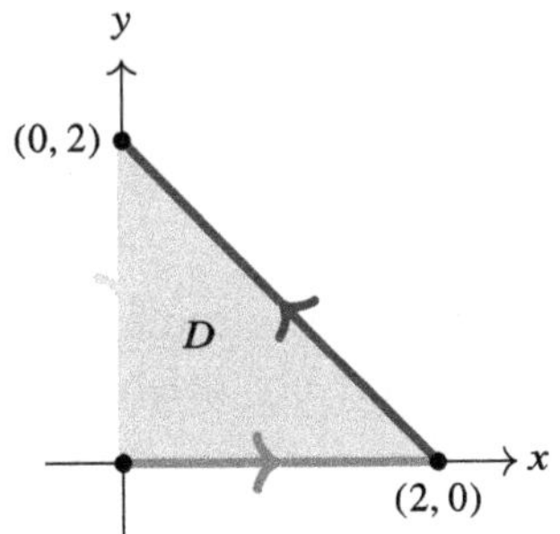

Solution 5.6.4. By Green's theorem, $\oint_C \vec{F}\cdot\vec{n}\,dS = \iint_D (2x+1)dA = \int_0^2 (2x+1)dx \int_0^{2-x} dy = \frac{14}{3}$.

Problem 5.6.5. Let $\vec{F}(x, y) = \langle x, x \rangle$, and let the curve C be the triangle with vertices at the origin and the points $(1,0)$ and $(0,1)$. Let D be the region on a plane that is bounded by C.

(a) Use the plot of the vector field $\vec{F}$ in the first quadrant to predict whether the net flux out of the region D, the interior of C, will be positive or negative.

(b) Compute the flux integral $\oint_C \vec{F} \cdot \vec{n}\, ds$ directly.

(c) Compute the flux integral $\oint_C \vec{F}\cdot\vec{n}\, ds$ using the appropriate double integral.

Solution 5.6.5. The figure depicts the images of the curve C, the region D, and the plot of the vector field $\vec{F}$.

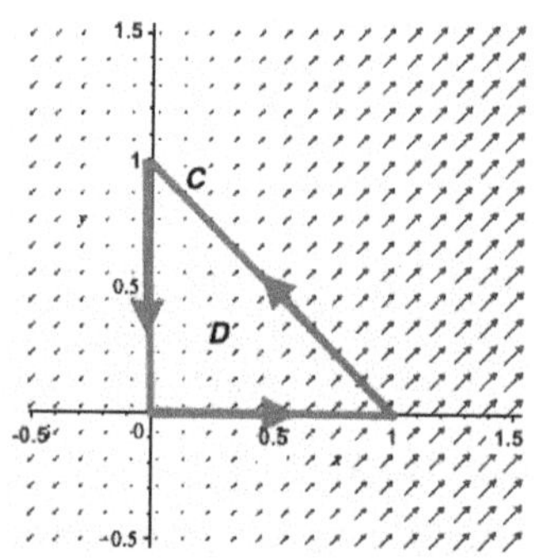

(a) The net flux out of D will be positive: as the image depicts, only a small amount of flux flows into D through the horizontal edge of the triangle.

(b) Since $C = C_1 \cup C_2 \cup C_3$, where $C_1 = \{(t,0) : t \in [0,1]\}$, $C_2 = \{(1-t,t) : t \in [0,1]\}$, and $C_3 = \{(0,1-t) : t \in [0,1]\}$, the flux is $\oint_C \vec{F} \cdot \vec{n}\, ds = \oint_C x\, dx + x\, dy = \int_0^1 t\, dt + \int_0^1 ((t-1) + (1-t))\, dt + \int_0^1 0\, dt = \frac{1}{2}$.

(c) By Green's theorem, $\oint_C \vec{F} \cdot \vec{n}\, ds = \iint_D dx\, dy = \frac{1}{2}$.

Problem 5.6.6. Evaluate the circulation of the vector field $\vec{a} = \langle 2y, 2x^2y \rangle$ along the positively oriented loop of the lemniscate $(x^2 + y^2)^2 = x^2 - y^2$, $x \geq 0$.

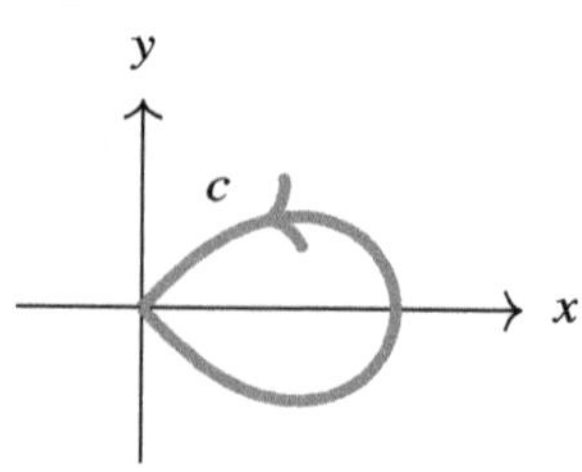

Solution 5.6.6. Let c be the loop of the lemniscate, and let D be the interior of the loop c. By Green's theorem, the circulation of the vector field $\vec{a}$ along c is $\oint_c \vec{a} \cdot d\vec{r} = \iint_D (4xy - 2)\, dxdy$. Since the loop of the lemniscate is symmetric with respect to the x-axis and since, for any $x \in \mathbb{R}$, the function $y \mapsto xy$ is odd, $\iint_D xy\, dxdy = 0$. It follows that

$$\oint_c \vec{a} \cdot d\vec{r} = -2 \iint_D dxdy = -2 \int_{-\frac{\pi}{4}}^{\frac{\pi}{4}} d\theta \int_0^{\sqrt{\cos 2\theta}} r\, dr = -1.$$

Problem 5.6.7. Let $A = \{(x,y) : y^2 < x\}$ and $B = \{(x,y) : x^2 < y\}$. Let c be the boundary of the region $A \cap B$. Is c a smooth closed curve? Evaluate $\oint_c y^2(x-1)\, dx + x^2 y\, dy$.

Answer 5.6.7. The curve c is closed and piecewise smooth. By Green's theorem, $\oint_c y^2(x-1)\, dx + x^2 y\, dy = \frac{3}{10}$.

Problem 5.6.8. Let c_1 be the oriented line segment with the initial point $A = (0,0)$ and the terminal point $B = (1,1)$, and let c_2 the oriented arc of parabola $y = x^2$ with the initial point B and the terminal point A. Evaluate $\oint_{c_1 \cup c_2} (x+y)^2\, dx - (x-y)^2\, dy$.

Solution 5.6.8. Observe that $c = c_1 \cup c_2$ is the negatively oriented boundary of the open region $D = \{(x,y) : x \in (0,1) \text{ and } y \in (x^2, x)\}$. By Green's theorem, $\oint_{c_1 \cup c_2} (x+y)^2\, dx - (x-y)^2\, dy = -\iint_D (-4x)\, dxdy = \frac{1}{3}$.

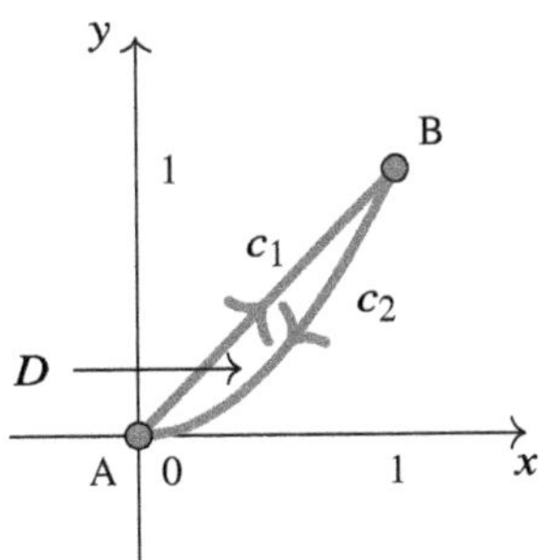

Problem 5.6.9. In each case, determine whether the line integral of the vector field along the oriented closed curve is positive or negative.

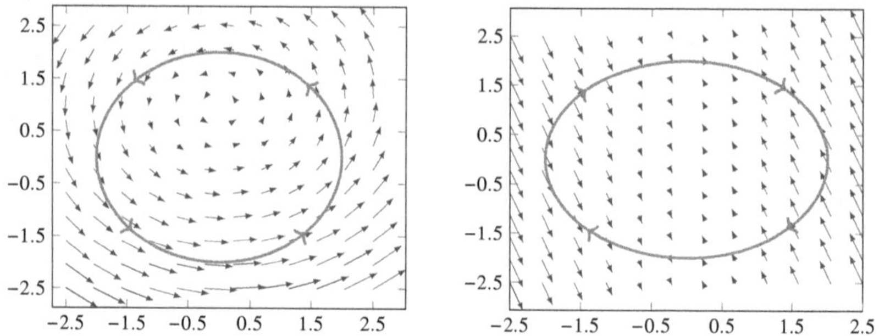

Solution 5.6.9. *Left* – positive. It appears that, at any point on the curve, the vector field and the tangent vector form an acute angle, which implies that their dot product is positive. Consequently, the line integral is positive. Alternatively, take $\vec{F}(x, y) = a\langle -y, x \rangle$, for some $a > 0$, and use Green's theorem.

Right – negative. It appears that, at any point on the curve, the vector field and the tangent vector, in the direction of the orientation of the curve, form an obtuse angle. This implies that their dot product is negative. Alternatively, take $\vec{F}(x, y) = x\langle a, b \rangle$, for some $a < 0$ and $b > 0$, and use Green's theorem. Consider the orientation of the curve.

5.7 Divergence Theorem

Problem 5.7.1. Determine if the flux of the vector field is positive or negative. In the left-side figure, the surface is oriented outward, while in the right-side figure, the surface is oriented in the positive z direction.

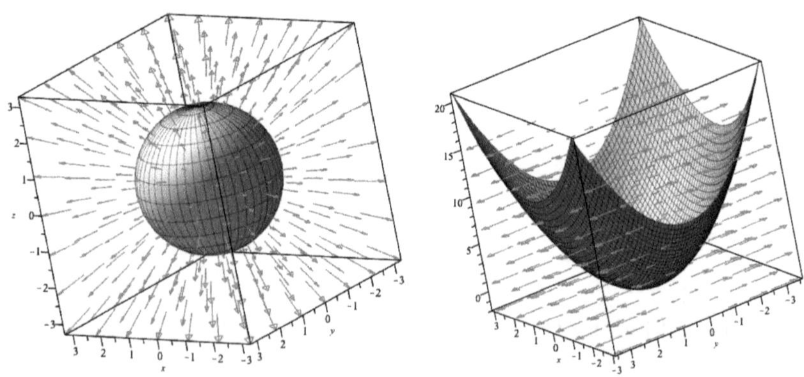

Solution 5.7.1. *Left* – positive. It appears that, at any point on the surface, the vector field and the outward-directed vector normal to the tangent plane are in the same direction, which implies that their dot product is positive. Consequently, the surface integral is positive. Alternatively, consider $\vec{F}(x, y, z) = \langle x, y, z \rangle$ and use the divergence theorem.

Right – negative. It appears that, at any point on the surface, the vector field and the upward-directed vector normal to the tangent plane form an obtuse angle. This implies that their dot product is negative. Alternatively, consider $\vec{F}(x, y, z) = \langle 0, y, 0 \rangle$ and the surface $z = -a + x^2 + y^2$, $a > 0$. Then, the normal vector is $\vec{N} = \langle -2x, -2y, 1 \rangle$ and $\vec{F} \cdot \vec{N} = -2y^2 < 0$, which determines the sign of the corresponding surface integral.

Problem 5.7.2. Evaluate the flux of the field $\vec{a} = xy\langle |x|, |y|, 0 \rangle$ through the outer side of the surface S determined in the spherical coordinates by $\rho = 2 \cos u \sin v$.

Solution 5.7.2. The surface S is a sphere with its centre at $(1, 0, 0)$ and radius 1. Since the function $t \mapsto t|t|$ is continuously differentiable, the vector field $\vec{a}$ is continuously differentiable. Since S is a closed smooth surface, by the divergence theorem, the flux is
$$\Phi = \iiint_V \operatorname{div} \vec{a} \, dV = 2 \iiint_V x(y + |y|) \, dV = \pi.$$

Problem 5.7.3.

(a) Determine a function $f : \mathbb{R} \to \mathbb{R}$ such that $f(0) = 0$ and that the field $\vec{a} = \langle (1 + x^2) f(x), 2xy f(x), -3z \rangle$ is solenoidal.
(b) Evaluate the flux of the solenoidal field obtained in Part (a) through the outer side of the sphere that has the hodograph of the vector function $\vec{r} = \langle \cos t, \sin t, 1 \rangle$, $t \in [0, 2\pi]$, as a great circle.

Solution 5.7.3.

(a) The field is solenoidal if $0 = \operatorname{div}\vec{a} = 2x f(x) + (1 + x^2) f'(x) + 2x f(x) - 3 = (1 + x^2) f'(x) + 4x f(x) - 3$, for all x in the domain of f. The function f is the solution of the initial value problem $y' + \frac{4xy}{1+x^2} = \frac{3}{1+x^2}$, $y(0) = 0$. It follows that $f(x) = \frac{x(3+x^2)}{(1+x^2)^2}$ and $\vec{a} = \left\langle \frac{x(3+x^2)}{(1+x^2)^2}, \frac{2x^2 y(3+x^2)}{(1+x^2)^2}, -3z \right\rangle$.
(b) By the divergence theorem, the flux through the outer side of the sphere S is zero.

Problem 5.7.4. Evaluate the flux of the vector field $\vec{F}(x, y, z) = \langle y^3 + \sin y, z^3 + \sin z, 5z + \sin xy \rangle$ through the outward-oriented surface S bounding the solid $R = \{(x, y, z) : x^2 \leq 4, y^2 \leq 4, z^2 \leq 4,$ and $x^2 + y^2 \geq 1\}$.

Solution 5.7.4. The solid R is obtained by removing the cylinder $\{(x, y, z) : x^2 + y^2 \leq 1\}$ from the cube centred at the origin, with its sides parallel to the coordinate planes and with an edge of length 4. Since, for all $(x, y, z) \in R$, div $\vec{F}(x, y, z) = 5$, by the divergence theorem, the flux is $5 \iiint_R dE = 5((\text{volume of the cube}) - (\text{volume of the cylinder})) = 5(4^3 - 4\pi) = 20(16 - \pi)$.

Problem 5.7.5. Evaluate the flux Φ of the vector field $\vec{a} = \langle x^3, y^3, x^3 \rangle$ through the outer side of the surface $x^2 + y^2 + z^2 = x$.

Answer 5.7.5. $\Phi = \frac{7\pi}{40}$.

Problem 5.7.6. Evaluate the flux Φ of the vector field $\vec{a} = \langle xz^2, y, -z^2 \rangle$ through the side of the surface $S = \{(x, y, z) \in \mathbb{R}^3 : y = 1 + \sqrt{2z - x^2 - z^2}\}$ for which the unit normal vector at the point $(0, 2, 1)$ is $\vec{j}$.

Solution 5.7.6. The surface S is a hemisphere with its centre at $(0, 1, 1)$ and radius 1, determined by $x^2 + (y - 1)^2 + (z - 1)^2 = 1$, $y \geq 1$. Let $S_1 = \{(x, 1, z) : x^2 + (z - 1)^2 \leq 1\}$ be a disc in the plane $y = 1$. Then, $\Sigma = S \cup S_1$ is a closed piecewise smooth surface. Let Φ_Σ be the flux of the field $\vec{a}$ through the outer side of Σ, and let Φ_1 be the flux of $\vec{a}$ through the side of S_1 determined by the normal vector $\vec{n} = -\vec{j}$. Since the vector $\vec{j}$ at the point $(0, 2, 1)$ determines the outer side of the surface Σ, it follows that $\Phi_\Sigma = \Phi + \Phi_1$ or, what is the same, $\Phi = \Phi_\Sigma - \Phi_1$. Let $V = \{(x, y, z) : x^2 + (y - 1)^2 + (z - 1)^2 \leq 1$ and $y \in [1, 2]\}$, i.e., let V be the region bounded by the surface Σ. By the divergence theorem, $\Phi_\Sigma = \iiint_V \text{div}\vec{a}\, dV = \iiint_V (z - 1)^2\, dV = \frac{2\pi}{15}$. Since $\Phi_1 = \iint_{S_1} \vec{a} \cdot (-\vec{j})\, dS = \iint_{S_1} (-y)\, dS = -\pi$, it follows that $\Phi = \frac{2\pi}{15} - (-\pi) = \frac{17\pi}{15}$.

Problem 5.7.7. Let R be the region in the first octant that is bounded by the coordinate planes, the cylinder $z = 1 - x^2$, and the plane $y + z = 2$. Evaluate the flux of the vector field $\vec{F}(x, y, z) = \langle \frac{x^2}{2}, \sin xz, xz + e^{x^2 y} \rangle$ through the outward-oriented surface S, the boundary of the region R.

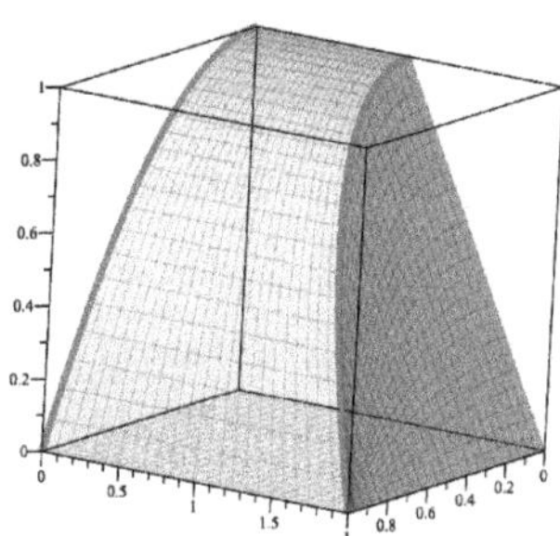

Solution 5.7.7. By the divergence theorem, $\iint_S \vec{F} \cdot \vec{n} \, dS = \iiint_R \operatorname{div} \vec{F} \, dE = \iiint_R 2x \, dE$. The projection of the region R into the xz-plane is the set $D = \{x, 0, z) : x \in [0, 1] \text{ and } 0 \le z \le 1 - x^2\}$. For each fixed point $(x, 0, z) \in D$, the point $(x, y, z) \in R$ if and only if $0 \le y \le 2 - z$. Therefore, $\iint_S \vec{F} \cdot \vec{n} \, dS = \iiint_R 2x \, dE = 2 \int_0^1 x \, dx \int_0^{1-x^2} dz \int_0^{2-z} dy = \frac{5}{6}$.

Problem 5.7.8. Evaluate the flux Φ of the vector field $\vec{a} = \langle x, xy, y \rangle$ through the side of the surface $S = \{(x, y, z) \in \mathbb{R}^3 : x = 1 + \sqrt{1 - y^2 - z^2}\}$, for which the unit normal vector at the point $(2, 0, 0)$ is $\vec{\imath}$.

Answer 5.7.8. $\Phi = \frac{31\pi}{12}$.

Problem 5.7.9. Let S be the upward oriented portion of the unit sphere, centred at the origin, that is inside the cone $z = \sqrt{x^2 + y^2}$. Evaluate $\iint_S \vec{F} \cdot d\vec{S}$, where $\vec{F}(x, y, z) = \langle xy + \cos z, -yx + x^2 + z^3, 2z^2 + x \rangle$.

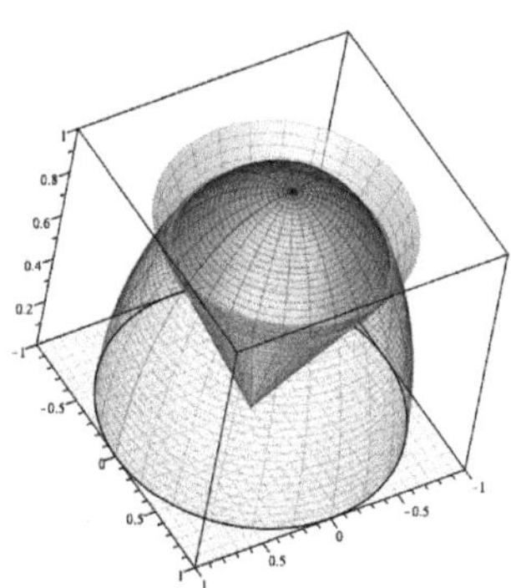

Solution 5.7.9. The boundary of the surface S is the circle $C = \{(x, y, \frac{1}{\sqrt{2}}) : x^2 + y^2 = \frac{1}{2}\}$. Let S' be the disc in the plane $z = \frac{1}{\sqrt{2}}$ bounded by C and oriented by the vector $-\vec{k}$. Then, $\iint_S \vec{F} \cdot d\vec{S} = \iint_{S \cup S'} \vec{F} \cdot d\vec{S} - \iint_{S'} \vec{F} \cdot d\vec{S}$. Let R be the solid bounded by $S \cup S'$. By the divergence theorem, $\iint_{S \cup S'} \vec{F} \cdot d\vec{S} = \iiint_R \operatorname{div} \vec{F}\, dE = \iiint_R (y - x + 4z)\, dE$. Because of the symmetry, $\iiint_R y\, dE = \iiint_R x\, dE$, which implies that $\iint_{S \cup S'} \vec{F} \cdot d\vec{S} = 4 \iiint_R z\, dE = 4 \int_0^{2\pi} d\theta \int_{\frac{1}{\sqrt{2}}}^1 z\, dz \int_0^{\sqrt{1-z^2}} r\, dr = 4\pi \int_{\frac{1}{\sqrt{2}}}^1 z(1 - z^2)\, dz = \frac{\pi}{4}$. This, together with $\iint_{S'} \vec{F} \cdot d\vec{S} = - \iint_{S'} (2z^2 + x)\, dS = - \int_0^{2\pi} d\theta \int_0^{\frac{1}{\sqrt{2}}} (1 + r\cos\theta) r\, dr = -\frac{\pi}{2}$, implies that $\iint_S \vec{F} \cdot d\vec{S} = \frac{\pi}{4} + \frac{\pi}{2} = \frac{3\pi}{4}$.

Problem 5.7.10. The surface S is determined by $y = 1 - x^2 - z^2$, $y \geq 0$. Evaluate $\iint_S \vec{F} \cdot d\vec{S}$ if $\vec{F}(x, y, z) = \langle e^{\sin y}, xz + 2, x^2 \sin y + z \rangle$ and the orientation of S is determined by the normal vector $\vec{j}$ at the point $(0, 1, 0)$.

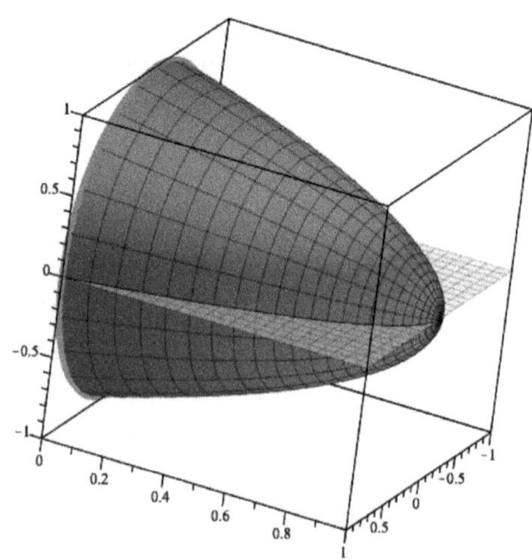

Solution 5.7.10. Let the disc $S_1 = \{(x, 0, z) : x^2 + z^2 \leq 1\}$ be oriented by $-\vec{j}$. Let R be the region bounded by the closed surface $S \cup S_1$. Then, $\iint_S \vec{F} \cdot d\vec{S} = \iint_{S \cup S_1} \vec{F} \cdot d\vec{S} - \iint_{S_1} \vec{F} \cdot d\vec{S} = \iiint_R \operatorname{div} \vec{F}\, dE - \iint_{S_1} \vec{F} \cdot d\vec{S}$. It follows that $\iiint_R \operatorname{div} \vec{F}\, dE = \iiint_R dE = \int_0^{2\pi} d\theta \int_0^1 r\, dr \int_0^{1-r^2} dy = 2\pi \int_0^1 r(1 - r^2)\, dr = \frac{\pi}{2}$. Also, since $\vec{n} = \langle 0, -1, 0 \rangle$ is the normal vector to the surface S_1 in the outward orientation

of $S \cup S_1$, $\iint_{S_1} \vec{F} \cdot d\vec{S} = \iint_{S_1} \vec{F} \cdot \vec{n} \, dS = -\iint_{S_1}(xz + 2) \, dS = -2\pi + \int_0^{2\pi} \sin\theta \cos\theta \, d\theta \int_0^1 r^3 dr = -2\pi$. Therefore, $\iint_S \vec{F} \cdot d\vec{S} = \frac{\pi}{2} - (-2\pi) = \frac{5\pi}{2}$.

Problem 5.7.11. Let $S = \{(x, y, z) : x^2 + y^2 = 4 \text{ and } x, y \geq 0 \text{ and } 0 \leq z \leq 4\}$, and let $\vec{F}(x, y, z) = \langle x, 0, 0 \rangle$.

(a) Sketch the surface S and the field $\vec{F}$.

(b) Evaluate $\iint_S \vec{F} \cdot \vec{n} \, dS$ through the side of S that is determined by the unit normal vector $\vec{N} = \frac{1}{\sqrt{2}}(\vec{\imath} + \vec{\jmath})$ at the point $(\sqrt{2}, \sqrt{2}, 1) \in S$.

(c) Let G be the region in the first octant bounded by S, the coordinate planes, and the plane $z = 4$. Use the divergence theorem to evaluate the flux of the field $\vec{F}$ through the outer side of the boundary $\delta(G)$ of the region G.

(d) The boundary of G comprises S together with four other faces. What is the outward flux through each of those four faces and why?

Solution 5.7.11.

(a) The surface S and the field $\vec{F}$ are depicted in the image.

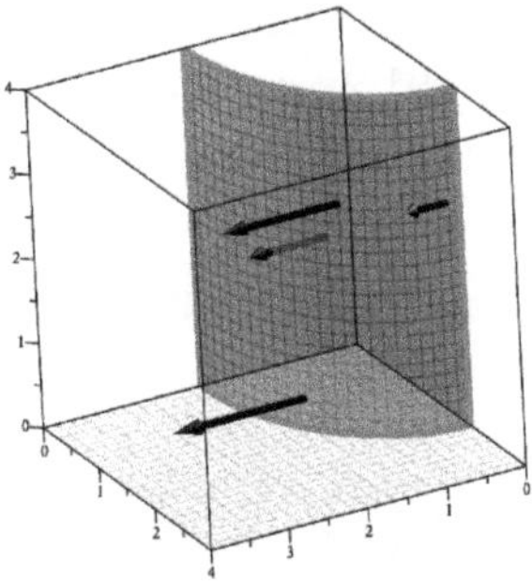

(b) The surface S is the hodograph of the vector function $\vec{r}(u, v) = \langle 2\cos u, 2\sin u, v \rangle$, $(u, v) \in D = \left[0, \frac{\pi}{2}\right] \times [0, 4]$.

$$\text{From } \vec{r}_u \times \vec{r}_v = \begin{vmatrix} \vec{\imath} & \vec{\jmath} & \vec{k} \\ -2\sin u & 2\cos u & 0 \\ 0 & 0 & 1 \end{vmatrix} = \langle 2\cos u, 2\sin u, 0 \rangle, \text{ it follows}$$

that $\iint_S \vec{F} \cdot \vec{n} \, dS = \iint_D \langle 2\cos u, 0, 0 \rangle \cdot \langle 2\cos u, 2\sin u, 0 \rangle \, dudv = 4\int_0^{\frac{\pi}{2}} \cos^2 u \, du \int_0^4 dv = 4\pi$.

(c) By the divergence theorem, $\iint_{\delta(G)} \vec{F} \cdot \vec{n} \, dS = \iiint_G dE = \frac{1}{4} \cdot 4\pi \cdot 4 = 4\pi$.

(d) The flux of $\vec{F}$ through each of four flat faces of G is zero.

- Since the face in the yz-plane is $S_{yz} = \{(0, y, z) : (y, z) \in [0, 2] \times [0, 4]\}$, $\iint_{S_{yz}} \vec{F} \cdot (-\vec{\imath})\, dS = \iint_{S_{yz}} \langle 0, 0, 0 \rangle \cdot \langle -1, 0, 0 \rangle\, dS = 0$.

- Since the face in the xz-plane is $S_{xz} = \{(x, 0, z) : (x, z) \in [0, 2] \times [0, 4]\}$, $\iint_{S_{xz}} \vec{F} \cdot (-\vec{\jmath})\, dS = \iint_{S_{xz}} \langle x, 0, 0 \rangle \cdot \langle 0, -1, 0 \rangle\, dS = 0$.

- Since the face in the xy-plane is $S_{xy} = \{(x, y, 0) : x \in [0, 2] \text{ and } 0 \le y \le \sqrt{4 - x^2}\}$, $\iint_{S_{xy}} \vec{F} \cdot (-\vec{k})\, dS = \iint_{S_{xy}} \langle x, 0, 0 \rangle \cdot \langle 0, 0, -1 \rangle\, dS = 0$.

- Since the face in the plane $z = 4$ is $S_{z=4} = \{(x, y, 4) : x \in [0, 2] \text{ and } 0 \le y \le \sqrt{4 - x^2}\}$, $\iint_{S_{z=4}} \vec{F} \cdot \vec{k}\, dS = \iint_{S_{z=4}} \langle x, 0, 0 \rangle \cdot \langle 0, 0, 1 \rangle\, dS = 0$.

Problem 5.7.12. Let $\vec{F}(x, y, z) = \langle -z, e^{x^2}, y^2 \rangle$. Let S be the outward-oriented (from the viewer's point of view) surface formed by glueing together the cylinder $x^2 + y^2 = 9$, $z \in [0, 9]$ and the paraboloid $5 + x^2 + y^2 = z$, $z \in [5, 9]$, and the ring in the plane $z = 9$ that is the hodograph of the vector function $\vec{r}(u, v) = \langle u\cos v, u\sin v, 9 \rangle$, $(u, v) \in [0, 2\pi] \times [2, 3]$. Evaluate $\iint_S \vec{F} \cdot d\vec{S}$.

Solution 5.7.12. The strategy is to close the surface with the disc $S' = \{(x, y, 0) : x^2 + y^2 \le 9\}$ and then use the divergence theorem. Let $\Sigma = S \cup S'$, and let R be the region bounded by the closed surface Σ. Then, $\iint_S \vec{F} \cdot d\vec{S} = \iint_\Sigma \vec{F} \cdot d\vec{S} - \iint_{S'} \vec{F} \cdot d\vec{S} = \iiint_R \operatorname{div}\vec{F}\, dE - \iint_{S'} \vec{F} \cdot d\vec{S}$.

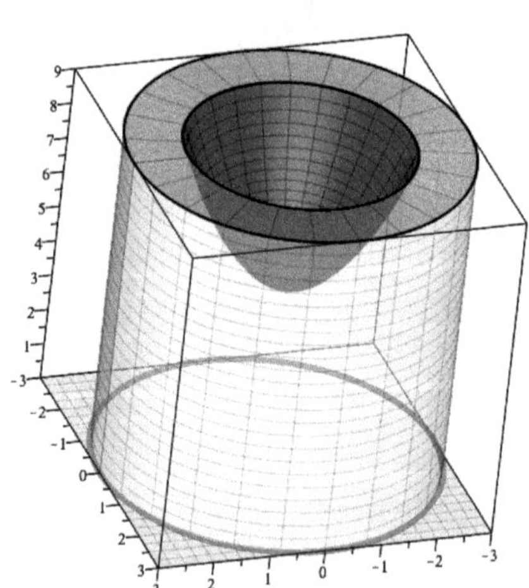

From div $\vec{F} = 0$, it follows that $\iiint_R \operatorname{div}\vec{F}\,dE = 0$ and $\iint_S \vec{F} \cdot d\vec{S} = -\iint_{S'} \vec{F} \cdot d\vec{S} = -\iint_{S'} \vec{F} \cdot (-\vec{k})\,dS = \int_0^{2\pi} \sin^2\theta\,d\theta \int_0^3 r^3\,dr = \frac{81\pi}{4}$.

Problem 5.7.13. Evaluate the flux Φ of the vector field $\vec{a} = \langle x, y, -z \rangle$ through the side of the surface $S = \{(x,y,z) : z \in [-1, 2]$ and $z^2 = x^2 + y^2\}$, for which the unit normal vector at the point $(1, 0, 1)$ is $\frac{1}{\sqrt{2}}\langle 1, 0, -1 \rangle$ and at the point $(\frac{1}{2}, 0, -\frac{1}{2})$ is $\frac{1}{\sqrt{2}}\langle 1, 0, 1 \rangle$.

Solution 5.7.13. The field $\vec{a}$ is continuously differentiable. The surface S is part of the cone with its vertex at the origin and the z-axis as its axis between the planes $z = -1$ and $z = 2$.

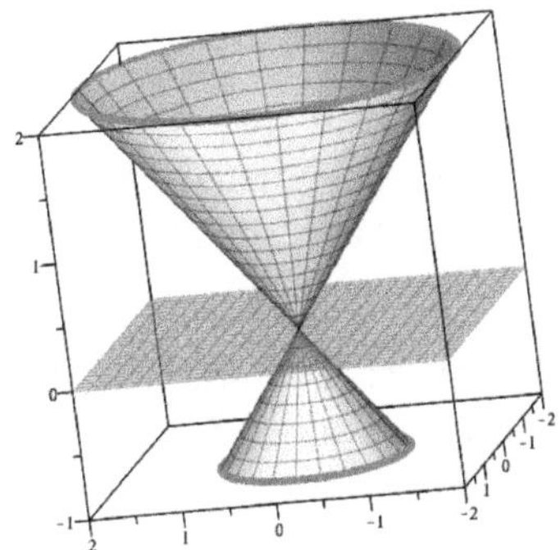

Let Φ_+ be the flux of the field $\vec{a}$ through S^+, the part of the surface S that is above the xy-plane, and let Φ_- be the flux of the field $\vec{a}$ through S^-, the part of the surface S that is below the xy-plane.

Let $S_1 = \{(x, y, 2) : x^2 + y^2 \leq 4\}$, and let $S_2 = \{(x, y, -1) : x^2 + y^2 \leq 1\}$. Let $\Sigma^+ = S^+ \cup S_1$, and let $\Sigma^- = S^- \cup S_2$. Each of the surfaces Σ^+ and Σ^- is closed and piecewise smooth. Let Φ^+ be the flux of $\vec{a}$ through the outer side of Σ^+, and let Φ^- be the flux of $\vec{a}$ through the outer side of Σ^-. Let Φ_1 be the flux of $\vec{a}$ through the side of S_1 determined by the vector $\vec{k}$, and let Φ_2 be the flux of $\vec{a}$ through the side of S_2 determined by the vector $-\vec{k}$. Since the vectors $\frac{1}{\sqrt{2}}\langle 1, 0, -1 \rangle$ and $\frac{1}{\sqrt{2}}\langle 1, 0, 1 \rangle$ determine the outer sides of Σ^+ and Σ^-, it follows that

$$\Phi = \Phi_+ + \Phi_- = (\Phi^+ - \Phi_1) + (\Phi^- - \Phi_2).$$

For $V^+ = \{(x, y, z) : x^2 + y^2 \leq z^2 \text{ and } x \in [0, 2]\}$ and $V^- = \{(x, y, z) : x^2 + y^2 \leq z^2 \text{ and } x \in [-1, 0]\}$, by the divergence theorem, $\Phi^+ = \iiint_{V^+} dV = \frac{8\pi}{3}$ and $\Phi^- = \iiint_{V^-} dV = \frac{\pi}{3}$. Since

$\Phi_1 = \iint_{S_1} \vec{a} \cdot \vec{k} \, dS = -8\pi$ and $\Phi_2 = \iint_{S_2} \vec{a} \cdot (-\vec{k}) \, dS = -\pi$, it follows that $\Phi = 12\pi$.

Problem 5.7.14. Evaluate the flux Φ of the vector field $\vec{a} = \langle x, y, z \rangle$ through the side of the surface S, determined by $(z-1)^2 = \frac{1}{4}(x^2+y^2)$, $z \in [0, 1]$, for which the unit normal vector at the point $(1, 0, \frac{1}{2})$ is $\frac{1}{\sqrt{5}} \langle 0, 1, 2 \rangle$.

Answer 5.7.14. $\Phi = 4\pi$.

Problem 5.7.15. Evaluate $I = \iint_S (x^2 \cos \alpha + y^2 \cos \beta + z^2 \cos \gamma) \, dS$ if the surface S is determined by $\frac{x^2}{a^2} + \frac{y^2}{b^2} - \frac{z^2}{b^2} = 0$, $z \in [0, b]$.

Answer 5.7.15. $I = \frac{ab^3 \pi}{2}$.

Problem 5.7.16. Evaluate the flux Φ of the radius vector $\vec{r} = \langle x, y, z \rangle$ through the side of the surface $S = \{(x, y, z) : x^2 + y^2 = 1$ and $z \in [0, 1]\}$, for which the unit normal vector at the point $(1, 0, \frac{1}{2})$ is $\vec{i}$.

Solution 5.7.16. *Solution 1*: The smooth surface S is part of the cylinder with the generating curve $x^2 + y^2 = 1$ between the planes $z = 0$ and $z = 1$. Let $S_1 = \{(x, y, 0) : x^2 + y^2 \le 1\}$ and $S_2 = \{(x, y, 1) : x^2 + y^2 \le 1\}$. The surface $\Sigma = S \cup S_1 \cup S_2$ is closed and piecewise smooth. Let Φ_Σ be the flux of the vector field $\vec{r}$ through the outer side of Σ, let Φ_1 be the flux of $\vec{r}$ through the side of S_1 determined by the vector $-\vec{k}$, and let Φ_2 be the flux of $\vec{r}$ through the side of S_2 determined by the vector $\vec{k}$. Since the vector $\vec{i}$ determines the outer side of Σ, it follows that $\Phi = \Phi_\Sigma - \Phi_1 - \Phi_2$. Let $V = \{(x, y, z) : x^2 + y^2 \le 1$ and $z \in [0, 1]\}$, i.e., let V be the region bounded by Σ. By the divergence theorem, $\Phi_\Sigma = \iiint_V 3 \, dV = 3\pi$. Since $\Phi_1 = \iint_{S_1} \vec{r} \cdot (-\vec{k}) \, dS = 0$ and $\Phi_2 = \iint_{S_2} \vec{r} \cdot \vec{k} \, dS = \pi$, it follows that $\Phi = 2\pi$.

Solution 2: Since the unit normal vector at the point $(x, y, z) \in S$ is $\vec{n} = \frac{x\vec{i} + y\vec{j}}{\sqrt{x^2 + y^2}} = x\vec{i} + y\vec{j}$, it follows that $\Phi = \iint_S \vec{r} \cdot \vec{n} \, dS = \iint_S (x^2 + y^2) \, dS = \iint_S dS = 2\pi$.

Problem 5.7.17. Evaluate the flux Φ of the field $\vec{a} = yz \langle 0, 1, 1 \rangle$ through the side of the surface $S = \{(x, y, z) : |x| + |y| = 1$ and $z \in [0, 1]\}$, for which the field of unit normal vectors is $\vec{n} = \frac{1}{\sqrt{2}} \langle \text{sign}\,(x), \text{sign}\,(y), 0 \rangle$.

Answer 5.7.17. $\Phi = 1$. The piecewise smooth surface S is part of the cylinder that has the square $\{(x, y, 0) : |x| + |y| = 1\}$ as its generating curve.

Problem 5.7.18. Evaluate the flux Φ of the field $\vec{a} = \langle x^3, 3yz^2, xy \rangle$ through the side of the surface S, obtained by rotation of the hodograph of the function $\vec{r}(t) = \langle t, t^2, 0 \rangle$, $t \in [0, 1]$, about the y-axis, for which the unit normal vector at the point $(0, 0, 0)$ is $-\vec{j}$.

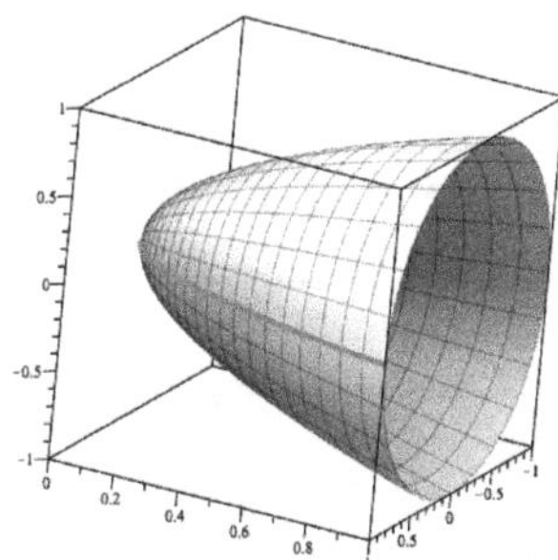

Solution 5.7.18. The vector field $\vec{a}$ is continuously differentiable. The surface S is part of the paraboloid $y = x^2 + z^2$ between the planes $y = 0$ and $y = 1$. Let $S_1 = \{(x, 1, z) : x^2 + z^2 \leq 1\}$.

The surface $\Sigma = S \cup S_1$ is closed and piecewise smooth. Let Φ_Σ be the flux of the vector field $\vec{a}$ through the outer side of Σ, and let Φ_1 be the flux of $\vec{a}$ through the side of S_1 determined by the vector $\vec{j}$. Then, $\Phi = \Phi_\Sigma - \Phi_1$. Let $V = \{(x, y, z) : x^2 + z^2 \leq y \text{ and } y \in [0, 1]\}$. By the divergence theorem, $\Phi_\Sigma = 3 \iiint_V (x^2 + z^2)\, dV = \frac{\pi}{2}$. Since $\Phi_1 = \iint_{S_1} \vec{a} \cdot \vec{j}\, dS = \frac{3\pi}{4}$, it follows that $\Phi = -\frac{\pi}{4}$.

Problem 5.7.19. Evaluate the flux Φ of the field $\vec{a} = \langle 2x^2, xy(y-2), -2xyz \rangle$ through the side of the surface S determined by $x^2 - y^2 - z^2 + 1 = 0$, $x \in [0, 1]$, for which the unit normal vector at the point $\left(\frac{1}{2}, \frac{\sqrt{5}}{2}, 0 \right)$ is $\frac{\langle 1, -\sqrt{5}, 0 \rangle}{\sqrt{6}}$.

Solution 5.7.19. The vector field $\vec{a}$ is continuously differentiable. The surface S is part of the one-sheeted hyperboloid, with its centre at the origin and the x-axis as its axis, between the planes $x = 0$ and $x = 1$. Let $S_1 = \{(1, y, z) : y^2 + z^2 \leq 2\}$, and let $S_2 = \{(0, y, z) : y^2 + z^2 \leq 1\}$. The surface $\Sigma = S \cup S_1 \cup S_2$ is closed and piecewise smooth.

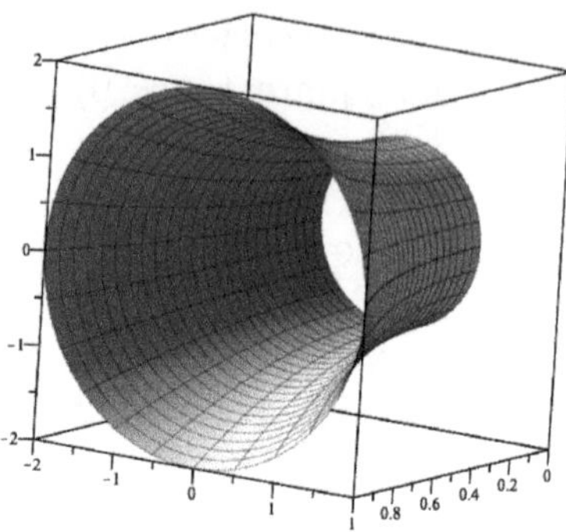

Let Φ_Σ be the flux of the vector field $\vec{a}$ through the outer side of Σ, let Φ_1 be the flux of $\vec{a}$ through the side of S_1 determined by the vector $\vec{\imath}$, and let Φ_2 be the flux of $\vec{a}$ through the side of S_2 determined by the vector $-\vec{\imath}$. Then, $\Phi = \Phi_\Sigma - \Phi_1 - \Phi_2$. Let $V_1 = \{(x,y,z) : y^2 + z^2 \leq 1 \text{ and } x \in [0,1]\}$, $V_2 = \{(x,y,z) : 1 \leq y^2 + z^2 \leq 2 \text{ and } \sqrt{y^2 + z^2 - 1} \leq x \leq 1\}$. The surface Σ is the boundary of the region $V = V_1 \cup V_2$. By the divergence theorem,

$$\Phi_\Sigma = 2 \iiint_V x \, dV = 2 \left(\iiint_{V_1} x \, dV + \iiint_{V_2} x \, dV \right)$$

$$= 2 \left(\frac{\pi}{2} + \frac{\pi}{4} \right) = \frac{3\pi}{2}.$$

Since $\Phi_1 = \iint_{S_1} \vec{a} \cdot \vec{\imath} \, dS = 4\pi$ and $\Phi_2 = \iint_{S_2} \vec{a} \cdot (-\vec{\imath}) \, dS = 0$, it follows that $\Phi = -\frac{5\pi}{2}$.

Problem 5.7.20. Evaluate the flux Φ of the field $\vec{r} = \langle x, y, z \rangle$ through the side of the surface S determined by $(x^2 + z^2)e^{2|y|} = 4$, $y \in [-1, 1]$, for which the unit normal vector at the point $(1, \ln 2, 0)$ is $\frac{\langle 1,1,0 \rangle}{\sqrt{2}}$ and the unit normal vector at the point $(1, -\ln 2, 0)$ is $\frac{\langle 1,-1,0 \rangle}{\sqrt{2}}$.

Solution 5.7.20. The vector field $\vec{r}$ is continuously differentiable. The surface S is obtained by rotation of the graph of the function $y \mapsto x = f(y) = 2e^{-|y|}$, $y \in [-1, 1]$ about the y-axis. Let $S_1 = \{(x, 1, z) : x^2 + z^2 \leq 4e^{-2}\}$, and let $S_2 = \{(x, -1, z) : x^2 + z^2 \leq 4e^{-2}\}$. The surface $\Sigma = S \cup S_1 \cup S_2$ is closed and piecewise smooth. Let Φ_Σ be the flux of the vector field $\vec{r}$ through the outer side of Σ, let Φ_1 be the flux of $\vec{r}$ through the side of S_1 determined by the vector $\vec{\jmath}$, and let Φ_2 be the flux of $\vec{r}$ through the side of S_2 determined by the vector $-\vec{\jmath}$. Then, $\Phi = \Phi_\Sigma - \Phi_1 - \Phi_2$. Let $V = \{(x,y,z) : (x^2 + z^2)e^{2|y|} \leq 4 \text{ and } y \in [-1,1]\}$. By the divergence theorem,

$\Phi_\Sigma = 3\iiint_V dV = 12\pi(1 - \frac{1}{e^2})$. Since $\Phi_1 = \iint_{S_1} \vec{r} \cdot \vec{j}\, dS = 4e^{-2}\pi$ and $\Phi_2 = \iint_{S_2} \vec{r} \cdot (-\vec{j})\, dS = 4e^{-2}\pi$, it follows that $\Phi = 4\pi(3 - \frac{5}{e^2})$.

Problem 5.7.21. Evaluate the flux Φ of the field $\vec{a} = \langle e^{xz} - xye^{yz},\ e^{xy} - yze^{xz},\ e^{yz} - zxe^{xy} + z\rangle$ through either side of the surface S determined by $z = 1 - \sqrt[3]{x^2} - \sqrt[3]{y^2}$, $z \geq 0$.

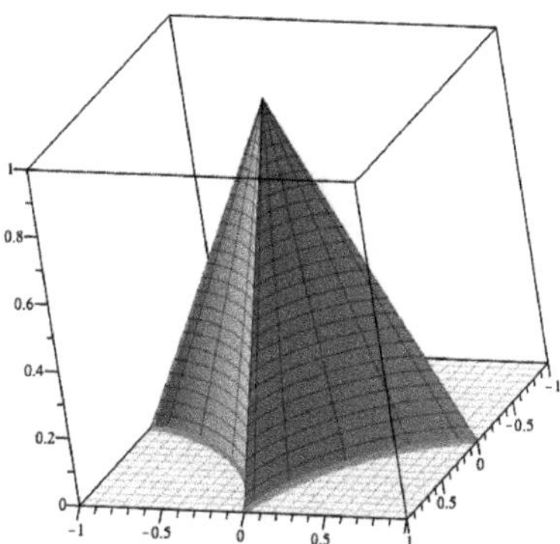

Solution 5.7.21. The vector field $\vec{a}$ is continuously differentiable. The figure depicts the surface S. Observe that the surface S is piecewise smooth and not closed. The intersection of the surface and the plane $z = 0$ is the astroid $\sqrt[3]{x^2} + \sqrt[3]{y^2} = 1$. For $S_1 = \{(x, y, 0) : \sqrt[3]{x^2} + \sqrt[3]{y^2} \leq 1\}$, the surface $\Sigma = S \cup S_1$ is closed and piecewise smooth.

Let Φ_Σ be the flux of the vector field $\vec{a}$ through the outer side of Σ, and let Φ_1 be the flux of $\vec{a}$ through the side of S_1 determined by the vector $-\vec{k}$. Let Φ be the flux of $\vec{a}$ through the side of surface S that matches the outer side of Σ. Then, $\Phi = \Phi_\Sigma - \Phi_1$. Let $V = \{(x, y, z) : 0 \leq 1 - \sqrt[3]{x^2} - \sqrt[3]{y^2}\}$. By the divergence theorem, $\Phi_\Sigma = \iiint_V dV = \frac{3\pi}{32}$. Since $\Phi_1 = \iint_{S_1} \vec{r} \cdot (-\vec{k})\, dS = -\iint_{S_1} dS = -\frac{3\pi}{8}$, it follows that $\Phi = \frac{15\pi}{32}$.

5.8 Stokes' Theorem

Problem 5.8.1. Let S be a region in the plane $x\cos\alpha + y\cos\beta + z\cos\gamma = 1$ bounded by a closed piecewise smooth curve Γ. Evaluate

$$I = \oint_\Gamma \begin{vmatrix} dx & dy & dz \\ \cos\alpha & \cos\beta & \cos\gamma \\ x & y & z \end{vmatrix},$$

where the orientation of Γ is induced by the vector $\vec{n} = \langle \cos\alpha, \cos\beta, \cos\gamma \rangle$, i.e., it is in agreement, in the sense of Stokes' theorem, with the orientation of S.[18]

Solution 5.8.1. Let $\vec{a} = \begin{vmatrix} \vec{i} & \vec{j} & \vec{k} \\ \cos\alpha & \cos\beta & \cos\gamma \\ x & y & z \end{vmatrix} = \vec{n} \times \vec{r}$. Then, by Stokes' theorem and using Problem 5.3.15, $I = \oint_\Gamma \vec{a} \cdot d\vec{r} = \iint_S \operatorname{curl} \vec{a} \cdot \vec{n}\, dS = 2\iint_S \vec{n} \cdot \vec{n}\, dS = 2\iint_S dS = 2 \cdot (\text{area of } S)$.

Problem 5.8.2. Let $\vec{F}(x, y, z) = \langle x\cos x, xy + z, e^z + y \rangle$, and let C be the square with the vertices $(1, 0, 0)$, $(1, 1, 0)$, $(1, 1, 1)$, and $(1, 0, 1)$, oriented clockwise when seen from the origin. Calculate $\oint_C \vec{F} \cdot d\vec{r}$.

Solution 5.8.2. Let S be the region in the plane $x = 1$ bounded by C. The vector $\vec{i}$ induces the orientation of S that is in agreement, in the sense of Stokes' theorem, with the orientation of C. By Stokes' theorem, $\oint_C \vec{F} \cdot d\vec{r} = \iint_S (\operatorname{curl} \vec{F}) \cdot \vec{i}\, dS$.

From $\operatorname{curl} \vec{F} = \begin{vmatrix} \vec{i} & \vec{j} & \vec{k} \\ \frac{\partial}{\partial x} & \frac{\partial}{\partial y} & \frac{\partial}{\partial z} \\ x\cos x & xy+z & e^z+y \end{vmatrix} = \langle 0, 0, y \rangle$, it follows that $(\operatorname{curl} \vec{F}) \cdot \vec{i} = 0$ and $\oint_C \vec{F} \cdot d\vec{r} = 0$.

Problem 5.8.3. Let $\vec{F}(x, y, z) = \langle 2z, -4x, 3y \rangle$. Let S be the part of the sphere $x^2 + y^2 + z^2 = 169$ that lies above the plane $z = 12$, and let the orientation of S be determined by the vector $\vec{k}$ at the point $(0, 0, 13)$.

(a) Evaluate directly the integral $\oint_C \vec{F} \cdot d\vec{r}$, where C is the boundary of S oriented counterclockwise relative to $\vec{k}$.

(b) Use Stokes' theorem to evaluate $\oint_C \vec{F} \cdot d\vec{r}$.

Solution 5.8.3.

(a) The boundary of the surface S is the circle $C = \{(5\cos\theta, 5\sin\theta, 12) : \theta \in [2\pi]\}$. It follows that $\oint_C \vec{F} \cdot d\vec{r} =$

---

[18] The curve Γ is oriented counterclockwise relative to the vector $\vec{n}$ normal to the surface, i.e., from the terminal point of the vector $\vec{n}$, the orientation of Γ is seen as counterclockwise. Equivalently, as a person *walks* around Γ on the side of S determined by $\vec{n}$, the region/surface S is on her left.

$$\int_0^{2\pi} \langle 24, -20\cos\theta, 15\sin\theta \rangle \quad \cdot \quad \langle -5\sin\theta, 5\cos\theta, 0 \rangle \qquad d\theta \qquad =$$
$$\int_0^{2\pi} (-120\sin\theta - 100\cos^2\theta)\, d\theta = -100\pi.$$

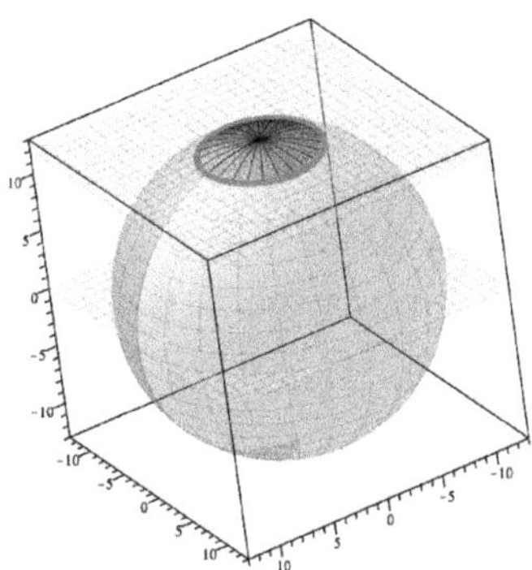

(b) The circle C is the boundary of the disc $S' = \{(x, y, 12) : x^2 + y^2 \le 5\}$ in the plane $z = 12$. Since $\operatorname{curl} \vec{F} = \begin{vmatrix} \vec{i} & \vec{j} & \vec{k} \\ \frac{\partial}{\partial x} & \frac{\partial}{\partial y} & \frac{\partial}{\partial z} \\ 2z & -4x & 3y \end{vmatrix} = \langle 3, 2, -4 \rangle$, by Stokes' theorem, $\oint_C \vec{F} \cdot d\vec{r} = \iint_{S'} \operatorname{curl} \vec{F} \cdot \vec{k}\, dS = \iint_{S'} \langle 3, 2, -4 \rangle \cdot \langle 0, 0, 1 \rangle\, dS = -4 \cdot 25\pi = -100\pi.$

Problem 5.8.4. Let S be the part of the plane $2x + y + 3z = 6$ that lies in the first octant, and let C be its boundary. Evaluate the circulation of the vector field $\vec{F}(x, y, z) = \langle 2z - x, -x + y + z, 2y - x \rangle$ along the curve C whose orientation is induced by an upward normal vector of the plane.

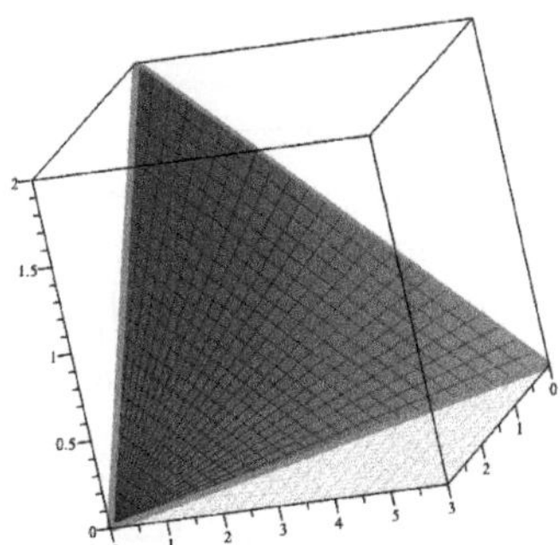

Solution 5.8.4. The projection of S into the xy-plane is $D = \{(x, y) : x \in [0, 3] \text{ and } 0 \le y \le 6 - 2x\}$ and $\operatorname{curl} \vec{F} = \langle 1, 3, -1 \rangle$. For $\vec{N} = \langle 2, 3, 6 \rangle$, by Stokes' theorem, $\oint_C \vec{F} \cdot d\vec{r} = \iint_S \operatorname{curl} \vec{F} \cdot d\vec{S} = \iint_D \langle 1, 3, -1 \rangle \cdot \langle 2, 1, 3 \rangle\, dx\,dy = 2 \iint_D dx\,dy = 18.$

Problem 5.8.5. Evaluate the circulation of the vector field $\vec{a} = \langle -\frac{z^2}{2}, -yz, 0 \rangle$ along the curve Γ, the intersection of the sphere $x^2 + y^2 + z^2 = 1$ and the plane $x + y = 1$. The orientation of Γ is induced by the vector $\vec{n} = \frac{1}{\sqrt{2}}(\vec{i} + \vec{j})$.

Solution 5.8.5. The circle Γ is the boundary of the disc $S = \{(x, y, z) : x + y = 1 \text{ and } x^2 + y^2 + z^2 \leq 1\}$. The projection of S into the xz-plane is $D = \{(x, 0, z) : 2(x - \frac{1}{2})^2 + z^2 \leq \frac{1}{2}\}$. By Stokes' theorem, the circulation is $\oint_\Gamma \vec{a} \cdot d\vec{r} = \iint_S \operatorname{curl} \vec{a} \cdot \vec{n} \, dS = \iint_S (y\vec{i} - z\vec{j}) \cdot \frac{1}{\sqrt{2}}(\vec{i} + \vec{j}) \, dS = \iint_D (1 - x - z) \, dxdz = \frac{\pi\sqrt{2}}{8}$.

Problem 5.8.6. Evaluate the circulation of the vector field $\vec{a} = \langle x^2, y^2, 0 \rangle$ along the boundary of the part of the ellipsoid $\frac{x^2}{a^2} + \frac{y^2}{b^2} + \frac{z^2}{c^2} = 1$ that lies in the first octant.

Solution 5.8.6. Since $\operatorname{curl} \vec{a} = \vec{0}$, by Stokes' theorem, the circulation is 0.

Problem 5.8.7. Consider the curve Γ that is the intersection of the ellipsoid $\frac{x^2}{a^2} + \frac{y^2}{b^2} + \frac{z^2}{c^2} = 2$ and the elliptic paraboloid $\frac{x^2}{a^2} + \frac{y^2}{b^2} - \frac{z}{c} = 0$, with a, b, $c > 0$.

(a) Establish that Γ is a planar curve.
(b) Evaluate the circulation of the vector field $\vec{a} = \langle a^2 y^3, -b^2 x^3, z^3 \rangle$ along Γ if the orientation of the curve is induced by the vector $\vec{n} = \vec{k}$.

Solution 5.8.7.

(a) The curve Γ lies in the plane $z = c$.
(b) Consider the set $S = \{(x, y, c) : b^2 x^2 + a^2 y^2 \leq a^2 b^2\}$, the region in the plane $z = c$ bounded by Γ, and its projection D into the xy-plane. By Stokes' theorem, $\oint_\Gamma \vec{a} \cdot d\vec{r} = \iint_S \operatorname{curl} \vec{a} \cdot \vec{k} \, dS = -3 \iint_D (b^2 x^2 + a^2 y^2) \, dxdy = -\frac{3}{2} a^3 b^3 \pi$.

Problem 5.8.8. Consider the curve Γ that is the intersection of the paraboloids $2z = x^2 + y^2$ and $2y = x^2 + z^2$.

(a) Establish that Γ is a planar curve.
(b) Evaluate the circulation of the vector field $\vec{a} = \langle xz, x(2z - y), z \rangle$ along Γ if the orientation of the curve is induced by the unit vector $\vec{n} = \frac{1}{\sqrt{2}}(\vec{j} - \vec{k})$.

Answer 5.8.8. (a) The curve lies in the plane $y = z$. (b) The circulation is $-\pi$.

Problem 5.8.9. Evaluate the circulation of the vector field $\vec{a} = \langle x, xy, y \rangle$ along the intersection Γ of the cylinders $x^2 + (y-2)^2 = 1$ and $z = 2 - \sqrt{1 - x^2}$ oriented either way.

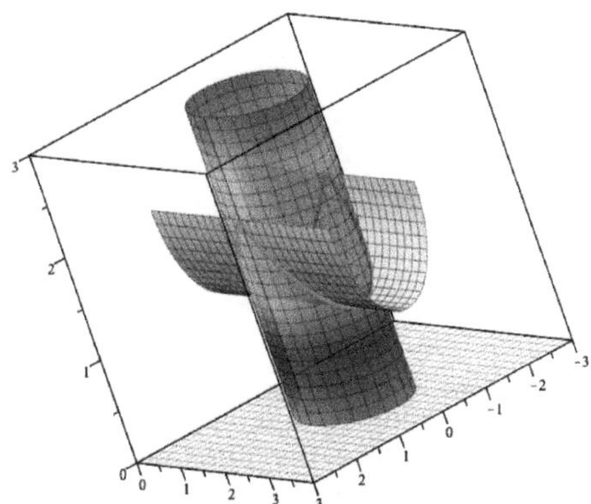

Solution 5.8.9. The projection of the curve Γ into the xy-plane is the circle that is the boundary of the disc $D = \{(x, y) : x^2 + (y-2)^2 \leq 1\}$. Let $S = \{(x, y, z) : (x, y) \in D$ and $z = 2 - \sqrt{1 - x^2}\}$, i.e., let S be the part of the cylinder $z = 2 - \sqrt{1 - x^2}$ that lies inside the cylinder $x^2 + (y-2)^2 = 1$.

The boundary of S is Γ. Let S^+ be the side of the surface S that corresponds to the field of unit normal vectors $\vec{n} = \langle x, 0, -\sqrt{1 - x^2} \rangle$.[19] Let the orientation of Γ be induced by the field $\vec{n}$. By Stokes' theorem, $\oint_\Gamma \vec{a} \cdot d\vec{r} = \iint_S \langle 1, 0, y \rangle \cdot \langle x, 0, -\sqrt{1 - x^2} \rangle \, dS = \iint_D \left(\frac{x}{\sqrt{1-x^2}} - y \right) dx dy = -2\pi$.

Note: It is possible to evaluate the required circulation without using Stokes' theorem. Observe that Γ is the hodograph of the vector function $\vec{R} = \langle \cos t, 2 + \sin t, 2 - |\sin t| \rangle$, $t \in [0, 2\pi]$.

Problem 5.8.10. Evaluate the circulation of the vector field $\vec{a} = \langle xy, x^2, z \rangle$ along the curve Γ that is the intersection of the cone $z + \sqrt{x^2 + y^2} = 1$ and the cylinder $|x| + |y| = 1$ if the orientation of the curve is induced by the vector $\vec{n} = \frac{1}{\sqrt{2}}(\vec{j} + \vec{k})$.

---

[19]In the figure, S^+ is the *bottom* side of the cylinder.

Solution 5.8.10. Let $D = \{(x,y) : |x| + |y| \leq 1\}$, and let $S = \{(x,y,z) : (x,y) \in D \text{ and } z + \sqrt{x^2+y^2} = 1\}$, i.e., let S be the part of the cone that lies inside the cylinder. The curve Γ is the boundary of S.

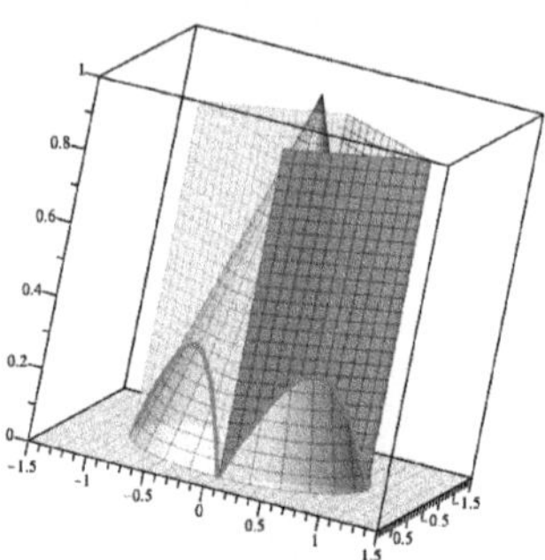

The side of S, which is in agreement, in the sense of Stokes' theorem, with the orientation of Γ, is determined by the vector field of unit vectors $\vec{N} = \frac{1}{\sqrt{2}}\langle \frac{x}{\sqrt{x^2+y^2}}, \frac{y}{\sqrt{x^2+y^2}}, 1 \rangle$.

By Stokes' theorem, $\oint_\Gamma \vec{a} \cdot d\vec{r} = \iint_S \frac{x}{\sqrt{2}}\, dS = \iint_D x\, dx dy = 0$.

Problem 5.8.11. Let $\vec{F}(x,y,z) = \langle yz, -xz, 1 \rangle$. Let S be the part of the paraboloid $z = 4 - x^2 - y^2$ which lies in the first octant. Let C, the boundary of S, be the closed curve $C = C_1 \cup C_2 \cup C_3$, where the curves C_1, C_2, and C_3 are the three curves formed by intersecting S with the xy-, yz-, and xz-planes, respectively. The orientation of the curve C is induced by the vector $\langle 2, 2, 1 \rangle$.

(a) Use Stokes' theorem to evaluate $\oint_C \vec{F} \cdot d\vec{r}$.

(b) Evaluate the integral $\oint_C \vec{F} \cdot d\vec{r}$ directly by parametrising each piece of the curve C.

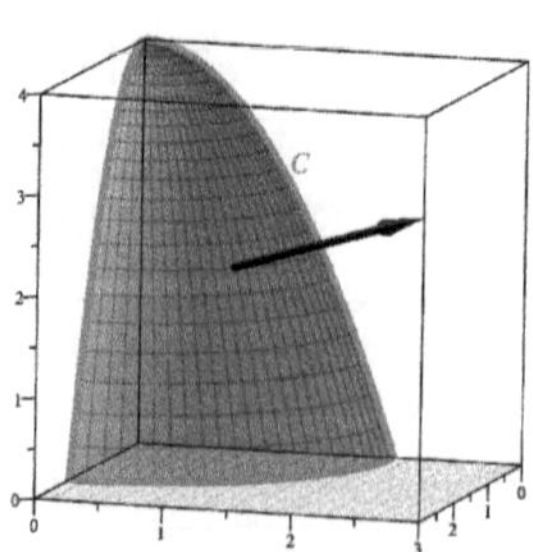

Solution 5.8.11. See the figure for a sketch of the surface S and the curve C. The projection of the surface into the xy-plane is the set $D = \{(x, y) : x^2 + y^2 \le 4 \text{ and } x \ge 0 \text{ and } y \ge 0\}$.

(a) The side of the surface, which is in agreement with the orientation of the curve, is determined by the field of normal vectors $\vec{n} = \langle 2x, 2y, 1 \rangle$. Since, for $(x, y, z) \in S$, $(\operatorname{curl} \vec{F}) \cdot \vec{n} = 2x^2 + 2y^2 - 2z = 2x^2 + 2y^2 - 2(4 - x^2 - y^2) = 4(x^2 + y^2 - 2)$, by Stokes' theorem, $\oint_C \vec{F} \cdot d\vec{r} = \iint_S \operatorname{curl} \vec{F} \cdot \frac{\vec{n}}{|\vec{n}|} dS = 4 \iint_D (x^2 + y^2 - 2) \, dx dy = 4 \int_0^{\frac{\pi}{2}} d\theta \int_0^2 (r^2 - 2) r dr = 0$.

(b) Keeping in mind the orientation, $C_1 = \{(2-t, \sqrt{4 - (2 - t)^2}, 0) : 0 \le t \le 2\}$, $C_2 = \{(0, 2 - t, 4 - (2 - t)^2) : 0 \le t \le 2\}$, and $C_3 = \{(t, 0, 4 - t^2) : 0 \le t \le 2\}$. From $\int_{C_1} \vec{F} \cdot d\vec{r} = \int_{C_1} yz \, dx - xz \, dy = 0$, $\int_{C_2} \vec{F} \cdot d\vec{r} = \int_{C_2} -xz \, dy + dz = \int_0^2 2(2 - y) \, dy = 4$, and $\int_{C_3} \vec{F} \cdot d\vec{r} = \int_{C_2} yz \, dx + dz = \int_0^2 (-2x) \, dx = -4$, it follows that

$$\oint_C \vec{F} \cdot d\vec{r} = \int_{C_1} \vec{F} \cdot d\vec{r} + \int_{C_2} \vec{F} \cdot d\vec{r} + \int_{C_3} \vec{F} \cdot d\vec{r} = 0.$$

Problem 5.8.12. Let the surface S be the hodograph of the vector function $\vec{r}(u, v) = \langle 2 - 2v^2, v \cos u, v \sin u \rangle$, $(u, v) \in D = [0, 2\pi] \times [0, 1]$.

(a) Which of these graphs represents the surface S?

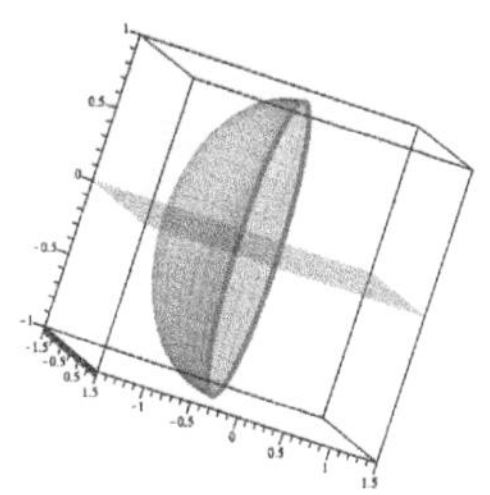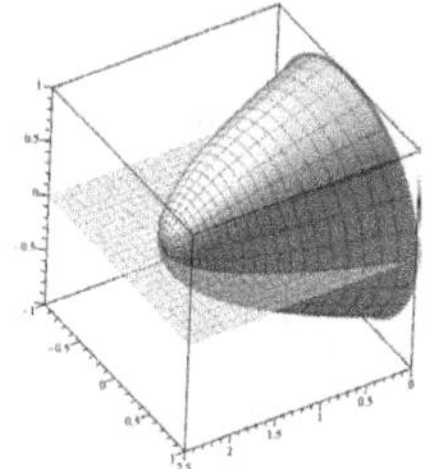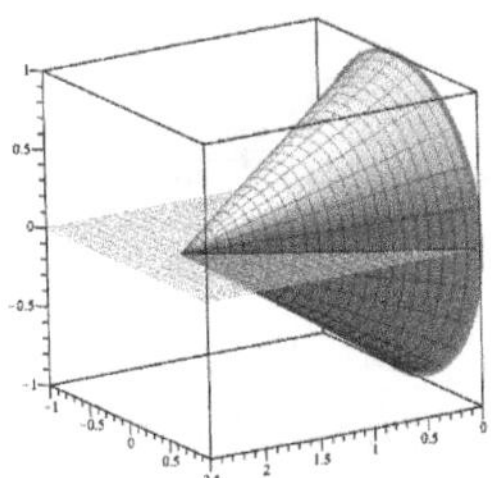

(b) For the vector field $\vec{F}(x, y, z) = \langle 0, -z, y \rangle$, evaluate $\iint_S (\operatorname{curl} \vec{F}) \cdot \vec{n} \, dS$ via the given parametrisation, where $\vec{n}$ is the normal vector field that points in the positive x-direction.

(c) Evaluate $\iint_S (\operatorname{curl} \vec{F}) \cdot \vec{n} \, dS$ by using Stokes' theorem.

Solution 5.8.12.

(a) The middle image. The left-side image is a hemisphere, while the one on the far right is a cone.

(b) From curl $\vec{F} = \langle 2, 0, 0 \rangle$ and $\frac{\partial \vec{r}}{\partial u} \times \frac{\partial \vec{r}}{\partial v} = \begin{vmatrix} \vec{\imath} & \vec{\jmath} & \vec{k} \\ 0 & -v\sin u & v\cos u \\ -4v & \cos u & \sin u \end{vmatrix} =$
$\langle -v, 4v^2 \cos u, -4v^2 \sin u \rangle$, it follows that

$$\iint_S (\operatorname{curl} \vec{F}) \cdot \vec{n}\, dS = -\iint_D \langle 2, 0, 0 \rangle \cdot \left(\frac{\partial \vec{r}}{\partial u} \times \frac{\partial \vec{r}}{\partial v} \right) dudv$$

$$= 2 \int_0^1 v\, dv \int_0^{2\pi} du = 2\pi.$$

(c) The boundary of S is the circle $C = \{(0, \cos t, \sin t) : t \in [0, 2\pi]\}$. The orientation of the circle, in the sense of increase of the variable t, is in agreement with the orientation of S. By Stokes' theorem, $\iint_S (\operatorname{curl} \vec{F}) \cdot \vec{n}\, d\vec{S} = \oint_C \vec{F} \cdot d\vec{r} = \int_0^{2\pi} \langle 0, -\sin t, \cos t \rangle \cdot \langle 0, -\sin t, \cos t \rangle\, dt = 2\pi$.

Problem 5.8.13. Let S be the hodograph of the vector function $\vec{r}(u, v) = \langle \cos u, \sin u, v \rangle$, $(u, v) \in [0, 2\pi] \times [0, 1]$.

(a) Which of these graphs represents the surface S?

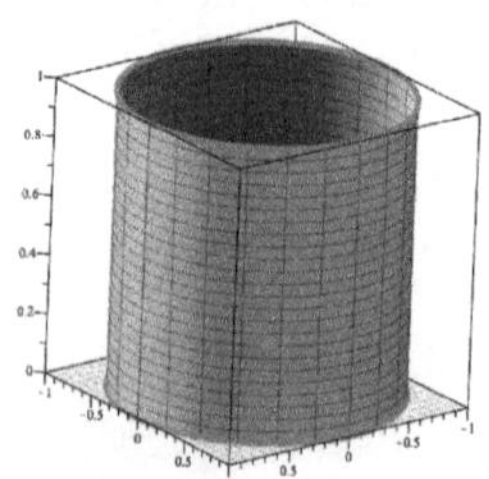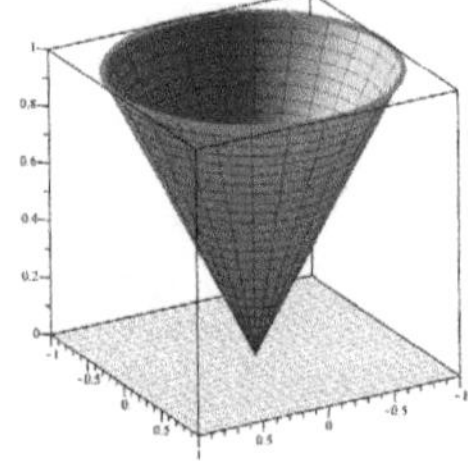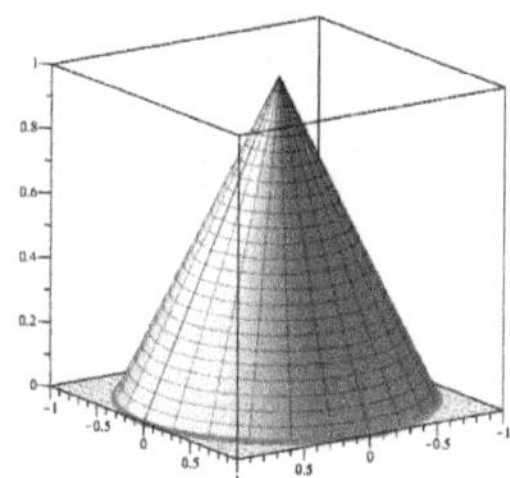

(b) For the vector field $\vec{F}(x, y, z) = \langle yz, -xz, 1 \rangle$, evaluate $\iint_S (\operatorname{curl} \vec{F}) \cdot \vec{n}\, dS$ via the given parametrisation, where $\vec{n}$ is the normal vector field pointing away from the z-axis.

(c) Evaluate $\iint_S (\operatorname{curl} \vec{F}) \cdot \vec{n}\, dS$ by using Stokes' theorem.

Solution 5.8.13.

(a) The left-side image; the surface S is the cylinder $x^2 + y^2 = \cos^2 u + \sin^2 u = 1$. The middle image is the cone parametrised by $\vec{r}(u, v) = \langle v\cos u, v\sin u, v \rangle$, $(u, v) \in [0, 2\pi] \times [0, 1]$, and the far-right image is the cone parametrised by $\vec{r}(u, v) = \langle v\cos u, v\sin u, 1 - v \rangle$, $(u, v) \in [0, 2\pi] \times [0, 1]$.

(b) From curl $\vec{F} = \langle x,\, y,\, -2z \rangle$ and $\frac{\partial \vec{r}}{\partial u} \times \frac{\partial \vec{r}}{\partial v} = \begin{vmatrix} \vec{i} & \vec{j} & \vec{k} \\ -\sin u & \cos u & 0 \\ 0 & 0 & 1 \end{vmatrix} =$

$\langle \cos u, \sin u, 0 \rangle$, it follows that $\iint_S (\text{curl } \vec{F}) \cdot \vec{n} \; dS =$ $\int_0^1 dv \int_0^{2\pi} \langle \cos u, \sin u, -2v \rangle \cdot \langle \cos u, \sin u, 0 \rangle du = 2\pi$.

(c) The boundary C of S is the union of two oppositely oriented circles: $C' = \{(\cos t, \sin t, 0) : t \in [0, 2\pi]\}$ and $C'' = \{(\cos t, -\sin t, 1) : t \in [0, 2\pi]\}$. By Stokes' theorem,

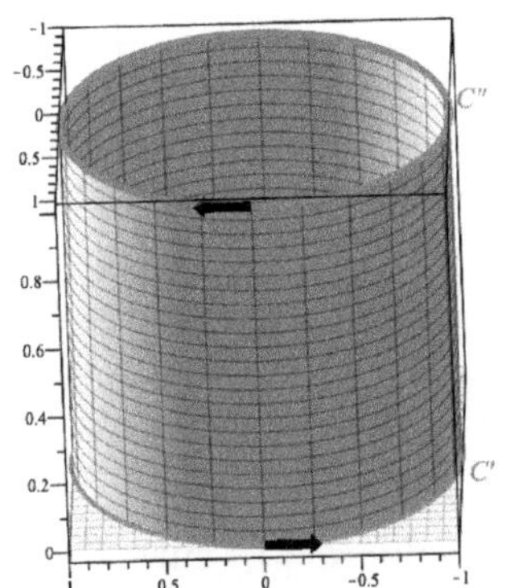

$$\iint_S (\text{curl } \vec{F}) \cdot \vec{n} \; dS = \int_{C' \cup C''} \vec{F} \cdot d\vec{r} = \oint_{C'} \vec{F} \cdot d\vec{r} + \oint_{C''} \vec{F} \cdot d\vec{r}$$

$$= \int_0^{2\pi} \langle 0, 0, 1 \rangle \cdot \langle -\sin t, \cos t, 0 \rangle \; dt$$

$$+ \int_0^{2\pi} \langle -\sin u, -\cos u, 1 \rangle \cdot \langle -\sin t, -\cos t, 0 \rangle \; dt = 2\pi.$$

Problem 5.8.14. Evaluate the flux of the curl of the field $\vec{F}(x, y, z) = \langle -y(x+1) + z^7 \cos z, \, z, \, z^{21} + z \cos xy \rangle$ through the surface S parametrised by $\vec{r}(t, s) = \langle s \cos t, s \sin t, (1-s)^2 \rangle$ with $0 \le t \le 2\pi$, $0 \le s \le 1$. The orientation of S is downward, i.e., the $\vec{k}$ component of the normal vector is negative.

Solution 5.8.14. The boundary of the surface S is the unit circle $C = \{(\cos t, \sin t, 0) : t \in [0, 2\pi]\}$. The orientation of the surface induces the negative orientation of C. By Stokes' theorem, $\iint_S (\text{curl } \vec{F}) \cdot \vec{n} \; dS = \oint_C \vec{F} \cdot d\vec{r} = -\int_0^{2\pi} \langle -\sin t (\cos t + 1), 0, 0 \rangle \cdot \langle -\sin t, \cos t, 0 \rangle \; dt = -\int_0^{2\pi} \sin^2 t (\cos t + 1) \; dt = -\pi$.

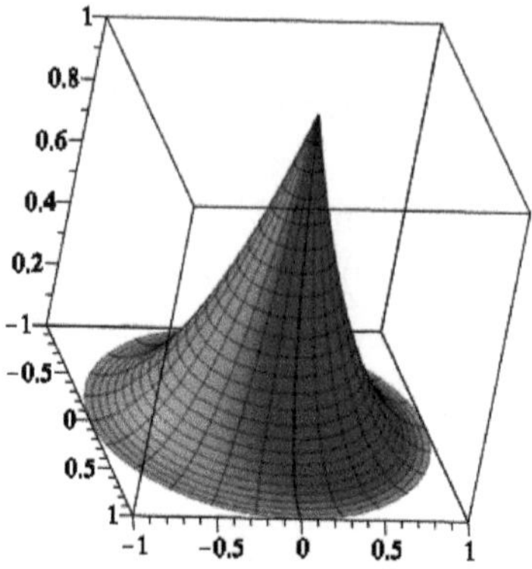

Problem 5.8.15. Consider the vector field $\vec{F}(x,y,z) = \langle y + \cos x, e^y \cos y, z^2 - x \rangle$. Let the surface S_1 be the hodograph of the vector function $\vec{r}(u,v) = \langle (2 + \sin v) \cos u, (2 + \sin v) \sin u, \cos v \rangle$, $(u,v) \in [0, \frac{\pi}{2}] \times [0, 2\pi]$, and let $S_2 = \{(x,0,z) : (x-2)^2 + z^2 \leq 1\}$. Let the orientation of the surface $S = S_1 \cup S_2$ be determined by the unit vectors $-\vec{j}$ at the point $(2,0,0)$ and $\vec{N} = \frac{\langle 1,1,0 \rangle}{\sqrt{2}}$ at the point $r(\frac{\pi}{4}, \frac{\pi}{2}) = (\frac{3}{\sqrt{2}}, \frac{3}{\sqrt{2}}, 0)$. Evaluate $\iint_S (\operatorname{curl} \vec{F}) \cdot dS$.

Solution 5.8.15. The surface S_1 is a quarter of a torus. The intersection of S_1 and the xz-plane is the circle $\{(x,0,z) : (x-2)^2 + z^2 = 1\}$, the boundary of S_2. We consider S_2 as a lid on one side of S_1. The circle $C = \{(0,y,z) : (y-2)^2 + z^2 = 1\}$ is the boundary of the surface S.

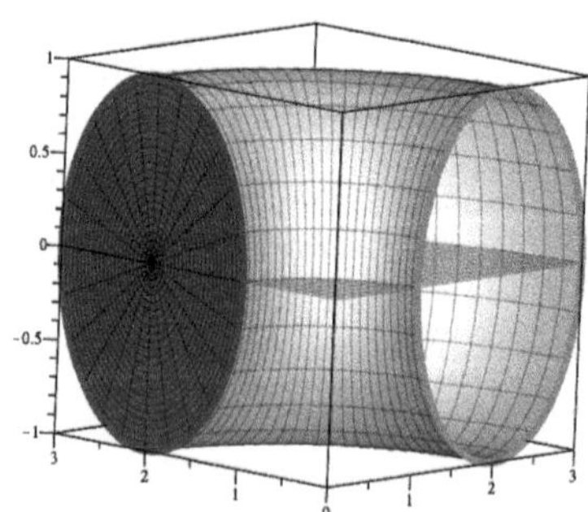

Let $S_3 = \{(0, y, z) : (y - 2)^2 + z^2 \leq 1\}$. By Stokes' theorem, $\iint_S \text{curl } \vec{F} \cdot d\vec{S} = \oint_C \vec{F} \cdot d\vec{r} = \iint_{S_3} (\text{curl } \vec{F}) \cdot d\vec{S}$.

From $\text{curl } \vec{F} = \begin{vmatrix} \vec{i} & \vec{j} & \vec{k} \\ \frac{\partial}{\partial x} & \frac{\partial}{\partial y} & \frac{\partial}{\partial z} \\ y+\cos x & e^y \cos y & z^2 - x \end{vmatrix} = \langle 0, 1, -1 \rangle$, it follows that $\text{curl } \vec{F} \cdot \langle -1, 0, 0 \rangle = 0$ and $\iint_{S_3} (\text{curl } \vec{F}) \cdot d\vec{S} = 0$.

References

[1] W. A. Adkins and M. G. Davidson. *Ordinary Differential Equations.* Springer, New York, NY, 1st edition, 2012.

[2] R. André. *Point-Set Topology with Topics: Basic General Topology For Graduate Studies.* World Scientific Publishing, Hackensack, New Jersey, NJ, 2024.

[3] T. A. Apostol. *Mathematical Analysis.* Addison Wesley, Boston, MA, 2nd edition, 1974.

[4] V. Barbu. *Differential Equations.* Springer, Cham, Switzerland, 1st edition, 2016.

[5] B. Barik. Lucas sequence, its properties and generalizations. Master's thesis, National Institute of Technology Rourkela, Odisha, India, May 2013.

[6] R. E. Bellman and R. S. Roth. *The Laplace Transform.* World Scientific Publishing, Hackensack, New Jersey, NJ, 1st edition, 1984.

[7] N. Borwein and V. Jungić. Experimental mathematics and the post-secondary visualization toolbox: Mathematics, history, and the visual. In B. Sriraman and Scott B. Lindstorm, editors, *Handbook of Visual, Experimental and Computational Mathematics – Bridges through Data.* Springer, New York, NY, 2025.

[8] W. E. Boyce, R. C. DiPrima, and D. B. Meade. *Elementary Differential Equations and Boundary Value Problems.* Wiley, Hoboken, New Jersey, NJ, 11th edition, 2017.

[9] K. Bryan. *Differential Equations: A Toolbox for Modeling the World.* SIMIODE, Chardon, OH, 2nd edition, 2025.

[10] M. A. Chaudry and S. M. Zubair. *On a Class of Incomplete Gamma Functions with Applications.* CRC Press, Boca Raton, FL, 1st edition, 2001.

[11]	V. Eiderman. *An Introduction to Complex Analysis and the Laplace Transform.* CRC Press, Boca Raton, FL, 1st edition, 2022.

[12]	S. Elaydi. *An Introduction to Difference Equations.* Springer, New York, NY, 3rd edition, 2005.

[13]	G. M. Fikhtengol'ts. *Kurs differentsialnogo i integralnogo ischislenia (in Russian),* volume 3. Lan, St. Petersburg, Russia, 16th edition, 2021.

[14]	P. Jones, M. Evans, and K. Lipson. *Essential Further Mathematics.* Cambridge University Press, Cambridge, UK, 4th edition, 2011.

[15]	V. Jungić. *Zbirka Zadataka iz Fourier-ove Analize, Diferencijalnih Jednačina i Vektorske Analize.* Elektrotehnički Fakultet, Banja Luka, Banja Luka, Yugoslavia, 1st edition, 1989.

[16]	W. Kaplan. *Advanced Calculus.* Addison-Wesley Press, Cambridge, MA, 5th edition, 2003.

[17]	W. G. Kelley and A. C. Peterson. *Difference Equations: An Introduction with Applications.* Academic Press, Cambridge, MA, 1st edition, 2000.

[18]	T. W. Körner. *Fourier Analysis.* Cambridge Mathematical Library. Cambridge University Press, Cambridge, UK, 2022.

[19]	T. Koshy. *Fibonacci and Lucas Numbers With Applications.* Wiley, Hoboken, New Jersey, NJ, 1st edition, 2001.

[20]	E. Kreyszig. *Advanced Engineering Mathematics.* John Wiley and Sons, Hoboken, New Jersey, NJ, 2011.

[21]	B. M. Makarov and A. N. Podkorytov. *Smooth Functions and Maps,* volume 7 of *Moscow Lectures.* Springer, Cham, Switzerland, 2021.

[22]	M. Manetti. *Topology.* Springer, Milano, Italy, 1st edition, 2014.

[23]	J. E. Marsden and A. Tromba. *Vector Calculus.* W. H. Freeman and Company, New York, NY, 6th edition, 2011.

[24]	R. E. Mickens. *Difference Equations: Theory and Applications.* Chapman and Hall/CRC, Boca Raton, FL, 2nd edition, 1991.

[25]	D. S. Mitrinović. *Matematika u obliku metodičke zbirke zadataka sa rešenjima, II.* Gradjevinska Knjiga, Belgrade, Serbia, 1st edition, 1967.

[26]	D. S. Mitrinović. *Matematika u obliku metodičke zbirke zadataka sa rešenjima, III.* Gradjevinska Knjiga, Belgrade, Serbia, 5th edition, 1990.

[27]	M. C. Potter, J. L. Lessing, and E. F. Aboufadel. *Advanced Engineering Mathematics.* Springer, Cham, Switzerland, 4th edition, 2019.

[28]	H. J. Ricardo. *A Modern Introduction to Differential Equations.* Academic Press, Cambridge, MA, 3rd edition, 2021.

[29]	R. Rockafellar. A property of piecewise smooth functions. *Computational Optimization and Applications,* (25):247–250, April 2003.

[30] W. Rudin. *Principles of Mathematical Analysis*. McGraw-Hill, New York, NY, 3rd edition, 1953.

[31] D. Slaughter. *Difference Equations to Differential Equations*. Orange Grove Texts Plus, Gainesville, FL, 1st edition, 2009.

[32] E. Stade. *Fourier Analysis*. Pure and Applied Mathematics. John Wiley and Sons, Hoboken, New Jersey, NJ, 2005.

[33] J. Stewart, D. Clegg, and S. Watson. *Calculus, Early Transcendentals*. Cengage, Boston, MA, 9th edition, 2021.

[34] R. Tahir-Kheli. *Ordinary Differential Equations*. Springer, Cham, Switzerland, 1st edition, 2018.

[35] T. Tao. *Analysis II*. Springer, Singapore, Singapore, 4 edition, 2022.

[36] D. Đ. Tošić. *Zbirka Rešenih Ispitnih Zadataka iz Matematike III*. Dobrilo Tošić, Belgrade, Serbia, 1st edition, 2002.

[37] G. van Dijk. *Distribution Theory - Convolution, Fourier Transform, and Laplace Transform*. De Gruyter, Berlin, Germany, 1st edition, 2013.

[38] B. Vick. *Applied Engineering Mathematics*. CRC Press, Boca Raton, FL, 1st edition, 2020.

[39] D. V. Winner. *The Laplace Transform*. Princeton University Press, Princeton, New Jersey, NJ, 2nd edition, 1946.

Index